Advances in
MICROBIAL ECOLOGY
Volume 14

ADVANCES IN MICROBIAL ECOLOGY

A Continuation Order Plan is available for this series. A continuation order will bring delivery of each new volume immediately upon publication. Volumes are billed only upon actual shipment. For further information please contact the publisher.

Advances in
MICROBIAL ECOLOGY

Volume 14

Edited by

J. Gwynfryn Jones

Freshwater Biological Association
Ambleside, Cumbria, England

PLENUM PRESS • NEW YORK AND LONDON

The Library of Congress cataloged the first volume of this title as follows:

Advances in microbial ecology. v. 1–
New York, Plenum Press ©1977–
v. ill. 24 cm.
Key title: Advances in microbial ecology, ISSN 0147-4863
1. Microbial ecology—Collected works.
QR100.A36   576'.15   77-649698

ISBN 0-306-45057-7

© 1995 Plenum Press, New York
A Division of Plenum Publishing Corporation
233 Spring Street, New York, N. Y. 10013

Contributors

D. J. Bagyaraj, Department of Agricultural Microbiology, University of Agricultural Sciences, Bangalore 560 065, India

Jacobus E. M. Beurskens, National Institute of Public Health and Environmental Protection (RIVM), NL-3720 BA Bilthoven, The Netherlands

D. E. Caldwell, Department of Applied Microbiology and Food Science, University of Saskatchewan, Saskatoon, Saskatchewan, Canada S7N 0W0

Ralf Conrad, Max Planck Institut für Terrestrische Mikrobiologie, D-35043 Marburg, Germany

David J. Des Marais, Ames Research Center, Moffett Field, California 94035-1000

Jan Dolfing, DLO–Research Institute for Agrobiology and Soil Fertility (AB-DLO), NL-9750 AC Haren, The Netherlands

Thomas Egli, Department of Microbiology, Swiss Federal Institute for Environmental Science and Technology (EAWAG), CH-8600 Dübendorf, Switzerland

D. R. Korber, Department of Applied Microbiology and Food Science, University of Saskatchewan, Saskatoon, Saskatchewan, Canada S7N 0W0

H. J. Laanbroek, Centre for Limnology, Netherlands Institute of Ecology, 3631 AC Nieuwersluis, The Netherlands

Claude Largeau, Laboratoire de Chimie Bioorganique et Organique Physique, UA CNRS D1381, Ecole Nationale Supérieure de Chimie de Paris, 75231 Paris Cedex 05, France

John R. Lawrence, National Hydrology Research Institute, Saskatoon, Saskatchewan, Canada S7N 3H5

Jan W. de Leeuw, Division of Marine Biogeochemistry, Netherlands Institute for Sea Research (NIOZ), 1790 AB Den Burg Texel, The Netherlands

Ajit Varma, School of Life Sciences, Jawaharlal Nehru University, New Delhi 110 067, India

J. W. Woldendorp, Department of Soil Biology, Centre for Terrestrial Ecology, Netherlands Institute of Ecology, 6666 ZG Heteren, The Netherlands

G. M. Wolfaardt, Department of Applied Microbiology and Food Science, University of Saskatchewan, Saskatoon, Saskatchewan, Canada S7N 0W0

Preface

There were many who joked when we took over *Advances in Microbial Ecology* at Volume 13; perhaps they should have reserved their expressions of superstition for Volume 14. As an example of British understatement, I think it would be fair to say that we have had a little bad luck. Never have I known a volume so bedeviled with misfortune, but we have been similarly fortunate in the patience exhibited by our authors, particularly those who were "first in line" with their chapters. It would be inappropriate to burden the reader with the catalogue of accidents and illnesses; suffice it to say that considerable experience has been gained in contingency planning.

We feel particularly delighted that the final product is a balanced volume, maintaining the tradition of *Advances in Microbial Ecology* in providing something for everyone. The chapters range from the strategies of growth to the role of microbes in maintaining sustainable agriculture, the significance of a single biochemical process to the complexities of coping with a wide range of substrates.

Of course, it is at this stage in the preparation of a volume that the editors' minds turn to the breadth of coverage and how this might be expanded. They question whether they are achieving the right mix and appealing to a wide enough audience. Microbial ecology covers an incredible range of subjects; the breadth of interaction between microbes demands a corresponding breadth in skills on the part of microbiologists. We are looking for a volume that could range from theoretical ecology to authoritative statements on a single species, from micro-habitat structures to the application of remote-sensing. We are, therefore, very receptive both to suggestions and offers from the readership. Symposia and conferences often act as fruitful grounds for the "hunter–gatherer" editor, but we also search for the spark of originality that identifies an area yet uncovered. At the time of this writing (May 1995), the chapters for Volume 15 are virtually agreed upon; we hope that the readership not only enjoys Volume 14, but is also willing to inform us about what it would like to see in Volume 16.

J. Gwynfryn Jones, Editor
Bernhard Schink
Warwick F. Vincent
David Ward

Contents

Chapter 1

Behavioral Strategies of Surface-Colonizing Bacteria

John R. Lawrence, D. R. Korber, G. M. Wolfaardt, and D. E. Caldwell

Chapter 2

Insoluble, Nonhydrolyzable, Aliphatic Macromolecular Constituents of Microbial Cell Walls

Claude Largeau and Jan W. de Leeuw

Chapter 3

Interaction between Arbuscular Mycorrhizal Fungi and Plants:
Their Importance in Sustainable Agriculture in Arid and
Semiarid Tropics

D. J. Bagyaraj and Ajit Varma

Chapter 4

The Microbial Logic and Environmental Significance of
Reductive Dehalogenation

Jan Dolfing and Jacobus E. M. Beurskens

Chapter 5

Soil Microbial Processes Involved in Production and Consumption of Atmospheric Trace Gases

Ralf Conrad

Chapter 6

The Biogeochemistry of Hypersaline Microbial Mats

David J. Des Marais

Chapter 7

Activity of Chemolithotrophic Nitrifying Bacteria under Stress in Natural Soils

H. J. Laanbroek and J. W. Woldendorp

Chapter 8

**The Ecological and Physiological Significance of the Growth of
Heterotrophic Microorganisms with Mixtures of Substrates**

Thomas Egli

1

Behavioral Strategies of Surface-Colonizing Bacteria

JOHN R. LAWRENCE, D. R. KORBER, G. M. WOLFAARDT, and D. E. CALDWELL

1. Introduction

Bacterial survival and reproductive success in many systems requires colonization of a surface and/or integration into a biofilm community. Success in a
community context requires morphological, physiological, and genetic attributes
that have only recently been explored. Previously, many aspects of microbial
behavior at interfaces have been explained in terms of physicochemical interactions. Indeed, Van Loosdrecht *et al.* (1989) concluded that virtually no studies
have shown a bacterial response to a surface. However, many studies past and
present have shown specific responses to the surface environment, including
chemoadherence, morphogenesis, gene induction, and variable rates of polymer
production (McCarter *et al.*, 1988, 1992; Vandevivere and Kirchman, 1993).
The induction of many genetic pathways has been shown to be surface-specific
phenomena. Ample evidence has also been provided for behavioral strategies
that only function during surface colonization and growth (Kjelleberg *et al.*,
1982; Lawrence *et al.*, 1987, 1991, 1992; Lawrence and Caldwell, 1987; Power
and Marshall, 1988; Lawrence and Korber, 1993). These strategies represent
essential elements for multicellular community growth and are expressed as
specific adaptations, life cycle alternation between attached and planktonic

JOHN R. LAWRENCE • National Hydrology Research Institute, Saskatoon, Saskatchewan, Canada S7N 3H5.   D. R. KORBER, G. M. WOLFAARDT, and D. E. CALDWELL • Department
of Applied Microbiology and Food Science, University of Saskatchewan, Saskatoon, Saskatchewan,
Canada S7N 0W0.

Advances in Microbial Ecology, Volume 14, edited by J. Gwynfryn Jones. Plenum Press, New York,
1995.

growth, as well as formation and maintenance of microcolonies, aggregates, and consortia.

The development of multicellular biofilm communities represents the interplay of many factors including specific cell–cell interactions (Shapiro and Hsu, 1989; Shapiro, 1991) and in many cases metabolic communication (Prosser, 1989). Microbial interactions enable a variety of microorganisms to coexist in environments in which individual organisms cannot survive. Typically, these communities consist of various microbial species with different metabolic activities and nutritional requirements. Particularly within a biofilm, temporal and spatial formation of chemical microzones, positioning of syntrophic partners, and establishment of complementary metabolic pathways may all occur. Interactions between different species and populations are often characterized by close but, in general, poorly understood interdependencies. A consequence of these interdependencies is that the spectrum of metabolic activities carried out by microbial communities is greater than the sum of those activities carried out by constituent members when cultured in isolation (Hamilton and Characklis, 1989). Thus, the existence of functional consortia and microcolonies within biofilms may present the most effective mechanism for satisfying the wide range of microbial growth requirements. In addition, community membership may improve the chance for reproductive success and survival.

Marshall (1993) indicated that microbial ecology, the discipline involving studies of living microorganisms in their natural environments, has three important goals: (1) to define population dynamics in communities, (2) to define physicochemical characteristics of microenvironments, and (3) to understand the metabolic processes carried out by microorganisms. In practice, much of microbial ecology has been process oriented and limited to the third of these three goals. To achieve the goals referred to above, it is essential to determine the roles of individual organisms, in the context of functioning communities in the environment (Caldwell, 1993).

Developments in a wide range of scientific disciplines have now provided an array of tools suitable for the analysis of biofilm communities in the ecological framework provided by Marshall. Among these developments are: molecular phylogenetic techniques (Amann *et al.*, 1990, 1992; Ward *et al.*, 1990, 1992), microelectrode analyses (Lens *et al.*, 1993; Lewandowski *et al.*, 1993), scanning confocal laser microscopy (SCLM), and environmentally sensitive fluorescent probes. Of these new methods, SCLM and microelectrode analyses have broad application for microbial studies (Geesey and White, 1990; Lawrence *et al.*, 1991; Caldwell *et al.*, 1992a,b, 1993; Assmus *et al.*, 1995; Korber *et al.*, 1993, 1994a,c; Lewandowski *et al.*, 1993; Lens *et al.*, 1993; Wolfaardt *et al.*, 1994a–c) and now allow characterization of microbial populations in a community context. Overall, these methods have provided evidence that alters our perception of microbial biofilms. For example, both pure culture and mixed-species biofilms are now widely accepted as spatially and temporally heterogeneous systems containing microscale variations in biofilm architecture and chemistry (Lawrence *et al.*, 1991, 1994; Korber *et al.*, 1993, 1994c; Lens *et al.*, 1993; Wolfaardt *et al.*, 1995a,b). These techniques have been used to define biofilm architecture, estimate cellular viability, identify microenvironmental chemical gradients, quantify molecular transport and molecular interactions, and phylogenetically differentiate individual biofilm members. Ultimately, it will be possible to characterize microbial populations in their niche and assess activity in the context of the natural community. The following discussion provides an overview of some of the surface colonization strategies employed by biofilm-forming bacteria that are believed to play a significant role in the development of attached communities.

2. Factors Influencing Initial Attachment of Bacteria to Surfaces

Integration of a microorganism into a community requires occupation of habitat and activity domains. Within biofilm communities this requires initial attachment, growth, and survival through a series of changing microenvironmental conditions. For the majority of microbiological ecosystems, factors that influence the success of bacterial attachment also play an important role in defining the population density and species diversity on newly exposed surfaces, thereby influencing the pathways of community development.

Numerous abiotic factors exist with demonstrable effects on bacterial attachment. Substratum microroughness, fluid dynamics, substratum chemistry, and solution ionic strength are typical examples of parameters known to influence the rates or success of microbial deposition and adhesion (Fletcher, 1977, 1988; Powell and Slater, 1983; Ho, 1986; Bryers, 1987; Sjollema *et al.*, 1988, 1990a,b). Microbial behaviors, including motile behaviors, chemotactic and sensory responses, and bacterial interactions with surfaces and other bacteria, have also been demonstrated to influence bacterial attachment (Korber *et al.*, 1990, 1994b; Lawrence *et al.*, 1989a, 1992). The following section examines some of the factors known to influence the initiation of stable microbiological films, particularly those that focus on bacterial adaptive behavior.

2.1. Physical Factors

2.1.1. Hydrodynamic Conditions

All microbiologically active surface microenvironments contain a constant or discontinuous water fraction, or phase, and as a consequence microorganisms have evolved mechanisms to optimize their growth and reproductive success at these interfaces. Although system hydrodynamics have been extensively discussed with regard to microbiological systems (Adamczyk and Van de Ven, 1981; Ho, 1986; Bryers, 1987; Characklis, 1990; Characklis *et al.*, 1990a,b), a brief review is relevant to any discussion of bacterial attachment because of the profound effect that the hydrodynamic regimen has during the attachment of suspended bacteria.

In static or quiescent systems, deposition of bacteria is governed primarily by various mechanisms of bacterial transport to surfaces, including simple Brownian diffusion, motility-enhanced diffusion, or cell settling via gravity (Walt *et al.*, 1985; Marmur and Ruckenstein, 1986; Sjollema *et al.*, 1988; Korber *et al.*, 1989, 1990). The majority of natural microbiological ecosystems possess a flow component when viewed spatially or temporally, the nature of which may be laminar or turbulent (Vogel, 1983). In either case, the velocity of flow decreases as the substratum is approached. This region is equally well-defined by the terms *viscous sublayer* or *hydrodynamic boundary layer*. All bacteria must traverse the hydrodynamic boundary layer in order to successfully colonize surfaces. In flowing systems, the attachment of bacteria is also under gravitational, diffusional, and behavioral control (Korber *et al.*, 1989, 1990, 1994b; Reynolds *et al.*, 1989; Camper *et al*, 1993); the magnitude of these contributing factors may be influenced by the velocity of flow as well as by the amount of advection that occurs (Vogel, 1983; Characklis *et al.*, 1990a,b).

2.1.2. Shear Effects

Flow velocity governs the hydraulic forces applied to attached or attaching bacteria, and has been mathematically defined in terms of shear stress (Rutter and Leech, 1980; Christersson *et al.*, 1988; Mittelman *et al.*, 1990; Characklis *et al.*, 1990a). The effect of shear forces on attached cells has been suggested as small by some workers; however, results from a number of experimental and theoretical studies evaluating the effect of shear force on bacterial attachment and desorption processes indicate otherwise (Christersson *et al.*, 1988; Characklis *et al.*, 1990a). For example, Powell and Slater (1983) demonstrated that the retention time for adhering *Bacillus cereus* cells to siliconized surfaces varied inversely with applied shear stress. Following a period of rapid cell deposition, it was observed that the number of attached cells achieved a maximum value, although the authors reported that this equilibrium appeared dynamic in that cells continually adhered and desorbed from the surface. Similarly, Korber *et al.* (1989) reported that increased flow rates decreased the attachment success of both motile and nonmotile *Pseudomonas fluorescens* suspensions to glass surfaces.

In contrast, some work suggests that an opposite relationship between flow or shear and bacterial attachment exists for other systems. Mittelman *et al.* (1990) observed that high shear forces resulted in the greatest rates of attachment for their *Pseudomonas atlantica* monoculture. Rutter and Leech (1980) determined that when flow velocity was increased, the deposition of both bacteria and latex microspheres increased. Interestingly, Rutter and Leech noted that cell deposition was enhanced under conditions of high fluid velocities when bacteria were grown at relatively low dilution rates (0.04 hr $^{-1}$), whereas when cells were grown under high dilution rates (> 0.2 hr $^{-1}$), adhesion decreased with increasing flow rates. Thus, the physiological state of the cell clearly defines the propensity for attachment in some systems.

2.1.3. Substratum Roughness

It has been observed that for flowing systems, bacterial adhesion may be locally increased by increasing surface "microroughness" (Geesey and Costerton, 1979; Baker, 1984). One result of surface roughness is increased advective transport of cells in the vicinity of the surface, with particles adhering more tenaciously once attached as a result of decreased shear force at the surface (Characklis *et al.*, 1990a). Although surface roughness may lead to increased cell deposition on surfaces, Baker (1984) observed that those cells that did attach were mainly found on flat regions rather than within crevices. Thus, any projection or irregularity present on an otherwise smooth surface may result in increased bacterial transport and attachment to surface sites in that region, though not specifically to the obstruction itself. In contrast, the observations of Notermans *et al.* (1991) indicated a preferential distribution of attached bacterial cells in association with crevices.

The effect of fluid hydrodynamics on the distribution of adhering particles has also been investigated with respect to site blocking by adsorbed particles rather than inherent surface roughness. Sjollema and Busscher (1989) compared the distribution of 0.8-μm diameter polystyrene particles that deposited on smooth surfaces within a laminar flow, parallel-plate system with a computer-simulated pattern of random particle adherence. Radial distributions between deposited particles indicated that a blocking effect inhibited the adherence of other particles over a range of 3 particle diameters, with a flow vector component resulting in stronger inhibition of particles adhering downstream of previously attached beads.

Interestingly, analysis of bacterial deposition patterns conflicted with previous information obtained using latex microspheres (Busscher *et al.*, 1990a). Positive cooperativity (the facilitation rather than inhibition of cell adsorption) of nonmotile *Streptococcus sanguis* cells to hydrophobic surfaces was demonstrated using laminar flow, where cells preferentially adsorbed downstream of previously attached cells. Further investigating the effects of flow velocity on bacterial adsorption, Sjollema *et al.* (1990a) demonstrated greater positive cooperativity for *S. sanguis* at high flow velocities, whereas studies using inert particles resulted in greater zones of inhibition. These observations clearly demonstrate the difficulty in using models based on abiotic observations or processes for predicting the behavior of living organisms.

Few studies have examined the role that fluid hydrodynamics or site blocking plays during the adsorption of motile bacteria capable of maneuvering within the surface microenvironment. Nearest-neighbor analysis conducted using motile and nonmotile strains of *P. fluorescens* demonstrated that motile bacteria released from growing microcolonies reattached randomly during surface recolonization in both high- and low-flow velocity environments (Korber *et al.*, 1989). In contrast, the distribution of nonmotile bacteria was approximately four times less random, with cells primarily reattaching immediately downstream of the parent microcolony. The observed patterns of nonmotile *P. fluorescens* cell reattachment were suggested to be the result of cells moving into a low-pressure region resulting from fluid flow over and around the parent microcolony from which the cells detached. The nonmotile strain was not capable of actively maneuvering over the surface prior to reattachment; detached cells simply reattached downstream of the parent colony or failed to reattach. It is important to note that in the case of initial attachment, bacteria would first have to traverse a relatively higher-pressure zone resulting from fluid flow over and around an obstruction prior to entering the low pressure area downstream or behind the obstruction.

2.1.4. Physicochemical and Thermodynamic Factors

The Derjaquin, Landau, Verwey, and Overbeek (DLVO, or double layer) theory has frequently been used to model the attractive and repulsive forces active at the bacterial–substratum interface, the sum of which dictates whether firm cell adhesion will occur (Van Loosdrecht *et al.*, 1987, 1989; Bellon-Fontaine *et al.*, 1990). Particle attraction occurs over shorter (≤ 1 nm) and longer distances (5–10 nm) (Ellwood *et al.*, 1982; Characklis *et al.*, 1990a) and are termed the primary minimum and secondary minimum, respectively. The DLVO theory predicts that between these two minima is a zone of maximum electrostatic repulsion, the force of which has been demonstrated theoretically and experimentally to decrease with solution electrolytic strength (Marshall *et al.*, 1971, 1989).

Van Loosdrecht *et al.* (1987), considering a large body of work examining DLVO theory and bacterial attachment, concluded that: (1) divalent cations increased attachment by adhering to the surfaces of bacteria or the substratum, thus decreasing the surface potential, and (2) increase in solution ionic strength or counter-ion valency increased attachment as a result of decreased electrostatic repulsion. Similar conclusions were reached by Absolom *et al.* (1983), who also attributed electrostatic interactions to cases where bacterial adhesion occurred at significant rates despite the setting of surface free energies at positive or zero values. These authors experimentally demonstrated that upon complexing divalent ions with a chelating agent, microbial adhesion was reduced to insignificant levels.

The microscale geometry of bacteria must be taken into account when using physicochemical models to describe microbial adhesion. Many cells possess appendages (e.g., fimbriae, pili, or flagella) that alter their effective diameter near the surface, and hence alter the diameter-dependent repulsive effects experienced between the primary and secondary minimum (Isaacson et al., 1977; Vesper and Bauer, 1986; Busscher et al., 1990b, 1992; Sjollema et al., 1990b). Complex mathematical derivations of DLVO theory that incorporate the geometry of the approaching bacterium and factors describing diffusion, convection, gravitational forces, and motile behaviors are likely necessary for general applicability of this approach (Adamczyk and Van de Ven, 1981; Sjollema et al., 1990b).

Hydrophobic interactions between cells and the substratum have also been demonstrated to play a role during bacterial adhesion (Rosenberg and Kjelleberg, 1986; Busscher et al., 1992). Free energy theory states that if the total free energy of a system is reduced by cell contact with a surface, then adsorption of the cell to the substratum is thermodynamically favorable (Absolom et al., 1983; Busscher et al., 1990a,b). Surface energy determinations involve measurement of the critical surface tensions, or hydrophobicities, of the surfaces involved and are not dependent on chemical measurements. Commonly, contact angles are measured microscopically between defined liquids and surfaces or surface preparations of microbial cells (Dexter et al., 1975; Fletcher and Loeb, 1979; Fletcher and Marshall, 1982; Fletcher, 1988; Rittle et al., 1990), the value of which may then be used to determine the critical surface tension for a surface after conditioning. Studies utilizing these criteria have demonstrated a range of results using various substrata, bacterial species, and assay methodology (Dexter, 1979; Fletcher and Loeb, 1979; Van Pelt et al., 1985; Rittle et al., 1990), with both increases and decreases in adsorption of bacteria using systems of various hydrophobicities. Much of the reported variability is largely due to the lack of procedural standards, as well as the heterogeneous, physicochemical nature of the bacterial cell (Fletcher, 1993).

2.2. Behavioral Factors

Various structures associated with the bacterial cell surface have been demonstrated or thought to be involved in the attachment of bacteria to surfaces. For instance, it was suggested that during the initial stages of attachment, adhesive polymers in the form of thin capsular coats, or thicker capsules (Marshall et al., 1989), may be involved (Geesey et al., 1977; Morris and McBride, 1984; Cowan *et al.*, 1991). McEldowney and Fletcher (1988) indicated that initial attachment may result from attractive interactions between polymers already present on the cell surface and the attachment surface without the requirement for the production of new adhesives. In contrast, Allison and Sutherland (1987) stated that little direct evidence has been presented to demonstrate the involvement of exopolymers in the initial adhesion process. These workers observed no differences in attachment rates between a capsulated bacterial strain and a nonmucoid mutant. However, subsequent colony formation was only observed for the capsulated and other slime-producing strains. This observation is in agreement with the findings from other studies indicating that the production of exopolymers provides a matrix in which biofilm cells are embedded (Allison and Sutherland, 1987; Marshall *et al.*, 1989; Vandevivere and Kirchman, 1993), with the possibility that the polymers have other functions in addition to adhesion (Neu and Marshall, 1991), and that they have an important role in the overall stability of biofilm communities (Uhlinger and White, 1983).

Other structures that have been suggested to play a role during bacterial attachment include fimbriae (fibrils) (Handley *et al.*, 1991), pili, and the holdfast associated with surface colonization by stalked bacteria such as *Caulobacter* species (Isaacson *et al.*, 1977; McSweegan and Walker, 1986; Vesper and Bauer, 1986; Allison and Sutherland, 1987; Merker and Smit, 1988). However, the most thoroughly studied and most visible facet of microbial behavior is flagellar motility. Motile behaviors have been demonstrated to influence a number of events associated with microbial surface colonization (Marshall *et al.*, 1971; Caldwell and Lawrence, 1986; Lawrence *et al.*, 1987, 1992; Marshall, 1988; Korber *et al.*, 1989, 1990, 1993, 1994b). These behaviors may be manifested in the form of increased mass transport of cells to surfaces, chemotactic movement toward attractant compounds, microbial positioning and recolonization behavior, and adhesion of the flagellar structure (Adler and Templeton, 1967; Adler, 1969, 1975; Lauffenburger *et al.*, 1981; Korber *et al.*, 1989; Reynolds *et al.*, 1989).

2.2.1. Motile Behavior and Initial Attachment

Motility has been shown to improve the success of many different bacteria during *in vitro* and *in situ* colonization events. For example, Stanley (1983) demonstrated that adsorption of *P. aeruginosa* cells decreased by 90% following mechanical removal of the flagellum. Similarly, the attachment success of motile *P. fluorescens* to glass surfaces decreased with increasing laminar flow velocities, but less than the decrease demonstrated for a nonmotile mutant of the same strain (Korber *et al.*, 1989). In addition, where the contribution of gravity to the deposition of these two strains was considered, the motile strain attached approximately 6 times more rapidly to lower surfaces than the nonmotile strain and approximately 77 times more rapidly to upper surfaces (Korber *et al.*, 1990).

This clearly demonstrated that gravity facilitated the attachment of nonmotile bacteria to surfaces in laminar systems and also provided evidence that otherwise physiologically equivalent motile cells possessed a significant competitive, surface-colonizing advantage in terms of attachment success and access to more restrictive or inaccessible sites.

Specialized flagellar systems have evolved for colonization of different surface environments. Vibrios possess lateral flagella (laf) that provide a mechanism for attachment to surfaces and dispersal in high-viscosity environments, as well as a polar flagellum active in low-viscosity environments (McCarter *et al.*, 1988, 1992; Lawrence *et al.*, 1992). A study by Korber *et al.* (1990) demonstrated that a polarly flagellated laf⁻ strain of *Vibrio parahaemolyticus* was poorly adherent even though the organism was motile and could effectively reach surface attachment sites at the same frequency as did the polarly and laterally flagellated wild-type strain. It appears that in the case of vibrios, mixed flagellation provides specific flagellar structures for both adherence and transport in the range of environments where vibrios are found (i.e., the open ocean and mucosal surfaces).

During competitive microbial attachment, likely the case in nature, the concept of monolayer attachment kinetics is important (Fletcher, 1977; Belas and Colwell, 1982; Korber *et al.*, 1994b). Organisms first gaining access to newly exposed surfaces may physically exclude a potential competitor organism, and by virtue of factors such as site blocking may enjoy significant advantages with respect to positioning and utilization of adsorbed nutrients. A number of different bacteria conform with monolayer attachment kinetics, with percent surface coverage (in the absence of significant growth) by various bacterial species ranging from 1 to 45% (Fletcher, 1977; Leech and Hefford, 1980; Belas and Colwell, 1982; Powell and Slater, 1983). Korber *et al.* (1990) determined that over a short-term time course (45 min) attachment rates for mot⁺ and fla⁻ *P. fluorescens* strains remained relatively constant, whereas attachment rates for two *V. parahaemolyticus* strains increased over time, even though multilayer attachment was not observed in either case. During a longer-term study, Korber *et al.* (1994b) found that attachment rates for mot⁺ and mot⁻ *P. fluorescens* did decrease over time, and even after equilibrium between attached and planktonic cell phases were reached, the surface area covered by bacteria did not exceed 28% for the more successful, motile *P. fluorescens* strain. The competitive significance of initial attachment and recolonization success was subsequently confirmed using dual-dilution continuous culture (DDCC), a system where dilution of both aqueous and particulate phases provided continuous selection pressure for rapidly colonizing bacteria. Nonmotile cells were consistently eliminated from the DDCC system before the motile strain at all aqueous rates of dilution (Fig. 1A,B), even though both strains of bacteria possessed identical attached and suspended growth rates (Korber *et al.*, 1989).

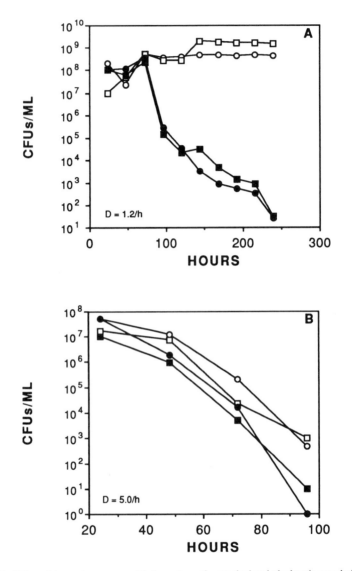

Figure 1. Effect of increasing aqueous dilution rate on the attached and planktonic populations of a wild-type *Pseudomonas fluorescens* and a fla⁻ nonmotile mutant of the same strain when grown in DDCC. (A) Note that at a dilution rate of 1.2 hr⁻¹, the motile strain maintained steady-state populations in both phases, whereas the nonmotile strain was eliminated from the system. (B) In contrast, when the dilution rate was increased to 5.0 hr⁻¹, both populations were unable to sustain themselves in either the attached or the planktonic phase. Reproduced with permission from Korber *et al.*, 1994b.

Advantages conferred to motile bacteria colonizing *in situ* surface microenvironments have also been examined experimentally and conceptually (Lauffenburger *et al.*, 1981; De Weger *et al.*, 1987; Haefele and Lindow, 1987; Kelly *et al.*, 1988). Haefele and Lindow (1987) showed that the spread of *Pseudomonas syringae* over bean leaf surfaces was facilitated by motility. Measuring the cell growth and competitive ability of these strains, Haefele and Lindow demonstrated that the loss of flagellar motility reduced the epiphytic fitness commonly observed in motile *P. syringae*, as indicated by the decreased resistance of motile cells to UV light irradiation and desiccation. The authors theorized that this decreased resistance was the result of the nonflagellated strain's inability to reach protected colonization sites. Guentzel and Berry (1983) and Gutierrex and Maddox (1987) also provided evidence that possession of a flagellum enhanced the pathogenesis of *Vibrio cholerae* in their model systems.

The role of motility during transport of motile and nonmotile bacteria has also been examined for porous soil/groundwater model systems (Jenneman *et al.*, 1985; Reynolds *et al.*, 1989). Such studies have shown that motile bacteria penetrate porous materials faster than nonmotile mutants. However, use of motile, chemotactic mutants demonstrated that chemotactic motility increased the breakthrough times in a porous model (Reynolds *et al.*, 1989). This suggests that in nature, penetration through porous materials may be less important, in terms of reproduction and survival, than attachment and colonization of surfaces. A study by Camper *et al.* (1993) revealed that for porous material with flow rates greater and less than the maximum rates of motility for the *P. fluorescens* strains, bacterial motility did not influence the timing of cell breakthrough, whereas adsorption of cells did. Benefits that motility provide bacteria in nature are often difficult to delineate because of the inherent complexity of most systems. To date, the use of carefully defined motility mutants has proved most useful in this regard (De Weger *et al.*, 1987; Haefele and Lindow, 1987; Korber *et al.*, 1989; Reynolds *et al.*, 1989; DeFlaun *et al.*, 1990; Camper *et al.*, 1993).

2.2.2. Chemosensory Behaviors during Attachment

The initial attachment of bacteria to surfaces may be described as a two-phase event that includes a reversible attachment phase and an irreversible attachment phase (Marshall *et al.*, 1971; Lawrence *et al.*, 1987; Marshall, 1988). Reversible attachment is exhibited by cells that attach to surfaces by a portion of the cell or flagellum while continuing to revolve (Meadows, 1971; Lawrence *et al.*, 1987; Malone, 1987; Marshall, 1988; Power and Marshall, 1988). This rotational behavior is then followed by either irreversible cell attachment or detachment.

Powell and Slater (1983) suggested that reversible attachment was predominant in nature. Marshall (1988), reaching a similar conclusion, reported that cells became irreversibly attached to surfaces only after a period of instability, during which time cells revolved around the axis of attachment and frequently emigrated from the attachment site rather than becoming irreversibly attached. This trend has subsequently been documented for pure cultures of bacteria and surface-colonizing bacteria from a number of natural and model systems (Meadows, 1971; Lawrence and Caldwell, 1987; Lawrence *et al.*, 1987, 1992; Malone, 1987; Korber *et al.*, 1989, 1990). It is clear from these and other results that not all cells that contact a surface become irreversibly attached. One explanation for bacterial spinning behavior in nature is that cells chemically evaluate potential attachment sites by binding, rotating, and sensing ambient conditions through chemoreceptors (Lawrence *et al.*, 1987).

The chemistry of acquired surface-conditioning films (Loeb and Niehof, 1975; Baier, 1980, 1984, 1985; Loeb, 1980; Rittle *et al.*, 1990) may thus play a central role in defining whether a surface is "viewed" as acceptable by bacteria exhibiting chemosensory behaviors. Bacterial chemotaxis to adsorbed organic films also provides an explanation for the directed movement and preferential positioning of many bacterial species on solid substrata (Berg and Brown, 1972; Adler, 1975; Kjelleberg *et al.*, 1982; Block *et al.*, 1983; Berg, 1985; Hermansson and Marshall, 1985).

Acquired and inherent surface chemistries have previously been demonstrated to affect the adsorption of natural epilithic bacterial communities, where preferential colonization of quartz by bacteria was observed relative to calcite (Mills and Maubrey, 1981). Ferris *et al.* (1989) found that variations in population densities of epilithic microorganisms present on different rocks in two rivers were inversely related to mineral substrate hardness. One question regarding the attachment of these cells is whether increased attachment was due to chemotaxis (active positioning of bacteria at specific surface sites) or stronger bonds between substratum and cell surface molecules. Indirect evidence that bacteria sense and respond to surface-bound nutrients was provided by Marshall (1988), who found the motile surface-association phase of *Pseudomonas* JD8 was dependent on the presence of bound stearic acid. Upon depletion of stearate, cells detached or remained on the surfaces but ceased to grow and divide. While changes in surface hydrophobicity were cited as a possible explanation for these observations, the events may represent a tactile response by the bacteria to substratum chemistry. A more compelling case for selective attachment based on substratum chemistry can be made for the thiobacilli. For example, the action of hydrophobicity as the selective mechanism during surface colonization by *Thiobacillus ferrooxidans* was generally inconsistent with the patterns of attachment described for pure sulfide minerals (Ohmura *et al.*, 1993), and thus was not likely to be the controlling factor. In addition, Ohmura and co-workers also established that electrostatic interactions could not explain the attachment to pyrite. It was suggested that the special interaction between *T. ferrooxidans* and sulfide minerals was based on sensing the presence of reduced iron in the substratum. In general, there is a lack of experimental evidence demonstrating chemotaxis to surface-bound attractant molecules; therefore, the preferential positioning of bacteria on surfaces coated by molecular films remains an area requiring further examination.

2.2.3. Motility at Interfaces

The motile attachment theory is based on the flow-independent movement of cells over surfaces in flowing microenvironments (Lawrence et al., 1987; Korber et al., 1989). During motile attachment, bacteria are thought to move along surfaces while remaining in a semiattached state, providing a mechanism by which cells may traverse or travel upstream against flow velocities exceeding the cells maximum rate of motility. Cells exhibiting motile attachment behavior have previously been reported by Lawrence et al. (1987) following observations of flow-independent behavior of P. fluorescens on surfaces where surface flow velocities were 200 μm sec^{-1}. Similarly, Rhizobium spp. have also been observed to travel upstream against flow rates higher than their maximum rates of motility (Malone, 1987). Korber et al. (1989) quantified the rates of backgrowth (upstream growth against flow) for mot$^+$ and mot$^-$ P. fluorescens strains grown within continuous-flow slide culture (CFSC) at flow velocities less than and greater than the rate of bacterial motility, and found that the mot$^+$ strain backgrowth at low flow was four times that of the mot$^-$ strain and at high flow backgrowth was two times greater than the mot$^-$ strain. The best explanation for the greater migration against flow by the motile strain was surface-associated motility. Lawrence and Caldwell (1987) also described a rolling behavior in which cells never became immobile but remained within the hydrodynamic boundary layer, growing and dividing. Gliding motility is a common surface-associated behavior as, for example, described for Myxococcus xanthus (Burchard, 1981; Kalos and Zissler, 1990; Shimkets, 1990). Overall, reports of movement of surface-associated cells are common, although the significance of such behaviors remains unclear. Explanations for motile attachment include bacterial positioning behaviors, nutrient capture strategies, and cellular migration within rapidly flowing systems.

2.2.4. Other Factors Influencing Initial Attachment

Nutrient status affects not only the size of bacteria but also the adhesiveness of the involved cells. For example, marine organisms grown under nutrient limitation have been shown to become more adhesive than when high-nutrient conditions exist (Rutter and Leech, 1980; Kjelleberg et al., 1987). Similarly, the adhesion of P. aeruginosa was related inversely to the rate of cell growth (Nelson et al., 1985). Under starvation conditions, enhanced adhesion has also been documented by a number of workers using a range of different organisms (Dawson *et al.*, 1981; Kjelleberg *et al.*, 1982; Camper *et al.*, 1993). However, this trend does not appear to be universal. Delaquis *et al.* (1989) demonstrated a decrease in the number of attached cells after nutrients were depleted in a batch culture system. Studies using starved bacteria suggest that cells at different phases of their life cycle (i.e., either growing or nongrowing) become physiologically distinct, sometimes resulting in an increase in adhesion of starved cells to surfaces (Wrangstadh *et al.*, 1989; Kjelleberg *et al.*, 1987). Van Loosdrecht *et al.* (1987) attempted to relate nutritional status of the cell to the physicochemistry of the cell surface. This study showed that cell surface hydrophobicity of *P. fluorescens* first increased and then decreased over time in batch culture. It was also shown that the surface hydrophobicity of unattached *P. fluorescens* increased as available substrate was depleted (Delaquis *et al.*, 1989). This increase was the result of emigration of cells from the surface to the bulk phase.

3. Growth and Behavior at Interfaces

Microorganisms express complex behavioral patterns during the colonization and occupation of surface environments. Behavior refers to any interaction between bacteria and their environment, including solution and surface chemistry, physical conditions, and other organisms occupying the same microhabitat (Lauffenberger *et al.*, 1984; Lawrence and Caldwell, 1987; Shapiro and Hsu, 1989; Lawrence *et al.*, 1992; Lawrence and Korber, 1993). A broad range of species-specific behaviors are included, proceeding through the series of events that include attachment (see Section 2), growth, microcolony formation, and redistribution/dispersal. These behaviors may be considered, regardless of the organism or the degree of morphological specialization, to represent a fundamental life cycle. A simple life cycle separates growth and reproduction from a dispersal phase. Thus, alternation between stalked mother cell and swarmer cell in *Caulobacter* spp., microcolony formation and recolonizing cell in *Pseudomonas* and *Vibrio* spp., or the complex life cycle of myxobacteria, all represent evolutionary solutions to growth and survival at interfaces.

Observations of bacterial growth and behavior at surfaces suggests that some organisms appear solitary in nature, whereas other behaviors are clearly colonial. It should be emphasized that some bacteria have undoubtedly been mistaken to exist solely in the solitary state due to the limitations inherent to many studies, including the inability to resolve surface-associated activity and behavior from within a complicated microbial community. The occurrence of "solitary" organisms such as the prosthecate bacteria or *Caulobacter* spp. within complex, high-nutrient systems such as sewage bioreactors, polluted streams, and intertidal zones, however, indicate successful adaptation to community life (Hirsch, 1968; Hirsch and Rheinheimer, 1968; Staley, 1971; Poindexter, 1981;

Semenov and Staley, 1992). For clarity and simplicity, a somewhat artificial distinction is drawn in the following discussion between those organisms that have been observed to exhibit a solitary life cycle component and those that associate or aggregate with other cells.

3.1. Solitary Surface Colonization Behaviors

Organisms that may be solitary in nature include the budding and shedding bacteria (Hirsch, 1974, 1984; Lawrence and Caldwell, 1987; Power and Marshall, 1988; Lawrence et al., 1989b). This includes organisms, which are not morphologically distinct, that alternate between the attached condition and a free-living swarmer cell, and have been reported to occur in a diverse range of natural systems (Henrici and Johnson, 1935; Helmsetter, 1969; Bott and Brock, 1970; Marshall and Cruickshank, 1973; Lawrence and Caldwell, 1987). A generalized illustration of shedding behavior, modeled after *Rhizobium*, is shown in Fig. 2. It also includes representatives of the budding bacteria, which encompass the prosthecate organisms such as *Pedomicrobium* and *Hyphomicrobium* (Hirsch, 1968; Lawrence et al., 1989b; Caldwell et al., 1992a).

Rhizobium spp. attach perpendicular to surfaces and release daughter cells into the bulk phase when growing in the free-living condition (i.e., not in a plant–microbe association) (Lawrence et al., 1989b). This growth pattern is similar to *Caulobacter* spp., which attaches via an adhesive, holdfast organelle, forms a stalk, and may develop planktonic aggregations or rosettes (Stove-Poindexter, 1964; Hirsch, 1984; Merker and Smit, 1988). Overall, the budding bacteria have extensively been reviewed in terms of morphology and growth habit (Hirsch, 1974; Dow et al., 1976; Moore, 1981; Poindexter, 1981; Semenov and Staley, 1992).

Kjelleberg et al. (1982) described the attachment of starved *Vibrio* DW1 adhering to a surface in a perpendicular position and following growth of the

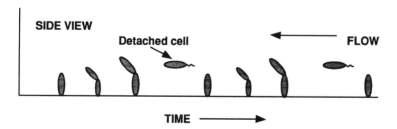

Figure 2. A schematic diagram illustrating shedding behavior as exhibited by numerous unidentified bacteria, including *Rhizobium* spp. grown in CFSC (continuous flow slide culture).

attached cells; motile daughter cells were released from these mother cells at regular intervals. In the case of *Rhizobium* spp., motile daughter cells were released that subsequently reattached at vacant surface sites and underwent growth and release of their own daughter cells. An additional variation on this behavioral theme was reported by Power and Marshall (1988), who described *Vibrio* MH3 that grew associated with the surface but completed cell division in the planktonic state.

Lawrence *et al.* (1989b) demonstrated that the growth and development of a *Pedomicrobium* sp. proceeded with the attachment of a pioneer cell, the formation and elongation of one to four hyphae from the central cell, and the production of daughter cells, which eventually were released, on the apical tips of hyphae. The growth of the mother and subsequently formed daughter cells occurred at distinct rates, with the daughter cell developing approximately 20 times more rapidly than the mother cell (doubling time = 24 hr for the mother and 1.2 hr for the daughter). It is significant that pedomicrobial nets, involving extensively intertwined hyphae and extracellular polysaccharides, may also form following extended surface colonization, a behavior clearly indicating a disposition for colonial growth under appropriate conditions (Hirsch, 1968; J. R. Lawrence, D. R. Korber, and D. E. Caldwell, unpublished data).

Solitary bacteria have also been demonstrated to exhibit various degrees of movement over surfaces. Rolling surface behavior, a simple although poorly understood growth habit, has been documented for members of stream communities by Lawrence and Caldwell (1987). These organisms slowly migrate over surfaces and divide without ever becoming firmly attached. It was suggested that this type of movement over surfaces was the result of sluggish rotation of the flagellum, while the cell remained within the secondary minimum of the diffuse double layer. Such a behavioral pattern may help disperse cells from the parent, facilitate nutrient scavenging, and prevent washout within flowing systems.

Gliding bacteria, including the genera *Cytophaga* and *Flexibacter,* provide further examples of single-cell behavior best observed on sparsely populated surfaces. Mechanisms of bacterial gliding are poorly understood; however, for *M. xanthus,* movement of the cell is achieved by the smooth movement of cells in the direction of their long axis without the use of flagella (Burchard, 1981; Kalos and Zissler, 1990). Locomotion for the various genera of gliding bacteria have been described as sluggish, often interrupted by spinning, flipping, flexing, or pendulumlike movements (Burchard, 1984; Godwin *et al.,* 1989). Hydrophobic interactions apparently play an important role during gliding behaviors. Burchard *et al.* (1990) demonstrated that adhesion of gliding bacteria was tenacious on hydrophobic surfaces and tenuous on hydrophilic surfaces. On highly hydrophobic surface sites, exclusion of water would facilitate adhesion, whereas on hydrophilic sites, a layer of water would result in gliding motility (Sorongon *et al.,* 1991). Burchard and co-workers (1990) hypothesized that the range of movements observed for gliding bacteria in their study may have reflected variations in substratum chemistry. The adaptive significance of the two distinct stages in the life cycle of these organisms include optimization of cell positioning and resource exploitation (Lawrence and Caldwell, 1987; Lawrence and Korber, 1993).

3.2. Colonial Surface Colonization Behaviors

Perhaps the most commonly observed colonization strategy employed by bacteria is the formation of colonies, or cell aggregates, from a single point of bacterial colonization. As shown for a variety *Pseudomonas, Vibrio, Klebsiella,* and *Escherichia* species, colonies develop from an individual attached pioneering cell that goes through a number of division cycles, eventually forming a bacterial microcolony. A range of variations on this main theme has been documented for pure culture and unidentified bacteria by a number of studies, the majority of which have been performed by manual microscopic observations (Marshall *et al.*, 1971; Hirsch, 1984; Lawrence and Caldwell, 1987; Lawrence *et al.*, 1987, 1989a,b, 1992; Korber *et al.*, 1989).

The spatial distribution of individual microcolony members varies with cell morphology, cell nutrient status, cell size, mode of cell wall growth, polymer production, and poorly understood events occurring at the moment of cell division. For example, *Corynebacterium* spp. undergo a snapping cell division, resulting in the formation of V-shaped daughter pairs comparable to Chinese letters (Hirsch, 1984). Similar snapping movements at the moment of separation of daughter cells also occurs in *P. fluorescens* and *Escherichia coli* (Lawrence *et al.*, 1987; Shapiro and Hsu, 1989). The cellular alignments that occur between *E. coli* individuals grown on agar surfaces were interpreted by Shapiro and Hsu (1989) as interactive behaviors.

Other unidentified bacteria attach to surfaces where they grow and divide, with progeny gradually but continuously moving away from other members of the spreading microcolony. The migration of "attached" cells has been reported by Power and Marshall (1988) using time-lapse video analysis to study the colonization behavior of starved *Pseudomonas* JD8. The authors determined that cells did not irreversibly attach to surfaces coated with fatty acids. Rather, bacteria grew and divided, with each daughter cell migrating away from the parent cell at slow rates ($0.04-0.19$ μm min^{-1}). This mode of surface colonization has also been documented by Lawrence and Caldwell (1987), who described daughter cells of an unidentified bacterium that moved apart during 14 hr of development, resulting in a loose association of attached cells. The mechanism proposed for this mode of surface growth was related to utilization of the surface-bound substrate stearic acid (Power and Marshall, 1988). When the nutrient concentration declined in the microenvironment surrounding the cell, it became reversibly attached and moved until encountering more substrate. This cycle was repeated until all substrate was utilized, at which time the cells returned to the

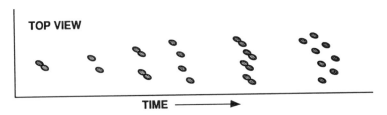

Figure 3. Schematic diagram showing the sequence of events involved in spreading colony formation.

planktonic state. Similar behaviors have been reported for an uncharacterized *Acinetobacter* sp. growing in CFSC (James *et al.*, 1995). A schematic diagram illustrating this mode of surface colonization is presented in Figure 3.

   Pseudomonas fluorescens CC-840406-E follows what has been termed "packing" microcolony formation behavior (Fig. 4), starting with attached individuals that grow and divide, forming a tightly packed monolayer of cells on the surface (Lawrence *et al.*, 1987; Lawrence and Korber, 1993). This packing behavior as exhibited by *P. fluorescens* is illustrated in Fig. 5. Other pseudomonads have remarkably different patterns of microcolony formation to that of *P. fluorescens*. Delaquis (1990) observed that *Pseudomonas fragi* microcolonies developed via a packing maneuver, which formed circular colonies consisting of a tightly packed monolayer of cells (Fig. 6).

   Vibrio parahaemolyticus and *Vibrio harveyi* similarly follow a packing colony formation pathway (Lawrence *et al.*, 1992; Lawrence and Korber, 1993), with events paralleling that of *P. fluorescens*. *Vibrio parahaemolyticus* grown on

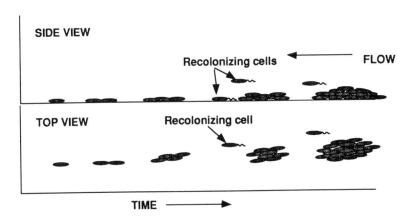

Figure 4. Schematic diagram illustrating the series of cell movements that occur during the formation of a packing colony, with a subsequent dispersal and reattachment/recolonization phase.

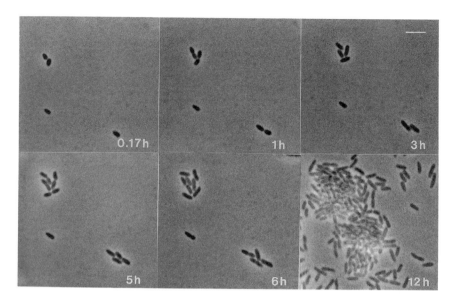

Figure 5. Phase-contrast photomicrographs showing packing colony formation during the development of *P. fluorescens* microcolonies (0.17 to 6 hr). Bar = 5μm.

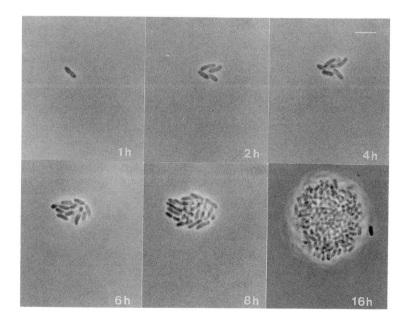

Figure 6. A series of phase-contrast photomicrographs showing the events resulting in the tightly packed microcolonies typical of *P. fragi*. Bar = 5 μm.

John R. Lawrence *et al.*

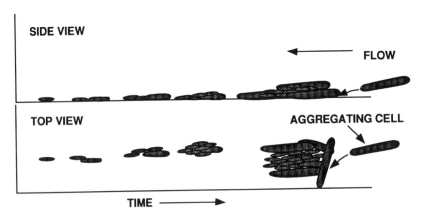

Figure 7. Diagram of colony formation by *Vibrio* spp. Initial colony formation occurred via a packing maneuver followed by dispersion of some colony members. Migrating cells subsequently reattached at the sites of original microcolonies, aggregating preferentially at these locations.

surfaces is capable of expressing two phenotypes. The opaque variant was expressed as a short, motile rod that was adapted to migration in low-viscosity environments. The translucent variant was expressed as an elongated, laterally flagellated variant demonstrated to be active during swarming in viscous systems up to 1000 cP as well as attachment. Following attachment of the opaque morphotype and during colonial growth, some colony members underwent phenotypic switching, assuming the translucent phenotype. During recolonization at approximately 4 hr, opaque cells frequently detached and recolonized at or near sites of existing colonies, facilitating the formation of a more complicated microcolony through aggregation. A schematic diagram illustrating this mode of microcolony formation is shown as Fig. 7. The translucent variant, in contrast, did not recolonize at low viscosity and remained attached. Under high-viscosity conditions, both types of cells detached from existing colonies; opaque cells were immobilized by the high-viscosity medium, whereas translucent cells emigrated or migrated continually, exhibiting a series of tactile chemosensory movements near and around existing microcolonies (Lawrence *et al.*, 1992). The mechanisms controlling the formation of both variants involve phenotypic switching at an unknown frequency. The induction of the lateral flagella gene system (McCarter *et al.*, 1992) in response to surfaces also plays a crucial role in adaptation of *V. parahaemolyticus* to surface environments.

3.3. Dispersal Phases

Each of the surface colonization patterns described in Section 3.2 features the critical phase of dispersal, involving detachment, migration, and in most,

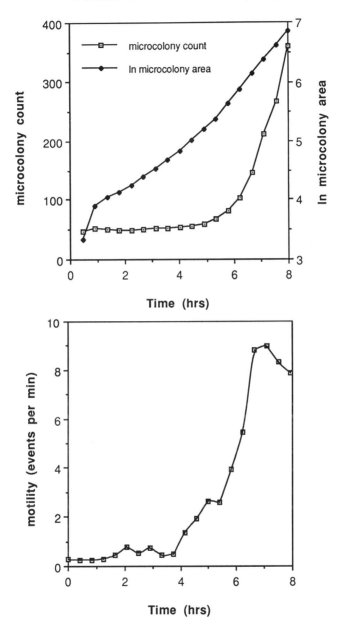

Figure 8. Digital image analyses of the change in number of attached *P. fluorescens* microcolonies and number of motile cells detected in the hydrodynamic boundary layer during growth in CFSC. Note the onset of the dispersal or recolonization phase at 4–5 hr as indicated by the increase in motility index and number of attached cells/microcolonies.

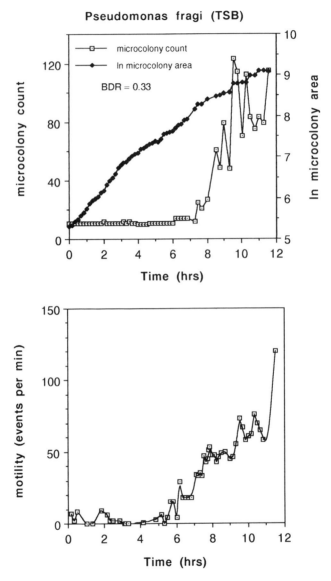

Figure 9. Digital image analysis of *P. fragi* growth in CFSC, showing the extended (relative to *P. fluorescens*) period of microcolony formation, followed by dispersal/recolonization phase. BDR, biofilm development rate.

although not all, cases subsequent reattachment at a suitable site. Lawrence *et al.*
(1987) termed this event *recolonization,* a phase that for *P. fluorescens* included
the emigration of attached cells from the surface to the suspended phase and the
subsequent reattachment of those cells to vacant surface sites. The timing of the
onset of this phase varied between species as well as with the conditions under
which the cells were growing (Lawrence and Korber, 1993). Attached *P. fluorescens* growing in CFSC recolonized following approximately 5 hr or after
microcolonies reach the 8- to 16-cell stage, during which time a number of cells
reattached in vacant areas between existing microcolonies. This event can result
in approximately a 600% increase in the number of microcolonies present in a
given area over an 8-hr period (Korber *et al.,* 1989). The extent of recolonization
has also been demonstrated to be under hydrodynamic and behavioral control.

A typical data set for *P. fluorescens* illustrating the growth phase, increase
in number of motile cells, and increase in microcolony number is shown in Fig.
8. A series of phase-contrast images showing the actual growth and development
of the microcolonies was shown previously as Fig. 5. In this instance, redistribu-

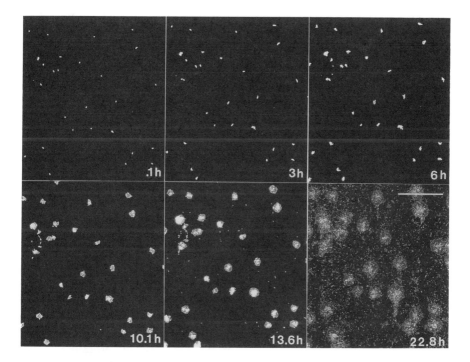

Figure 10. Darkfield photographs illustrating the events occurring during *P. fragi* growth in CFSC,
illustrating microcolony development and redistribution of cells on the surface after dispersal. Bar =
10 μm.

tion occurred following 4–5 hr of surface-associated growth. In contrast, *P. fragi* exhibited delayed recolonization (Fig. 9) and a distinctive microcolony morphology when compared with *P. fluorescens,* as shown in Fig. 10. In addition, *V. parahaemolyticus* grown in 2216 marine broth (75% w/v) was determined to recolonize after 4-hr surface growth, whereas *V. harveyi* grown under identical conditions recolonized after only 2 hr (Lawrence *et al.,* 1992; Lawrence and Korber, 1993).

Variation in the timing of cell redistribution represents an important behavioral characteristic that appears to be species-specific. Recolonization provides a mechanism whereby cells may emigrate from heavily colonized regions, where growth factors may be depleted, to regions more amenable to rapid growth. However, in detailed studies of individual cells within microcolonies, no evidence for a change in growth rate as a trigger has been detected (J. R. Lawrence and D. R. Korber, unpublished data). It is therefore likely that this phenomenon represents variable expression of traits, spontaneous variants, or other mechanisms that generate diversity, allowing organisms to adapt to changing environments, and that provide at least two states optimal for different circumstances. This latter mechanism appears to operate in the control of recolonization of *V. parahaemolyticus* in which two variants spontaneously arise: a translucent variant with a functional laf system, and an opaque variant that functions solely for planktonic dispersal at low viscosity (Lawrence *et al.,* 1992).

3.4. Aggregation and Coaggregation

Behaviors such as aggregation play an important role in the development of microcolonies and, subsequently, biofilm architecture. For example, nearly all human oral bacteria participate in intergeneric coaggregation (Komiyama *et al.,* 1987; Kolenbrander, 1989, 1993). The term *cooperative effects* among oral bacteria designates a microbial process by which adhering cells can modify their surrounding environment into a more favorable one for further attachment (Sjollema *et al.,* 1990a; Busscher *et al.,* 1992). Surveys of more than 700 strains of oral bacteria concluded that partner recognition during bacterial coaggregation is very specific (Kolenbrander, 1989). Coaggregation may occur intergenerically, intragenerically, and multigenerically, and these interactions appear to be extremely important during the colonization and occupation of substrata. These phenomena have been reviewed in detail by Kolenbrander (1989) and Kolenbrander and London (1992).

Observations of the growth and development of *P. fragi* microcolonies indicated that after onset of migration an extensive layer of vertically attached cells (a palisade layer) developed on the surface of the initial microcolonies (Fig. 11). This layer has also been observed in biofilms of *P. fluorescens,* suggesting that some cells may aggregate to form specialized structural elements during

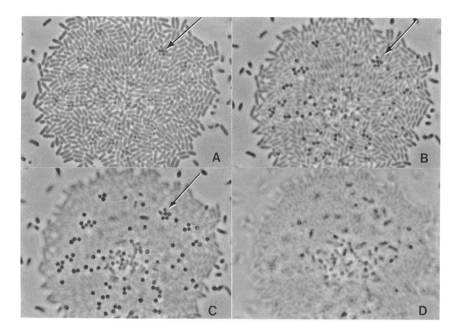

Figure 11. Phase-contrast micrographs showing the development of the palisade layer on the surface of *P. fragi* microcolonies during and after recolonization. These vertically attached cells are thought to play a role in the further development of biofilm structure. Bar = 5 μm.

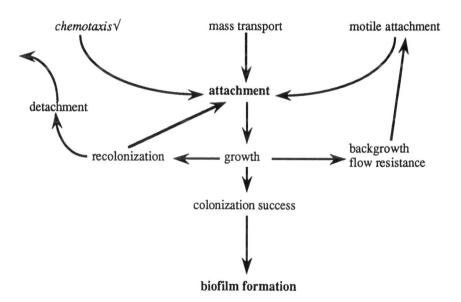

Figure 12. A schematic diagram illustrating the distinct phases of surface colonization behaviors exhibited in whole or in part by various bacterial species.

biofilm development (Korber *et al.*, 1993; Lawrence and Korber, 1993). Similarly, extensive aggregation during the development of *V. parahaemolyticus* microcolonies has been described (Fig. 7) (Lawrence *et al.*, 1992). Cell aggregates were still evident in the form of distinct microcolonies attached to the surface at the base of the *V. parahaemolyticus* biofilms after 12 and 24 hr of growth (Lawrence *et al.*, 1991). The architecture of these biofilms was also influenced by cell migration, a phenomenon apparent during the development of a diffuse basal layer with numerous channels.

The nature of cellular interactions in biofilm formation may be further investigated through the use of nonmotile mutants. Studies of this nature have been carried out using wild-type and mot⁻ mutants of *P. fluorescens* (Korber *et al.*, 1989, 1993). These latter studies have concluded that many of the structures that develop in microcolonies and within biofilms require active, motility dependent cell–cell interactions. The phases of surface colonization exhibited by various bacterial species are summarized in Fig. 12.

3.5. Environmental Effects on Microcolony Formation and Dispersal

Surface colonization pathways are influenced by a number of factors, including flow rates, viscosity, nutrient status, starvation, and light intensity (Hirsch, 1968, 1974; Kjelleberg *et al.*, 1982; Marshall, 1988; Power and Marshall, 1988; Korber *et al.*, 1989, 1990; Lawrence *et al.*, 1989a, 1992). However, in some cases, no effect of the environment has been detected. For example, the developmental pathway of *P. fluorescens* strain CC840406-E was unaffected by factors such as initial cell density, nutrition, and laminar flow velocity (Korber *et al.*, 1989).

In general, behavioral sequences do vary with respect to environmental conditions, possibly through the regulation and induction of specific genes. Nutritional status affects the sequence of events that occur during surface-associated growth by *P. syringae,* where the formation of microcolonies in CFSC was inhibited by growth in 10% trypticase soy medium. Under these conditions, *P. syringae* initially adhered to the surface and then detached without a growth phase or microcolony development. In contrast, *P. syringae* cells cultured using defined glucose minimal salts media grew to form microcolonies, without significant dispersal (Fig. 13). Digital-image analysis was used to quantify rates of change in colony number, frequency of motile cells, and attached biomass (Fig. 14 A,B). It is likely that these patterns represent the behavior of at least two variants that exist in the population of *P. syringae.*

Kjelleberg *et al.* (1982) and Marshall (1989) have documented that surface overcrowding resulted in nutrient depletion, cessation of growth, and induction of a starvation-survival phase in *Vibrio* DW1. James *et al.* (1995) described a nutrient-dependent, behavioral response for an *Acinetobacter* isolate, where cells

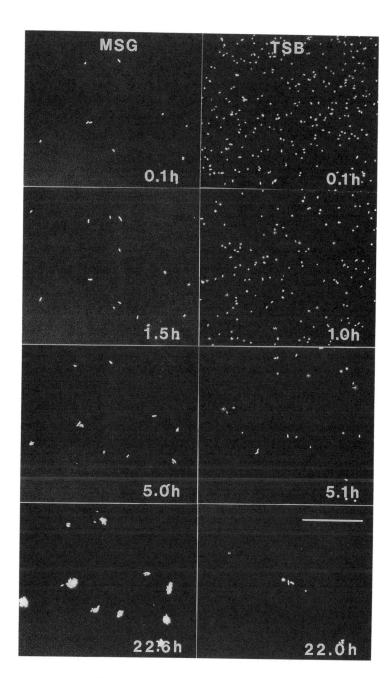

Figure 13. A series of darkfield images showing the effect of carbon source on the attachment and microcolony formation pattern of *P. syringae*. In this instance, the bacteria attached in high numbers in the presence of 10% trypticase soy broth medium (TSB) but subsequently detached over time, whereas in the presence of glucose-mineral salts medium (MSG), initial attachment was reduced and the attached cells formed microcolonies without exhibiting a dispersal phase. Bar = 10 μm.

Pseudomonas syringae (MS 1 g/L glucose)

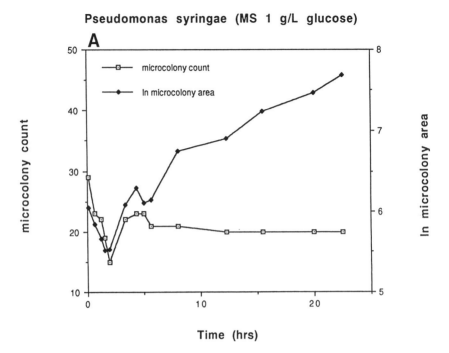

Time (hrs)

Figure 14. Digital image analysis of the two surface colonization patterns exhibited by *P. syringae*, showing the kinetics of detachment and microcolony formation in the presence of either (A) glucose-mineral salts medium or (B) 10% trypticase soy broth.

formed a spreading microcolony under high-nutrient conditions, whereas packed microcolonies were formed under low-nutrient conditions. In a study of natural stream communities, Lawrence and Caldwell (1987) observed that one colony type responded to the cessation of flow by immediately emigrating from the surface. The microcolony was stable over the preceding 10–12 hr, indicating a possible response to changes in the microenvironment as a consequence of cessation of flow.

Thus, bacteria have a variety of strategies for adapting to life in a surface habitat and also for releasing progeny from the attached to the planktonic state. Behavioral adaptations are an integral part of surface colonization and represent: (1) adaptations for colonization and occupation of the substratum, (2) optimization of substrate availability, (3) strategies for reproductive success, (4) balancing of intra- and interspecific competition, and (5) responses to environmental stress. However, more information is required regarding the quantitative relationships between growth rate, reproductive success, resistance to stress, and the

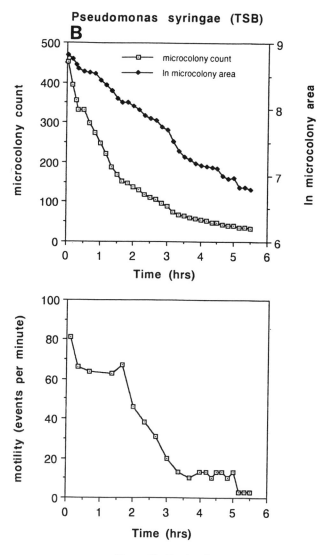

Figure 14. Continued

relative merits of specific behavioral patterns. In addition, it is clear that cell–cell interactions play a role in the establishment of the structural matrix of microcolonies and biofilms. The role of these behaviors in community development will require further studies involving direct, continuous microscopic observation.

4. Formation of Heterogeneous Biofilm Communities

Biofilm development begins when a solid surface is immersed in an aqueous environment and organic molecules adsorb, forming a macromolecular conditioning film. Attachment of a variety of bacteria to the surface occurs rapidly with growth and production of exopolymers (EPs). The production of extracellular polymers is pivotal to community development on surfaces because it provides an interface between the cell and the external environment, influencing rates of chemical exchange, availability of nutrients, and facilitating the creation of microniches. In addition, the presence of EPs influences the susceptibility of bacteria to stress and facilitates cell–cell interaction. Biofilm organisms, which frequently represent a range of physiological/functional groups, may benefit from the attached mode of growth through protection provided by the polymeric matrix against outside perturbations (Fletcher, 1984). Fletcher (1984) also suggested that bacterial cells that are embedded in a polymeric matrix may be capable of interactions.

Bacteria live in a diffusion-dominated environment, and as individuals are unable to alter the environment in which they live (Koch, 1991). As a consequence of microbial activity and diffusion limitation, the development of specialized microniches or microenvironments occurs. The microorganisms respond to the changing environment through activation or inactivation of genes, for example, in response to increasing carbon dioxide (Goodman *et al.*, 1993a). In biofilms, the close proximity of organisms to one another may promote interaction and genetic exchange between different species (Bouwer, 1989). The larger genetic pool of the microbial community therefore improves the chances for the development of new metabolic pathways and strategies to adapt to changing conditions.

Although many studies over the last decade have provided a better understanding of the structure and function of biofilms, very little has been done to study the spatial organization within these communities. For example, while the molecular basis of degradation and the evolution of new genes by microorganisms to facilitate the degradation of xenobiotic compounds have been investigated by many to obtain a better understanding of biodegradation (Furukawa *et al.*, 1989; Rothmel *et al.*, 1989), the ways that these genotypic changes are expressed in degradative biofilms are largely unknown. Therefore, our current knowledge of biofilm communities is based on observations of pure culture biofilms or is extrapolated from simplified model systems.

4.1. Concept of the Microenvironment

The ambient conditions associated with clean surfaces are temporally and spatially modified by the colonization, growth, and activity of bacteria. Microbial consumption of nutrients, production of wastes, and synthesis of cellular and extracellular materials all act in concert to physicochemically define what is known as the microbial microenvironment (Hamilton, 1987). Biofilm communities are likely to be characterized by steep, continuously shifting, physicochemical gradients within the biofilm. In addition to these physical gradients, the community is also spatially heterogeneous in terms of cell physiological state and genetically heterogeneous in terms of cell phenotype and genotype. Recently, Costerton *et al.* (1994) discussed these concepts in terms of biofilms being a customized microniche. These microenvironments persist over time due to diffusion limitations in combination with sustained metabolic activity. The presence of concentration gradients can influence bacterial population diversity and spatial distribution as well as microbial metabolic activity within biofilms, allowing the survival of other organisms with specific, and often stringent, growth requirements. The microenvironment is therefore believed to be the foundation for much of the microbial diversity observed in nature.

Although the microniche is a fundamental concept in biofilm community development, its presence has proved difficult to directly verify. However, there is extensive direct and indirect information on physical and chemical spatial heterogeneity in biofilms and includes: chemical measurements performed using microelectrodes; visualization of biofilm architecture and chemical gradients using SCLM and environmentally sensitive fluorescent probes; estimation of diffusion rates from fluor penetration studies; examination of biofilm surface and EPs chemistry using lectin and charge-sensitive probes; and rRNA oligonucleotide studies of biofilm population phylogeny (Lawrence *et al.*, 1991, 1994; Assmus *et al.*, 1995; Lens *et al.*, 1993; Wolfaardt *et al.*, 1994a–c; 1995a,b).

Architectural analyses of biofilms have demonstrated the presence of a high degree of spatial heterogeneity (Lawrence *et al.*, 1991; Stewart *et al.*, 1993; Korber *et al.*, 1993, 1994a,c). SCLM analyses have shown variations in arrangements of cells, EPs, and space within pure culture biofilms of *V. parahaemolyticus, P. fluorescens,* and *P. aeruginosa* (Lawrence *et al.*, 1991). These arrangements result in the development of channels and pores through the biofilm matrix that allow penetration of high-molecular-weight probes (e.g., 2×10^6 mol. wt. dextrans) (Lawrence *et al.*, 1994). Observation of inert particles and nuclear magnetic resonance (NMR) visualization of flow patterns have also shown the presence of interconnected channels within biofilms (Costerton *et al.*, 1994). These observations and others have provided evidence that the scale of analysis for biofilms must be sufficient to encompass this heterogeneity (i.e., $> 1 \times 10^5$ µm^2) (Korber *et al.*, 1993).

Analyses have also shown mixed-species biofilms to be quite variable (Robinson *et al.*, 1984; Wolfaardt *et al.*, 1994a). Lawrence *et al.* (1994) demonstrated that extensive pore and channel networks also developed in a multispecies biofilm community. Keevil and Walker (1992) reported that natural biofilms observed with differential interference contrast (DIC) and epifluorescence microscopy consisted of a basal cell layer approximately 5- to 10-μm thick, with stacks of polymer-associated cells extending from the surface into the bulk aqueous phase at regular intervals. These structures were presumed to facilitate the flux of nutrients to the basal biofilm layers, thereby supporting the growth of aerobic heterotrophs while creating chemical microenvironments that would support the growth of anaerobic organisms as well.

The application of pH-sensitive probes in conjunction with scanning confocal microscopy has also provided information on microenvironments, indicating the presence of 1- to 3-μm zones of reduced fluorescence surrounding *V. parahaemolyticus* cells and microcolonies within biofilms (Caldwell *et al.*, 1992a, 1993). Use of microelectrode techniques to study bacteria on surfaces and in aggregates also provides a wide range of data on activities within bacterial associations (Alldredge and Cohen, 1987; Revsbech, 1989; Lens *et al.*, 1993; Lewandowski *et al.*, 1989, 1993), and has confirmed the existence of steep

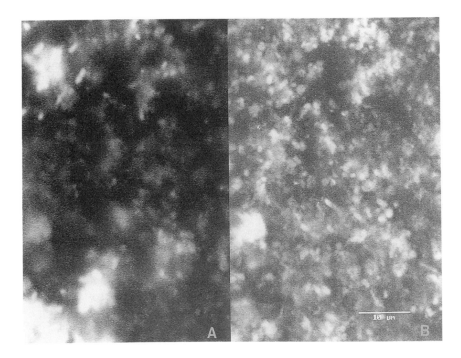

Figure 15. A panel of fluor-conjugated lectins was used to assess the location of specific residues within the matrix of a biofilm community. In this instance the relative binding of two lectins is shown in confocal laser optical thin sections using dual channel imaging: (A) the TRITC-*Lycopersicon* lectin, which binds to (D-glcNAC)₃ and (B) the FITC-*Canavalia* lectin, which binds to α-D-mannose and α-D-glucose. Note that the binding pattern of *Lycopersicon* (A) is limited in distribution. Bar = 10 μm.

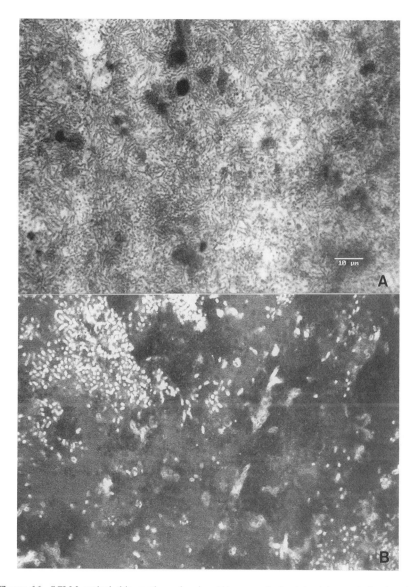

Figure 16. SCLM optical thin sections showing (A) a negatively stained image indicating the distribution of all cells, and (B) the localized binding pattern of a 70-kDa molecular weight dextran with a net polyanionic charge. These images show that although biofilms may exhibit a net negative charge, considerable spatial variability exists in the nature of the charge and charge density within the matrix. Bar = 10 μm.

gradients through biofilms and bioaggregates. Lewandowski *et al.* (1991) demonstrated that dissolved oxygen levels decreased at the biofilm–bulk phase interface and declined further with depth. Following microelectrode studies by Lens *et al.* (1993), it was concluded that the outer regions of sludge granules appeared to be the sites of acetogen activity (from pH profiles), whereas granule centers appeared to be the sites of methanogenic activity, as evidenced by a decreased pH in this region. McLean *et al.* (1991) indirectly demonstrated the development of high-pH microzones in *Proteus mirabilis* biofilms by studying the formation of struvite crystals in the biofilm matrix and showing that they resisted dissolution.

The microenvironment is also significantly influenced by the chemical nature of the exopolymeric matrix. Wolfaardt *et al.* (1995b) used fluorescently conjugated lectin probes and charged dextrans to demonstrate that different members of a community expressed unique EP chemistries (Figs. 15 and 16). These variations in EP chemistry likely play an important role in the development of chemical microzones through their influence on sorption/desorption and diffusion characteristics of the matrix. The nature of these polymers has also been characterized by monitoring the mobility of various fluorescently labeled chemical species through the biofilm matrix and calculation of effective diffusion coefficients and porosity values for the matrix (Lawrence *et al.*, 1994). This study showed that diffusion coefficients (D_e) were highly variable depending on location within the biofilm, and that D_e was also dependent on the hydrodynamic radius of the diffusing molecule. An example of spatial variability during a time course of the penetration of fluorescently labeled dextrans is shown in Fig. 17. These types of studies have proved valuable for defining the spatial variability of the biofilm EPs matrix.

One of the best examples of temporal/spatial modification of biofilm chemistry results from the activity of aerobic heterotrophic bacteria. Within a newly formed biofilm, rapid utilization of available nutrients leads to an increase in microbial biomass with a concurrent increase in the demand for oxygen. Thus, anaerobic zones form due to metabolic oxygen depletion and diffusion limitation, potentially allowing proliferation of anaerobic microorganisms while limiting the success of aerobes. Korber *et al.* (1993) indirectly demonstrated the presence of low O_2 microenvironments following observations of cell filaments within the lower regions of 24-hr *P. fluorescens* biofilms (Fig. 18). Earlier studies have demonstrated that filament formation in aerobic organisms is a physiological response to conditions of oxygen limitation (Jensen and Woolfolk, 1985; Wright *et al.*, 1988).

Recently, Amann *et al.* (1992) conducted studies using natural populations of sulfate-reducing bacteria (SRB) within multispecies biofilms, demonstrating that two types of cell distributions were observed when biofilms were stained with SRB-specific oligonucleotides. One probe distribution pattern indicated a microcolony growth habit for *Desulfovibrio vulgaris,* whereas the *Desulfovibrio*

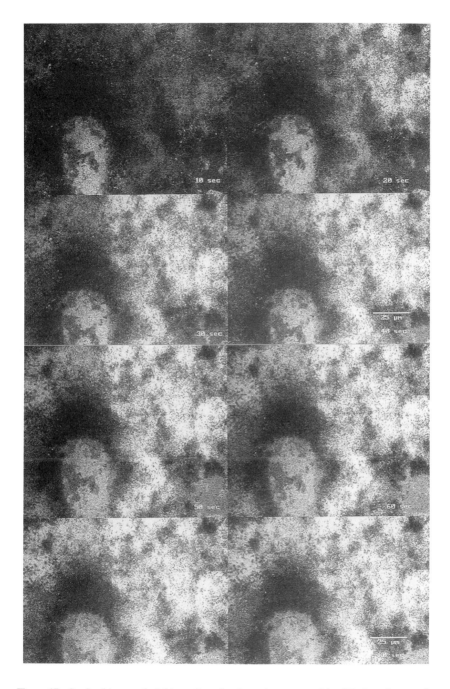

Figure 17. Confocal laser optical thin sections showing a time course of the diffusion of a neutrally charged 40 - kDa molecular weight dextran into a biofilm matrix. The penetration pattern indicated the inherent spatial variability of diffusion within the matrix. Bar = 10 μm.

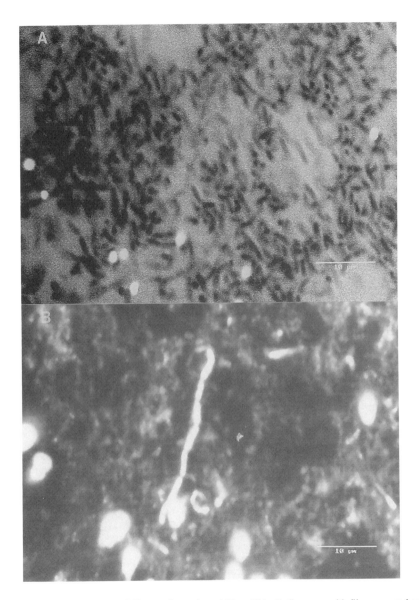

Figure 18. The occurrence of filament formation within a 48 hr *P. fluorescens* biofilm was used to indicate microzones of reduced oxygen tension. (A) Negatively stained region of the biofilm. Note positions of fluorescent marker beads (0.97-μm diameter). (B) The same location stained with FITC. The FITC preferentially bound to filaments rather than the predominant rod shaped cells. Bar = 10 μm. From Korber *et al.*, 1993, with permission.

desulfuromonas was more randomly distributed. Time-course investigations using these techniques may provide important information regarding the temporal success of organisms that have specific chemical requirements. For example, development of anaerobic zones may be found to correlate with the occurrence of SRB microcolonies. In pure laboratory cultures, the cultivation of strict anaerobes involves prereducing the growth medium and eliminating any exposure to oxidized equivalents during transfer, growth, or subculture. Within the biofilm microenvironment, these same organisms flourish because the conditions are met within microniches created chemically and biologically through the activity of other community members.

4.2. Interactions within Heterogeneous Biofilm Communities

Within a habitat is a particular set of conditions allowing proliferation of a species or a group of species. For spatially organized communities, the concept of domain (Wimpenny, 1981, 1992) is a useful construct. As defined by Wimpenny, domains represent either passive physicochemical conditions (i.e., habitat domains) or they encompass the influence of the organism on its environment, or the activity domain. Whenever some overlap of domains exist, interaction between the organisms is likely to occur. These interactions between microorganisms with compatible and/or incompatible habitat and activity domains play a significant role in the development of highly structured communities such as those occurring at interfaces.

4.2.1. Bacterial Interactions

Bacterial interactions include the complete range of potential interactions, from direct predation to commensalism. Typically, these interactions are defined in terms of substrate utilization. For example, introduction of a new, nonmetabolizable compound into the environment may result in the development of stress conditions (Slater, 1980). Evolution of a degradative mechanism to relieve this stress by the mineralization of the contaminant is important for two reasons: (1) for optimizing the utilization of available carbon and energy sources, and (2) for maintaining the stability and character of the environment. Slater (1980) suggested that microbial communities may be important in developing novel degradative pathways to mineralize xenobiotic compounds. Häggblom (1990) discussed the probability that the degradation of these compounds may be facilitated by a sequence of fortuitous reactions catalyzed by enzymes produced by more than one organism.

The coexistence of different individuals or groups of individuals in natural ecosystems is regulated by a range of metabolic and genetic interactions. It has been demonstrated that synergistic metabolism and cometabolic processes are involved in the transformation of many compounds, including pesticides (Walker, 1976). Häggblom (1990) pointed out that although a xenobiotic compound may be degraded by a single organism, these compounds are usually degraded via a sequence of reactions catalyzed by more than one organism. For instance, it was impossible to isolate a pure culture capable of using chlorinated propionamide as sole carbon and nutrient source from the soil (Slater and Hartman, 1982). However, the same workers isolated a mixture of organisms capable of doing so. Pothuluri *et al.* (1990) made reference to a number of previous studies by other workers showing that the degradation of the pesticides alachlor and metolachlor were cometabolic processes.

Fogarty and Tuovinen (1991) described the important role of interactions between members in microbial consortia during the degradation of pesticides. These authors referred to primary degraders as organisms able to produce metabolites that support the growth of other bacteria that cannot use the parent compound. A similar phenomenon was observed during the metabolism of polychlorinated biphenyls (Kröckel and Focht, 1987), furfural (Knight *et al.*, 1990), and various other contaminants. In the case of furfural, a *Desulfovibrio* species was responsible for the primary degradation of furfural to acetic acid, while a methanogenic bacterium was responsible for the conversion of acetate and methanol to methane. Schmidt *et al.* (1992) described three organisms subsisting as commensals on metabolic by-products of *Bacillus cereus* when 2,4-dinitrophenol was provided as the sole carbon source. Furthermore, there is also evidence that the growth kinetic properties of mixed populations are more favorable than those of the axenic cultures, and that the overall rate of biodegradation of compounds may be significantly greater by mixtures of several organisms compared with the sum of the rates of transformation by individual organisms.

Caldwell (1993) suggested that the range of conditions under which a bacterial species can survive in a natural habitat is restricted by its limited genetic and physiological capabilities, and that reproductive success is often dependent on associations with other species. These associations occur within communities, and are characterized by multiple trophic levels, nutrient exchange, spatial organization, and the development of internal microenvironments and steady-state conditions. For example, anaerobic waste digestion is only efficient when biofilms, sludge granules, or flocs are present (Bochem *et al.*, 1982). Thiele *et al.* (1988) referred to syntrophic relationships, facilitated by the formation of consortia, in which the participating organisms share a common spatial microniche. Interspecies hydrogen transfer is an example of such a spatially dependent syntrophic relationship found in anaerobic digesters (Wilson *et al.*, 1985; Boone *et al.*, 1989; MacLeod *et al.*, 1990). Although the importance of microbial consortia in degrading environmental contaminants was recognized by many workers (Lang *et al.*, 1992; Rozgaj and Glancer-Soljan, 1992), and use of microorganisms for treatment of toxic chemicals is now commonplace (Schmidt *et al.*, 1992), very little is known about the interrelationships between the organisms involved. In addition, the majority of work has involved studies of pure (mostly planktonic) cultures, making it difficult to assess the role of complementary metabolic activities among degradative populations on surfaces.

Predatory interactions between bacterial species may also play a significant role in community development. The research of Casida and co-workers (Byrd *et al.*, 1985; Casida, 1992) has revealed a range of obligate and nonobligate predatory interactions that occur between bacteria and may be significant in regulating cell numbers within communities. For example, in soil the nonobligate predator N-1 produces an initiation factor that stimulates the growth of *Agromyces ramosus* and other bacteria. After the initiation of growth, *A. ramosus* advances toward N-1 and kills some of them. N-1 then kills all of the *A. ramosus* cells in an attack–counterattack phenomenon. Other bacterial interactions such as these including *Bdellovibrio, Ensifier adherens,* and others may also be important in defining the development of biofilm communities. However, these interactions have not been extensively studied from this perspective.

4.2.2. Bacterial–Algal Interactions

Algal–bacterial associations in bioaggregates and in biofilm communities are commonplace, with algae representing both sites of attachment and sources of nutrients. The ability of heterotrophic bacteria to utilize extracellular products of phytoplankton is generally recognized and is based on evidence from experiments carried out by various workers (Nalewajko *et al.*, 1976). In these studies, bacterial–algal interactions were determined in terms of the utilization of algal extracellular products by bacteria as a source of carbon. In many cases, attention was focused on the effect of bacteria on the kinetics of algal extracellular release. When grown in axenic culture, the net algal extracellular release was linear with time, but followed a plateau-type curve in the presence of bacteria, suggesting consumption of the extracellular products by bacteria after a lag period (Nalewajko *et al.*, 1976, 1980). Bacterial utilization of algal extracellular products as a source of organic carbon was also discussed by Bowen *et al.* (1993). Their studies focused on the clustering of chemotactic bacteria around phytoplankton cells to improve their exposure to the carbon exuded by phytoplankton. They also discussed observations by other workers that the uptake of glucose by bacteria remained unsaturated at concentrations orders of magnitude higher than background levels in their natural environment, and they attributed this enhanced uptake to multiphasic assimilation.

Further evidence for bacterial utilization of extracellular substances produced by other organisms such as algae came from investigations not directly related to ecological studies. Examples include a study by Von Riesen (1980), who focused on the search for novel biochemical activities of nonfastidious nonfermentative gram-negative bacilli, and a study by Hansen *et al.* (1984), who described a *Bacillus* species. These studies indicated that many bacterial species isolated from soil, fresh water, and marine systems have the ability to grow on complex polysaccharides such as cellulose, chitin, and algin (sodium alginate). The latter compound, which is produced by various algal species, is degraded by bacterial enzymes responsible for hydrolysis of the guluronide and mannuronide linkages of algin. One can expect that bacteria producing enzymes specific for polysaccharides exuded by marine or freshwater organisms such as algae would predominantly be isolated from those habitats. However, the two bacterial species studied by Von Riesen (1980) were terrestrial isolates. It was therefore suggested that these aquatic polysaccharides may exist in modified forms in terrestrial environments.

Bacteria associated with aggregates of *Aphanizomenon* and *Anabaena* are capable of forming reduced microzones in O_2-saturated waters. The site-selective colonization of heterocysts and adjoining cells appear to enhance N_2-fixation by these cyanobacteria (Paerl, 1980). Schieffer and Caldwell (1982) also demonstrated that synergistic interactions that occurred during growth of *Anabaena* and *Zooglea* spp. in CO_2-limited continuous culture.

Studies examining algal–bacterial interactions, and others on this topic, have demonstrated that algal exudates serve as an important growth substrate for heterotrophic bacteria (Nalewajko *et al.*, 1980). However, the role of these exudates in the bacterial mineralization of xenobiotic compounds has not been extensively investigated. Wolfaardt *et al.* (1994c) demonstrated that algal exudates may serve as the primary substrate supporting the growth of microorganisms and also as a stimulant that induced production of enzymes with low specificity that also act on a cometabolized substrate (xenobiotic compound). In this case, degradation of the herbicide diclofop was enhanced by the presence of the algae, which increased diclofop degradation and removal by 36%. Batch culture experiments with ^{14}C-labeled diclofop confirmed algal involvement during the mineralization of diclofop, as there was no significant difference in the amount of [14]CO_2 evolved by the bacterial consortium with and without the algal activity when cultivated in the dark, while 11% more [14]CO_2 was evolved in the light by the algal–bacterial consortium. Existence of these community-based mechanisms is of significance for the determination of bacterial diversity and community structure.

4.2.3. Bacterial–Predator Interactions

The role of protozoa in regulating population numbers in microbial communities is well recognized. For instance, a study by Clarholm (1981) demonstrated that amoebas were able to cause a 60% reduction of bacterial numbers in soil, while Sanders and Porter (1986) suggested that grazing by Protista on bacteria in marine and freshwater systems may account for the relative constancy of bacterial numbers over a broad range of trophic states. Similar findings were made by Vargas and Hattori (1990). Protozoan predation has also been shown to have a significant effect on biofilm development in marine systems (Pederson, 1982). In addition, the grazing effects of protozoa can be species-specific and mediated by external recognition factors such as exopolymer chemistry. Snyder (1990) has shown the importance of extracellular components in prey recognition by the ciliate *Pseudocolloembus marinus*. Lawrence *et al.* (1993) indicated that the presence of protistan grazers markedly altered the EP chemistry and architecture of a biofilm community. A transmission electron micrograph of an amoeba "caught" grazing in a microbial biofilm community is presented in Fig. 19.

Predation may also affect microbial activity. Hunt *et al.* (1984) reported previous experiments that demonstrated increased mineralization of carbon, phosphorus, and nitrogen as a result of predation. A few studies have investigated the effect of predation during the mineralization of xenobiotic compounds by bacterial populations. Schmidt *et al.* (1992) cited a short list of papers by others who have studied the role of protozoa in sewage treatment systems and in natural systems polluted with toxic organic chemicals. In their own study, Schmidt *et al.* (1992) analyzed the population dynamics of protozoa in sewage and toxic waste treatment systems, demonstrating that the presence of grazing

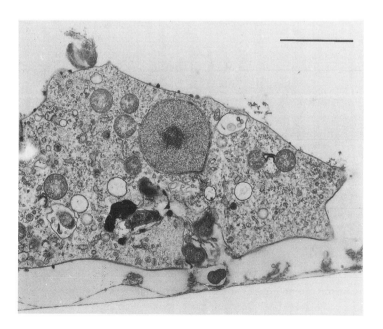

Figure 19. Transmission electron micrograph of an amoeba grazing in a degradative microbial community. Bar = 2 μm. (Photo provided by S. J. Caldwell.)

protozoa increased mineralization rates. It was suggested that the bacteria were maintained in a state of high metabolic activity, resulting in increased rates of assimilation of organic materials. Similar observations were made by Zanyk (1993), who demonstrated increased bacterial mineralization of ^{14}C-labeled diclofop to [14]CO_2 in microcosms when predators were present. Protozoa may further affect microbial activity by the regeneration of nutrients (Güde, 1985), which may lead to cometabolism of contaminants, and by the excretion of growth factors that stimulate only certain bacterial populations (Huang *et al.*, 1981). Thus, protozoans may act as a link between different bacterial populations and between bacterial populations and organisms positioned higher on the food chain (Clarholm, 1984).

Bacteriophages are extremely abundant in natural systems, and as noted by Bratbak *et al.* (1993), almost any disturbance of natural microbial communities may result in cell lysis and release of viruses. These viruses have been shown to be important in algal and bacterial population dynamics and diversity. Although the bacteriophages may represent an important controlling mechanism, their activity within complex systems is still poorly understood.

In general, despite the range of investigations conducted on predator–prey interactions, few studies of predation pressure effects on biofilm communities and their activity have been conducted.

4.2.3. Competition and Succession in Biofilms

In natural systems, there exists a high degree of cellular interaction and competitive behavior (Fredrickson, 1977). Much of this interactive behavior results from the competition of cells for available resources, whether those resources be attachment sites or growth substrates. Higher organisms may also influence the outcome of microbial surface colonization, and events such as predatory grazing have the potential to limit the success of one strain while having relatively little effect on other community members. Few studies have focused on the effect of competitive pressure during the development of microbial biofilms. Cowan *et al.* (1991) utilized image analysis to measure the extent of surface coverage during colonization of pure cultures and mixed-species over a 27-day period; in the presence of *Xanthomonas maltophilia*, greater surface colonization was observed during coculture experiments than when *X. maltophilia* was absent. The researchers also noted the limitations of their system in that three-dimensional cell aggregates could not effectively be discriminated from two-dimensional growth and development using phase-contrast image analysis. Also noted was the inability to discriminate between bacterial species; thus determination of dominance by any particular colonizing species within the system was difficult.

Radiochemical labeling of bacteria has also been used to evaluate competitive behavior in biofilm cocultures (Banks and Bryers, 1991). Discrimination between the two strains used during this study was possible because the *Pseudomonas putida* strain metabolized glucose exclusively, whereas the *Hyphomicrobium* sp. metabolized only methanol. Replicate culture systems were used to cultivate either mixed-species biofilms grown in the presence of [^{14}C]glucose and unlabeled methanol, or [^{14}C]methanol and unlabeled glucose. Success of coculture members was determined by comparison of radioactivity accumulated over time, sacrificing flow cells for analytical purposes. When both strains were simultaneously added to flow cells, the *P. putida* isolate outnumbered the *Hyphomicrobium* sp. at 24 and 48 hr. Similarly, when the *P. putida* strain was added to an established *Hyphomicrobium* sp. biofilm, the *P. putida* cells became the dominant biofilm species after only 48 hr. In both cases, the *P. putida* strain outnumbered the *Hyphomicrobium* sp. by approximately 5 to 1 following 50 hr of biofilm development. The authors attributed these results to the differences in growth rates of the two species, whereas the effect of microbial colonization behavior was not addressed.

Korber *et al.* (1994b) utilized DDCC analysis to study the dynamics of competition between motile and nonmotile *P. fluorescens* strains. During this study, enhanced cell transport, flow resistance, and improved recolonization success led to the dominance of the motile strain even though the growth rates of the two strains were identical. This demonstrated the pivotal role of the flagellum during competitive surface colonization in flowing microenvironments.

Current limitations to the study of complex microbial communities lie in the inability to visually discriminate between similarly shaped consortium members. Use of SCLM, in conjunction with fluorescently labeled antibodies, has potential in this area of study, as laser optical thin sectioning may then be used to help elucidate the position of fluorescent antibody-stained bacteria present within the biofilm matrix (James *et al.*, 1993).

Observations of adherent populations in natural habitats provided evidence that these systems are in constant flux (Characklis *et al.*, 1990a). A sequence of events occurs during surface fouling, resulting in a succession of surface-colonizing species that dominate at various times following immersion of clean surfaces in natural waters. In a process termed *epibiosis*, the initial colonizing species are subsequently replaced before other organisms with evolutionarily defined requirements become involved with the surface consortium (Wahl, 1989). The succession is based on a sequence of physical and biological events, starting with the adsorption of organic films and closely followed by surface colonization by bacterial species (Fig. 20). The time frame for these two initial phases of surface colonization are relatively short, with the molecular film forming over a period of minutes and with significant bacterial colonization occurring within 24 hr. Lawrence, Korber, and Caldwell (unpublished results) observed that only brief periods were required for the onset of bacterial deposition from a

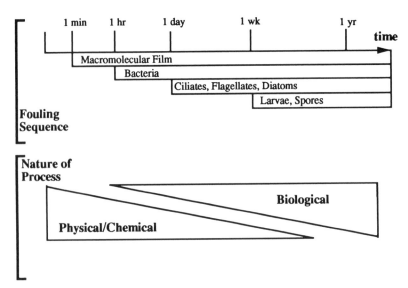

Figure 20. Diagram illustrating the generalized series of successional events that occur at a surface during the development of a biofilm community. Adapted from Wahl, 1989.

simulated marine microcosm, and that common marine species (*Hyphomicrobium, Pedomicrobium, Pseudomonas, Caulobacter* spp.) were primarily involved.

While bacteria continue to adsorb to exposed surfaces, larger fouling organisms (including amoeba, flagellates, ciliates, diatoms, and larvae) subsequently colonize the surface over a period of days (for predatory eukaryotes and diatoms) or weeks (for larval and spore deposition) (Wahl, 1989; Decho, 1990). This successional pattern of surface colonization has been observed during long-term observation of immersed surfaces in flowing or static systems. Researchers have reported that rod-shaped bacteria are often the primary colonizers following exposure of clean surfaces to natural aqueous systems (Marshall *et al.*, 1971; Marszalek *et al.*, 1979), followed by stalked bacteria such as *Caulobacter* spp. (Dempsey, 1981). Subsequent colonization by filamentous algae, diatoms, and larvae has reportedly been observed in stream, river, and oceanic environments, with eventual attraction of predators that feed on the biofilm (Marszalek *et al.*, 1979; Baier, 1985; Wahl, 1989; Rittle *et al.*, 1990).

The majority of studies describing community change or community succession are not based on direct assessment of the fate of species within the community. These studies have primarily been limited to traditional culture techniques with their inherent limitations. Direct analysis of biofilm development is possible using SCLM. Studies of a nine-member degradative community showed that distinctive spatial arrangements between community members occurred during growth on the herbicide diclofop (Wolfaardt *et al.*, 1994a). The structures included conical projections rising 30 μm above the attachment surface, grapelike clusters surrounding these conical structures, radially arranged bacilli, and other highly specific patterns. These relationships did not develop when the community was grown on labile carbon sources. The general pattern of community growth is described in a series of laser micrographs (Fig. 21) and also in a schematic representation (Fig. 22).

The development of molecular phylogenetic techniques has provided the tools to study the activity of microbial populations in their natural communities (Stahl, 1993). These probes have been used to identify or localize bacterial types or strains within biofilms or attached to surfaces (Amann *et al.*, 1992; Assmus *et al.*, 1995). However, their application for studies of changes in species abundance and composition have been relatively few. Stahl *et al.* (1988) used species- and group-specific probes to enumerate strains of *Bacteroides succinogenes* and *Lachnospira multiparus*-like organisms in the bovine rumen. They noted that the

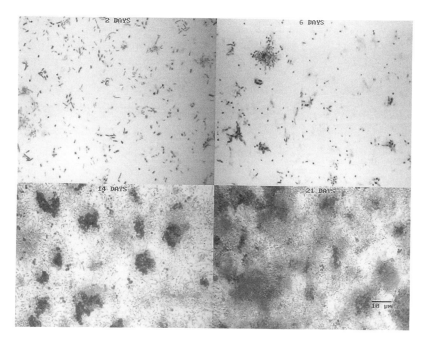

Figure 21. A series of confocal laser images showing development of a degradative community grown with the herbicide diclofop (14 μg ml⁻¹) as sole source of carbon and energy, after (A) 2, (B) 6, (C) 14, and (D) 21 days. Bar = 10 μm.

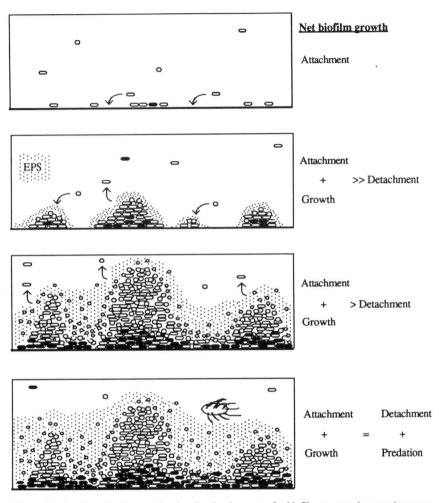

Figure 22. A schematic diagram showing the development of a biofilm community growing on a chlorinated aromatic substrate. The dark cells represent bacteria involved in reductive dehalogenation, a step critical for subsequent cleavage of the aromatic rings. Creation of (anaerobic) microzones that allow these cells to proliferate and the subsequent release of ring-carbon, likely control colonization of other species. Protistan grazers shown in the diagram also appear to play an important role in the development of the community influencing EP chemistry and the fate of the chlorinated aromatics bound to EPs. The events shown occur over approximately 21 days as shown in Fig. 21.

antibiotic momensin depressed numbers of *L. multiparus* strains with a transient five- to tenfold increase after cessation of treatment. Significant changes in the populations of *B. succinogenes* strains were also observed. The authors concluded that the observed changes in ruminal fermentation patterns occurred not as the result of altered populations, but rather as the result of altered microbial metabolism. In the future, application of these probes should greatly enhance studies of microbial populations *in situ*.

4.3. Adaptive Strategies

4.3.1. Role of Genetic Integration and Regulation

Bacteria exhibit a prodigious capacity to adapt to changes in their environment. Much of bacterial adaptation stems from phenotypic plasticity founded on altered or variable gene expression. An organism may successfully integrate into a community through switching genes on or off, or by optimizing the level of expression of appropriate genetic pathways that enhance its ability to survive and reproduce. In addition to this inherent plasticity, bacteria have also derived numerous mechanisms for generating variation within populations. This variation provides the basis for survival under unpredictable changing conditions, and thus enhances long-term evolutionary success. For example, an organism may improve its chances for survival by being a recipient or a donor in genetic exchange within the community or through spontaneous production of variants. All of these processes are critical in the determination of the long-term success of a biofilm community.

4.3.1.a. Short Time Frame Responses. In order to sustain growth at an interface and succeed in a developing community, bacteria must initially respond to the surface and adapt to changing conditions. Essentially all microorganisms are capable of sensing changes in their environment and responding by altering patterns of gene expression (Stock *et al.*, 1990). The microbial cell can be viewed as being in a dynamic state (Roszak and Colwell, 1987), and as a result of this flexibility can readily adapt to changes in its environment by means of genotypic and phenotypic adjustments.

Studies of *V. parahaemolyticus* represent the most comprehensive investigations of surface-induced gene expression (McCarter *et al.*, 1992). The genetic control of *V. parahaemolyticus* lateral flagellum (laf) production has been shown to be physical (i.e., increasing the viscosity of the medium was sufficient to induce the laf phenotype within 30–60 min). In this instance, the polar flagellum has been shown to act as a tactile sensory device that informs the bacterium about its ability to move. Additional studies have shown that iron deficiency is also required to initiate production of swarmer cells (laterally flagellated cells) (McCarter *et al.*, 1988; McCarter and Silverman, 1989).

Dagostino *et al.* (1991) reported that genes related to other specific functions may only be switched on at surfaces and not while the bacteria are growing planktonically. For example, Goodman *et al.* (1993a) showed that the marine *Pseudomonas* S9 has a gene that was only switched on at a hydrophobic polystyrene surface. Extracellular polysaccharide and glycoprotein production is often derepressed (Deretic *et al.*, 1989). Davies *et al.* (1993) demonstrated that *Alg-c* genes of *P. aeruginosa* were up-regulated by adhesion to the surface and growth in the biofilm compared with planktonic cells in the liquid medium. Extracellular enzyme induction has also been linked to the biofilm mode of growth (Giwercman *et al.*, 1991). All of the above observations suggest that some surface-colonizing bacteria respond to a broad set of environmental cues, including concentrations of nutrients, micronutrients, O_2, CO_2, and so forth. It is also possible that substances such as pheromones or signal proteins similar to those active in myxobacteria (Nealson, 1977; Kim and Kaiser, 1990) may be involved in controlling some surface growth-related phenomena.

    4.3.1b. Variable Time Frame Responses. Gene flux in microbial communities is of fundamental importance to the study of microbial ecology. Reflecting this importance, a large number of extensive reviews on the subject have been published (Reanney *et al.*, 1983; Slater, 1985; Bale *et al.*, 1987, 1988; Levy and Marshall, 1988; Levy and Miller, 1989; Fry and Day, 1990; Marshall and Levy, 1990; Stotzky *et al.*, 1990; Veal *et al.*, 1992). In addition to responding to the initial conditions at a surface microenvironment, the bacterial cell is required to respond to changing conditions within the biofilm. Bacteria have recently been shown to induce genes in response to reduced oxygen tensions (Lee and Falkow, 1990), increased CO_2 levels (Goodman *et al.*, 1993a), changes in pH (Slonczewski, 1992), and carbon starvation (Ostling *et al.*, 1991).

    Despite a limited genetic complement, bacteria obtain a high degree of genetic flexibility through high rates of mutation, plasmid-borne genes, and haploidy (Day, 1987). The phenomenon of and mechanisms for generation of intraclonal polymorphism in bacteria have been reviewed by Rainey *et al.* (1993). These properties of bacteria facilitate their survival, and proliferation in changing environments and genetic interactions within microbial communities are believed to be complex. Thus, genetic transfer is an additional mechanism by which bacteria within a biofilm community may interact. Indeed, the individual members of a community may be considered to be a pool of metabolic processes linked by their ability to be genetically altered by gene transfer (Palenik *et al.*, 1989). Slater (1985) characterized communities as either tight, or those that are highly interdependent, or loose, or those that utilize plasmid transfer as a dominant mechanism for interaction rather than metabolic interdependence. The topics of maintenance of genes and gene flux in microbial communities have been reviewed recently by Veal *et al.* (1992), and include examinations of selection, periodic selection, cryptic and mutator genes, and directed mutation.

    Microcosms have been used in many studies to demonstrate the possibility of gene exchange (Veal *et al.*, 1992). The use of artificial media, dense populations, high concentrations of substrates, and the enhancement of cell transformation may make these studies of little relevance for the study of gene adaptation in natural communities. Gene transfer has been shown to occur frequently within a number of natural environments (Fry and Day, 1990). Notably, interfacial regions that contain biofilm communities, such as the rhizoplane, phyloplane, and epilithion, are sites of elevated rates of gene transfer. These regions are likely to be sites of high population density, and the state of the bacteria is conducive to gene exchange (Veal et al., 1992). For example, high rates of gene transfer have been reported for the epilithion (Rochelle et al., 1989) and the rhizosphere (Van Elsas et al., 1988).

Genetic exchange mechanisms have been used to explain the enhanced biodegradation of pesticides in soil communities previously exposed to pesticides, where a change in degradative efficiency was attributed to adaptation by the microorganisms involved in the degradation process (McCuster et al., 1988; Mueller et al., 1988). Lee and Ward (1985) used the term acclimated microorganisms to describe organisms capable of enhanced degradation as a result of prior exposure to contaminants. Wilson et al. (1985) and Wiggins et al. (1987) described adaptation as a phenomenon caused by several mechanisms, including the growth of active organisms, enzyme induction, mutation, or genetic changes, which resulted in an increase in the rate of biotransformation. Madsen (1991) noted that the mechanisms responsible for metabolic adaptation are seldom studied.

The role of specific genetic elements has also been addressed. Genetic elements encoding for the degradation of aromatic and chloroaromatic compounds are often carried on conjugative plasmids, and are thus potentially mobile in microbial communities. For instance, Wickham and Atlas (1988) studied short-term effects of environmental stress, caused by exposure to contaminants, on the occurrence and frequency of plasmids within the bacterial population of a soil microbial community. For most chemical stress factors there was an increase in plasmid frequency. However, in mercury-chloride stressed soils, there was a decrease in plasmid frequency, despite an increase of overall mercury resistance, indicating that this resistance was chromosome encoded. In contrast to this observation, Wyndham et al. (1994) demonstrated the significance of transposable elements in the mobilization of catabolic genes and discussed the importance of genetic rearrangements during adaptation. Tam et al. (1987), as well as others, also discussed the role of plasmids and plasmid transfer on the degradative capability of organisms. Catabolic plasmids do show broad distributions, as demonstrated for the TOL plasmids (Chatfield and Williams, 1986). The distribution and abundance of Mer genes in microbial communities have also been related to the presence of toxic mercury species in the environment (Barkay et al., 1991). Thus, these mechanisms may be significant in natural biofilm communities.

The temporal nature and physicochemical requirements for genetic transfer have also been a topic for conjecture. The transfer of genetic elements in a degradative community growing in freshwater mesocosms supplemented with 3-chlorobiphenyl was demonstrated in the work of Fulthorpe and Wyndham (1991, 1992). However, plasmid transfer only occurred in the presence of 3-chlorobiphenyl selection pressure and only after months of exposure. In contrast,

Goodman *et al.* (1993a) reported that transconjugant organisms arose within biofilms after only 24 hr of exposure and in the absence of selection pressure for the plasmid. In addition, the transfer frequency was highest among attached cells within biofilms. In addition, Fernandez-Astorga *et al.* (1992) and Jones *et al.* (1991) demonstrated that plasmid transfer by conjugation occurred between *E. coli* and *P. aeruginosa* strains in freshwater microcosms with no added nutrients. Similarly, Goodman *et al.* (1993b) reported that conjugative plasmid transfer between bacteria proceeded under oligotrophic marine conditions. Thus, it is clear that community adaptation can be facilitated by mechanisms such as enzyme induction, gene exchange, and mutation (Wilson *et al.*, 1985). In addition, microbial communities do have an important role to play during the development of new metabolic pathways in natural systems (Slater and Somerville, 1979).

4.3.2. Starvation/Survival Mechanisms

It has been established that nongrowth behaviors, or starvation survival, is a widespread phenomenon in bacteria with important evolutionary and ecological significance. During the life of any naturally occurring bacterium, there will be periods of growth at various rates, starvation, recovery, and regrowth (Kjelleberg, 1980, 1984; Morita, 1982, 1986; Kjelleberg *et al.*, 1987; Roszak and Colwell, 1987). Morphological, physiological, and molecular biological variations have all been studied in relation to the onset of starvation in a variety of microorganisms.

Survival strategies of bacteria in their natural environment have been identified (Roszak and Colwell, 1987). The phenomenon of multiple divisions leading to the formation of ultramicrobacteria (<0.3 μm diameter) has been reported (Novitsky and Morita, 1976, 1977, 1978; Morita, 1982). These specialized ultramicrobacterial forms have been suggested to be exogenously dormant cells responding to unfavorable environmental conditions (Roszak and Colwell, 1987). Jannasch (1967) indicated that two strategies may be observed: (1) the ability to grow at low substrate concentrations, and (2) the ability to become temporarily inactive but able to survive. The mechanism of low metabolic rate is widespread in soil and marine bacteria (Roszak and Colwell, 1987). Brown and Gilbert (1993) noted that cells respond to nutrient deprivation by: (1) reducing the requirement for the deficient nutrient by limiting cell components containing the element by using alternate substrates or by reorganizing cellular metabolic pathways, (2) altering the cell surface and inducing higher-affinity transport systems, and (3) reducing cellular growth rate.

Many bacterial cells undergoing starvation do alter their basal macromolecular pools and change their key metabolic capabilities (Geesey and Morita, 1979; Guckert *et al.*, 1986; Hood *et al.*, 1986). In addition, starving bacteria produce a range of specific starvation or stress proteins. During starvation periods, the marine *Vibrio* S14 induces the synthesis of a complete, time-dependent series of proteins. The multiple starvation stimulation of strain 14 consists of approximately 66 proteins that are induced sequentially during simultaneous starvation for carbon, nitrogen, and phosphorus (Nystrom *et al.*, 1990). This starvation response has been noted to be generally similar to the classical differentiation program of spore formation (Kjelleberg *et al.*, 1993). The similarity in the responses of spore-forming and non–spore-forming bacteria is likely due to the possession of the stringent response gene by both groups of organisms (Gilbert *et al.*, 1990). In general, any nutrient-deprived bacterium shows a sequential pattern involving a complex turn-on/turn-off pattern of protein synthesis (Kjelleberg *et al.*, 1993).

One consequence of growth on surfaces or within biofilms is the potential for development of limiting conditions, particularly for the growth of aerobic organisms. For example, Caldwell and Lawrence (1986) demonstrated flow-dependent growth of individual *P. fluorescens* cells in microcolonies at low nutrient concentrations (<100 mg liter^{-1} glucose) and flow-independent growth at high nutrient concentrations (1 gm liter^{-1} glucose), ascribing limitations to nutrients rather than oxygen. Flow-dependent growth has similarly been demonstrated using low concentrations of flowing nutrient (Confer and Logan, 1991). In thick microbial biofilms, however, oxygen has been identified as a primary factor for the formation and maintenance of microbial biofilms consisting of aerobic bacteria (Sanders, 1966; Mack *et al.*, 1975; Howell and Atkinson, 1976; Characklis *et al.*, 1990a).

Starving bacteria have been shown to be able to utilize surface-bound substrates, leading to cellular growth and reproduction. Hermansson and Marshall (1985) showed that both *Pseudomonas* JD8 and *Vibrio* spp. (MH3 and DW1) utilized bound fatty acids. Kefford *et al.* (1982) demonstrated efficient surface scavenging of stearic acid from surfaces by a *Leptospira biflexa*. In contrast, bacterial utilization of substrates requires that they be available for assimilation. For example, amino acids may be bound to clays and become unavailable for microbial utilization (Stotzky and Burns, 1982).

Consequently, bacteria within biofilms are likely to express a wide range of metabolic responses (a range of growth responses) to their ambient environment, a phenomenon that has previously been discussed in the context of inherent mechanisms of resistance to the effects of antimicrobial agents (Brown *et al.*, 1988; Evans *et al.*, 1990; Gilbert *et al.*, 1990; Eng *et al.*, 1991; Brown and Gilbert, 1993). While little direct evidence is available demonstrating slow growth as a mechanism of resistance for attached microbial populations, Korber *et al.* (1994a) found that in established *P. fluorescens* biofilms exposed to the fluoroquinolone, fleroxacin, a gradient of cell elongation was observed from the biofilm–liquid interface (where maximum cell elongation occurred) to the base of the biofilm (where the least cell elongation occurred). Quinolone compounds, which inhibit DNA–gyrase activity, have previously been used to estimate growth rates of cells by measuring the amount of cell elongation that occurs following treatment with the antimicrobial agent (Bottomley and Maggard, 1990). Thus, the depth-dependent range of cell elongation response from within *P. fluorescens* biofilms suggests that cells at different depths in the biofilm were growing at different rates at the time of fleroxacin exposure.

Microbial cells that are quiescent are thought to play a significant role as reservoirs of diversity in natural communities and to be of major importance during nutrient cycling events (Lewis and Gattie, 1991). Thus, greater emphasis and study of this phenomenon is required for understanding community processes.

4.3.3. The Role of EPs in Biofilm Communities

Attached microbial communities are characterized by the presence of an extensive exopolymeric network. Observations of adherent microbial cells indicate that virtually all cells in biofilms are surrounded by extracellular polymer (Costerton, 1984; White, 1984; Costerton *et al.*, 1987), either as capsules attached to the cell wall or as mucoid secretions in the extracellular environment (Troy, 1979; Shen *et al.*, 1993). Rose (1976) estimated that 60% of cell dry weight may be EPs and Christensen and Characklis (1990) indicated that EPs may constitute many times the carbon found in cell biomass. Characterization of the components of bacterial EPs has revealed that these polymers consist of a large variety of chemical structures (Kenne and Lindberg, 1983). The nature and chemistry of exopolysaccharides have been extensively reviewed (Dudman, 1977; Kenne and Lindberg, 1983; Sutherland, 1985; Allison and Sutherland, 1987; Decho, 1990).

The importance of EPs in microbial communities is clear, and includes a variety of functions that improve microbial competitiveness and reproductive success. These functions include: (1) attachment of cells to surfaces (Marshall *et al.*, 1989); (2) formation of microcolonies (Allison and Sutherland, 1987); (3) increased resistance of attached cells to environmental stress (Uhlinger and White, 1983; Brown and Gilbert, 1993); (4) facilitation of trophic interactions; and (5) functioning for nutrient accumulation and storage.

4.3.1a. Attachment and Microcolony Formation. Bacteria may adhere to surfaces and grow as sessile biofilm communities in response to oligotrophic conditions (Kjelleberg, 1984; Parkes and Senior, 1988). In general, attached cells are usually associated with greater amounts of EPs than planktonic cells (Wrangstadh *et al.*, 1989; Bengtsson, 1991). Although low nutrient concentrations may stimulate production of EPs by both sessile and planktonic cells (Bengtsson, 1991), Vandevivere and Kirchman (1993) demonstrated a fivefold greater production of extracellular polysaccharides by attached cells than free-living cells of the same bacterium. Mittelman *et al.* (1990) observed that in addition to increased attachment rates, irreversibly attached cells exhibited an increased carbohydrate : protein ratio. The authors concluded that an observed shift in the 1130 cm^{-1} Fourier transform infrared band resulted from shear-induced production of EPs. Extracellular polymers play a critical role in microcolony development. For example, Allison and Sutherland (1987) showed that a freshwater bacterium and a non–EP-producing mutant colonized surfaces equally well, but only the former could produce microcolonies.

4.3.1b. Protection from Stress.   In most environments, bacteria are subject to predation by protozoa, slime molds, and bdellovibrios. The exopolymeric secretions of the bacterial cell provides an interface that regulates the interaction between the attached cell and surface-grazing protozoans. The EP clearly functions as a protective barrier against predators such as amoeba, bacteriophage, and *Bdellovibrio* (Venosa, 1975). The presence of EPs has also been shown to influence grazing of the cell biomass; attached cells have been noted to have enhanced resistance to ingestion (Caron, 1987) or digestion (Fenchel and Jorgenson, 1977). Flagellates may either be mechanically excluded from the biofilm or graze selectively within aggregates (Biddanda and Pomeroy, 1988; Sibbald and Albright, 1988). Lawrence *et al.* (1993) demonstrated selective feeding of protozoans on a biofilm community, based on changes in EP chemistry observed with and without the activity of predators.

Resistance to antimicrobial agents may occur through several mechanisms, including simple reduction of diffusion coefficient, accumulation of protective enzymes within the EP (Giwercman, 1991), or sorption of the agent or agents to the EPs (Brown and Gilbert, 1993). These mechanisms of protection from xenobiotic compounds likely also operate for environmentally relevant stresses. The EPs have also been shown to substantially increase bacterial resistance to desiccation (Ophir and Gutnick, 1994).

4.3.1c. The Role of EPs in Microbial Nutrition.   It has also been established that the EPs of certain microbial species accumulate metals and it has been suggested that extracellular polymers play a similar role in the concentration of nutrients (Rudd *et al.,* 1983; Uhlinger and White, 1983; Costerton, 1984; Mittelman and Geesey, 1985; Decho, 1990). Metal binding is a common phenomenon in waste water treatment, and involves complexation of EPs with metals (Brown and Lester, 1982a,b; Rudd *et al.,* 1983) and ion exchange, mostly via carboxyl groups on uronic acids (Decho, 1990). Based on the observation that dissolved organic material competed with trace metals for binding sites on organic ligands, it was suggested by others that the binding mechanisms of organic compounds and metals to microbial EPs were similar (Decho, 1990). Christensen and Characklis (1990) reasoned that although most EPs have cation exchange properties that may enable bacteria to use EPs as a nutrient trap under oligotrophic conditions, it is unlikely that they can function effectively as a trap for soluble nutrients in environments where divalent cations are abundant. This is in contrast with Costerton (1984) and others (Costerton *et al.,* 1981; Hansen *et al.,* 1984; Decho,

1990) who proposed that the EP matrix of biofilms can act as an ion-exchange resin active in the accumulation of nutrients. However, these studies did not present direct evidence for the mechanisms involved. The relative importance of extracellular sorption to EPs in relation to absorption into cells or adsorption onto cells is also not known (Bellin and Rao, 1993).

Decho (1990) suggested that it is more likely for high-molecular-weight compounds to be accumulated by EPs than low-molecular-weight compounds, because the latter can be directly taken up by the cell without the need for substantial modification prior to crossing the cell wall. Alternately, high-molecular-weight compounds first have to be hydrolyzed by exoenzymes. However, the same author, as well as others (La Motta, 1976; Matson and Characklis, 1976), provided evidence for the polymer-based accumulation of low-molecular-weight compounds.

Based on the arguments presented by Decho (1990) and Dudman (1977), it seems reasonable that EPs can function as an effective nutrient trap when the concentration of these nutrients is below the threshold values required to keep cells in a viable state. It is also possible that the increased production of EPs by attached cells improves nutrient-trap efficiency of biofilms. Exopolymers may play an additional role during the utilization of trapped nutrients by the localization of extracellular enzymes that hydrolyze high-molecular-weight organic matter into a more labile form before it enters the cell (Decho, 1990). The extent to which any nutrients accumulated in EPs can be utilized during periods of nutrient limitation is not known. However, studies of the accumulation of organic chemicals indicate that EPs may also have a role in their sorption and subsequent mineralization by bacteria. A recent study by Wolfaardt *et al.* (1995a) indicated that chlorinated ring compounds bound to EPs in a microbial community were mineralized to CO_2 during periods of starvation.

Some authors have postulated that the production of EPs may be for the long-term storage of carbon and energy. Patel and Gerson (1974) demonstrated that EPs could function in this manner for a *Rhizobium*. In general, EP is unlikely to function for storage because most EP-producing bacteria lack the enzymes utilize their EPs (Dudman, 1977). However, bacteria have been shown to utilize the EPs of other bacteria (Decho, 1990). For example, Obayashi and Gaudy (1973) found that several genera of bacteria produced polymers that could be used by other bacteria as growth substrates. Thus, cross-feeding may occur through EP utilization between adjacent microbial populations, although this possibility has not been investigated.

4.3.1d. Role of EPs in Trophic Interactions. Microbial biofilms represent the lowest trophic level of the food chain, and as such constitute a significant portion of the carbon assimilated by organisms in higher trophic levels. Clearly, bacteria and their polymers play an important role during nutrient cycling and energy flow processes (Decho, 1990; Schmidt *et al.*, 1992). The vast quantities of EP produced by the microbial community constitute a major food resource for higher trophic levels, including protozoa (Lock *et al.*, 1984; Decho, 1990; Decho and Lopez, 1993). Sherr (1988) demonstrated the ingestion of fluorescently labeled polymers (dextrans) by both freshwater and marine protozoans. An additional argument for the significance of EP as a carbon source for higher trophic levels is the observation that cellular biomass of bacteria is insufficient to support the grazer populations. Consequently, EP has been suggested as the logical nutritional substitute (Moriarty, 1982; Decho, 1990; Schmidt *et al.*, 1992; Decho and Lopez, 1993). If that is the case, then EPs may play an important role in nutrient cycling in natural food chains. This possibility has not been studied and discussions on it are mostly speculative (Decho, 1990). However, it was demonstrated that EPs that accumulated copper could serve as a route of entry for this metal into environmental food chains (Mittelman and Geesey, 1985). This phenomenon has also been discussed by Patrick and Loutit (1976).

It is also possible that the chemically diverse exopolymeric matrix in which biofilm cells are embedded is in a similar way responsible for the accumulation of significant amounts of organic contaminants and the subsequent transfer of these compounds to higher trophic levels in natural ecosystems. Recently, Wolfaardt *et al.* (1994b) demonstrated trophic transfer of the herbicide diclofop methyl from the polymers of a biofilm community to the flagellate, amoeba, and ciliate grazers. However, information on the sites where sorption occur, the relative importance of sorption to EPs or absorption into cells, and the ecological significance of this phenomenon require further study (Uhlinger and White, 1983; Bell and Tsezos, 1987; Tsezos and Bell, 1988; Decho, 1990; Bellin and Rao, 1993).

4.3.1e. Characterization of EPs in a Biofilm Community. For a mechanistic understanding of the role of EPs in microbial processes, it is necessary to characterize the EPs in terms of chemical composition, charge, density, and other parameters. Conventionally, the chemical characterization of EPs was performed by colorimetric or chromatographic analysis of preparations obtained from a series of steps, including harvesting and concentration of biomass, resuspension, and separation of EPs from the cell fraction (Bartell *et al.*, 1970; Christensen *et al.*, 1985; Henningson and Gudmestad, 1992). Although the chemical composition of EPs and the relative abundance of components can be accurately assessed following this approach, it does not allow characterization of the nature and spatial arrangement of the components within attached microbial communities. It would therefore be difficult to understand the mechanisms involved in EP function, such as the selective accumulation of nutrients, if these traditional techniques are utilized. For this purpose, the spatial arrangement of the different EP components within the biofilm matrix must be determined, preferentially by nondestructive techniques without the need to alter the biofilm structure. Because of the lack of appropriate, nondestructive methods for analyzing biofilms, little is known about spatial variability in the chemical composition of biofilms. As a result, the importance of chemically defined regions in the functioning of biofilms has not been established.

A series of studies by Wolfaardt *et al.* (1994b,d,e) have attempted to address the role of EPs in a biofilm community by utilizing a variety of nondestructive assays, including a panel of Fluorescein Isothiocyanate (FITC)- and Tetramethyl Rhodamine Isothiocyanate (TRITC)-conjugated probes, in conjunction with SCLM. This approach allowed detection and quantification of lectin binding to a variety of EP residues including α-L-fucose, α-D-glucose, α-D-mannose, and sialic acid. It was concluded that the relative abundance of these components varied between 0 and 67% of the biofilm area at any depth. Lectin-binding sites were distributed nonuniformly, both horizontally and vertically within the >30-μm thick biofilms when the herbicide diclofop methyl was provided as the sole carbon source. However, a more uniform distribution of lectin-binding sites was observed in biofilms formed by the same consortium but grown on a labile growth medium (Wolfaardt *et al.*, 1995b).

In the diclofop-grown biofilms, extracellular polymers, present as either dense capsules surrounding individual cells or as a diffuse matrix filling intercellular spaces, accumulated the fluorescent diclofop residues. There was a near one-to-one relationship in the distribution of these regions that accumulated diclofop (and other chlorinated ring compounds) and regions with binding sites for the α-L-fucose-specific, *Ulex europaeus* I lectin. These regions also bound polyanionic and cationic fluor-conjugated dextrans and a hydrophobic-specific dye, demonstrating the nonuniform distribution of charged and hydrophobic areas in the biofilm matrix. Hydrolytic enzymes, some selected for their specificity against residues identified by the lectin assay, had no effect on the structural integrity and diclofop-binding capacity of the biofilms. The variable chemical composition and charge density observed in biofilms grown on chlorinated ring compounds indicated a high degree of spatial organization and differentiation in this degradative community. An example of the variability inherent to the EPs of this community is shown in Figs. 15 and 16, which illustrate variability in lectin-binding sites and charge distribution in the biofilm. Wolfaardt *et al.* (1995a) were also able to demonstrate that the bound diclofop and its metabolites could be utilized as a source of carbon and energy during periods of carbon limitation. In addition (as discussed in Section 4.3.1), the EPs acted as a mechanism for transfer between bacterial and protistan populations. Thus, the EPs represent the key to community development in prokaryotes, providing the matrix, the interface, and a fundamental currency of carbon and energy.

5. Advantages of Biofilm Communities

Recent publications have stressed the need for evolutionary- and community-based research in the field of microbial ecology (Caldwell, 1993;

Marshall, 1993; Stahl, 1993; Woese, 1994). It has been argued that for pro-karyotes, the reproductive success of a single cell line may be tightly linked to the success of the group as a whole, and that the microbial community is the unit of selection rather than individual microbial species (Sonea and Panisset, 1983; Caldwell, 1993). Microbial communities can be seen as highly organized, with keystone or principal species, exhibiting temporal positioning in successional developmental patterns. Despite their overriding importance to the evolution and continuance of life on earth, few if any communities have been well studied. Consequently, it is not clear how they arise, what their boundaries are, how resistant they are to change, and how they evolve with time.

The development of microbial communities may be viewed as "unlimited"; for example, most species are available to most habitats, abiotic factors are relatively unimportant, and therefore biotic interactions are the driving force behind microbial community development and structure. In contrast, community membership may be limited to a small fraction of those individuals available that are capable of coexistence, interaction, and persistence in the environment. McCormick and Cairns (1991) outlined a series of paradigms for development of some microbial communities; (1) microbial species exhibit extremely high rates of dispersal among aquatic habitats such that external transport processes do not influence the structure of individual communities; (2) because most microbes are tolerant of a wide range of environmental conditions, communities of these organisms are relatively insensitive to abiotic habitat conditions; and (3) rather than merely representing random collections of species regulated by autecologi-cal factors (i.e., dispersal rates and fundamental niches), microbial communities have a structure that is controlled by endogenous biotic forces. These paradigms were proposed for planktonic communities, and the development of biofilm communities is more likely to be regulated by factors such as (1) recruitment to the surface, (2) development of microenvironmental conditions, and (3) endoge-nous biotic forces. During the development of a biofilm community, there is also a gradual shift over time from an initial phase, with low competition/interactions and high growth rates (an r strategist-dominated community) to increased popu-lation densities with increased competition/interactions (a k strategist-dominated community). In the final stage, population density may be high with intense competition/interactions and a k strategy-dominated community (Marshall, 1989).

From a variety of studies, microbial ecologists have gained insight into some of the factors that govern successful integration of microorganisms into microbial communities. For example, inoculants have been observed to fail for a variety of reasons (Goldstein *et al.,* 1985): (1) the concentration of growth substrates may be to low to support growth; (2) the environment may contain substances inhibiting the growth or activity of the organism; (3) the growth rate may be lower than the rate of predation; and (4) the organism may fail to migrate to locate a suitable microhabitat. Thus, to achieve active community membership, a colonizing organism must overcome these limitations. For structured communities, as outlined by Wimpenny (1981, 1992), there must exist a suitable habitat domain and activity domain. The habitat domain reflects the relationship between a species and its environment, and the activity domain represents the effect of the individual on the environment. Within a community are a series of interacting populations, the presence of which provides the basis for community structure. Where habitat domains overlap, there will be potential for interactions to occur. Many of these interactions are optimized if the populations or individuals form physical associations such as those that occur at interfaces. The range of interactions possible between populations can be considered on a continuum, as described by Lewis (1985).

It may be inferred from a variety of observations (Bouwer and McCarty, 1982; Namkung and Rittman, 1987; Bouwer, 1989; Characklis and Marshall, 1990) that the development of biofilm communities should impart several advantages. Growth of bacteria in communities may (1) enhance the adaptation processes by permitting a high diversity of microorganisms to develop in the presence of a broad range of microenvironmental conditions; (2) promote interactions and genetic exchange between different species; (3) favor sequential cometabolic transformation of compounds; (4) allow sorption of the primary and secondary carbon and energy sources, sequestering them from the cells but releasing them for subsequent degradation as solution concentrations fall in the microenvironment; (5) sequestration may allow communities to tolerate and degrade higher concentrations of toxic organics than pure cultures; (6) bioconcentration of carbon and energy in exopolymers (see Wolfaardt *et al.*, 1995a) may also provide functional advantages at extremely low environmental concentrations; (7) EP matrices may provide for controlled cross-feeding between populations; and (8) growth in biofilms also provides refuge from environmental stresses (UV, predation, pH nutrient limitation). In addition, these communities may be stable and competent over substantial time periods, providing long-term access for bacteria to zones of proliferation and survival.

6. Conclusions

Our understanding of microbial community development has been hampered by the spatial scale of measurements relative to the scale of the microenvironment, the plasticity of bacterial phenotypes, dominance of pure culture studies, and the lack of complexity in microbial communities examined in laboratory studies. Few studies have addressed the role of specific adaptations in relation to reproductive success in natural or even model systems. Thus, although our knowledge of the regulation of a specific process may be detailed, our understanding of its role in bacterial survival and proliferation in natural systems is limited. Interpreting how heterotrophic bacteria respond to and benefit from community growth in the natural environment, as well as the underlying molecular biological mechanisms, awaits application of the range of techniques now available. However, they must be applied within a philosophical framework that emphasizes the natural relationships and evolutionary connections within the prokaryotes. In addition, the concepts of multicellularity, whereby bacteria have evolved a variety of strategies based on cell–cell interactions as an alternative to their unicellular nature, must be incorporated in studies of biofilm communities.

References

Absolom, D. R., Lamberti, F. V., Policova, Z., Zingg, W., Van Oss, C. J., and Neumann, A. W., 1983, Surface thermodynamics of bacterial adhesion, *Appl. Environ. Microbiol.* **46**:90–97.

Adamczyk, Z., and Van de Ven, T. G. M., 1981, Deposition of particles under external forces in laminar flow through parallel-plate and cylindrical channels, *J. Colloid Interface Sci.* **80**:340–357.

Adler, J., 1969, Chemoreceptors in bacteria, *Science* **166**:1588–1597.

Adler, J., 1975, Chemotaxis in bacteria, *Annu. Rev. Biochem.* **44**:341–355.

Adler, J., and Templeton, B., 1967, The effect of environmental conditions on the motility of *Escherichia coli, J. Gen. Microbiol.* **46**:175–184.

Alldredge, A. L., and Cohen, Y., 1987, Can microscale chemical patches persist at sea? Microelectrode study of marine snow, fecal pellets, *Science* **235**:689–691.

Allison, D. G., and Sutherland, I. W., 1987, The role of exopolysaccharides in adhesion of freshwater bacteria, *J. Gen. Microbiol.* **133**:1319–1327.

Amann, R. I., Krumholz, L., and Stahl, D. A., 1990, Fluorescent oligonucleotide probing of whole cells for determinative phylogenetic, and environmental studies in microbiology, *J. Bacteriol.* **172**:762–770.

Amann, R. I., Stromley, J., Devereux, R., Key, R., and Stahl, D. A., 1992, Molecular and microscopic identification of sulfate-reducing bacteria in multispecies biofilms, *Appl. Environ. Microbiol.* **48**:614–623.

Assmus, B., Hutzler, P., Kirchhof, G. Amann, R., Lawrence, J. R., and Hartmann, A., 1995, *In situ* localization of *Azospirillum brasilense* in the rhizosphere of wheat with fluorescently labeled, rRNA-targeted oligonucleotide probes and scanning confocal laser microscopy, *Appl. Environ. Microbiol.* **61**:1013–1019.

Baier, R. E., 1980, Substrate influence on adhesion of microorganisms and their resultant new surface properties, in: *Adsorption of Microorganisms to Surfaces* (G. Bitton and K. C. Marshall, eds.), Wiley-Interscience, New York, pp. 59–104.

Baier, R. E., 1984, Initial events in microbial film formation, in: *Marine Biodeterioration: An Interdisciplinary Approach* (J. D. Costlow and R. C. Tipper, eds.), E. and F. N. Spon Ltd., London, pp. 57–62.

Baier, R. E., 1985, Adhesion in the biologic environment, *Biomater. Med. Devices Artif. Organs* **12**:133–159.

Baker, J. H., 1984, Factors affecting the bacterial colonization of various surfaces in a river, *Can. J. Microbiol.* **30**:511–515.

Bale, M. J., Fry, J. C., and Day, M. J., 1987, Transfer and occurrence of large mercury resistance plasmids in river epilithion, *Appl. Environ. Microbiol.* **54**:972–978.

Bale, M. J., Day, M. J., and Fry, J. C., 1988, Novel method for studying plasmid transfer in undisturbed river epilithon, *Appl. Environ. Microbiol.* **54**:2756–2758.

Banks, M. K., and Bryers, J. D., 1991, Bacterial species dominance within a binary culture biofilm, *Appl. Environ. Microbiol.* **57**:1874–1979.

Barkay, T., Turner, R., Vandenbrook, A., and Liebert, C., 1991, The relationships of Hg (II) volatilization from a freshwater pond to the abundance of mer genes in the gene pool of the indigenous microbial community, *Microb. Ecol.* **21**:151–161.

Bartell, P. F., Orr, T. E., and Chudio, B., 1970, Purification and chemical composition of the protective slime antigen of *Pseudomonas aeruginosa, Infect. Immun.* **2**:543–548.

Belas, M. R., and Colwell, R. R., 1982, Adsorption kinetics of laterally and polarly flagellated *Vibrio, J. Bacteriol.* **151**:1568–1580.

Bell, J. P., and Tsezos, K., 1987, Removal of hazardous organic pollutants by biomass adsorption, *J. Water Poll. Control Fed.* **59**:191–198.

Bellin, C. A., and Rao, P. S. C., 1993, Impact of bacterial biomass on contaminant sorption and transport in a subsurface soil, *Appl. Environ. Microbiol.* **59**:1813–1820.

Bellon-Fontaine, M.-N., Mozes, N., van der Mei, H. C., Sjollema, J., Cerf, O., Rouxhet, P. G., and Busscher, H. J., 1990, A comparison of thermodynamic approaches to predict the adhesion of dairy microorganisms to solid substrata, *Cell. Biophys.* **17**:93–106.

Bengtsson, G., 1991, Bacterial exopolymer and PHB production in fluctuating groundwater habitats, *FEMS Microbiol. Lett.* **86**:15–24.

Berg, H. C., 1985, Physics of bacterial chemotaxis, in: *Sensory Perception and Transduction in Aneural Organisms* (G. Colombetti, F. Linci, and P-S. Song, eds.), Plenum Press, London, pp. 19–30.

Berg, H. C., and Brown, D. A., 1972, Chemotaxis in *Escherichia coli* analyzed by three-dimensional tracking, *Nature* **239**:500–504.

Biddanda, B. A., and Pomeroy, I. R., 1988, Microbial aggregation and dehydration of phytoplankton-derived detritus in seawater. I. Microbial succession, *Mar. Ecol. Prog. Ser.* **42**:79–88.

Block, S. M., Segall, J. E., and Berg, H. C., 1983, Adaptation kinetics in bacterial chemotaxis, *J. Bacteriol.* **154**:312–324.

Bochem, H. P., Schoberth, S. M., Sprey, B., and Wengler, P., 1982, Thermophilic biomethanation of acetic acid: Morphology and ultrastructure of a granular consortium, *Can. J. Microbiol.* **28**:500–510.

Boone, D. R., Johnson, R. L., and Liu, Y., 1989, Diffusion of the interspecies electron carriers H_2 and formate in methanogenic ecosystems and its implications in the measurement of K_m for H_2 or formate uptake, *Appl. Environ. Microbiol.* **55**:1735–1741.

Bott, T. L., and Brock, T. D., 1970, Growth and metabolism of periphytic bacteria: Methodology, *Limnol. Oceanogr.* **20**:191–197.

Bottomley, P. J., and Maggard, S. P., 1990, Determination of viability within serotypes of a soil population of *Rhizobium leguminosarum* bv. *trifolii, Appl. Environ. Microbiol.* **56**:533–540.

Bouwer, E. J., 1989, Transformations of xenobiotics in biofilms, in: *Structure and Function of Biofilms* (W. G. Characklis and P. H. Wilderer, eds.), Wiley and Sons, Toronto, Canada, pp. 251–267.

Bouwer, E. J., and McCarty, P. L., 1982, Removal of trace chlorinated organic compounds by activated carbon and fixed film bacteria, *Environ. Sci. Technol.* **16**:836–843.

Bowen, J. D., Stolzenbach, K. D., and Chrisholm, S. W., 1993, Simulating bacterial clustering around phytoplankton cells in a turbulent ocean, *Limnol. Oceanogr.* **38**:36–51.

Bratbak, G., Heldal, M., Naess, A., and Roeggen, T., 1993, Viral impact on microbial communities, in: *Trends in Microbial Ecology* (R. Guerrero and C. Pedrós-Alío, eds.), Spanish Society for Microbiology, Barcelona, Spain, pp. 299–302.

Brown, M. J., and Lester, J. N., 1982a, Role of bacterial extracellular polymers in metal uptake in pure bacterial culture and activated sludge—I, *Water Res.* **16:**1539–1548.

Brown, M. J., and Lester, J. N., 1982b, Role of bacterial extracellular polymers in metal uptake in pure bacterial culture and activated sludge—II: Effects of mean cell retention time, *Water Res.* **16:**1549–1560.

Brown, M. R. W., and Gilbert, P., 1993, Sensitivity of biofilms to antimicrobial agents, in: *Microbial Cell Envelopes: Interactions and Biofilms* (L. B. Quesnel, P. Gilbert, and P. S. Handley, eds.), Blackwell Scientific, Oxford, England, pp. 87S–97S.

Brown, M. R. W., Allison, D. G., and Gilbert, G., 1988, Resistance of bacterial biofilms to antibiotics: A growth-rate related effect? *J. Antimicrob. Chemother.* **22:**777–780.

Bryers, J. D., 1987, Biologically active surfaces: Processes governing the formation and persistence of biofilms, *Biotechnol. Prog.* **3:**57–68.

Burchard, R. P., 1981, Gliding motility of prokaryotes: Ultrastructure, physiology, and genetics, *Annu. Rev. Microbiol.* **35:**497–529.

Burchard, R. P., 1984, Inhibition of *Cytophaga* sp. strain U67 gliding motility by inhibitors of peptide synthesis, *Arch. Microbiol.* **139:**248–254.

Burchard, R. P., Rittschof, D., and Bonaventura, J., 1990, Adhesion and motility of gliding bacteria on substrata with different surface free energies, *Appl. Environ. Microbiol.* **56:**2529–2534.

Busscher, H. J., Bellon-Fontaine, M.-N., Mozes, N, Van Der Mei, H. C., Sjollema, J., Cerf, O., and Rouxhet, P. G., 1990a, Deposition of *Leuconostoc mesenteroides* and *Streptococcus thermophilus* to solid substrata in a parallel plate flow cell, *Biofouling* **2:**55–63.

Busscher, H. J., Bellon-Fontaine, Sjollema, J., and Van Der Mei, H. C., 1990b, Relative importance of surface free energy as a measure of hydrophobicity in bacterial adhesion to solid surfaces, in: *Microbial Cell Surface Hydrophobicity* (R. J. Doyle and M. Rosenberg, eds.), American Society for Microbiology Press, Washington, D.C., pp. 335–359.

Busscher, H. J., Cowan, M. M., and Van Der Mei, H. C., 1992, On the relative importance of specific and non-specific approaches to oral microbial adhesion, *FEMS Microbiol. Rev.* **88:**199–210.

Byrd, J. J., Zeph, L. R., and Casida, Jr., L. E., 1985, Bacterial control of *Agromyces ramosus* in soil, *Can. J. Microbiol.* **31:**157–1163.

Caldwell, D. E., 1993, The microstat: Steady-state microenvironments for subculture of steady-state consortia, communities, and microecosystems, in: *Trends in Microbial Ecology* (R. Guerrero and C. Pedrós-Alío, eds.), Spanish Society for Microbiology, Barcelona, Spain, pp. 123–128.

Caldwell, D. E., and Lawrence, J. R., 1986, Growth kinetics of *Pseudomonas fluorescens* microcolonies within the hydrodynamic boundary layers of surface microenvironments, *Microb. Ecol.* **12:**299–312.

Caldwell, D. E., Korber, D. R., and Lawrence, J. R., 1992a, Confocal laser microscopy and digital image analysis in microbial ecology, in: *Advances in Microbial Ecology,* Vol. 12 (K. C. Marshall, ed.), Plenum Press, New York, pp. 1–67.

Caldwell, D. E., Korber, D. R., and Lawrence, J. R., 1992b, Imaging of bacterial cells by fluorescence exclusion using scanning confocal laser microscopy, *J. Microbiol. Methods* **15:**249–261.

Caldwell, D. E., Korber, D. R., and Lawrence, J. R., 1993, Analysis of biofilm formation using 2-D versus 3-D digital imaging, in: *Microbial Cell Envelopes: Interactions and Biofilms* (L. B. Quesnel, P. Gilbert, and P. S. Handley, eds.), Blackwell Scientific, Oxford, England, pp. 52S–66S.

Camper, A. K., Hayes, J. T., Sturman, P. J., Jones, W. L., and Cunningham, A. B., 1993, Effects of motility and adsorption rate coefficient on transport of bacteria through saturated porous media, *Appl. Environ. Microbiol.* **59:**3455–3462.

Caron, D. A., 1987, Grazing of attached bacteria by heterotrophic microflagellates, *Microb. Ecol.* **13**:203–218.

Casida, L. E., Jr., 1992, Competitive ability and survival in soil of *Pseudomonas* strain 679-2, a dominant, nonobligate bacterial predator of bacteria, *Appl. Environ. Microbiol.* **58**:32–37.

Characklis, W. G., 1990, Biofilm processes, in: *Biofilms* (W. G. Characklis and K. C. Marshall, eds.), Wiley and Sons, New York, pp. 195–231.

Characklis, W. G. and Marshall, K. C. (eds.), 1990, *Biofilms*, Wiley-Interscience, New York.

Characklis, W. G., McFeters, G. A., and Marshall, K. C., 1990a, Physicological ecology in biofilm systems, in: *Biofilms* (W. G. Characklis and K. C. Marshall, eds.), Wiley and Sons, New York, pp. 341–393.

Characklis, W. G., Turakhia, M. H. and Zelver, N., 1990b, Transfer and interfacial transport phenomena, in: *Biofilms* (W. G. Characklis and K. C. Marshall, eds.), Wiley and Sons, New York, pp. 265–340.

Chatfield, L. K., and Williams, P. A., 1986, Naturally occurring TOL plasmids in *Pseudomonas* strains carry either two homologous or two nonhomologous catechol 2,3-oxygenase genes, *J. Bacteriol.* **168**:878–885.

Christensen, B. E., and Characklis, W. G., 1990, Physical and chemical properties of biofilms, in: *Biofilms* (W. G. Characklis and K. C. Marshall, eds.), Wiley and Sons, New York, pp. 93–130.

Christensen, B. E., Kjosbakken, J., and Smidsrød, O., 1985, Partial chemical and physical characterization of two extracellular polysaccharides produced by marine, periphytic *Pseudomonas* sp. Strain NCMB 2021, *Appl. Environ. Microbiol.* **50**:837–845.

Christersson, C. E., Glantz, P-O. J., and Baier, R. E., 1988, Role of temperature and shear forces on microbial detachment, *Scand. J. Dent. Res.* **96**:91–98.

Clarholm, M., 1981, Protozoan grazing of bacteria in soil-impact and importance, *Microb. Ecol.* **7**:343–350.

Clarholm, M., 1984, Heterotrophic, free-living protozoa: Neglected microorganisms with an important task in regulating bacterial populations, in: *Current Perspectives in Microbial Ecology* (M. J. Klug and C. A. Reddy, eds.), American Society for Microbiology Press, Washington, D. C., pp. 321–326.

Confer, D. R., and Logan, B. E., 1991, Increased bacterial uptake of macromolecular substrates with fluid shear, *Appl. Environ. Microbiol.* **57**:3093–3100.

Costerton, J. W., 1984, The formation of biocide-resistant biofilms in industrial, natural and medical systems, *Dev. Ind. Microbiol.* **25**:363–372.

Costerton, J. W., Irvin, R. T., and Cheng, K. J., 1981, The bacterial glycocalyx in nature and disease, *Annu. Rev. Microbiol.* **35**:299–324.

Costerton, J. W., Cheng, K-J., Geesey, G. G., Ladd, T. I., Nickel, N. C., Dasgupta, M., and Marrie, T. J., 1987, Bacterial biofilms in nature and disease, *Annu. Rev. Microbiol.* **41**:435–464.

Costerton, J. W., Lewandowski, Z., Caldwell, D. E., Korber, D. R., and James, G., 1994, Biofilms: The customized microniche, *J. Bacteriol.* **176**:2137–2142.

Cowan, M. M., Warren, T. M., and Fletcher, M., 1991, Mixed-species colonization of solid surfaces in laboratory biofilms, *Biofouling* **3**:23–34.

Dagostino, L., Goodman, A. E., and Marshall, K. C., 1991, Physiological responses induced in bacteria adhering to surfaces, *Biofouling* **4**:113–19.

Davies, D. G., Chakrabarty, A. M., and Geesey, G. G., 1993, Exopolysaccharide production in biofilms: Substratum activation of alginate gene expression by *Pseudomonas aeruginosa*, *Appl. Environ. Microbiol.* **59**:1181–1186.

Dawson, M. P., Humphrey, B. A., and Marshall, K. C., 1981, Adhesion: A tactic in the survival strategy of a marine vibrio during starvation, *Curr. Microbiol.* **6**:195–201.

Day, M. J., 1987, The biology of plasmids, *Sci. Prog.* **71**:203–220.

Decho, A. W., 1990, Microbial exopolymer secretions in ocean environments: Their role(s) in food webs and marine processes, *Oceanogr. Mar. Biol. Annu. Rev.* **28:**73–153.

Decho, A. W., and Lopez, G. R., 1993, Exopolymer microenvironments of microbial flora: Multiple and interactive effects on trophic relationships, *Limnol. Oceanogr.* **38:**1633–1645.

DeFlaun, M. F., Tanzer, A. S., McAteer, A. L., Marshall, B., and Levy, S. B., 1990, Development of an adhesion assay and characterization of an adhesion-deficient mutant of *Pseudomonas fluorescens, Appl. Environ. Microbiol.* **56:**112–119.

Delaquis, P. J., Caldwell, D. E., Lawrence, J. R., and McCurdy, A. R., 1989, Detachment of *Pseudomonas fluorescens* from biofilms on glass surfaces in response to nutrient stress, *Microb. Ecol.* **18:**199–210.

Delaquis, P. J., 1990, *Colonization of Model and Meat Surfaces by Pseudomonas fragi and Pseudomonas fluorescens,* University of Saskatchewan, Saskatoon, Canada, Ph.D. thesis.

Dempsey, M. J., 1981, Marine bacterial fouling: A scanning electron microscope study, *Mar. Biol.* **61:**305–315.

Deretic, V., Dikshit, R., Konyecsni, W. M., Chakrabarty, A. M., and Misra, T. K., 1989, The algR gene, which regulates mucoidy in *Pseudomonas aeruginosa,* belongs to a class of environmentally responsive genes, *J. Bacteriol.* **171:**1278–1283.

De Weger, L. D., Van Der Vlugt, C. I. M., Wijfjes, A. H. M., Bakker, P. A. H. M., Schippers, B., and Lugtenberg, B., 1987, Flagella of a plant-growth-stimulating *Pseudomonas fluorescens* strain are required for colonization of potato roots, *J. Bacteriol.* **169:**2769–2773.

Dexter, S. C., 1979, Influence of substratum critical surface tension on bacterial adhesion—*in situ* studies, *J. Colloid Interface Sci.* **70:**346–353.

Dexter, S. C., Sullivan Jr., J. D., Williams, J., and Watson, S. W., 1975, Influence of substrate wetability on the attachment of marine bacteria to various surfaces, *Appl. Microbiol.* **30:**298–308.

Dow, C. S., Westmacott, D., and Whittenbury, R., 1976, Ultrastructure of budding and prosthecate bacteria, in: *Microbial Ultrastructure* (R. Fuller and D. W. Loverlock, eds.), Academic Press, New York, pp. 187–221.

Dudman, W. F., 1977, The role of surface polysaccharides in natural environments, in: *Surface Carbohydrates of the Prokaryotic Cells* (I. W. Sutherland, ed.), Academic Press, New York, pp. 357–414.

Ellwood, D. C., Keevil, C. W., Marsh, P. D., Brown, C. M., and Wardell, J. N., 1982, Surface-associated growth, *Phil. Trans. R. Soc. Lond. B* **297:**517–532.

Eng, R. H. K., Padberg, F. T., Smith, S. M., Tan, E. N., and Cherubin, C. E., 1991, Bactericidal effects of antibiotics on slowly growing and nongrowing bacteria, *Antimicrob. Agents Chemother.* **35:**1824–1828.

Evans, D. J., Brown, M. R. W., Allison, D. G., and Gilbert, P., 1990, Susceptibility of bacterial biofilms to tobramycin: Role of specific growth rate and phase in division cycle, *J. Antimicrob. Chemother.* **25:**585–591.

Fenchel, T., and Jorgensen, B. B., 1977, Detritus food chains of aquatic ecosystems: The role of bacteria, in: *Advances in Microbial Ecology,* Vol. 1 (M. Alexander, ed.), Plenum Press, New York, pp. 3–37.

Fernandez-Astorga, A., Muela, A., Cisterna, R., Iriberri, J., and Barcina, I., 1992, Biotic and abiotic factors affecting plasmid transfer in *Escherichia coli* strains, *Appl. Environ. Microbiol.* **58:**392–398.

Ferris, F. G., Fyfe, W. S., Witten, T., Schultze, S., and Beveridge, T. J., 1989, Effect of mineral substrate hardness on the population density of epilithic microorganisms in two Ontario rivers, *Can. J. Microbiol.* **35:**744–747.

Fletcher, M., 1977, The effects of culture concentration and age, time, and temperature on bacterial attachment to polystyrene, *Can. J. Microbiol.* **23:**1–6.

Fletcher, M., 1984, Comparative physiology of attached and free-living bacteria, in: *Microbial Adhesion and Aggregation* (K. C. Marshall, ed.), Springer Verlag, New York, pp. 223–232.

Fletcher, M., 1988, Attachment of *Pseudomonas fluorescens* to glass and influence of electrolytes on bacterium-substratum separation distance, Abstracts of the 88th Annual Meeting of the American Society for Microbiology, Miami Beach, Fla.

Fletcher, M., 1993, Physicochemical aspects of surface colonization, in: *Trends in Microbial Ecology* (R. Guerrero and C. Pedrós-Alío, eds.), Spanish Society for Microbiology, Barcelona, Spain, pp. 109–112.

Fletcher, M., and Loeb, G. I., 1979, Influence of substratum characteristics on the attachment of a marine pseudomonad to solid surfaces, *Appl. Environ. Microbiol.* **37:**67–72.

Fletcher, M., and Marshall, K. C., 1982, Bubble contact angle method for evaluating substratum interfacial characteristics and its relevance to bacterial attachment, *Appl. Environ. Microbiol.* **44:**184–192.

Fogarty, A. M., and Tuovinen, O. H., 1991, Microbiological degradation of pesticides in yard waste composting, *Microbiol. Rev.* **55:**225–233.

Fredrickson, A. G., 1977, Behavior of mixed cultures of microorganisms, *Annu. Rev. Microbiol.* **33:**63–87.

Fry, J. C., and Day, M. J. (eds.), 1990, *Bacterial Genetics in Natural Environments,* Chapman and Hall, London, England.

Fulthorpe, R. R., and Wyndham, R. C., 1991, Transfer and expression of the catabolic plasmid pBRC60 in wild bacterial recipients in a freshwater ecosystem, *Appl. Environ. Microbiol.* **57:**1546–1553.

Fulthorpe, R. R., and Wyndham, R. C., 1992, Involvement of a chlorobenzoate-catabolic transposon, Tn5271, in community adaptation to chlorobiphenyl, chloroaniline and 2,4-dichlorophenoxyacetic acid in a freshwater ecosystem, *Appl. Environ. Microbiol.* **58:**314–325.

Furukawa, K., Taira, K., and Hayase, N., 1989, Molecular organization of chromosomal genes coding for biphenyl/PCB catabolism in various soil bacteria, in: *Recent Advances in Microbial Ecology* (T. Hattori, Y. Ishida, Y. Maruyama, R. Y. Morita and A. Uchida, eds.), Japan Scientific Society Press, Tokyo, pp. 611–616.

Geesey, G. G., and Costerton, J. W., 1979, Microbiology of a northern river: Bacterial distribution and relationship to suspended sediment and organic carbon, *Can. J. Microbiol.* **25:**1058–1062.

Geesey, G. G., and Morita, R. Y., 1979, Capture of arginine at low concentrations by a marine psychrophilic bacterium, *Appl. Environ. Microbiol.* **38:**1092–1097.

Geesey, G. G., and White, D. C., 1990, Determination of bacterial growth and activity at solid–liquid interfaces, *Annu. Rev. Microbiol.* **44:**579–602.

Geesey, G. G., Richardson, W. T., Yoemans, H. G., Irvin, R. T., and Costerton, J. W., 1977, Microscopic examination of natural bacterial populations from an alpine stream, *Can. J. Microbiol.* **23:**1733–1736.

Gilbert, P., Collier, P. J., and Brown, M. R. W., 1990, Influence of growth rate on susceptibility to antimicrobial agents: Biofilms, cell cycle, dormancy, and stringent response, *Antimicrob. Agents Chemother.* **34:**1856–1868.

Giwercman, B., Jensen, E. T., Hoiby, N., Kharazmi, A., and Costerton, J. W., 1991, Induction of β-lactamase production in *Pseudomonas aeruginosa* biofilms, *Antimicrob. Agents Chemother.* **35:**1008–1010.

Godwin, S. L., Fletcher, M., and Burchard, R. P., 1989, Interference reflection microscopic study of sites of association between gliding bacteria and glass substrata, *J. Bacteriol.* **171:**4589–4594.

Goldstein, R. M., Mallory, L. M., and Alexander, M., 1985, Reasons for possible failure of inoculation to enhance biodegradation, *Appl. Environ. Microbiol.* **50:**977–983.

Goodman, A. E., Angles, M. L., and Marshall, K. C., 1993a, Genetic responses of bacteria in biofilms, in: *Trends in Microbial Ecology* (R. Guerrero and C. Pedrós-Alío, eds.), Spanish Society for Microbiology, Barcelona, Spain, pp. 119–122.

Goodman, A. E., Hild, E., Marshall, K. C., and Hermansson, M., 1993b, Conjugative plasmid transfer between bacteria under simulated marine oligotrophic conditions, *Appl. Environ. Microbiol.* **59:**1035–1040.

Guckert, J. B., Hood, M. A., and White, D. C., 1986, Phospholipid ester-linked fatty acid profile changes during nutrient deprivation of *Vibrio cholerae;* increases in the *trans/cis* ratio and proportions of cyclopropyl fatty acids, *Appl. Environ. Microbiol.* **52:**794–801.

Güde, H., 1985, Influence of phagotrophic processes on the regeneration of nutrients in two-stage continuous culture systems, *Microb. Ecol.* **11:**193–204.

Guentzel, M. N., and Berry, L. J., 1983, Motility as a virulence factor for *Vibrio cholerae, Infect. Immun.* **11:**890–897.

Gutierrex, N. A., and Maddox, I. S., 1987, Role of chemotaxis in solvent production by *Clostridium acetobutylicum, Appl. Environ. Microbiol.* **53:**1924–1927.

Haefele, D. M., and Lindow, S. E., 1987, Flagellar motility confers epiphytic fitness advantages upon *Pseudomonas syringae, Appl. Environ. Microbiol.* **53:**2528–2533.

Häggblom, M., 1990, Mechanisms of bacterial degradation and transformation of chlorinated mono-aromatic compounds, *J. Basic Microbiol.* **30:**115–141.

Hamilton, W. A., 1987, Biofilms: Microbial interactions and metabolic activities, in: *Ecology of Microbial Communities* (M. Fletcher, T. R. G. Gray, and J. G. Jones, eds.), Cambridge University Press, Cambridge, England, pp. 361–385.

Hamilton, W. A., and Characklis, W. G., 1989, Relative activities of cells in suspension and in biofilms, in: *Structure and Function of Biofilms* (W. G. Characklis and P. A. Wilderer, eds.), Wiley and Sons, Toronto, Canada, pp. 199–219.

Handley, P. S., Hesketh, L. M., and Moumena, R. A., 1991, Charged and hydrophobic groups are localized in the short and long tuft fibrils on *Streptococcus sanguis* strains, *Biofouling* **4:**105–111.

Hansen, J. B., Doubet, R. S., and Ram, J., 1984, Alginase enzyme production by *Bacillus circulans, Appl. Environ. Microbiol.* **47:**704–709.

Helmsetter, C. E., 1969, Sequence of bacterial reproduction, *Annu. Rev. Microbiol.* **23:**223–238.

Henningson, P. J., and Gudmestad, N. C., 1992, Comparison of exopolysaccharides from mucoid and nonmucoid strains of *Clavibacter michiganensis* subspecies *sepedonicus, Can. J. Microbiol.* **39:**291–296.

Henrici, A. T., and Johnson, D. E., 1935, Studies of freshwater bacteria II. Stalked bacteria, a new order of Schizomycetes, *J. Bacteriol.* **30:**61–93.

Hermansson, M., and Marshall, K. C., 1985, Utilization of surface localized substrate by non-adhesive marine bacteria, *Microb. Ecol.* **11:**91–105.

Hirsch, P., 1968, Biology of budding bacteria, *Arch. Mikrobiol.* **60:**201–216.

Hirsch, P., 1974, Budding bacteria, *Annu. Rev. Microbiol.* **28:**391–433.

Hirsch, P., 1984, Microcolony formation and consortia, in: *Microbial Adhesion and Aggregation* (K. C. Marshall, ed.), Springer-Verlag, New York, pp. 373–393.

Hirsch, P., and Rheinheimer, G. 1968, Biology of budding bacteria. V. Budding bacteria in aquatic habitats: Occurrence, enrichment and isolation, *Arch. Mikrobiol.* **62:**289–306.

Ho, C. S., 1986, An understanding of the forces in the adhesion of micro-organisms to surfaces, *Proc. Biochem.* **21:**148–152.

Hood, M. A., Guchert, J. B., White, D. C., and Deck, F., 1986, Effect of nutrient deprivation on lipid, carbohydrate, RNA, DNA, and protein levels in *Vibrio cholerae, Appl. Environ. Microbiol.* **52:**788–793.

Howell, J. A., and Atkinson, B., 1976, Sloughing of microbial film in trickling filters, *Water Res.* **10:**307–315.

Huang, T.-C., Chang, M.-C., and Alexander, M., 1981, Effect of protozoa on bacterial degradation of an aromatic compound, *Appl. Environ. Microbiol.* **41:**229–232.

Hunt, H. W., Coleman, D. C., Cole, C. V., Ingham, R. E., Elliott, E. T., and Woods, L. E., 1984, Simulation model of a food web with bacteria, amoebae, and nematodes in soil, in: *Current Perspectives in Microbial Ecology* (M. J. Klug and C. A. Reddy, eds.), American Society for Microbiology Press, Washington, D.C., pp. 346–352.

Isaacson, R. E., Nagy, B., and Moon, H. W., 1977, Colonization of porcine small intestine by *Escherichia coli:* Colonization and adhesion factors of pig enteropathogens that lack K88, *J. Infect. Dis.* **135:**531–538.

James, G. A., Caldwell, D. E., and Costerton, J. W., 1993, Spatial relationships between bacterial species within biofilms, Abstract, CSM/SIM annual meeting, Toronto, Canada.

James, G. A., Korber, D. R., Caldwell, D. E., and Costerton, J. W., 1995, Digital image analysis of growth and starvation responses of a surface-colonizing *Acinetobacter* sp. *J. Bacteriol.* **177:**905–915.

Jannasch, H. W., 1967, Growth of marine bacteria at limiting concentrations of organic carbon in seawater, *Limnol. Oceanogr.* **12:**264–271.

Jenneman, G. E., McInerney, M. J., and Knapp, R. M., 1985, Microbial penetration through nutrient-saturated Berea sandstone, *Appl. Environ. Microbiol.* **50:**383–391.

Jensen, R. H., and Woolfolk, C. A., 1985, Formation of filaments by *Pseudomonas putida*, *Appl. Environ. Microbiol.* **50:**364–372.

Jones, G. W., Baines, L., and Genthner, F. J., 1991, Heterotrophic bacteria of the freshwater neuston and their ability to act as plasmid recipients under nutrient deprived conditions, *Microb. Ecol.* **22:**15–25.

Kalos, M., and Zissler, J. F., 1990, Defects in contact-stimulated gliding during aggregation by *Myxococcus xanthus, J. Bacteriol.* **172:**6476–6493.

Keevil, C. W., and Walker, J. T., 1992, Nomarski DIC microscopy and image analysis of biofilms, *Binary* **4:**93–95.

Kefford, B., Kjelleberg, S., and Marshall, K. C., 1982, Bacterial scavenging: Utilization of fatty acids localized at a solid–liquid interface, *Arch. Microbiol.* **133:**257–260.

Kelly, F. X., Dapsis, K. J., and Lauffenburger, D. A., 1988, Effect of bacterial chemotaxis on dynamics of microbial competition, *Microb. Ecol.* **16:**115–131.

Kenne, L., and Lindberg, B., 1983, Bacterial polysaccharides, in: *The Polysaccharides* (G. O. Aspinall, ed.), Academic Press, New York, pp. 287–363.

Kim, S. K., and Kaiser, D., 1990, C-factor; a cell–cell signalling protein required for fruiting body morphogenesis of *M. xanthus, Cell* **61:**19–26.

Kjelleberg, S., 1980, Effects of interfaces on survival mechanisms of copiotrophic bacteria in low-nutrient environments, in: *Microbial Adhesion to Surfaces* (R. C. W. Berkeley, J. M. Lynch, J. Melling, P. R. Rutter, and B. Vincent, eds.), Horwood, Chichester, England, pp. 151–159.

Kjelleberg, S., 1984, Effects of interfaces on survival mechanisms of copiotrophic bacteria in low-nutrient habitats, in: *Current Perspectives in Microbial Ecology* (M. J. Klug and C. A. Reddy, eds.), American Society for Microbiology Press, Washington, D.C., pp. 151–159.

Kjelleberg, S., Humphrey, B. A., and Marshall, K. C., 1982, The effect of interfaces on small, starved marine bacteria, *Appl. Environ. Microbiol.* **43:**1166–1172.

Kjelleberg, S., Hermansson, M., Marden, P., and Jones, G. W., 1987, The transient phase between growth and non-growth of heterotrophic bacteria with emphasis on the marine environment, *Annu. Rev. Microbiol.* **41:**25–49.

Kjelleberg, S., Ostling, J., Holmquist, L., Flardh, K., Svenblad, B., Jouper-Jann, A., Weichart, D., and Albertson, N., 1993, Starvation and recovery of *Vibrio*, in: *Trends in Microbial Ecology* (R. Guerrero and C. Pedrós-Alío, eds.), Spanish Society for Microbiology, Barcelona, Spain, pp. 169–174.

Knight, E. V., Novick, N. J., Kaplan, D. L., and Meeks, J. R., 1990, Biodegradation of 2-fur-aldehyde under nitrate-reducing and methanogenic conditions, *Environ. Toxicol. Chem.* **9**:725–730.

Koch, A. L., 1991, Diffusion: The crucial process in many aspects of the biology of bacteria, in: *Advances in Microbial Ecology*, Vol. 11 (K. C. Marshall, ed.), Plenum Press, New York, pp. 37–70.

Kolenbrander, P. E., 1989, Surface recognition among oral bacteria: Multigeneric coaggregations and their mediators, *Crit. Rev. Microbiol.* **17**:137–159.

Kolenbrander, P. E., 1993, Coaggregation of human oral bacteria: Potential role in the accretion of dental plaque, in: *Microbial Cell Envelopes: Interactions and Biofilms* (L. B. Quesnel, P. Gilbert, and P. S. Handley, eds.), Blackwell Scientific, Oxford, England, pp. 79S–86S.

Kolenbrander, P. E., and London, J., 1992, Ecological significance of coaggregation among oral bacteria, in: *Advances in Microbial Ecology*, Vol. 12 (K. C. Marshall, ed.), Plenum Press, New York, pp. 183–217.

Komiyama, K., Habbick, B. F., and Gibbons, R. J., 1987, Interbacterial adhesion between *Pseudomonas aeruginosa* and indigenous oral bacteria isolated from patients with cystic fibrosis, *Can. J. Microbiol.* **33**:27–32.

Korber, D. R., Lawrence, J. R., Sutton, B., and Caldwell, D. E., 1989, The effect of laminar flow on the kinetics of surface recolonization by mot$^+$ and mot$^-$ *Pseudomonas fluorescens, Microb. Ecol.* **18**:1–19.

Korber, D. R., Lawrence, J. R., Zhang, L., and Caldwell, D. E., 1990, Effect of gravity on bacterial deposition and orientation in laminar flow environments, *Biofouling* **2**:335–350.

Korber, D. R., Lawrence, J. R., Hendry, M. J., and Caldwell, D. E., 1993, Analysis of spatial variability within mot$^+$ and mot$^-$ *Pseudomonas fluorescens* biofilms using representative elements, *Biofouling* **7**:339–358.

Korber, D. R., James, G. A., and Costerton, J. W., 1994a, Evaluation of fleroxacin activity against established *Pseudomonas fluorescens* biofilms, *Appl. Environ. Microbiol.* **60**:1663–1669.

Korber, D. R., Lawrence, J. R., and Caldwell, D. E., 1994b, Effect of motility on surface colonization and reproductive success of *Pseudomonas fluorescens* in dual-dilution continuous culture and batch culture systems. *Appl. Environ. Microbiol.* **60**:1421–1429.

Korber, D. R., Hanson, K. G., Lawrence, J. R., Caldwell, D. E., and Costerton, J. W., 1994c, The effect of environmental laminar flow velocities on the architecture of *Pseudomonas fluorescens* biofilms, in: *Abstracts of the ASM 94th Annual Meeting* (Las Vegas, Nevada), American Society of Microbiology.

Kröckel, L., and Focht, D. D., 1987, Construction of chlorobenzene-utilizing recombinants by progenitive manifestation or a rare event, *Appl. Environ. Microbiol.* **53**:2470–2475.

La Motta, E. J., 1976, Internal diffusion and reaction in biological films, *Environ. Sci. Technol.* **10**:765–769.

Lang, E., Viedt, H., Egestorff, J., and Hanert, H. H., 1992, Reaction of the soil microflora after contamination with chlorinated aromatic compounds and HCH, *FEMS Microbiol. Ecol.* **86**:275–282.

Lauffenburger, D., Aris, R., and Keller, K. H., 1981, Effects of random motility on growth of bacterial populations, *Microb. Ecol.* **7**:207–227.

Lauffenburger, D., Grady, M., and Keller, K. H., 1984, An hypothesis for approaching swarms of myxobacteria, *J. Theoret. Biol.* **110**:257–274.

Lawrence, J. R., and Caldwell, D. E., 1987, Behavior of bacterial stream populations within the hydrodynamic boundary layers of surface microenvironments, *Microb. Ecol.* **14**:15–27.

Lawrence, J. R., and Korber, D. R., 1993, Aspects of microbial surface colonization behavior, in: *Trends in Microbial Ecology* (R. Guerrero and C. Pedrós-Alío, eds.), Spanish Society for Microbiology, Barcelona, Spain, pp. 113–118.

Lawrence, J. R., Delaquis, P. J., Korber, D. R., and Caldwell, D. E., 1987, Behavior of *Pseu-*

domonas fluorescens within the hydrodynamic boundary layers of surface microenvironments, *Microb. Ecol.* **14**:1–14.

Lawrence, J. R., Korber, D. R., and Caldwell, D. E., 1989a, Computer-enhanced darkfield microscopy for the quantitative analysis of bacterial growth and behavior on surfaces, *J. Microbiol. Methods* **10**:123–138.

Lawrence, J. R., Malone, J. A., Korber, D. R., and Caldwell, D. E., 1989b, Computer image enhancement to increase depth of field in phase contrast microscopy, *Binary* **1**:181–185.

Lawrence, J. R., Korber, D. R., Hoyle, B. D., Costerton, J. W., and Caldwell, D. E., 1991, Optical sectioning of microbial biofilms, *J. Bacteriol.* **173**:6558–6567.

Lawrence, J. R., Korber, D. R., and Caldwell, D. E., 1992, Behavioral analysis of *Vibrio parahaemolyticus* variants in high- and low-viscosity microenvironments by use of digital image processing, *J. Bacteriol.* **174**:5732–5739.

Lawrence, J. R., Wolfaardt, G. M., and Snyder, R. A., 1993, Influence of microbial trophic interactions on the architecture and exopolymer chemistry of a biofilm community, Abstract, Canadian Society for Microbiologists and Society for Industrial Microbiology Meeting, Toronto, Ontario.

Lawrence, J. R., Wolfaardt, G. M., and Korber, D. R., 1994, Determination of diffusion coefficients in biofilms using confocal laser microscopy, *Appl. Environ. Microbiol.* **60**:1166–1173.

Lee, C. A., and Falkow, S., 1990, The ability of salmonella to enter mammalian cells is affected by bacterial growth state, *Proc. Natl. Acad. Sci. USA* **87**:4304–4308.

Lee, M. D., and Ward, C. H., 1985, Biological methods for the restoration of contaminated aquifers, *Environ. Toxicol. Chem.* **4**:743–750.

Leech, R., and Hefford, R. J. W., 1980, The observation of bacterial deposition from a flowing suspension, in: *Microbial Adhesion to Surfaces* (R. C. W. Berkley, J. M. Lynch, J. Melling, P. R. Rutter, and D. Vincent, eds.), Ellis Horwood, Chichester, England, pp. 544–545.

Lens, P. N. L., De Beer, D., Cronenberg, C. C. H., Houwen, F. P., Ottengraf, S. P. P., and Verstraete, W. H., 1993, Heterogeneous distribution of microbial activity in methanogenic aggregates: pH and glucose microprofiles, *Appl. Environ. Microbiol.* **59**:3803–3815.

Levy, S. B., and Marshall, B. M., 1988, Genetic transfer in the natural environment, in: *Release of Genetically Engineered Microorganisms* (M. Sussman, C. H. Collins, F. A. Skinner, and D. E. Stewart-Tull, eds.), Academic Press, New York, pp. 61–76.

Levy, S. B., and Miller, R. V. (eds.), 1989, *Gene Transfer in the Environment*, McGraw-Hill, New York.

Lewandowski, Z., Lee, W. C., Characklis, W. G., and Little, B., 1989, Dissolved oxygen and pH microelectrode measurements at water immersed metal surfaces, *Corrosion* **45**:92–98.

Lewandowski, Z., Walser, G., and Characklis, W. G., 1991, Reaction kinetics in biofilms, *Biotechnol. Bioeng.* **38**:877–882.

Lewandowski, Z., Altobelli, S. A., and Fukushima, E., 1993, NMR and microelectrode studies of hydrodynamics and kinetics in biofilms, *Biotechnol. Prog.* **9**:40–45.

Lewis, D. H., 1985, Symbiosis and mutualism: Crisp concepts and soggy semantics, in: *The Biology of Mutualism* (D. H. Boucher, ed.), Oxford University Press, New York, pp. 29–39.

Lewis, D. L., and Gattie, D. K., 1991, The ecology of quiescent microbes, *Am. Soc. Microbiol. News* **57**:27–32.

Lock, M. A., Wallace, R. R., Costerton, J. W., Ventullo, R. M., and Charlton, S. E., 1984, River epilithion: Toward a structural–functional model, *Oikos* **42**:10–22.

Loeb, G. I., 1980, Measurement of microbial marine fouling films by light section microscopy, *Mar. Technol. Soc. J.* **14**:17–23.

Loeb, G. I., and Neihof, R. A., 1975, Marine conditioning films, *Adv. Chem. Ser.* **145**:319–335.

Mack, W. N., Mack, J. P., and Ackerson, A. O., 1975, Microbial film development in a trickling filter, *Microb. Ecol.* **2**:215–226.

Madsen, E. L., 1991, Determining *in situ* biodegradation, *Environ. Sci. Technol.* **25:**1663–1673.

MacLeod, F. A., Guiot, S. R., and Costerton, J. W., 1990, Layered structure of bacterial aggregates produced in an upflow anaerobic sludge bed reactor, *Appl. Environ. Microbiol.* **56:**1598–1607.

Malone, J. A., 1987, *Colonization of Surface Microenvironments by Rhizobium* spp., University of Saskatchewan, Saskatoon, Canada, M.Sc. thesis.

Marmur, A., and Ruckenstein, E., 1986, Gravity and cell adhesion, *J. Colloid Interface Sci.* **114:**261–266.

Marshall, B., and Levy, S. B., 1990, Gene exchange in the natural environment, in: *Advances in Biotechnology* (E. Heseltine, ed.), AB Boktryck HBG, Stockholm, pp. 131–143.

Marshall, K. C., 1988, Adhesion and growth of bacteria at surfaces in oligotrophic habitats, *Can. J. Microbiol.* **34:**503–506.

Marshall, K. C., 1989, Growth of bacteria on surface-bound substrates: Significance on biofilm development, in: *Recent Advances in Microbial Ecology* (T. Hattori, Y. Ishida, Y. Maruyama, R. Y. Morita, and A. Uchida, eds.), Japan Scientific Society Press, Tokyo, pp. 146–150.

Marshall, K. C., 1993, Microbial ecology: Whither goest thou? in: *Trends in Microbial Ecology* (R. Guerrero and C. Pedrós-Alió, eds.), Spanish Society for Microbiology, Barcelona, Spain, pp. 5–8.

Marshall, K. C., and Cruickshank, R. H., 1973, Cell surface hydrophobicity and the orientation of certain bacteria at interfaces, *Arch. Mikrobiol.* **91:**29–40.

Marshall, K. C., Stout, R., and Mitchell, R., 1971, Mechanisms of the initial events in the sorption of marine bacteria to solid surfaces, *J. Gen. Microbiol.* **68:**337–348.

Marshall, P. A., Loeb, G. I., Cowan, M. M., and Fletcher, M., 1989, Response of microbial adhesives and biofilm matrix polymers to chemical treatments as determined by interference reflection microscopy and light section microscopy, *Appl. Environ. Microbiol.* **55:**2827–2831.

Marszalek, D. S., Gerchakov, S. M., and Udey, L. R., 1979, Influence of substrate composition on marine microfouling, *Appl. Environ. Microbiol.* **38:**987–995.

Matson, J. V., and Characklis, W. G., 1976, Diffusion into microbial aggregates, *Water Res.* **10:**877–881.

McCarter, L. L., and Silverman, M., 1989, Iron regulation of swarmer cell differentiation of *Vibrio parahaemolyticus, J. Bacteriol.* **171:**731–736.

McCarter, L. L., Hilmen, M., and Silverman, M., 1988, Flagellar dynamometer controls swarmer cell differentiation of *Vibrio parahaemolyticus, Cell* **54:**345–351.

McCarter, L. L., Showalter, R. E., and Silverman, M. R., 1992, Genetic analysis of surface sensing in *Vibrio parahaemolyticus, Biofouling* **5:**163–175.

McCormick, P. V., and Cairns, J. C., 1991, Limited versus unlimited membership in microbial communities: Evaluation and experimental tests of some paradigms, *Hydrobiologia* **21:**77–91.

McCuster, V. W., Skipper, H. D., Zublena, J. P., and Gooden, D. T., 1988, Biodegradation of carbamothioates in butylate-history soils, *Weed Sci.* **36:**818–823.

McEldowney, S., and Fletcher, M., 1988, Effect of pH, temperature, and growth conditions in the adhesion of a gliding bacterium and three nongliding bacteria to polystyrene, *Microb. Ecol.* **16:**183–195.

McLean, R. J. C., Lawrence, J. R., Korber, D. R., and Caldwell, D. E., 1991, *Proteus mirabilis* biofilm protection against struvite crystal dissolution and its implications in struvite urolithiasis, *J. Urol.* **146:**1138–1142.

McSweegan, E., and Walker, R. I., 1986, Identification and characterization of two *Campylobacter jejuni* adhesins for cellular and mucous substrates, *Infect. Immun.* **53:**141–148.

Meadows, P. S., 1971, The attachment of bacteria to solid surfaces, *Arch. Mikrobiol.* **85:**374–381.

Merker, R. I., and Smit, J., 1988, Characterization of the adhesive holdfast of marine and freshwater Caulobacters, *Appl. Environ. Microbiol.* **54:**2078–2085.

Mills, A. L., and Maubrey, R., 1981, Effect of mineral composition on bacterial attachment to submerged rock surfaces, *Microb. Ecol.* **7:**315–322.

Mittelman, M. W., and Geesey, G. G., 1985, Copper-binding characteristics of expolymers from a freshwater-sediment bacterium, *Appl. Environ. Microbiol.* **49:**846–851.

Mittelman, M. W., Nivens, D. E., Low, C., and White, D. C., 1990, Differential adhesion, activity, and carbohydrate : protein ratios of *Pseudomonas atlantica* monocultures attaching to stainless steel in a linear shear gradient, *Microb. Ecol.* **19:**269–278.

Moore, R. L., 1981, The biology of *Hyphomicrobium* and other prosthecate, budding bacteria, *Annu. Rev. Microbiol.* **35:**567–594.

Moriarty, D. J. W., 1982, Feeding of the holothurians on bacteria and organic matter, *Aust. J. Mar. Freshwater Res.* **33:**255–263.

Morita, R. Y., 1982, Starvation–survival of heterotrophs in the marine environment, *Adv. Microb. Ecol.* **6:**171–198.

Morita, R. Y., 1986, Starvation–survival: The hormonal mode of most bacteria in the ocean, in: *Proceedings of the IVISME* (F. Mequsar and M. Gantar, eds.), Slovene Society for Microbiology, Ljubljana, Yugoslavia, pp. 242–248.

Morris, E. J., and McBride, B. C., 1984, Adherence of *Streptococcus sanguis* to saliva-coated hydroxyapatite: Evidence for two binding sites, *Infect. Immun.* **43:**656–663.

Mueller, J. G., Skipper, H. D., and Kline, E. L., 1988, Loss of butyrate-utilizing ability by a *Flavobacterium, Pest. Biochem. Phys.* **32:**189–196.

Nalewajko, C., Dunstall, T. G., and Shear, H., 1976, Kinetics of extracellular release in axenic algae and in mixed algal–bacterial cultures: Significance in estimation of total (gross) phytoplankton excretion rates, *J. Phycol.* **12:**1–5.

Nalewajko, C., Lee, K., and Fay, P., 1980, Significance of algal extracellular products to bacteria in lakes and in cultures, *Microb. Ecol.* **6:**199–207.

Namkung, E., and Rittman, B. E., 1987, Modelling substrate removal by biofilms, *Biotech. Bioeng.* **29:**269–278.

Nealson, K. H., 1977, Autoinduction of bacterial luciferase: Occurrence, mechanism and significance, *Arch. Microbiol.* **112:**73–79.

Nelson, C. H., Robinson, J. A., and Characklis, W. A., 1985, Bacterial adsorption to smooth surfaces: Rate, extent, and spatial pattern, *Biotech. Bioeng.* **27:**1662–1667.

Neu, T. R., and Marshall, K. C., 1991, Microbial "footprints"—a new approach to adhesive polymers, *Biofouling* **3:**101–112.

Notermans, S., Dormans, J. A. M. A., and Mead, G. C., 1991, Contribution of surface attachment to the establishment of micro-organisms in food processing plants: A review, *Biofouling* **5:**21–36.

Novitsky, J. A., and Morita, R. Y., 1976, Morphological characterization of small cells resulting from nutrient starvation of a psychrophilic marine vibrio, *Appl. Environ. Microbiol.* **32:**617–622.

Novitsky, J. A., and Morita, R. Y., 1977, Survival of a psychrophilic marine vibrio under long-term nutrient starvation, *Appl. Environ. Microbiol.* **33:**635–641.

Novitsky, J. A., and Morita, R. Y., 1978, Possible strategy for the survival of marine bacteria under starvation conditions, *Mar. Biol.* **48:**289–295.

Nystrom, T., Flardh, K., and Kjelleberg, S., 1990, Responses to multiple nutrient starvation in marine *Vibrio* sp. strain CCUG 15956, *J. Bacteriol.* **172:**7085–7097.

Obayashi, A. W., and Gaudy, A. F., Jr., 1973, Aerobic digestion of extracellular microbial polysaccharides, *J. Water Pollut. Control Fed.* **45:**1584–1594.

Ohmura, N., Kitamura, K., and Saiki, H., 1993, Selective adhesion of *Thiobacillus ferrooxidans* to pyrite, *Appl. Environ. Microbiol.* **59:**4044–4050.

Ophir, T., and Gutnick, D. L., 1994, A role for expolysaccharides in the protection of microorganisms from desiccation, *Appl. Environ. Microbiol.* **60:**740–745.

Ostling, J., Goodman, A. E., and Kjelleberg, S., 1991, Behaviour in IncP-1 plasmids and a miniMu transposon in a marine *Vibrio* sp.: Isolation of starvation inducible lac operon fusions, *FEMS Microbiol. Ecol.* **86**:83–96.

Paerl, H. W., 1980, Attachment of microorganisms to living and detrital surfaces in freshwater systems, in: *Adsorption of Microorganisms to Surfaces* (G. Bitton and K. C. Marshall, eds.), Wiley and Sons, New York, pp. 375–402.

Palenik, B., Block, J.-C., Burns, R. G., Characklis, W. G., Christensen, B. E., Ghiorse, W. C., Gristina, A. G., Morel, F. M. M., Nichols, W. W., Tuovinen, O. H., Tuschewitzki, G.-J., and Videls, H. A., 1989, Group report biofilms: Properties and processes, in: *Structure and Function of Biofilms* (W. G. Characklis and P. A. Wilderer, eds.), Wiley and Sons, Toronto, Canada, pp. 351–366.

Parkes, R. J., and Senior, E., 1988, Multistage chemostats and other models for studying anoxic environments, in: *Handbook of Laboratory Model Systems for Microbial Ecosystems* (J. W. T. Wimpenny, ed.), CRC Press, Boca Raton, Fla., pp. 51–71.

Patel, J. J., and Gerson, T., 1974, Formation and utilization of carbon reserves by *Rhizobium, Arch. Microbiol.* **101**:211–220.

Patrick, F. M., and Loutit, M., 1976, Passage of metals in influents, through bacteria to higher organisms, *Water Res.* **10**:333–335.

Pedersen, K., 1982, Factors regulating microbial biofilm development in a system with slowly flowing seawater, *Appl. Environ. Microbiol.* **44**:1196–1204.

Poindexter, J. S., 1981, The caulobacters: Ubiquitous unusual bacteria, *Microbiol. Rev.* **45**:123–179.

Pothuluri, J. V., Moorman, T. B., Obenhuber, D. C., and Wauchope, R. D., 1990, Aerobic and anaerobic degradation of alachlor in samples from a surface-to-groundwater profile, *J. Environ. Qual.* **19**:525–530.

Powell, M. S., and Slater, N. K. H., 1983, The deposition of bacterial cells from laminar flows onto solid surfaces, *Biotechnol. Bioeng.* **25**:891–900.

Power, K., and Marshall, K. C., 1988, Cellular growth and reproduction of marine bacteria on surface-bound substrates, *Biofouling* **1**:163–174.

Prosser, J. I., 1989, Modeling nutrient flux through biofilm communities, in: *Structure and Function of Biofilms* (W. G. Characklis and P. A. Wilderer, eds.), Wiley and Sons, Toronto, Canada, pp. 239–250.

Rainey, P. B., Moxon, E. R., and Thompson, I. P., 1993, Intraclonal polymorphism in bacteria, in: *Advances in Microbial Ecology,* Vol. 13 (J. A. Jones, ed.), Plenum Press, New York, pp. 263–300.

Reanney, D. C., Gowland, P. C., and Slater, J. H., 1983, Genetic interactions among communities, in: *Microbes in the Natural Environment* (J. H. Slater, R. Whittenbury, and J. W. T. Wimpenny, eds.), Cambridge University Press, Cambridge, England, pp. 379–421.

Revsbech, N. P., 1989, Diffusion characteristics of microbial communities determined by use of oxygen microsensors, *J. Microbiol. Methods* **49**:111–122.

Reynolds, P. J., Sharma, P., Jenneman, G. E., and McInerney, M. J., 1989, Mechanisms of microbial movement in subsurface materials, *Appl. Environ. Microbiol.* **55**:2280–2286.

Rittle, K. H., Helmstetter, C. E., Meyer, A. E., and Baier, R. E., 1990, *Escherichia coli* retention on solid surfaces as functions of substratum surface energy and cell growth phase, *Biofouling* **2**:121–130.

Robinson, R. W., Akin, D. E., Nordstedt, R. A., Thomas, M. V., and Aldrich, H. C., 1984, Light and electron microscopic examinations of methane-producing biofilms from anaerobic fixed-bed reactors, *Appl. Environ. Microbiol.* **48**:127–136.

Rochelle, P. A., Fry, J. C., and Day, M. J., 1989, Plasmid transfer between *Pseudomonas* spp. within epilithic films in a rotating disc microcosm, *FEMS Microbiol. Ecol.* **62**:127–136.

Rose, A. H., 1976, *Chemical Microbiology: An Introduction to Microbial Physiology,* Plenum Press, New York, pp. 295–299.

Rosenberg, M., and Kjelleberg, S., 1986, Hydrophobic interactions: Role in bacterial adhesion, *Adv. Microb. Ecol.* **9**:353–393.

Roszak, D. B., and Colwell, R. R., 1987, Survival strategies of bacteria in the natural environment, *Microbiol. Rev.* **51**:365–379.

Rothmel, R. K., Haugland, R. A., Coco, W. M., Sangodkar, U. M. X., and Chakrabarty, A. M., 1989, Natural and directed evolution: Microbial degradation of synthetic chlorinated compounds, in: *Recent Advances in Microbial Ecology* (T. Hattori, Y. Ishida, Y. Maruyama, R. Y. Morita, and A. Uchida, eds.), Japan Scientific Society Press, Tokyo, pp. 605–610.

Rozgaj, R., and Glancer-Soljan, M., 1992, Total degradation of 6-aminonaphthalene-2-sulphonic acid by a mixed culture consisting of different bacterial genera, *FEMS Microbiol. Ecol.* **86**:229–235.

Rudd, T., Sterritt, R. M., and Lester, J. N., 1983, Mass balance of heavy metal uptake by encapsulated cultures of *Klebsiella aerogenes, Microb. Ecol.* **9**:261–272.

Rutter, P., and Leech, R., 1980, The deposition of *Streptococcus sanguis* NCTC 7868 from a flowing suspension, *J. Gen. Microbiol.* **120**:301–307.

Sanders, R. W., and Porter, K. G., 1986, Use of metabolic inhibitors to estimate protozooplankton grazing and bacterial production in a monomictic eutrophic lake with an anaerobic hypolimnion, *Appl. Environ. Microbiol.* **52**:101–107.

Sanders, W. M., 1966, Oxygen utilization by slime organisms in continuous-culture, *J. Air Water Poll.* **10**:253–276.

Schieffer, G. E., and Caldwell, D. E., 1982, Synergistic interaction between *Anabaena* and *Zooglea* spp. in carbon dioxide limited continuous cultures, *Appl. Environ. Microbiol.* **44**:84–87.

Schmidt, S. K., Smith, R., Sheker, D., Hess, T. F., Silverstein, J., and Radehaus, P. M., 1992, Interactions of bacteria and microflagellates in sequencing batch reactors exhibiting enhanced mineralization of toxic organic chemicals, *Microb. Ecol.* **23**:127–142.

Semenov, A., and Staley, J. T., 1992, Ecology of the polyprosthecate bacteria, *Adv. Microb. Ecol.* **12**:339–382.

Shapiro, J. A., 1991, Multicellular behavior of bacteria, *Am. Soc. Microbiol. News* **57**:247–253.

Shapiro, J. A., and Hsu, C., 1989, *Escherichia coli* K-12 cell–cell interactions seen by time-lapse video, *J. Bacteriol.* **171**:5963–5974.

Shen, C. F., Kosaric, N., and Blaszczyk, R., 1993, The effect of heavy metals (Ni, Co, and Fe) on anaerobic granules and their extracellular substance, *Water Res.* **27**:25–33.

Sherr, E. B., 1988, Direct use of high molecular weight polysaccharide by heterotrophic flagellates, *Nature* **335**:348–351.

Shimkets, L. J., 1990, Social and developmental biology of the myxobacteria, *Microbiol. Rev.* **54**:473–501.

Sibbald, M. J., and Albright, L. J., 1988, Aggregated and free bacteria as food sources for heterotrophic microflagellates, *Appl. Environ. Microbiol.* **54**:613–616.

Sjollema, J., and Busscher, H. J., 1989, Deposition of polystyrene particles in a parallel plate flow cell. 2. Pair distribution functions between deposited particles, *Colloids Surfaces* **47**:337–352.

Sjollema, J., Busscher, H. J., and Weerkamp, A. H., 1988, Deposition of oral streptococci and polystyrene latices onto glass in a parallel plate flow cell, *Biofouling* **1**:101–112.

Sjollema, J., Van der Mei, H. C., Uyen, H. M., and Busscher, H. J., 1990a, Direct observations of cooperative effects in oral streptococcal adhesion to glass by analysis of the spatial arrangement of adhering bacteria, *FEMS Microbiol. Let.* **69**:263–270.

Sjollema, J., Van der Mei, H. C., Uyen, H. M. W., and Busscher, H. J., 1990b, The influence of collector and bacterial cell surface properties on the deposition of oral streptococci in a parallel plate flow cell, *J. Adhes. Sci. Technol.* **4**:765–777.

Slater, J. H., 1980, Physiological and genetic implications of mixed population and microbial community growth, in: *Microbiology—1980* (D. Schlessinger, ed.), American Society for Microbiology Press, Washington, D.C., pp. 314–316.

Slater, J. H., 1985. Gene transfer in microbial communities, in: *Engineered Organisms in the Environment* (H. O. Halverson, D. Pramer, and M. Rogal, eds.) American Society for Microbiology Press, Washington, D.C., pp. 89–98.

Slater, J. H., and Hartman, D. J., 1982, Microbial ecology in the laboratory: Experimental systems, in: *Experimental Microbial Ecology* (R. G. Burns and J. H. Slater, eds.), Blackwell Scientific Publications, Oxford, England, pp. 255–274.

Slater, J. H., and Somerville, H. J., 1979, Microbial aspects of waste treatment with particular attention to the degradation of organic compounds, *Symp. Soc. Gen. Microbiol.* **29**:221–261.

Slonczewski, J. L., 1992, pH regulated genes in enteric bacteria, *Am. Soc. Microbiol. News* **58**:140–144.

Snyder, R. A., 1990, Chemoattraction of a bactivorous ciliate to bacteria surface compounds, *Hydrobiologia* **215**:205–213.

Sonea, S., and Panisset, M., 1983, *A New Bacteriology,* Jones and Bartlett, Boston, Mass.

Sorongon, M. L., Bloodgood, R. A., and Burchard, R. P., 1991, Hydrophobicity, adhesion, and surface-exposed proteins of gliding bacteria, *Appl. Environ. Microbiol.* **57**:3193–3199.

Stahl, D. A., 1993, The natural history of microorganisms, *Am. Soc. Microbiol. News* **59**:609–613.

Stahl, D. A., Flesher, B., Mansfield, H. R., and Montgomery, L., 1988, Use of phylogenetically based hybridization probes for studies of ruminal microbial ecology, *Appl. Environ. Microbiol.* **54**:1079–1084.

Staley, J. T., 1971, Incidence of prosthecate bacteria in a polluted stream, *J. Appl. Microbiol.* **22**:496–502.

Stanley, P. M., 1983, Factors affecting the irreversible attachment of *Pseudomonas aeruginosa* to stainless steel, *Can. J. Microbiol.* **29**:1493–1499.

Stewart, P. S., Peyton, B. M., Drury, W. J., and Murga, R., 1993, Quantitative observations of heterogeneities in *Pseudomonas aeruginosa* biofilms, *Appl. Environ. Microbiol.* **59**:327–329.

Stock, J. B., Stock, A. M., and Mottonen, J. M., 1990, Signal transduction in bacteria, *Nature* **344**:395–400.

Stotzky, G., and Burns, R. G., 1982, The soil environment: Clay–humus–microbe interactions, in: *Experimental Microbial Ecology* (R. G. Burns and J. H. Slater, eds.), Blackwell Scientific, Oxford, England pp. 105–133.

Stotzky, G., Devana, M. A., and Zeph, L. R., 1990, Methods for studying bacterial gene transfer in soil by conjugation and transduction, *Adv. Appl. Microbiol.* **35**:57–169.

Stove-Poindexter, J., 1964, Biological properties and classification of the *Caulobacter* group, *Bacteriol. Rev.* **28**:231–95.

Sutherland, I. W., 1985, Biosynthesis and composition of gram-negative bacterial extracellular and wall polysaccharides, *Annu. Rev. Microbiol.* **39**:243–270.

Tam, A. C., Behki, R. M., and Khan, S. U., 1987, Isolation and characterization of an *s*-ethyl-*N,N*-dipropylthiocarbamate-degrading *Arthrobacter* strain and evidence for plasmid-associated *s*-ethyl-*N,N*-dipropylthiocarbamate degradation, *Appl. Environ. Microbiol.* **53**:1088–1093.

Thiele, J. H., Chartrain, M., and Zeikus, J. G., 1988, Control of interspecies electron flow during anaerobic digestion: Role of floc formation in syntrophic methanogenesis, *Appl. Environ. Microbiol.* **54**:10–19.

Troy, F. A., 1979, The chemistry and biosynthesis of selected bacterial capsular polymers, *Annu. Rev. Microbiol.* **33**:519–560.

Tsezos, K., and Bell, J. P., 1988, Significance of biosorption for the hazardous organics removal efficiency of a biological reactor, *Water Res.* **22**:391–394.

Uhlinger, D. J., and White, D. C., 1983, Relationship between physiological status and formation of extracellular polysaccharide glycocalyx in *Pseudomonas atlantica, Appl. Environ. Microbiol.* **45**:64–70.

Vandevivere, P., and Kirchman, D. L., 1993, Attachment stimulates exopolysaccharide synthesis by a bacterium, *Appl. Environ. Microbiol.* **59**:3280–3286.

Van Elsas, J. D., Trevors, J. T., and Starodub, M. E., 1988, Bacterial conjugation between *Pseudomonas* in the rhizosphere of wheat, *FEMS Microbiol. Ecol.* **53**:299–306.

Van Loosdrecht, M. C. M., Norde, W., and Zehnder, A. J. B., 1987, Influence of cell surface characteristics on bacterial adhesion to solid surfaces, *Proc. Eur. Congr. Biotechnol.* **4**:575–580.

Van Loosdrecht, M. C. M., Lyklema, J., Norde, W., and Zehnder, A. J. B., 1989, Bacterial adhesion: A physiochemical approach, *Microb. Ecol.* **17**:1–15.

Van Pelt, A. W. J., Weerkamp, A. H., Uyen, M. H. W. J. C., Busscher, H. J., de Jong, H. P., and Arends, J., 1985, Adhesion of *Streptococcus sanguis* CH3 to polymers with different surface free energies, *Appl. Environ. Microbiol.* **49**:1270–1275.

Vargas, R., and Hattori, T., 1990, The distribution of protozoa among soil aggregates, *FEMS Microbial. Ecol.* **74**:73–78.

Veal, D. A., Stokes, H. W., and Daggard, G., 1992, Genetic exchange in natural microbial communities, in: *Advances in Microbial Ecology,* Vol. 12 (K. C. Marshall, ed.), Plenum Press, New York, pp. 383–430.

Venosa, A. D., 1975, Lysis of *Spaerotilus natans* swarms by *Bdellovibrio bacteriovirus, Appl. Environ. Microbiol.* **29**:702–705.

Vesper, S. J., and Bauer, W. D., 1986, Role of pili (fimbriae) in attachment of *Bradyrhizobium japonicum* to soybean roots, *Appl. Environ. Microbiol.* **52**:134–141.

Vogel, S., 1983, *Life in Moving Fluids: The Physical Biology of Flow,* Princeton University Press, Princeton, N.J.

Von Riesen, V. L., 1980, Digestion of algin by *Pseudomonas maltophilia* and *Pseudomonas putida, Appl. Environ. Microbiol.* **39**:92–96.

Wahl, M., 1989, Marine epibiosis. I. Fouling and antifouling: Some basic aspects, *Mar. Ecol. Prog. Ser.* **58**:175–189.

Walker, N., 1976, Microbial degradation of plant protection chemicals, in: *Soil Microbiology* (N. Walker, ed.), Wiley and Sons, Toronto, Canada, pp. 181–192.

Walt, D. R., Smulow, J. B., Turesky, S. S., and Hill, R. G., 1985, The effect of gravity on initial microbial adhesion, *J. Colloid Interface Sci.* **107**:334–336.

Ward, D. M., Weller, R., and Bateson, M. M., 1990, 16S rRNA sequences reveal numerous uncultured microorganisms in a natural community, *Nature* **453**:63–65.

Ward, D. M., Bateson, M. M., Weller, R., and Ruff-Roberts, A. L., 1992, Ribosomal RNA analysis of microorganisms as they occur in nature, in: *Advances in Microbial Ecology,* Vol. 12 (K. C. Marshall, ed.), Plenum Press, New York, pp. 219–286.

White, D. C., 1984, Chemical characterization of films, in: *Microbial Adhesion and Aggregation* (K. C. Marshall, ed.), Springer Verlag, New York, pp. 159–176.

Wickham, G. S., and Atlas, R. M., 1988, Plasmid frequency fluctuations in bacterial populations from chemically stressed soil communities, *Appl. Environ. Microbiol.* **54**:2192–2196.

Wiggins, B. A., Jones, S. H., and Alexander, M., 1987, Explanations for the acclimation period preceding the mineralization of organic chemicals in aquatic environments, *Appl. Environ. Microbiol.* **53**:791–796.

Wilson, J. T., McNabb, J. F., Cochran, J. W., Wang, T. H., Tomson, M. B., and Bedient, P. B., 1985, Influence of microbial adaptation on the fate of organic pollutants in ground water, *Environ. Toxicol. Chem.* **4**:721–726.

Wimpenny, J. W. T., 1981, Spatial order in microbial ecosystems, *Biol. Rev.* **56**:295–342.

Wimpenny, J. W. T., 1992, Microbial systems: Patterns in time and space, in: *Advances in Microbial Ecology,* Vol. 12 (K. C. Marshall, ed.), Plenum Press, New York, pp. 469–522.

Woese, C. R., 1994, There must be a prokaryote somewhere: Microbiology's search for itself *Microbiol. Rev.* **58**:1–9.

Wolfaardt, G. M., Lawrence, J. R., Robarts, R. D., Caldwell, S. J., and Caldwell, D. E., 1994a, Multicellular organization in a degradative biofilm community, *Appl. Environ. Microbiol.* **60**:434–446.

Wolfaardt, G. M., Lawrence, J. R., Headley, J. V., Robarts, R. D., and Caldwell, D. E., 1994b, Microbial exopolymers provide a mechanism for bioaccumulation of contaminants, *Microb. Ecol.* **27**:279–291.

Wolfaardt, G. M., Lawrence, J. R., Robarts, R. D., and Caldwell, D. E., 1994c, The role of interactions, sessile growth and nutrient amendment on the degradative efficiency of a bacterial consortium, *Can. J. Microbiol.* **40**:331–340.

Wolfaardt, G. M., Lawrence, J. R., Robarts, R. D., and Caldwell, D. E., 1995a, Bioaccumulation of the herbicide diclofop in extracellular polymers and its utilization by a biofilm community during starvation, *Appl. Environ. Microbiol.* **61**:152–158.

Wolfaardt, G. M., Lawrence, J. R., Robarts, R. D., and Caldwell, D. E., 1995b, *In situ* characterization of biofilm exopolymers involved in the accumulation of chlorinated organics, *Appl. Environ. Microbiol.,* submitted.

Wrangstadh, M., Conway, P. L., and Kjelleberg, S., 1989, The role of an extracellular polysaccharide produced by the marine *Pseudomonas* sp. S9 in cellular detachment during starvation, *Can. J. Microbiol.* **35**:309–312.

Wright, J. B., Costerton, J. W., and McCoy, W. F., 1988, Filamentous growth of *Pseudomonas aeruginosa*, *J. Indust. Microbiol.* **3**:139–146.

Wyndham, R. C., Nakatsu, C., Peel, M., Cashore, A., Ng, J., and Szilagyi, F., 1994, Distribution of the catabolic transposon Tn5271 in a groundwater bioremediation system, *Appl. Environ. Microbiol.* **60**:86–93.

Zanyk, B. N., 1993, *Degradation and Mobility of Diclofop Methyl in Model Groundwater Systems,* University of Saskatchewan, Saskatoon, Canada, M.Sc. thesis.

2

Insoluble, Nonhydrolyzable, Aliphatic Macromolecular Constituents of Microbial Cell Walls

CLAUDE LARGEAU and JAN W. DE LEEUW

1. Introduction

The recognition of nonhydrolyzable, highly aliphatic biomacromolecules stems from the analysis of organic matter present in sediments varying in age from very Recent to hundreds of millions years old. The far greater part of sedimentary organic matter, i.e., over 95%, is high molecular in nature, nonhydrolyzable and insoluble in water and organic solvents, and commonly referred to as kerogen. The total amount of kerogen present in the earth's crust is presently estimated 30 $\times$ 10^{21} g (Hedges, 1992) and is by far the largest pool of organic carbon on our planet (for comparison: organic carbon in living marine organisms is estimated only 2 $\times$ 10^{15} g C). It has long been known that kerogen-rich deposits act as source rocks of crude oil and gas after burial over eons at elevated temperatures (Tissot and Welte, 1984). As a consequence of its complexity, insolubility, and high-molecular-weight nature, kerogen is one of the most problematic organic substances to characterize on a molecular level. Until a few years ago, it was generally accepted that kerogen resulted from a random condensation process of small amounts of nonmineralized amino acids and monosaccharides, resulting from microbially induced depolymerization of proteins and polysaccharides, respectively, and lipid components (Tissot and Welte, 1984). Such a random condensation was supposed to generate humic type substances in Recent sedi-

CLAUDE LARGEAU • Laboratoire de Chimie Bioorganique et Organique Physique, UA CNRS D1381, Ecole Nationale Supérieure de Chimie de Paris, 75231 Paris Cedex 05, France.   JAN W. DE LEEUW • Division of Marine Biogeochemistry, Netherlands Institute for Sea Research (NIOZ), 1790 AB Den Burg Texel, The Netherlands.

Advances in Microbial Ecology, Volume 14, edited by J. Gwynfryn Jones. Plenum Press, New York, 1995.

ments, which upon further burial underwent chemical transformations finally resulting in kerogen. The absence of recognizable entities in most kerogens when studied by light microscopic techniques seemed to support this kerogen formation hypothesis. Recent developments in analytical chemistry have enabled a more detailed analysis of kerogens. Fourier transform infrared (FTIR) and solid-state ^{13}C-labeled nuclear magnetic resonance (NMR) spectroscopy gave clues to the presence or absence of specific functional groups in kerogens in a semiquantitative fashion (Largeau *et al.*, 1984; Barwise *et al.*, 1984; Derenne *et al.*, 1987; Kister *et al.*, 1990; Landais *et al.*, 1993). Based on these spectroscopic results in combination with elemental analyses, it became clear that the so-called aliphaticity of many kerogens was much higher than originally expected and that polymethylene chains were probably relatively abundant. Convincing evidence for the relatively abundant presence of long, unbranched alkyl moieties in kerogens came from analytical pyrolysis methods. Pyrolysis of kerogens under well-controlled and reproducible conditions in combination with gas chromatographic separation of the pyrolysis products and their structural identification by mass spectrometry revealed the abundant presence of n-alkanes and n-alkenes with carbon chain lengths ranging from C_6 to C_{35}. Figure 1 shows the total ion current trace obtained by gas chromatography–mass spectrometry of a pyrolysate of kerogen from the 50-million-year-old Messel oil shale in Germany obtained by Curie point pyrolysis of approximately 30 μg of this material. The peaks indicated by open and filled circles represent homologous series of alkanes and alkenes, respectively. The elemental analysis (H/C ratio, approx. 1.6) and the solid-state [^{13}C]-NMR spectrum (Fig. 2) were in full agreement with these pyrolysis results (Goth *et al.*, 1988; Collinson *et al.*, 1994). Homologous series of alkanes and alkenes were also reported earlier by Largeau *et al.* (1984, 1986) as major products obtained by heating under an inert atmosphere of other highly aliphatic kerogens called torbanites. At the time these pyrolytic results were obtained, the only aliphatic biopolymers known were cutins and suberins, both naturally occurring polyesters present in higher plants. However, pyrolysis experiments with these polyesters isolated from several plants unambiguously indicated that such compounds could not be the precursors of the above-mentioned homologous series of alkanes and alkenes, since their pyrolysates contained exclusively C_{16} and C_{18} fatty acids and hydroxy fatty acids generated via six-membered rearrangement reactions during pyrolysis (Tegelaar *et al.*, 1989a). Hence, the pyrolytic results to some extent seemed to support the depolymerization–condensation pathway of kerogen formation, assuming that a relatively large proportion of lipids would have been involved. However, electron microscopic observations of the torbanites and the Messel oil shale kerogen already mentioned indicated the overwhelming presence of algal cell walls of *Botryococcus braunii* and *Tetrahedron minimum,* respectively (Largeau *et al.*, 1986, 1990a; Goth *et al.*, 1988). Because of these observations, it was decided to

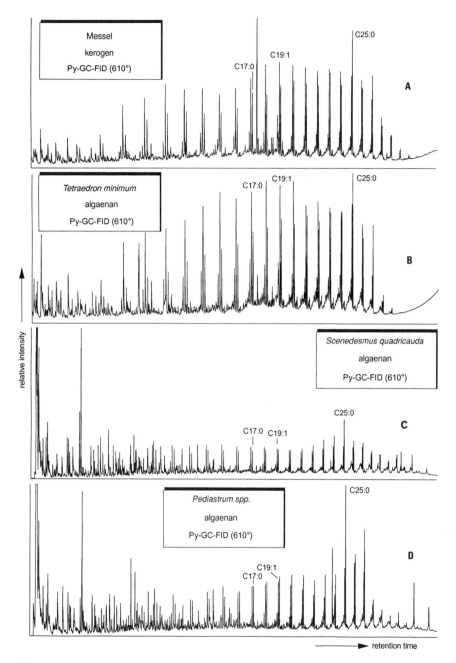

Figure 1. Total ion current (TIC) trace of flash pyrolysates of (A) Messel oil shale kerogen, (B) algaenans isolated from *Tetraedron minimum,* (C) *Scenedesmus quadricauda,* and (D) *Pediastrum* spp. Numbers indicate chain lengths of alkene/alkane doublets.

30.76

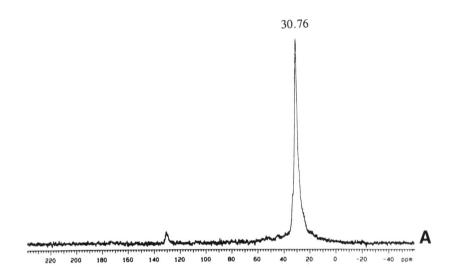

30.76

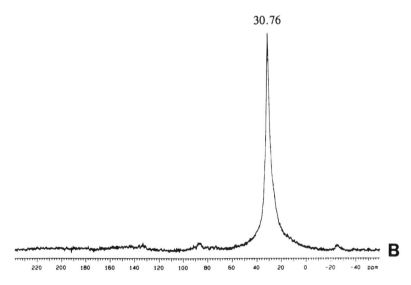

Figure 2. Solid-state [¹³C]-NMR spectra of algaenan isolated from (A) *T. minimum* and (B) kerogen of the Messel oil shale. (After Collinson *et al.*, 1994.)

investigate the cell walls of cultured *B. braunii* and *T. minimum* strains isolated after lipid extraction followed by drastic base and acid hydrolysis (Berkaloff *et al.*, 1983; Kadouri *et al.*, 1988; Goth *et al.*, 1988). It was demonstrated that the pyrolysates of these cell walls derived from the cultured species also generated abundant homologous series of *n*-alkenes and *n*-alkanes. The conclusion based on these results was rather straightforward: Obviously the major chemical component building up the thick cell walls of these green microalgae is highly aliphatic in nature but does not represent a polyester. A new type of aliphatic biopolymers was thus recognized. Further research has indicated that nonhydrolyzable aliphatic biomacromolecules are not restricted to green algae but also occur in bacteria (Chalansonnet *et al.*, 1988; Le Berre *et al.*, 1991), higher plant cuticles, bark tissue, spore and pollen walls, plant resins, and in the inner layers of seed coats of water plants* (de Leeuw *et al.*, 1991; de Leeuw and Largeau, 1993). The recognition of these new types of aliphatic biopolymers as major constituents of kerogens indicate their selective preservation and thus relative enhancement during microbial decay of organic debris in the environment, witnessing their relatively high resistant nature. Moreover, the concept of kerogen formation has drastically changed since the recognition of these aliphatic and other resistant biomacromolecules. Presently, it can be stated that kerogens consist, for a large part, of selectively preserved resistant biomacromolecules, and that crude oil formation can be regarded in many cases as the genesis of *n*-alkanes by "natural pyrolysis" of predominantly resistant, nonhydrolyzable highly aliphatic biomacromolecules over geological times at enhanced temperatures in the subsurface (Tegelaar *et al.*, 1989b,c).

In this chapter, we limit ourselves to highly aliphatic macromolecules in microorganisms. Such constituents have been detected so far in microalgae and bacteria and are termed *algaenans* and *bacterans,* respectively. Their structures are discussed in relation to their possible biosynthesis when known. Whether the low-molecular-weight components in these microorganisms can be considered as potential building blocks will be discussed also. Furthermore, the possible role of these biomacromolecules in algae and bacteria will be discussed.

2. Occurrence, Chemical Structure, and Biosynthesis of Algaenans and Bacterans

2.1. Microalgae

2.1.1. Botryococcus braunii

The first systematic study of nonhydrolyzable macromolecular constituents in microbial cell walls was concerned with the freshwater colonial Chlorophycea

---

*The occurrence of insoluble and nonhydrolyzable constituents was also observed in some fungal spores (Furch and Gooday, 1978).

(green microalga), *B. braunii*. This ubiquitous freshwater species exhibits unusual features regarding (1) the high level and nature of its lipid constituents, (2) its colony and cell wall organization, and (3) the production of algaenan (Metzger *et al.*, 1991). The lipid components of most *B. braunii* strains are dominated by hydrocarbons that account for up to 60% of total dry biomass, whereas very low amounts of hydrocarbons usually occur in microalgae. Further studies revealed important, strain-dependent differences in the nature of *B. braunii* hydrocarbons and resulted in the recognition of three distinct chemical races (Metzger *et al.*, 1991). These races show similar morphological features but biosynthesize sharply different types of hydrocarbons:

- straight chain, odd-carbon-numbered, C_{25} to C_{31} alkenes (**1, 2,** in Appendix) in the A race.
- C_{30} to C_{37} isoprenoid hydrocarbons (**3**), termed botryococcenes, in the B race.
- a lycopadiene (C_{40} isoprenoid hydrocarbon) (**4**) in the L race.

Regarding morphological features, *B. braunii* cells are surrounded by a classic polysaccharidic wall, but they also comprise a thick outer wall (up to 2 μm) (Plain *et al.*, 1993) where the basal parts of the cells are deeply embedded (Fig. 3). Following division, the daughter cells, with their own outer wall, remain embedded in the basal wall of the mother cell. The closely associated thick outer walls, corresponding to successive divisions, thus built up the matrix of *B. braunii* colonies. Similar basal walls and colony organization have not been observed so far in other species of microalgae. *B. braunii* outer walls are also a major site of lipid biosynthesis and storage (Largeau *et al.*, 1980a,b; Metzger *et al.*, 1987). Thus, for example, approximately 95% of A race hydrocarbons are located in the basal walls, along with a complex mixture of other lipids, including high-molecular-weight compounds, and the last steps of the biosynthesis leading to these hydrocarbons and high-molecular-weight lipids take place in the outer walls.

The skeleton of *B. braunii* basal outer walls was shown to be composed of algaenans termed PRB A, PRB B, and PRB L, in the races A, B, and L, respectively (Berkaloff *et al.*, 1983; Kadouri *et al.*, 1988; Derenne *et al.*, 1989). These insoluble, nonhydrolyzable, highly aliphatic macromolecular compounds are major constituents of *B. braunii*. PRB accounts for up to 30% of total biomass in the L race, and values around 10% were observed in samples from the A and B races. The isolation of algaenans via drastic basic and acid hydrolyses of extracted, lipid-free biomass does not affect the morphological features of the basal outer walls. Accordingly, the general organization of *B. braunii* colonies is thus preserved, while cell contents and polysaccharide inner walls are entirely eliminated.

Pronounced differences in chemical structure were observed between PRB

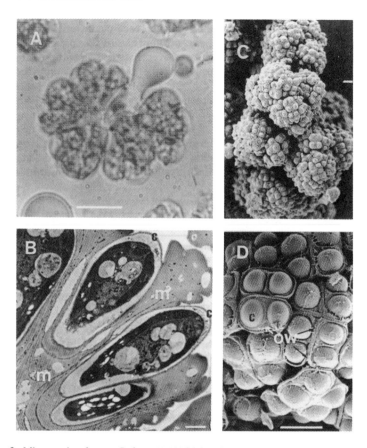

Figure 3. Micrographs of extant *B. braunii.* (A) Light microscopy; a refringent globule of lipids is excreted from the colony by pressure on the coverglass. (B) Transmission electron microscopy; the longitudinal section shows numerous lipid inclusions within the cells; the successive outer walls form a dense matrix (*m*). (C and D) Scanning electron microscopy: (C) whole colony; (D) the basal part of the cells is embedded in thick outer walls (*ow*) when the apical part is covered by a thin cap (*c*). Scale bars: A, C, and D = 10 μm; B = 1 μm. (After Metzger *et al.,* 1991.)

A and PRB B, on the one hand, and PRB L on the other hand. The structures of the former are based on a macromolecular network of long $(CH_2)_n$ chains, with n up to 31 (Berkaloff *et al.,* 1983; Kadouri *et al.,* 1988), whereas PRB L is chiefly built up from C_{40} isoprenoid chains (Derenne *et al.,* 1989, 1990). Various functionalities, including hydroxyl groups, double bonds, and ether and ester groups, were shown to occur in the three PRBs. In addition, PRB A and PRB B comprise carbonyl groups that possibly can be ascribed to α-unsaturated aldehydes. Ether bridges are likely to play an important role in the cross-linking of the long hydrocarbon chains that build up the macromolecular structure of PRBs. The three-dimensional network thus generated affords an efficient steric protection against hydrolysis reagents. Although esters are known to be cleaved easily under mild acid or basic conditions, the ester functions occurring in PRBs are unaffected, even after drastic acid and basic treatments. Similarly, ether bridges in PRBs can withstand prolonged acid attacks, thus preventing the destruction of the macromolecular structures. These protected ester and ether functions can only be cleaved upon thermal stress (pyrolysis), resulting in the production of fatty acids and alkylketones, respectively.

The biosynthesis of PRBs and the nature of their building blocks were recently examined, especially in the A race. In fact, a complex mixture of lipids was shown to be stored in the basal outer walls of *B. braunii*. These "external" lipids are easily extracted by a short contact with hexane, while intracellular lipids are only recovered after prolonged extractions with mixtures of polar solvents (Largeau *et al.*, 1980a). In addition to hydrocarbons (**1, 2**) (see Appendix) and common lipids like triacylglycerols, numerous series of compounds, e.g., (**5**) to (**7**), including high-molecular-weight ones, were identified in the external lipids of the A race. These series comprise several families of lipids so far not observed in any living organism (Metzger *et al.*, 1991). The series of high-molecular-weight lipids identified in the A race are probably derived, as shown in Fig. 4 for (**5**), from condensation reactions via the formation of ether bridges through opening of epoxides. Additional series of complex lipids resulting from further condensations of basic units, such as (**5**) to (**7**), likely occur also in the external lipids of the A race, though their chemical structures have not been firmly established yet.

The ester-bound lipids released after saponification of the residue obtained after the extraction of the external lipids are mainly composed of fatty acids (Fig. 5). In the A race, oleic acid is very abundant, though mono- and diunsaturated fatty acids ranging in chain length from C_{22} up to C_{32}, maximizing at C_{28} and C_{30}, are clearly present. In the L race, the monounsaturated C_{28} fatty acid is even more abundant than oleic acid, while the distribution of the esterified fatty acids in the B race is similar to that in the A race. It is worthy of note that relatively high amounts of long-chain fatty acids, in particular C_{28} mono- and diunsaturated fatty acids, are biosynthesized by these algae. Fatty acids with chain lengths between C_{24} and C_{32} are commonly observed in tissues of higher plants. Their occurrence in algae is rather unexpected, though their role in the formation of PRBs seems unimportant.

A rubbery material, soluble in chloroform, was also isolated from the external lipid fraction of the A race. This "rubber" corresponds to a polymer probably produced by condensation of C_{32} dialdehydic units (Fig. 6). As emphasized above, the chemical structure of PRB A is based on a network of long polymethylenic chains bearing various functionalities, including hydroxyl group, esters, and possibly α-unsaturated aldehydes. Taking into account such features and the nature of the external lipids, PRB A might be derived from further condensations of rubber and high-molecular-weight lipids. Recent data resulting from RuO$_4$ oxidations of PRB A firmly supported this (Schouten *et al.*, in preparation); Fig. 7 shows the gas chromatogram of the compounds released (corresponding to approx. 10% of the original amount of PRB A). Based on RuO$_4$ oxidations of model compounds indicating that under the conditions used double bonds and carbon–carbon bonds between carbon atoms bearing ether moieties are oxidized, it could be concluded that the distribution pattern of the major dicarboxylic acids released reflected the positions of the functional groups in PRB A, assuming the prolonged condensation of the dialdehydes (Fig. 7). These condensations and increasing cross-linking will finally lead to the

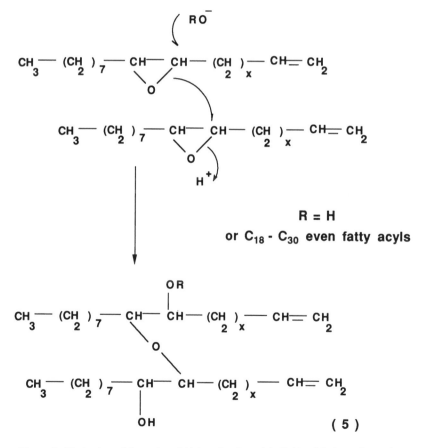

Figure 4. Mechanism of formation of high-molecular-weight lipids of *B. braunii* race A.

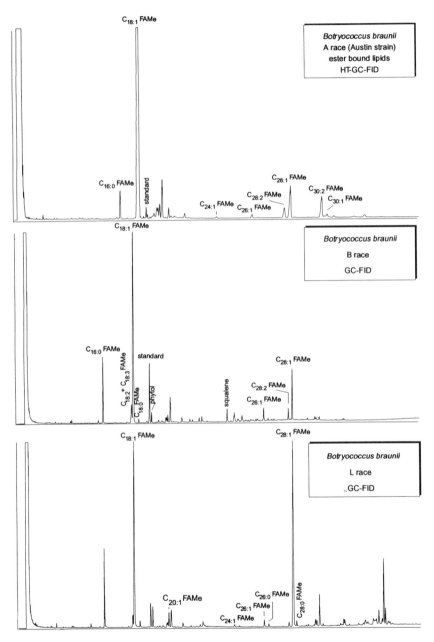

Figure 5. Gas chromatography traces of ester-bound lipids obtained from extracted residues of *B. braunii* race A (upper trace), *B. braunii* race B (middle trace), and *B. braunii* race L (lower trace). Fatty acids were analyzed as their methylester derivatives (FAMe).

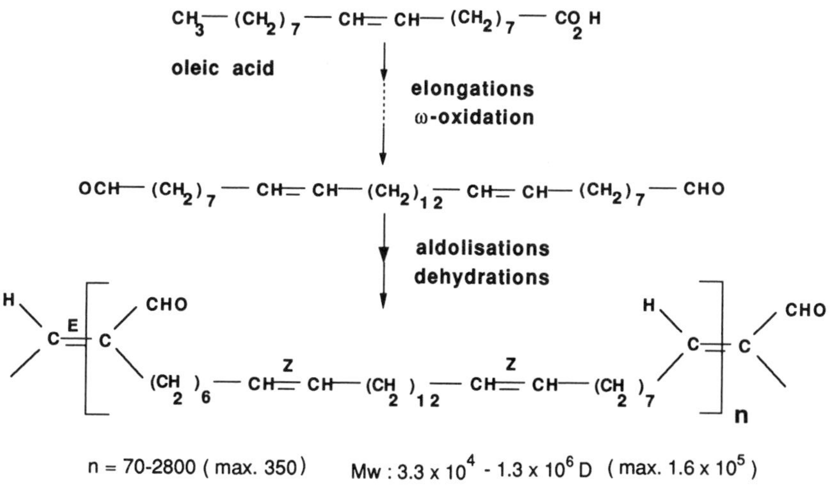

$$CH_3-(CH_2)_7-CH=CH-(CH_2)_7-CO_2H$$

oleic acid

elongations
ω-oxidation

$$OCH-(CH_2)_7-CH=CH-(CH_2)_{12}-CH=CH-(CH_2)_7-CHO$$

aldolisations
dehydrations

n = 70-2800 (max. 350)   Mw : 3.3 x 10^4 - 1.3 x 10^6 D (max. 1.6 x 10^5)

Figure 6. Structure of the "rubber" of *B. braunii* race A and of its precursor, C_{32} dialdehyde. (After Metzger *et al.*, 1991.)

formation of PRB A, i.e., an insoluble and nonhydrolyzable polymer. Such a type of structure and mechanism of formation was also fully supported by recent pyrolytic studies (Gelin *et al.*, 1994a).

Feeding experiments with ^{14}C-labeled substrates were also carried out with cultures of A race (Laureillard *et al.*, 1986, 1988; Templier *et al.*, 1992a). These experiments revealed that the biosynthetic pathways leading to the long alkyl chains of PRB A and to long-chain external lipids [including hydrocarbons, complex lipids like (5) to (7), and aldehydic rubber], all start from oleic acid elongation. In addition, feeding with [U-14C]-*n*-alkadienes confirmed (1) that such hydrocarbons are precursors, via formation of epoxides, of the *n*-alkenyl chains with a terminal unsaturation occurring in condensation products like (5) to (7), and (2) that a part of the long alkyl chains building up the macromolecular network of PRB A is derived from these dienes, as well. Inhibition experiments were also performed, with a cinnolinyl acid derivative that strongly reduces PRB A formation without affecting the primary metabolism of *B. braunii* (Templier *et al.*, 1992b, 1993). The associated changes in the abundance of the different families of long-chain external lipids confirmed precursor–product relationships between some of these lipids and PRB A. Conspicuous changes in outer wall morphology and rigidity were also observed along with PRB A inhibition.

All these observations are consistent with tight biosynthetic relationships between PRB A, on the one hand, and long-chain external lipids, including hydrocarbons, condensation products, and "rubber," on the other hand. Based on this mode of formation and on the nature of its precursors, PRB A is considered

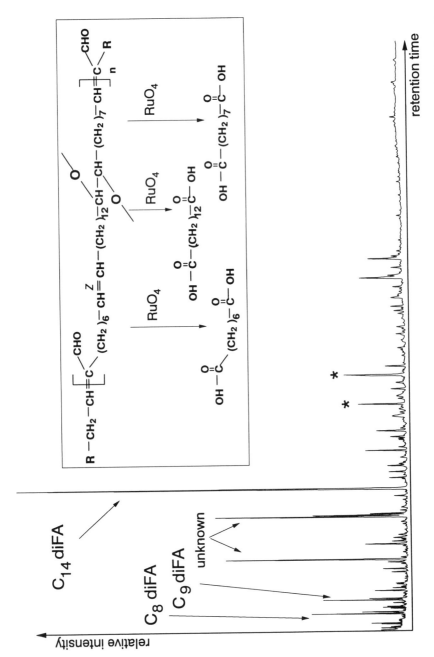

Figure 7. Gas chromatography of trace compounds obtained from PRB A after RuO$_4$ oxidation; * indicates contaminants; diFA, dicarboxylic acids.

to be a "macromolecular lipid," and is characterized by a highly aliphatic nature. Several series of high-molecular-weight ether lipids, for example, (8) and (9), comprising phenoxy bridges were recently identified (Metzger, 1994; Metzger and Largeau, 1994) along with the already mentioned alkoxy ether lipids like (5) to (7), in the external lipids of the A race. In sharp contrast, the former phenolic compounds were not detected, even in trace amounts, in the case of the B race. These phenoxy ether lipids were shown to be produced in very large amounts in some A race strains and to contribute to PRB A formation. Nevertheless, the corresponding phenol units are systematically associated with long poly-methylenic chains, and the latter always appear as the major structural elements of PRB A.

Regarding the B race, the chemical features of PRB B pointed, in sharp contrast to the situation with the A race, to a complete lack of relationship between hydrocarbons and the macromolecular constituents of the outer walls. Indeed, as already discussed, the macromolecular structure of PRB B is based on a network of polymethylenic chains, while botryococcenes are highly unsaturated isoprenoid hydrocarbons. This lack of relationship was confirmed by feeding experiments (Laureillard *et al.*, 1986), since (1) no radioactivity was incorporated into PRB B following incubations with [2-^{14}C]mevalonate, whereas botryococcenes were labeled, as expected, from this universal precursor of isoprenoid compounds; and (2) the alkyl chains of the PRB B network were shown to originate from oleic acid elongation. The biosynthetic pathways leading to hydrocarbons and to PRB are therefore entirely distinct in the case of the B race (Fig. 8). The external lipids of this race are composed, in addition to botryococcenes, of various series of high-molecular-weight lipids. A biosynthetic pathway similar to that leading to PRB A, i.e., starting with oleic acid elongation followed by condensation of complex lipids, is thus suggested for the formation of PRB B. However, no direct evidence of the involvement of such a pathway has been so far obtained in the B race, and no information about the chemical structure of its high-molecular-weight lipids has been reported yet.

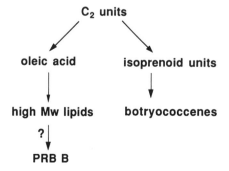

Figure 8. Schematic biosynthetic pathways of hydrocarbons and PRB in *B. braunii* race B.

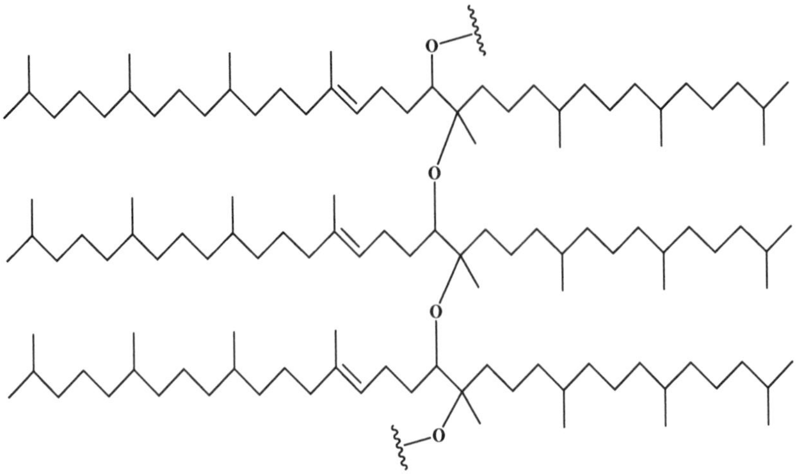

Figure 9. Proposed structure of PRB L. (After Gelin *et al.*, 1994b.)

The basic structure of PRB L, with its network of long isoprenoid chains, suggested a relationship with the C_{40} isoprenoid hydrocarbon, a lycopadiene (**4**), produced by the L race. This structural correlation was confirmed by the predominant presence of C_{40} isoprenoid hydrocarbons and ketones in pyrolysates of PRB L (Derenne *et al.*, 1990, 1994; Behar *et al.*, 1995). These C_{40} compounds show either the same typical skeleton as lycopane (i.e., a central tail-to-tail linkage between two phytyl-type units) or skeletons derived from lycopene moieties via cyclization and subsequent aromatization. A detailed structure for PRB L was recently proposed based on Curie point flash pyrolysis gas chromatography–mass spectrometry (GC–MS) analyses of various model compounds, by which detailed knowledge of mechanisms of pyrolysis of di- and polyalkenyl ethers was obtained (Gelin *et al.*, 1993, 1994b). The specific flash pyrolysis products of PRB L firmly indicated the sites of the ether cross-links of the lycopane units, so that a full structure of PRB L could be reconstructed from these pyrolysis products (Fig. 9).

In conclusion, the comparison of chemical structures and biosynthetic pathways indicated that PRBs originate from the condensation of long-chain lipids in the three races of *B. braunii*. These insoluble and nonhydrolyzable constituents of outer walls can therefore be considered as "macromolecular lipids," and they exhibit a highly aliphatic nature. Important differences in the nature of the chains building up the macromolecular network are noted, however, between PRB A and PRB B, on the one hand, and PRB L on the other hand, with the former composed of polymethylene chains, whereas PRB L network has an isoprenoid nature. In addition, it appears that tight biosynthetic relationships occur between hydrocarbons and PRBs in the A and L races, while a complete lack of such a correlation is noted for the B race.

2.1.2. Tetrahedron minimum

This freshwater Chlorophycea shows relatively thick, approximately 200 nm, algaenan-composed outer walls. Contrary to *B. braunii*, however, *T. minimum* does not form colonies, and each cell is completely surrounded by a resistant outer wall. Flash pyrolytic studies (Goth *et al.*, 1988) on isolated algaenan of *T. minimum* indicated a highly aliphatic structure and a macromolecular network to some extent similar to the one observed in PRB A and PRB B, i.e., based on long $(CH_2)_n$ chains, with *n* up to 31 (Fig. 1). It is noteworthy that the clusters of pyrolysis products eluting before the alkane/alkene doublets consist of the same compounds as observed in flash pyrolysates of PRB A and PRB B, although their relative intensities are significantly lower (Gatellier *et al.*, 1993). The nature and distribution of the esterified lipids of *T. minimum* are, however, different from those of *B. braunii* (De Leeuw *et al.*, in preparation). Polyunsaturated C_{16} and C_{18} fatty acids as well as hexadecanoic acid are the major fatty acids present, though small amounts of saturated long-chain fatty acids maximizing at C_{28} are present in this alga as well (Fig. 10). The sterol composition of this alga is also unique in that exclusively $\Delta7$ C_{28} and $-C_{29}$ sterols are biosynthesized (Nes and Mckean, 1977; De Leeuw *et al.*, in preparation). A series of monounsaturated and saturated ω-hydroxy acids with chain lengths of 30, 32, and 34 carbon atoms are noteworthy. The location of the double bond was determined by dimethyldisulfide derivatization and subsequent GC–MS analysis of the methylthio adducts and turned out to be at the ω9 position (De Leeuw *et al.*, in preparation). Based on RuO_4 oxidation of the algaenan isolated from this alga, it was assumed that the ω9-ω-hydroxy fatty acids are probably important building blocks of this biopolymer (Schouten *et al.*, in preparation). The dominance of dicarboxylic fatty acids with 9, 10, 21, 23, and 25 carbon atoms probably reflects the position of the functional groups in a polymer built up from these hydroxy fatty acids or derivatives thereof (Fig. 11). A highly speculative structure of *T. minimum* algaenan may thus be proposed.

2.1.3. Scenedesmus quadricauda, Pediastrum *spp., and other Microalgae with Resistant Trilaminar Sheath*

Thin, algaenan-composed outer walls were observed in the vegetative cells of a number of green microalgae (Derenne *et al.*, 1992a) (Table I). Such outer walls, only 10- to 30-nm thick, show a typical trilaminar organization with two electron-dense layers sandwiching an electron-lucent one when examined by transmission electron microscopy (TEM). This type of outer wall structure, or trilaminar sheath (TLS) (Atkinson *et al.*, 1972), is commonly observed in green

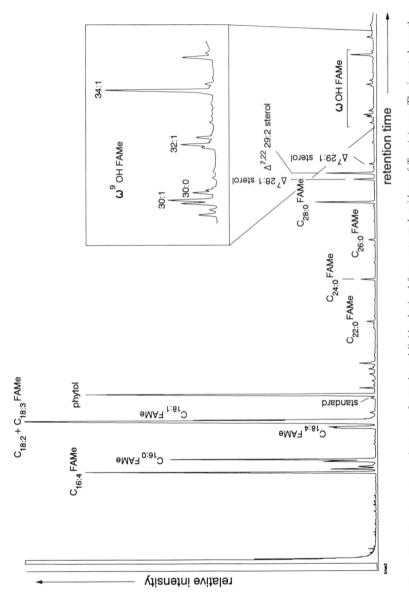

Figure 10. Gas chromatography trace of ester-bound lipids obtained from extracted residues of *T. minimum*. The insert shows the distribution of the monounsaturated ω-hydroxy fatty acids with a double bond at the ω⁹ position.

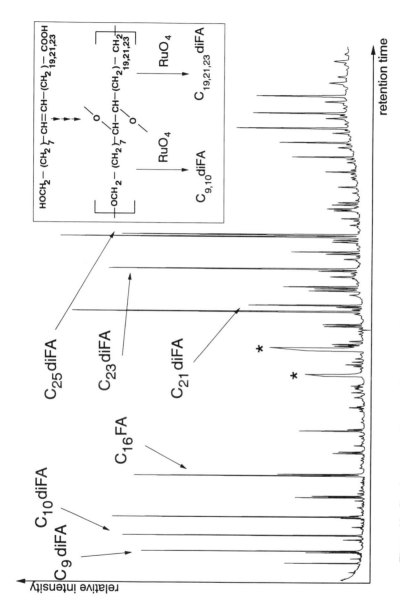

Figure 11. Gas chromatography trace of compounds obtained from *T. minimum* algaenan after RuO$_4$ oxidation.

**Table I. Microalgal Species Known to Comprise Algaenan-
Composed, Thin, Trilaminar Outer Walls** [a,b]

Ankistrodesmus braunii (1)	*Pediastrum duplex* (1)
Chlorella fusca (6)	*Phycopeltis epiphyton* (1)
Chlorella sp. (2)	*Prototheca portoricensis* (1)
Chlorella nana (1)	*Prototheca wickerhamii* (1)
Chlorella pyrenoidosa (1)	*Prototheca zopfii* (1)
Coccomyxa glaronensis (1)	*Pseudodidymocystis fina* (1)
Coccomyxa tiroliensis (1)	*Pseudodidymocystis planctonica* (1)
Coelastrum proboscidum (1)	*Scenedesmus longus* (1)
Elliptochloris bilobata (1)	*Scenedesmus obliquus* (2)
Elliptochloris sp. (1)	*Scenedesmus quadricauda* (2)
Nanochlorum eucaryotum (1)	

[a] After Derenne *et al.* (1992a).
[b] Only species comprising resistant TLS in vegatative cells are listed in this table. Regarding chlorophycean zygospores and cysts, the presence of acetolysis-resistant layers in their cell walls was noted in some cases (Devries *et al.*, 1983; Hull *et al.*, 1985; Zarsky *et al.*, 1985; Margulis *et al.*, 1988). It seems, however, that no information on the chemical composition of such layers has been reported so far. Values in parentheses correspond to the number of strains in which resistant TLS were detected. All the algae of this table, except *C. nana* and *N. eucaryotum*, are freshwater species. The presence of TLS is often considered to be an important evolutionary and taxonomic feature (Komarek, 1987; Kalina and Puncocharova, 1987). In this connection, it shall be noted that all the Chlorophyceae so far shown to comprise resistant TLS, but *P. epiphyton*, all belong to the same order, Chlorococcales. Electron microscopy observations revealed the presence of TLS in a number of other species of Chlorophyceae, e.g., *Chlorella minutissima* (Dempsey *et al.*, 1980) and *Scenedesmus verrucosus* (Hegewald and Schnepf, 1974), but the chemical resistance of these wall layers was not tested; accordingly, the corresponding species are not considered in this table.

microalgae. However, the occurrence of TLS in a given species does not alone provide definite evidence for the presence of algaenan. Indeed, a complete lack of insoluble and nonhydrolyzable macromolecular constituents was noted in several microalgae, following drastic chemical treatments, although TLS was clearly observed by TEM (Brunner and Honegger, 1985). Accordingly, only the species where the presence of algaenan-composed TLS was ascertained by chemical tests are listed in Table I, and flash pyrolysates of some are shown in Fig. 1. Resistant TLS thus appears to be of a quite widespread occurrence in green microalgae, and 21 species, corresponding to 11 different genera, have been shown to be composed of such outer walls so far.

In accordance with the thickness of TLS, the above species contain quite low levels of algaenan, and contents of 1 to 8%, relative to the total dry biomass, are generally obtained (Derenne *et al.*, 1992a). It must be noted, however, that following cell divisions in these microalgae, the TLS of the mother cell bursts and is released in the medium along with the freed daughter cells. In sharp contrast, the resistant basal outer walls corresponding to successive divisions remain tightly associated, in the case of *B. braunii,* and build up the matrix of colonies. Accordingly, the difference in algaenan production, between *B. braunii,* on the one hand, and microalgae with resistant TLS, on the other hand, is not as high as suggested when algaenan levels determined from living cells are directly compared. Indeed, actively growing populations of species comprising resistant TLS will produce rather large amounts of algaenans that will accumulate in the growth medium as "mother" TLS.

Although the typical trilaminar structure of these thin, resistant, microalgal outer walls is retained, after drastic basic and acid hydrolyses, the TLS morphology appears to be highly distorted (Fig. 12). As a result, in contrast with the thick resistant outer walls of *B. braunii* and *T. minimum,* the initial shape of such cells is no longer recognizable following algaenan isolation.

The chemical structures of the algaenans building up the resistant TLS of several species of microalgae, including *Scenedesmus quadricauda, Chlorella*

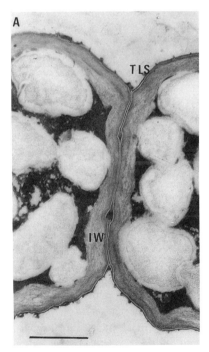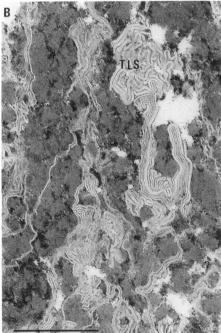

Figure 12. Transmission electron microscopy. (A) Portion of *Scenedesmus quadricauda* colony showing two adjacent cells; each cell is surrounded by a thick polysaccharidic inner wall (IW) and a thin trilaminar outer wall (TLS). (B) Resistant material obtained after the final acid treatment; the algaenan-composed TLS are still easily recognized, but they appear highly distorted and associated into small bundles. Scale bars = 1 μm.

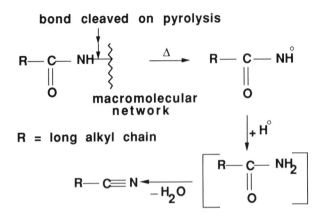

Figure 13. Mechanism of formation of *n*-alkylnitriles upon pyrolysis of algaenans isolated from species with resistant TLS.

fusca, Pediastrum spp., and *Nanochlorum eucaryotum,* were recently examined (Derenne *et al.*, 1991, 1992a; De Leeuw *et al.*, in preparation). These algaenans exhibit the same basic structure as PRB in the A and B races of *B. braunii;* that is, a macromolecular network based on long polymethylenic chains cross-linked by ether bridges. Identification of pyrolysis products, however, revealed an important difference: *n*-alkylnitriles are generated upon pyrolysis of the algaenans isolated from species with resistant TLS, whereas such compounds are not formed, even not in trace amounts, from *B. braunii* and *T. minimum* algaenans. The important morphological differences observed between thick resistant outer walls, on the one hand, and resistant TLS, on the other hand, are thus associated with a substantial difference in chemical composition regarding nitrogenous compounds. Spectroscopic features suggested that the above nitriles could originate from the thermal cleavage of amide groups, associated with polymethylenic chains in the macromolecular structure, followed by a fast dehydration step (Fig. 13). This assumption concerning the presence of amide groups was fully confirmed by growth of *Scenedesmus quadricauda* on a medium with $^{15}NO_3$ as sole nitrogen source, followed by solid-state [^{15}N]-NMR examination of the isolated algaenan (Derenne *et al.*, 1993). It is well documented that amide groups usually exhibit a high susceptibility to hydrolysis, and amide-based macromolecules, like proteins, are thus easily cleaved. The very high resistance of the amide groups occurring in "TLS-type" algaenans therefore reveals, as indicated in Section 2.1.1 for esters in PRBs, an efficient steric protection within the macromolecular network. This type of algaenan thus provides the first example of nonhydrolyzable amide functions in biomacromolecules. Other features of the

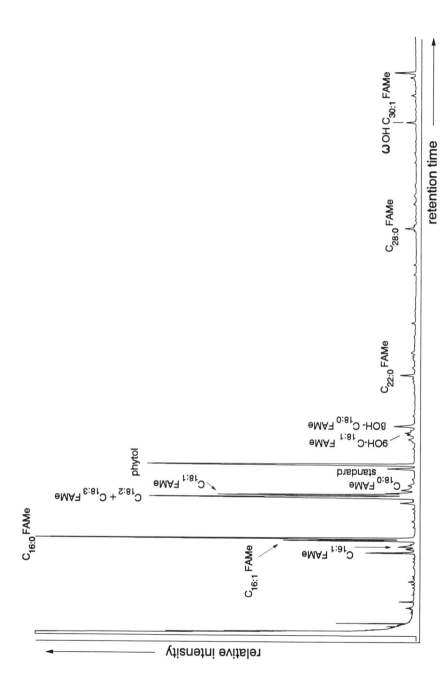

Figure 14. Gas chromatography trace of ester-bound lipids obtained from extracted residues of *S. quadricauda*.

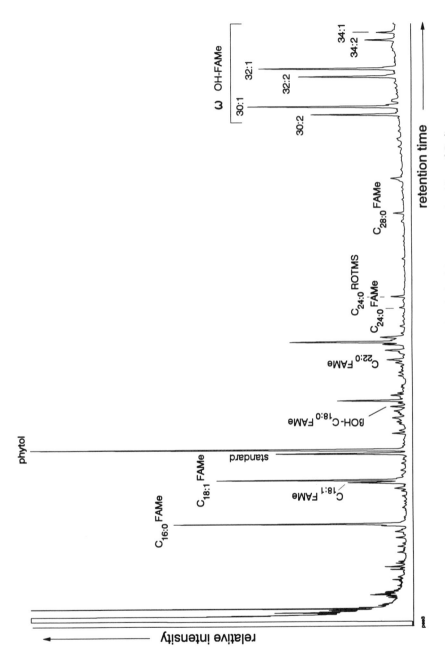

Figure 15. Gas chromatography trace of ester-bound lipids obtained from extracted residues of *Pediastrum* spp.

algaenan structure of these algae are revealed by the presence of specific lipids in *S. quadricauda* and *Pediastrum* spp.

The ester-bound lipids of *Scenedesmus quadricauda* and *Pediastrum* spp. have been analyzed in some detail (De Leeuw *et al.*, in preparation). These lipids in *S. quadricauda* are dominated by C_{16} and C_{18} unsaturated fatty acids. It should be noted that C_{28} fatty acids as well as saturated and monounsaturated C_{26} and C_{28} alcohols are present in minor amounts (Fig. 14). Among the ester-bound lipids in *Pediastrum*, highly specific compounds present in relatively high amounts are noticed apart from the fatty acids and phytol (Fig. 15). GC-MS analysis indicated that these lipids are mono- and diunsaturated ω-hydroxy fatty acids with chain lengths of 30, 32, and 34 carbon atoms. Catalytic hydrogenation of these compounds resulted in the formation of their saturated counterparts, thus strengthening their structural identification. As in *T. minimum*, it can be speculated that these compounds serve as building blocks for their algaenans. This is further supported by the virtual identical flash pyrolysates of *T. minimum*, *Pediastrum* spp., and *S. quadricauda* (Fig. 1).

The selective preservation of the algaenans from microalgal TLS results in the formation of the thin lamellar structures, termed *ultralaminae*, recently discovered in a number of kerogens by TEM observations (Raynaud *et al.*, 1989; Largeau *et al.*, 1990b) and shown to exhibit the same chemical features, including *n*-alkylnitrile formation upon pyrolysis, as the above algaenans (Derenne *et al.*, 1991, 1992b,c; Gillaizeau *et al.*, 1995).

2.2. Bacteria

Pioneering studies by Strohl *et al.* (1977) pointed to the presence of nonhydrolyzable macromolecular constituents in the vegetative cells of four species of *Myxococcus,* whereas a complete lack of this type of material was observed for eight species of other genera (Table II). No information was provided, however, about the location and the chemical structure of the resistant compounds of these *Myxococcus* species.

The first systematic study concerning the occurrence of resistant macromolecular constituents in Eubacteria was carried out via examination of several groups of filamentous and coccoid cyanobacteria (Chalansonnet *et al.*, 1988). The presence of cyanobacterans was thus observed in five of six species (Table III). Such constituents were shown to be located in cell walls, since cell contents are entirely eliminated following saponification, while several wall layers are still observed. In sharp contrast with algaean-containing species, however, the morphological features of the walls were completely lost following the final acid hydrolysis. The isolated cyanobacterans thus appeared as an amorphous material when observed by TEM at high magnifications. Accordingly, the precise location of these cyanobacterans could not be established among the different cell wall

Claude Largeau and Jan W. de Leeuw

**Table II. Occurence of
Acetolysis-Resistant Organic
Constituents in Myxobacterial
Cell Walls[a,b]**

Myxococcus xanthus (myxospores)
M. xanthus (vegetative cells)
M. fulvus (myxospores)
M. fulvus (vegetative cells)
M. stipitatus (myxospores)
M. stipitatus (vegetative cells)
M. virescens (myxospores)
M. virescens (vegetative cells)

[a]After Strohl *et al.* (1977).
[b]No resistant material was obtained from *Cytophaga aquatilis, Flexibacter* sp., *Escherichia coli, Micrococcus roseus, M. luteus, Bacillus cereus, B. firmis,* and *Cellulomonas flavogena.*

layers retained after saponification. Contrary to microalgal outer walls, none of the layers occurring in the cell walls of the above cyanobacteria therefore comprise a skeleton entirely built up by a nonhydrolyzable macromolecular material.

Substantial levels of cyanobacterans were observed in the five species listed in Table III. Regarding the chemical structure, only the bulk chemical features of the nonhydrolyzable macromolecular material of *Schizothrix* sp. have been examined so far (Chalansonnet *et al.*, 1988). This cyanobacteran exhibits pronounced differences when compared to algaenans: a lower contribution of polymethylenic chains is observed, whereas methyl branching, unsaturations, and oxygen-containing functional groups are relatively more abundant.

**Table III. Cyanobacterial Species Tested for the
Occurence of Cyanobacterans[a,b]**

Species		Abundance[c]
Anacystis montana (c)	+	2.1
Calothrix anomala (f)	+	0.9
Fischerella muscicola (f)	+	1.2
Gloeocapsa alpicola (c)	−	
Oscillatoria rubescens (f)	+	2.9
Schizothrix sp. (f)	+	5.6

[a]After Chalasonnet *et al.* (1988).
[b]+, Presence; −, absence; c, coccoid; f, filamentous.
[c]As percent of the dry weight of the initial untreated biomass.

Several species of bacteria belonging to different families were recently examined for the occurrence of bacteran (Le Berre *et al.*, 1991; Le Berre, 1992). The species tested (Table IV) comprised both Eubacteria (four ubiquitous aerobic species; *Rhodomicrobium vannielii*, a photosynthetic bacterium; *Beggiatoa alba*, a hydrogen sulfide oxidizer) and Archaebacteria (two methanogens and one halophile). Bacterans were detected in six of nine species, including three ubiquitous aerobic species, the two methanogens and the halophilic Archaebacterium (Table IV).

All the above observations are therefore consistent with a rather widespread occurrence of insoluble nonhydrolyzable macromolecular constituents in cyanobacteria, other groups of Eubacteria and in Archaebacteria. The examined Archaebacteria exhibit, however, substantially lower levels of resistant constituents when compared to Eubacteria (Tables III and IV).

Electron microscopy observations revealed that in the three bacteran-containing Eubacteria, the same morphological features as already discussed for cyanobacteria occur: resistant material located in cell walls and that becomes amorphous once isolated by drastic hydrolyses (Fig. 16). It is well documented that eubacterial cell walls are chiefly based on a peptidoglycan skeleton. In addition, peptidoglycans are known to withstand drastic basic hydrolyses but are cleaved under acid conditions. TEM examinations of *Mycobacterium smegmatis*, *Rhodococcus rhodochrous* (*Nocardia opaca*), and *Corynebacterium glutamicum*, following lipid extraction and drastic saponification, showed that only the peptidoglycan layer of the cell wall is retained at this stage. Bacterans in the above species should therefore be linked covalently to the peptidoglycan skeleton. The final acid treatment results in a complete collapse of this skeleton via peptidoglycan cleavage; hence, the amorphous nature of the isolated bacterans.

The cell walls of the halophilic Archaebacterium *Halococcus morrhuae* consist

Table IV. Bacterial Species Tested for the Occurence of Bacterans [a,b]

	Species		Abundance[c]
Ubiquitous aerobic	*Pseudomonas putida* (E)	−	.
	Mycobacterium smegmatis (E)	+	2
	Nocardia opaca (E)	+	2.2
	Corynebacterium glutamicum (E)	+	1.6
Methanogens	*Methanosarcina thermophila* (A)	+	*ca* 0.2
	Methanosarcina barkeri (A)	+	*ca* 0.2
Halophilic	*Halococcus morrhuae* (A)	+	*ca* 0.2
Photosynthetic	*Rhodomicrobium vannielii* (E)	−	
H$_2$S oxidizer	*Beggiatoa alba* (E)	−	

[a]From Le Berre *et al.* (1991).
[b]+, Presence; −, absence; E, Eubacteria; A, Archaebacteria.
[c]As percent of the dry weight of the intitial untreated biomass.

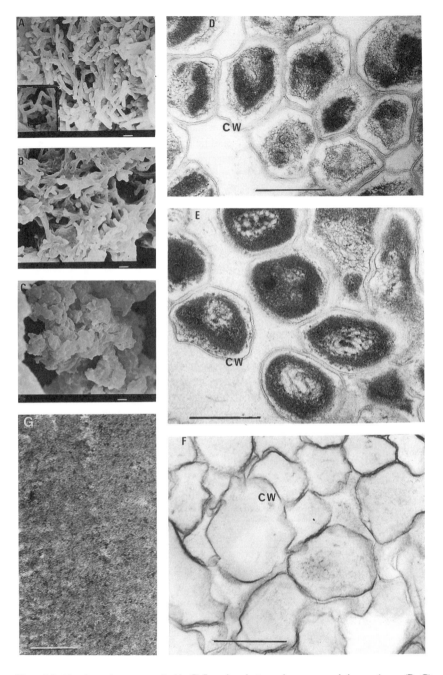

Figure 16. *Mycobacterium smegmatis*. (A–C) Scanning electron microscopy; scale bars = 1 μm. (D–G) Transmission electron microscopy; scale bars = 0.5 μm. (A and D) Lyophilized biomass, insert in A shows an Y-shaped cell; CW = cell walls. (B and E) After lipid extraction, cell shape and cell wall morphology are

mainly of polysulfated heteroglycans (Larsen, 1984). TEM observations after saponification and acid hydrolysis (Fig. 17) indicate that *H. morrhuae* bacteran should be covalently linked to this heteroglycan skeleton. Hence, as before, the bacteran of *H. morrhuae* is amorphous in nature.

Chemical structures of bacterans were examined by spectroscopic and pyrolytic methods in the case of *M. smegmatis, N. opaca,* and *C. glutamicum* (Le Berre, 1992; Flaviano *et al.,* 1994). These three species belong to closely related genera characterized by the formation of specific lipids and glycolipids (Lederer *et al.,* 1975; Asselineau and Asselineau, 1978; Daffé *et al.,* 1981; Minnikin, 1982; Couderc *et al.,* 1988; Brennan, 1988). These typical constituents include mycolic acids (**10**), trehalose esters/fatty acyl esters (**12**), mycolates (**11**), and trehalose-containing lipooligosaccharides. In fact, the chemical features of the three bacterans are consistent with some contribution of moieties derived from such lipids and glycolipids. For example, in the case of *M. smegmatis* bacteran, the FTIR and solid-state $[^{13}C]$-NMR spectra point to the presence of sugar units and polymethylenic chains. In addition, the $[^{13}C]$-NMR spectrum shows peaks corresponding to the typical resonances of C_α, C_β, and C_γ carbon atoms in mycolic acids and derived compounds (Fig. 18).

The relationship between *M. smegmatis* bacteran and some specific lipids and glycolipids was also supported by pyrolysis. First, the fatty acid fraction isolated from the pyrolysate of *M. smegmatis* bacteran shows an unusual distribution, with a marked predominance of *n*-tetracosanoic acid, thus pointing to contribution of mycolate units in this bacteran. Indeed, it is well documented that (1) heating of mycolic acids and derivatives results in a selective cleavage of the C_α–C_β bond (Fig. 19), and (2) the mycolic acids of *M. smegmatis* all comprise a normal saturated C_{22} chain; hence, the formation of *n*-tetracosanoic acid, upon pyrolysis, from any structure containing esters of such mycolic acids. Second, the pyrolysis products of *M. smegmatis* bacteran comprise, in addition to *n*-alkanes and *n*-alk-1-enes, some pyrolysis products not observed in algaenan pyrolysates. Such compounds include a series of linear, saturated, C_{14} to C_{32} ethyl esters, dominated by the C_{18}, C_{20}, and C_{26} homologues and two C_{16} and C_{18} "aldehydic esters" (Fig. 20). Based on their chemical structures and distributions, the above products probably originate from the thermal cracking of moieties derived from the trehalose esters and the trehalose-containing lipooligosaccharides of *M. smegmatis,* respectively (Fig. 20). Similar relationships between bacterans on the one hand and some specific lipids and glycolipids, especially mycolate derivatives, on the other hand were also observed after examination of *N. opaca* and *C. glutamicum* bacterans.

---

not significantly altered. (C and F) After saponification, *M. smegmatis* cells are massed together (C), cell contents are almost entirely eliminated, cell walls are thinner, and the trilayered organization is no longer observed (F). (G) After the final acid treatment, the isolated algaenan appears as an amorphous material.

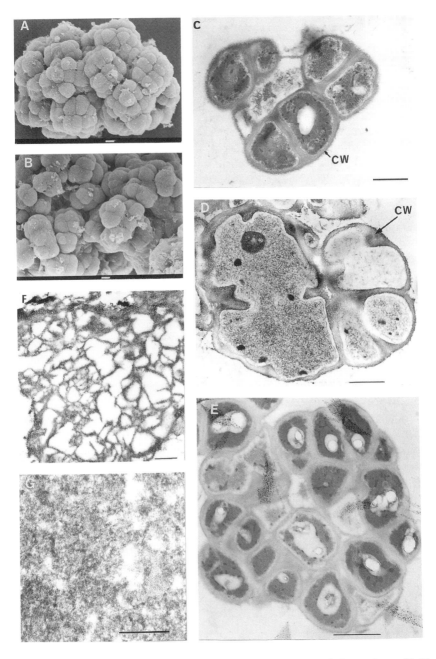

Figure 17. *Halococcus morrhnae*. (A, B) Scanning electron microscopy; scale bars = 1 μm. (C–G) Transmission electron microscopy; scale bars = 0.5 μm. (A, C, and D) Lyophilized biomass, cells occur in

Because of the similarities between these moieties in bacterans and the specific complex lipids in these bacteria, we had to consider the possibility of artifact formation. Such a formation could occur through condensation of nonextracted mycolic-type lipids, induced by the base and/or acid treatments used for bacteran isolation. Hence, blank experiments were undertaken to test this hypothesis. A sample of *M. smegmatis* biomass was separated in two parts, which were extracted by the procedure used in algaenan isolation and by the classic method used for extracting mycolic-type lipids (Asselineau *et al.*, 1984). The extracts were then subjected to base and acid treatments in exactly the same way as for bacteran isolation. However, no formation of "bacteran" was observed. Furthermore, the residual bacterial biomass obtained after the two extraction procedures was treated for bacteran isolation and no significant differences either in the amount or in the bulk chemical structure of the final insoluble material were noted. Accordingly, the latter material should actually correspond to insoluble macromolecules originally occurring in *M. smegmatis*, and artifact formation can be ruled out.

As already emphasized, algaenans originate from the condensation of high-molecular-weight lipids, and the resulting macromolecular structures exhibit, owing to steric protection, a very strong resistance to hydrolyses. Based on the chemical features discussed above, a similar process is likely to be involved in the formation of the bacterans of *M. smegmatis, N. opaca,* and *C. glutamicum* and would also account for their resistance to drastic basic and acid treatments. These bacterans, along with algaenans, would therefore belong to a new family of biomacromolecules and they can be regarded as "macromolecular lipids." Due to the different nature of the lipids involved and the lack of glycolipid contributions in algaenans, markedly different chemical structures are observed between bacterans and algaenans.

The cell walls of *M. smegmatis, N. opaca,* and *C. glutamicum* are known to comprise, like all the bacteria belonging to these three genera, a "mycoloyl–arabinogalactan–peptidoglycan" macromolecular complex (Minnikin, 1982; McNeil *et al.*, 1990, 1991). The lipid constituents of cell walls are loosely associated with this skeleton and they are removed by solvent extraction. Basic hydrolyses result in the elimination of the mycoloyl–arabinogalactan portion of the complex, via cleavage of the mycoloyl–arabinogalactan ester links and of the arabinogalactan–peptidoglycan phosphodiester bridges. In fact, the morphological and chemical features of the three bacterans indicate that the macromolecular complex of *M. smegmatis, N. opaca,* and *C. glutamicum* cell walls also comprise insoluble and nonhydrolyzable units. These

---

pairs, and tetrads associated in irregular clusters (A and C); groups of cells which contents are partly lyzed and cell walls eliminated in contact areas are also observed (D); cell walls (CW) appear homogeneous and relatively thick. (B and E) After lipid extraction, cell shape and wall morphology are not significantly altered. (F) After saponification, cell contents are almost entirely eliminated; cell walls are thinner and somewhat distorted. (G) After the final acid treatment, the isolated algaenan appears as an amorphous material.

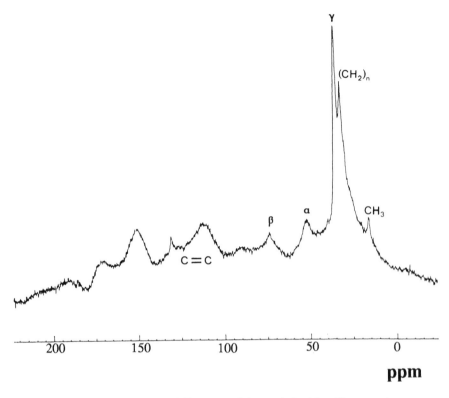

Figure 18. Solid-state [^{13}C]-NMR spectrum of algaenan isolated from *M. smegmatis.*

bacteran units would be covalently linked to the peptidoglycan portion of the complex. In sharp contrast with peptidoglycan, such lipid-derived macromolecules cannot be cleaved by hydrolysis, and they are retained as an insoluble amorphous material following the final acid treatment.

Regarding geochemical implications, the selective preservation of this type of bacterans should therefore result in the formation of amorphous kerogens. Indeed,

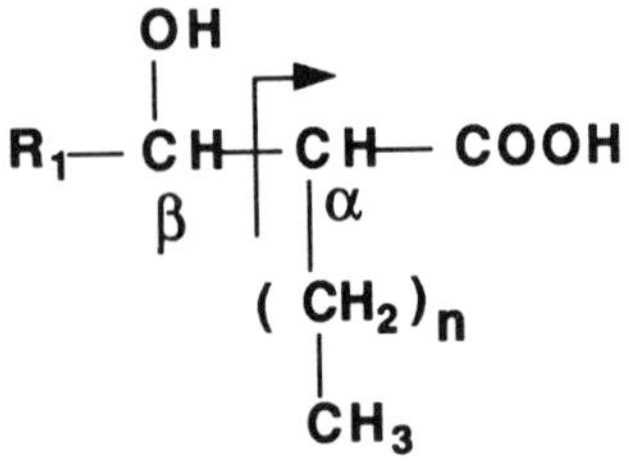

$$R_1 - \underset{\beta}{\overset{\overset{\displaystyle OH}{|}}{C}H} \!\!-\!\! \underset{\underset{(CH_2)_n}{\overset{|}{\underset{CH_3}{|}}}}{\overset{|}{C}H} \!\!-\!\! COOH$$

Figure 19. Preferential pyrolytic cleavage of mycolic acid derivatives; $n = 21$ for *M. smegmatis* mycolates.

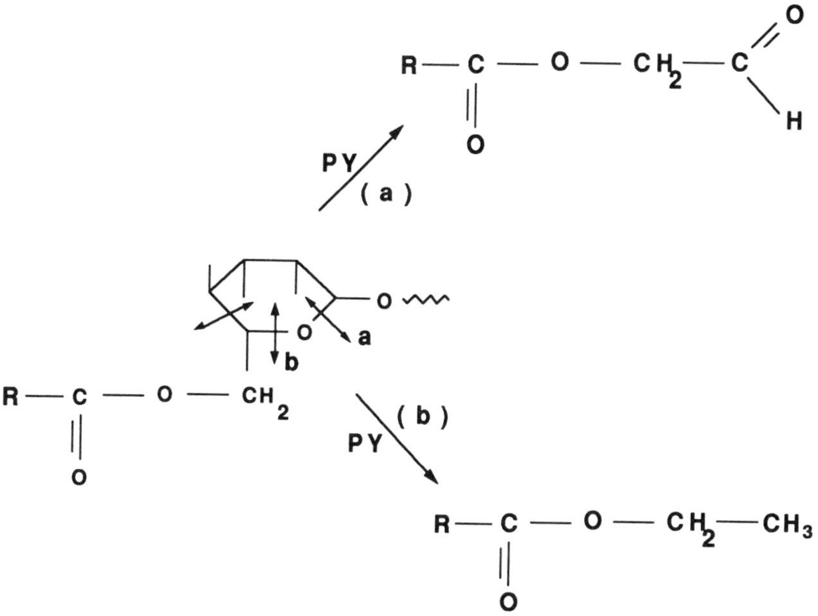

Figure 20. Assumed formation of (a) "aldehydic esters" and (b) ethylesters during pyrolysis of bacteran isolated from *M. smegmatis.*

recent comparative studies with the pyrolysis products of three amorphous kerogens (Flaviano *et al.*, 1994; Boussafir *et al.*, 1995) indicated some chemical correlations and point to bacteran contribution to their formation.

Only bulk chemical information was obtained, by FTIR spectroscopy, on the insoluble residues of *H. morrhuae, M. thermophila,* and *M. barkeri* (Le Berre, 1992). These nonhydrolyzable macromolecular constituents of archaebacterial cell walls show, when compared to the bacterans isolated from the three Eubacteria mentioned above, (1) a lower content of oxygenated groups, (2) a higher contribution of alkyl chains, and (3) a higher level of branching by methyl groups. Archaebacterial lipids are known to exhibit a typical structure based on polyisoprenoid chains (Woese *et al.*, 1978). Accordingly, the bulk chemical features of such "archaebacterans" are also consistent with formation via condensation of high-molecular-weight lipids. Circumstantial evidence for the possible existence of archaebacterans comes from the chemical degradation of immature kerogens by BBr_3 (Michaelis and Albrecht, 1979). The products isolated are clearly related to the high-molecular-weight, ether-bound membrane lipids typical of Arachaebacteria. Since these compounds were released from kerogens, i.e., insoluble sedimentary organic matter, one may assume that they are derived from insoluble macromolecular constituents originally present in Archaebacteria.

3. Biological Role of Algaenans and Bacterans

No extensive studies so far have been performed regarding the function of algaenans and bacterans. The available information, however, suggests that the presence of nonhydrolyzable aliphatic macromolecular constituents in cell walls has important consequences for living cells. With regard to microalgae, the thin, resistant trilaminar outer walls (TLS) occurring in a number of species (Table I) seem to play a major protective role. Thus,

1. Various high-molecular-weight compounds, including enzymes, cannot enter cells surrounded by TLS (Syrett and Thomas, 1973); indeed, the first study of such wall structures stemmed from the inability to obtain protoplasts, via enzymatic treatments, from a strain of *Chlorella fusca* (Atkinson *et al.*, 1972).

2. Close relationships were observed between the ability of various strains of *Chlorella* and *Scenedesmus* to survive high concentrations of detergents and the presence of a TLS; such correlations were noted from growth determination (Biedlingmaier *et al.*, 1987) and photosynthetic and respiratory activity measurements (Corre *et al.*, submitted).

3. Protection of phycobionts against fungal penetration in some lichens is also likely to be provided by TLS (Honegger and Brunner, 1981).

4. The high resistance of the parasitic alga *Prototheca wickerhamii* to drugs and lysosomes could be related to a thin resistant trilaminar outer wall in this species (Atkinson *et al.*, 1972; Puel *et al.*, 1987).

5. TLS could also afford some protection against dessication since *Phycopeltis epiphyton*, a subaerial microalga growing on leaves, is surrounded by several TLSs (Good and Chapman, 1987) instead of a single trilaminar outer wall as for the other species mentioned in Table I that all live in normal water habitats. This multilayered outer wall of *P. epiphyton* may act as the cuticle in higher plants.

In the case of *B. braunii,* the thick algaenan-composed outer walls seem to be important in adaptation to salinity changes. Recent observations with the A and B races (Derenne *et al.*, 1992d; Sabelle *et al.*, 1993) indicated that growth on media of increasing salinity is associated with pronounced morphological changes and that, under saline conditions, the apical part of cells is entirely covered by the thick outer wall. In addition, chemical analyses on the A race revealed higher levels of phenolic moieties in PRB A with increasing medium salinity. Studies on the susceptibility of various microalgae and cyanobacteria to decomposition by heterotrophic bacteria are consistent with a much higher resistance in species with cell walls comprising nonhydrolyzable aliphatic macromolecular constituents (Shilo, 1970; Serruya *et al.*, 1974; Gunnison and Alexander, 1975; Yamamoto and Suzuki, 1990). Such walls could therefore act as an efficient barrier to biodegradative agents.

Regarding the bacteria, it was suggested (Strohl *et al.*, 1977) that the presence of nonhydrolyzable macromolecular compounds in vegetative cells of several *Myxococcus* could be related to the resistance of myxobacteria to environmental stress. Finally, it can be noted that a limited survey of different families of bacteria (Table IV) revealed the presence of bacteran in the three tested Actinomycetes, including *Mycobacterium smegmatis*. The latter genus is well known to comprise a number of pathogenic species, like *M. tuberculosis* and *M. leprae*. Accordingly, future studies will have to examine whether bacterans also occur in these species and whether a relationship could be established between their resistance to some drugs and the presence of such macromolecular constituents in their cell walls.

4. Conclusions

Insoluble, nonhydrolyzable, aliphatic macromolecular constituents have been recently shown in cell walls of several groups of microalgae (algaenans), bacteria (bacterans and cyanobacterans), and possibly of some Archaebacteria. It should be noted that these types of constituents have also been recognized recently in various higher plant tissues (cuticles, barks, seed coats). The limited number of ubiquitous green microalgae tested so far all contain relatively higher amounts of algaenans. It must be emphasized, however, that until now only a few marine species have been investigated.

The high molecular weight and insoluble nature of these aliphatic constituents have seriously hampered their recognition and make it difficult to elucidate their structures. It was only through a combined analytical approach using spectroscopic methods (FTIR, [^{13}C]-NMR), pyrolysis, chemolysis, and electron microscopy that these macromolecules could be identified as an unprecedented family of biomacromolecules derived from condensation of complex lipids.

Pronounced differences in algaenan structures were observed for the algae studied so far. These differences reflect, in most cases, the structural differences of the complex lipids from which they are derived. Among the three races of *B. braunii,* major differences concerning the nature of the long alkyl chains are even observed between the A and B races on the one hand and the L race on the other. Substantial differences are also observed between the algaenans and the bacterans as well as between the bacterans and the cyanobacterans. Accordingly, these types of biomacromolecules may be useful for chemotaxonomical purposes in the future.

Their location in the cell walls or at the periphery of the organism, as well as their highly resistant nature, strongly point to an important protective function for these insoluble macromolecular lipids. If they are indeed biosynthesized as a protection against saprophytic organisms, it becomes quite understandable that these constituents are selectively preserved during deposition and subsequent fossilization; hence, their relatively high contribution to fossil organic matter and their importance in oil formation.

ACKNOWLEDGMENTS. We gratefully acknowledge Mrs. M. R. S. Baas, W. I. C. Rijpstra, S. Derenne, and J. Templier, and Mr. S. Schouten, P. Metzger, F. Gelin, and Y. Pouet, for analyses, assistance, and helpful discussions. This is NIOZ Division of Marine Biogeochemistry contribution No. 353. Dr. J. Connan (Société Elf-Aquitaine, Géochimie Organique, Pau) is acknowledged for providing a part of the bacterial biomass used for bacteran studies by the Paris group; financial support to the latter studies was provided by CNRS and Société Elf-Aquitaine.

5. Appendix

$$CH_3 - (CH_2)_7 \overset{E \text{ or } Z}{- CH= CH} - (CH_2)_x - CH= CH_2$$

$$\omega 9 \quad \omega 1 0$$

(1) x = odd 11 - 19

$$CH_3 - (CH_2)_5 - \overset{Z}{CH= CH} - \overset{Z}{CH= CH} - (CH_2)_x - CH= CH_2$$

$$\omega 7 \qquad \omega 9$$

(2) x=15, 17

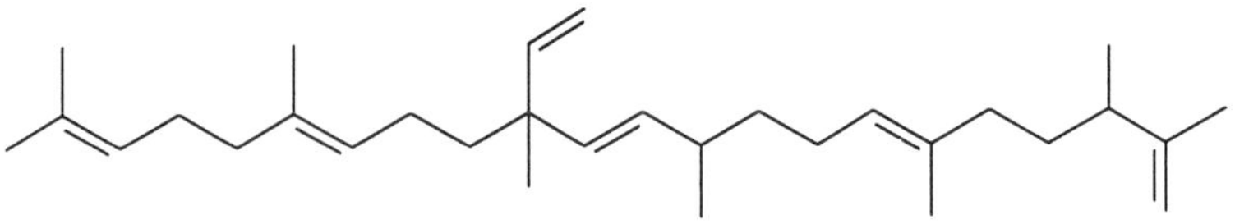

(3) C_{31} botryococcene

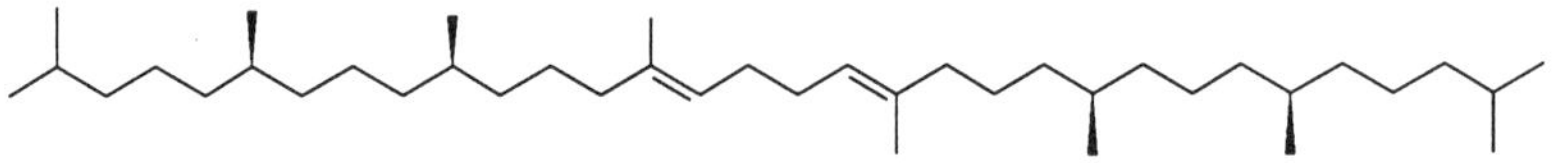

(4)

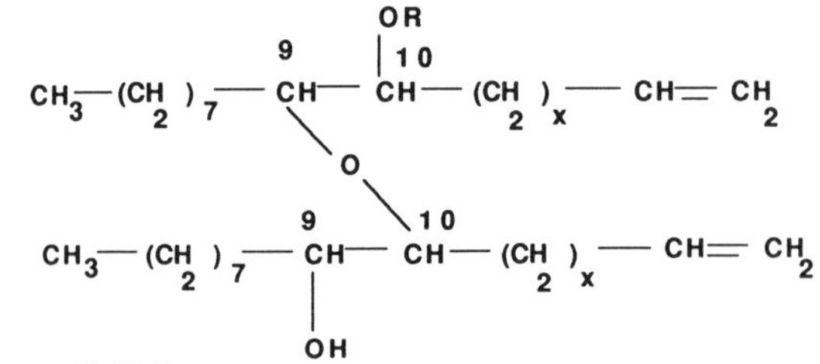

(5)

R = H

R = C_{18} to C_{30}

evenfatty acyls

x = 15,17,19

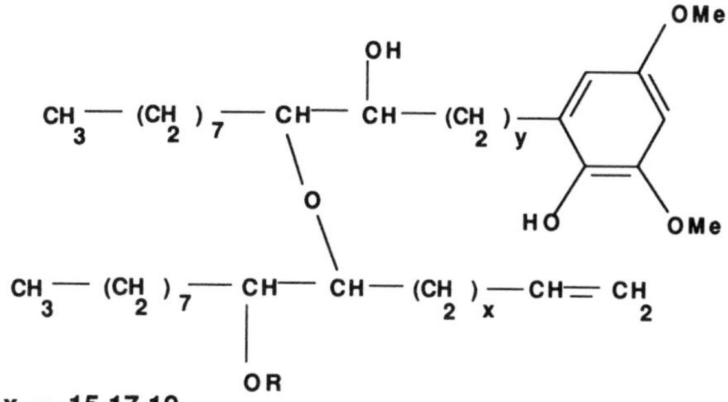

(6) R as in (5)

x = 15,17,19
y = 17,19,21

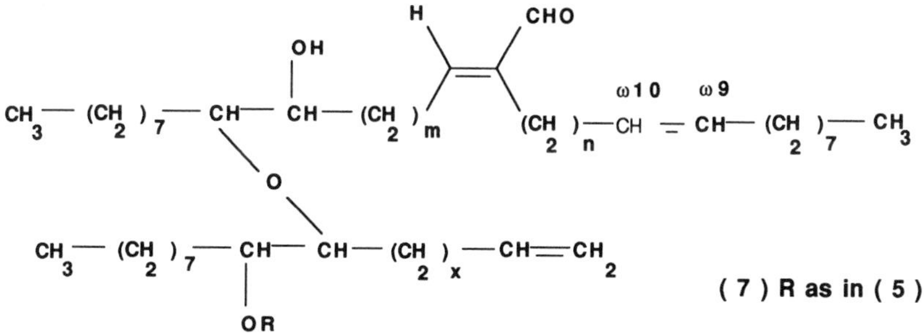

(7) R as in (5)

m = odd 15 to 21
x = 15,17,19

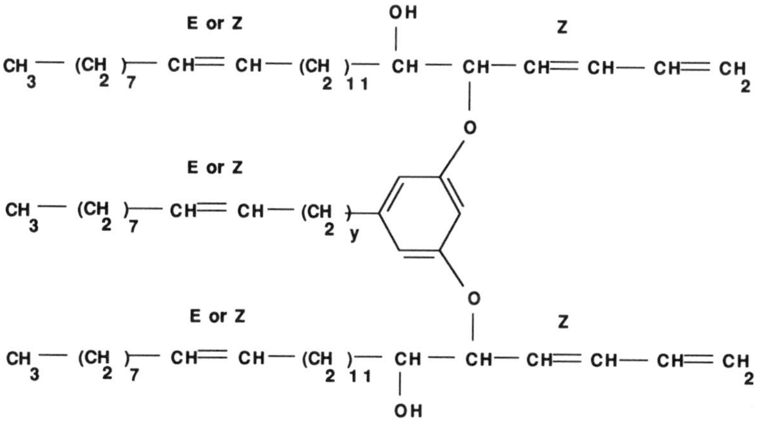

(8)

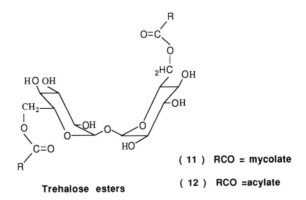

$$CH_3 - (C_nH_{2n-2}) - \underset{\beta}{CH} - \underset{C_{22}H_{45}}{\overset{\alpha}{CH}} - COOH$$

n up to 62

(10)

typical mycolic acid of M. smegmatis

(11)   RCO = mycolate

(12)   RCO = acylate

Trehalose esters

References

Asselineau, C., and Asselineau, J., 1978, Trehalose-containing glycolipids, *Prog. Chem. Fats Other Lipids* **16**:59–99.

Asselineau, C., Daffé, M., David, H. L., Lanéelle, M. A., and Rastogi, N., 1984, Lipids as taxonomic markers for bacteria derived from leprosy infection, *Acta Leprol.* **95**:86–92.

Atkinson, A. W., Gunning, B. E. S., and John, P. C. L., 1972, Sporopollenin in the cell wall of *Chlorella* and other algae: Ultrastructure, chemistry and incorporation of ^{14}C acetate, studied in synchronous cultures, *Planta* **107**:1–32.

Barwise, A. J. G., Mann, A. L., Eglinton, G., Gowar, A. P., Wardroper, A. M. K., and Gutteridge, C. S., 1984, Kerogen characterization by ^{13}C NMR spectroscopy and pyrolysis-mass spectrometry, *Org. Geochem.* **6**:343–349.

Behar, F., Derenne, S., and Largeau, C., 1995, Closed pyrolyses of the isoprenoid algaenan of *Botryococcus braunii, L* race. Geochemical implications for derived kerogens, *Geochim. Cosmochim. Acta,* **59**:2983–2997.

Berkaloff, C., Casadevall, E., Largeau, C., Metzger, P., Peracca, S., and Virlet, J., 1983, The resistant polymer of the walls of the hydrocarbon-rich alga *Botryococcus braunii, Phytochemistry* **22**:389–397.

Biedlingmaier, S., Wanner, G., and Schmidt, A., 1987, A correlation between detergent tolerance and cell wall structure in green algae, *Z. Naturforsch. C* **42**:245–250.

Boussafir, M., Gelin, F., Lallier-Vergés, E., Derenne, S., Bertrand, P., and Largeau, C., 1995, Electron microscopy and pyrolysis of kerogens from the Kimmeridge Clay Formation (U.K.). Source organisms, preservation processes and origin of microcycles, *Geochim. Cosmochim. Acta,* **59**:in press.

Brennan, P. J., 1988, Mycobacterium and other actinomycetes, in: *Microbial Lipids,* Vol. 1 (C. Ratledge and S. G. Wilkinson, eds.), Academic Press, London, pp. 203–298.

Brunner, U., and Honegger, R., 1985, Chemical and ultrastructural studies on the distribution of sporopollenin-like biopolymers in six genera of lichen phycobionts, *Can. J. Bot.* **63**:2221–2230.

Chalansonnet, S., Largeau, C., Casadevall, E., Berkaloff, C., Peniguel, G., and Couderc, R., 1988, Cyanobacterial resistant biopolymers. Geochemical implications of the properties of *Schizothrix sp.* resistant material, *Org. Geochem.* **13**:1003–1010.

Collinson, M.E., van Bergen, P. F., Scott, A., and de Leeuw, J. W., 1994, The oil-generating potential of plants from coal and coal-bearing strata through time: A review with new evidence from carboniferous plants, in: *Coal and Coal-bearing Strata as Oil-prone Source Rocks* (A. C. Scott and A. J. Fleet, eds.), Geol. Soc. Special Publ. No. 77, Geological Society, London, 31–70.

Corre, G., Templier, J., and Largeau, C., Influence of cell wall composition on the resistance of microalgae to detergents. 1. Comparative studies of photosynthesis in a TLS-comprising and a TLS-devoid species of *Chorella* (Chlorophyceae) exposed to Dodecyl benzene sulfonate and Triton X-100, *J. Phycology,* submitted.

Couderc, F., Aurelle, H., Promé, D., Savagnac, A., and Promé, J. C., 1988, Analysis of fatty acids by negative ion gas chromatography/tandem mass spectrometry: Structural correlations between α-mycolic acid chain and Δ-5-monounsaturated fatty acids from *Mycobacterium phlei, Biomed. Environ. Mass Spectr.* **16**:317–321.

Daffé, M., Lanéelle, M. A., Puzo, G., and Asselineau, C., 1981, Acide mycolique époxydique: Un nouveau type d'acide mycolique, *Tet. Lett.* **22**:4515–4516.

de Leeuw, J. W., and Largeau, C., 1993, A review of macromolecular organic compounds that comprise living organisms and their role in kerogen, coal, and petroleum formation, in: *Organic Geochemistry, Principles and Applications* (M. H. Engel and S. A. Macko, eds.), Plenum Press, New York, pp. 23–72.

de Leeuw, J. W., van Bergen, P. F., van Aarssen, B. G. K., Gatellier, J-P. L. A., Sinninghe Damsté, J.

S., and Collinson, M. E., 1991, Resistant biomacromolecules as major contributors to kerogen, *Philos. Trans. R. Soc. Lond. Biol.* **333**:329–337.

de Leeuw, J. W., Rijpstra, W. I. C., Baas, M., and van de Ende, H., 1995, Lipid profiles of several algaenan-producing freshwater green microalgae, in preparation.

Dempsey, G. P., Lawrence, D., and Cassie, V., 1980, The ultrastructure of *Chlorella minutissima* Fott et Novahova (Chlorophyceae, Chlorococcales), *Pycologia* **19**:13–19.

Derenne, S., Largeau, C., Casadevall, E., and Laupretre, F., 1987, Mise au point—The quantitative analysis of coals and kerogens by ^{13}C CP/MAS n.m.r., *J. Chim. Phys.* **10**:1231–1238.

Derenne, S., Largeau, C., Casadevall, E., and Berkaloff, C., 1989, Occurrence of a resistant biopolymer in the L race of *Botryococcus braunii*, *Phytochemistry* **28**:1137–1142.

Derenne, S., Largeau, C., Casadevall, E., and Sellier, N., 1990, Direct relationship between the resistant biopolymer and the tetraterpenic hydrocarbon in the lycopadiene-race of *Botryococcus braunii*, *Phytochemistry* **29**:2187–2192.

Derenne, S., Largeau, C., Casadevall, E., Berkaloff, C., and Rousseau, B., 1991, Chemical evidence of kerogen formation in source rocks and oil shales via selective preservation of thin resistant outer walls of microalgae: Origin of ultralaminae, *Geochim. Cosmochim. Acta* **55**:1041–1050.

Derenne, S., Largeau, C., Berkaloff, C., Rousseau, B., Wilhelm, C., and Hatcher, P., 1992a, Non-hydrolysable macromolecular constituents from outer walls of *Chlorella fusca* and *Nanochlorum eucaryotum*, *Phytochemistry* **31**:1923–1929.

Derenne, S., Largeau, C., and Hatcher, P. G., 1992b, Structure of *Chlorella fusca* algaenan: Relationships with ultralaminae in lacustrine kerogens; species- and environment-dependent variations in the composition of fossil ultralaminae, *Org. Geochem.* **18**:417–422.

Derenne, S., Le Berre, F., Largeau, C., Hatcher, P., Connan, J., and Raynaud, J. F., 1992c, Formation of ultralaminae in marine kerogens via selective preservation of thin resistant outer walls of microalgae, *Org. Geochem.* **19**:345–350.

Derenne, S., Metzger, P., Largeau, C., van Bergen, P. F., Gatellier, J. P., Sinninghe Damsté, J. S., and de Leeuw, J. W., 1992d, Similar morphological and chemical variations of *Gloeocapsomorpha prisca* in Ordovician sediments and cultured *Botryococcus braunii* as a response to changes in salinity, *Org. Geochem.* **19**:299–313.

Derenne, S., Largeau, C., and Taulelle, F., 1993, Occurrence of non-hydrolysable amides in the macromolecular constituent of *Scenedesmus quadricauda* cell wall as revealed by ^{15}N NMR. Origin of *n*-alkylnitriles in pyrolysates of ultralamine-containing kerogens, *Geochim. Cosmochim. Acta* **57**:851–857.

Derenne, S., Largeau, C., and Behar, F., 1994, Low polarity pyrolysis products of Permian to Recent *Botryococcus*-rich sediments: First evidence for the contribution of an isoprenoid algaenan to kerogen formation, *Geochim. Cosmochim. Acta* **58**:3703–3711.

Devries, P. J. R., Simmons, J., and Van Beern, A. P., 1983, Sporopollenin in the spore wall of *Spirogyra* (Zygnemataceae, Chlorophyceae), *Acta Bota. Neerl.* **32**:25–28.

Flaviano, C., Le Berre, F., Derenne, S., Largeau, C., and Connan, J., 1994, First indications of the formation of kerogen amorphous fractions by selective preservation. Role of non-hydrolysable macromolecular constituents of Eubacterial cell walls, *Org. Geochem.* **22**:759–771.

Furch, B., and Gooday, G. W., 1978, Sporopollenin in *Phycomyces blakesleeanus*, *Trans. Br. Mycol. Soc.* **70**:307–309.

Gatellier, J-P. L. A., de Leeuw, J. W., Sinninghe Damsté, J. S., Derenne, S., Largeau, C., and Metzger, P., 1993, A comparative study of macromolecular substances of a Coorongite and cell walls of the extant alga *Botryococcus braunii*, *Geochim. Cosmochim. Acta* **57**:2053–2068.

Gelin, F., Gatellier, J-P. L. A., Sinninghe Damsté, J. S., Metzger, P., Derenne, S., Largeau, C., and de Leeuw, J. W., 1993, Mechanisms of flash pyrolysis of ether lipids isolated from the green microalga *Botryococcus braunii* race A, *J. Anal. App.. Pyrol.* **27**:155–168.

Gelin, F., de Leeuw, J. W., Sinninghe Damsté, J. S., Derenne, S., Largeau, C., and Metzger, P., 1994a, The similarity of chemical structures of soluble aliphatic polyaldehyde and insoluble algaenan in the green microalga *Botryococcus braunii* race A as revealed by analytical pyrolysis, *Org. Geochem.* **21**:423–435.

Gelin, F., de Leeuw, J. W., Sinninghe Damsté, J. S., Derenne, S., Largeau, C., and Metzger, P., 1994b, Scope and limitations of flash pyrolysis–gas chromatography–mass spectrometry as revealed by the thermal behaviour of high-molecular-weight lipids derived from the green microalga *Botryococcus braunii*, *J. Anal. Appl. Pyrol.* **28**:183–204.

Gillaizeau, B., Derenne, S., Behar, F., Berkaloff, C., and Largeau, C., 1995, Main source organisms and mode of formation of the Goynuk oil shale (Turkey), *Org. Geochem.*, submitted.

Good, B. H., and Chapman, R. L., 1978, The ultrastructure of *Phycopeltis* (Chroolepidaceae: Chlorophyta). I. Sporopollenin in cell walls, *Am. J. Bot.* **65**:27–33.

Goth, K., de Leeuw, J. W., Püttmann, W., and Tegelaar, E. W., 1988, Origin of Messel oil shale kerogen, *Nature* **336**:759–761.

Gunnison, D., and Alexander, M., 1975, Basis for the resistance of several algae to microbial decomposition, *Appl. Microbiol.* **1975**:729–738.

Hedges, J. I., 1992, Global biogeochemical cycles: Progress and problems, *Mar. Chem.* **39**:67–93.

Hegewald, E., and Schnepf, E., 1974, Contribution to the knowledge of the green alga *Scenesdesmus verrucosus*. Roll, *Arch. Hydrobiol.* **46**(suppl):151–162.

Honegger, R., and Brunner, U., 1981, Sporopollenin in the cell walls of *Coccomyxa* and *Myrmecia* phycobionts of various lichens: An ultrastructural and chemical investigation, *Can. J. Bot.* **59**:2713–2734.

Hull, H. M., Hoshaw, R. W., and Wang, J. C., 1985, Interpretation of zygospore wall structure and taxonomy of *Spirogyra* and *Sirogonium* (Zygnemataceae, Chlorophyta), *Phycologia* **24**:231–239.

Kadouri, A., Derenne, S., Largeau, C., Casadevall, E., and Berkaloff, C., 1988, Resistant biopolymer in the outer walls of *Botryococcus braunii* B Race, *Phytochemistry* **27**:551–557.

Kalina, T., and Puncocharova, M., 1987, Taxonomy of the subfamily Scotiellocystoidaea Fott 1976 (Chlorellaceae, Chlorophyceae), *Arch. Hydrobiol.* **73**(suppl):473–521.

Kister, J., Guiliano, M., Largeau, C., Derenne, S., and Casadevall, E., 1990, Characterization of chemical structure, degree of maturation and oil potential of Torbanites (type I kerogens) by quantitative FTIR spectroscopy, *Fuel* **69**:1356–1361.

Komarek, J., 1987, Species concept of coccal green algae, *Arch. Hydrobiol.* **73**(suppl.):437–471.

Landais, P., Rochdi, A., Largeau, C., and Derenne, S., 1993, Chemical characterization of Torbanites by transmission micro-FTIR spectroscopy. Origin and extent of compositional heterogeneities, *Geochim. Cosmochim. Acta.* **57**:2529–2539.

Largeau, C., Casadevall, E., Berkaloff, C., and Dhamelincourt, P., 1980a, Sites of accumulation and composition of hydrocarbons in *Botryococcus braunii*, *Phytochemistry* **19**:1043–1051.

Largeau, C., Casadevall, E., and Berkaloff, C., 1980b, The biosynthesis of long-chain hydrocarbons in the green alga *Botryococcus braunii*, *Phytochemistry* **19**:1081–1085.

Largeau, C., Casadevall, E., Kadouri, A., and Metzger, P., 1984, Formation of *Botryococcus braunii* kerogens. Comparative study of immature Torbanite and of the extant alga *Botryococcus braunii*, *Org. Geochem.* **6**:327–332.

Largeau, C., Derenne, S., Casadevall, E., Kadouri, A., and Sellier, N., 1986, Pyrolysis of immature Torbanite and of the resistant biopolymer (PRB A) isolated from extant alga *Botryococcus braunii*. Mechanism of formation and structure of Torbanite, *Org. Geochem.* **10**:1023–1032.

Largeau, C., Derenne, S., Clairay, C., Casadevall, E., Raynaud, J. F., Lugardon, B., Berkaloff, C., Corolleur, M., and Rousseau, B., 1990a, Characterization of various kerogens by scanning electron microscopy (SEM) and transmission electron microscopy (TEM)—Morphological relationships with resistant outer walls in extant microorganisms, *Meded. Rijks Geol. Dienst.* **45**:91–101.

Largeau, C., Derenne, S., Casadevall, E., Berkaloff, C., Corolleur, M., Lugardon, B., Raynaud, J. F., and Connan, J., 1990b, Occurrence and origin of "ultralaminar" structures in "amorphous" kerogens of various source rocks and oil shales *Org. Geochem.* **16:**889–895.

Larsen, H., 1984, *Halococcus* Schoop 1935, in: *Bergey's Manual of Systematic Bacteriology,* Vol. 1 (N. R. Krieg and J. G. Holt, eds.), Williams and Williams, Baltimore, pp. 266–267.

Laureillard, J., Largeau, C., Waeghmaeker, F., and Casadevall, E., 1986, Biosynthesis of the resistant polymer in the alga *Botryococcus braunii.* Studies on the possible direct precursors, *J. Nat. Prod.* **49:**794–799.

Laureillard, J., Largeau, C., and Casadevall, E., 1988, Oleic acid in the biosynthesis of the resistant biopolymers of *Botryococcus braunii, Phytochemistry* **27:**2095–2098.

Le Berre, F., 1992, *Formation de Kérogènes par Préservation Sélective de Biopolymères résistants (PR) de Parois de Microorganisms,* Université Pierre et Marie Curie, Paris, Ph.D. thesis.

Le Berre, F., Derenne, S., Largeau, C., Connan, J., and Berkaloff, C., 1991, Occurrence of non-Hydrolysable, macromolecular, wall constituents in bacteria. Geochemical implications, in: *Organic Geochemistry Advances and Applications in Energy and the Nature Environment* (D. A. C. Manning, ed.), University Press, Manchester, England, pp. 428–431.

Lederer, E., Adam, A., Ciorbaru, F., Petit, J. F., and Weitzerbin, J., 1975, Cell walls of Mycobacteria and related organisms: Chemistry and immunostimulant properties, *Mol. Cell. Biochem.* **7:**87–104.

Margulis, L., Hinkle, G., McKhann, H., and Moynihan, B., 1988, *Mychonastes desiccatus* Brown sp. nova (Chlorococcales, Chlorophyta)—an intertidal alga forming achlorophyllous dessication-resistant cysts, *Arch. Hydrobiol.* **78**(suppl.)**:**425–446.

McNeil, M., Daffé, M., and Brennan, P., 1990, Evidence for the nature of the link between the arabinogalactan and peptidoglycan of mycobacterial cell walls, *J. Biol. Chem.* **265:**18200–18206.

McNeil, M., Daffé, M., and Brennan, P., 1991, Location of the mycolyl ester substituents in the cell walls of Mycobacteria, *J. Biol. Chem.* **265:**13217–13223.

Metzger, P., 1994, Phenolic ether lipids with an *n*-alkenylresorcinol moiety from a bolivian strain of *Botryococcus braunii (A race), Phytochemistry* **36:**195–212.

Metzger, P., and Largeau, C., 1994, A new type of ether lipids comprising phenolic moieties in *Botryococcus braunii.* Chemical structure and abundance, geochemical implications, *Org. Geochem.* **22:**801–814.

Metzger, P., David, M., and Casadevall, E., 1987, Biosynthesis of triterpenoid hydrocarbons in the B race of the green alga *Botryococcus braunii.* Sites of production and nature of the methylating agent, *Phytochemistry* **26:**129–134.

Metzger, P., Largeau, C., and Casadevall, E., 1991, Lipids and macromolecular lipids of the hydrocarbon-rich microalga *Botryococcus braunii.* Chemical structure and biosynthesis. Geochemical and Biotechnological importance, in: *Progress in the Chemistry of Organic Natural Products,* Vol. 57 (W. Herz, G. W. Kirby, W. Steglich, and C. Tamm, eds.), Springer-Verlag, Wien, New-York, pp. 1–70.

Michaelis, W., and Albrecht, P., 1979, Molecular fossils of Archaebacteria in kerogen, *Naturwissenschaften* **66:**420–422.

Minnikin, D. E., 1982, Lipids: Complex lipids, their chemistry, biosynthesis and roles, in: *The Biology of the Mycobacteria,* Vol. 2 (C. Ratledge and J. Stanford, eds.), Academic Press, London, pp. 95–184.

Plain, N., Largeau, C., Derenne, S., and Couté, A., 1993, Variabilité morphologique de *Botryococcus braunii* (Chlorococcales, Chlorophyta): Corrélations avec les conditions de croissance et la teneur en lipides, *Phycologia* **32:**259–265.

Puel, F., Largeau, C., and Giraud, G., 1987, Occurrence of a resistant biopolymer in the outer walls of the parasitic alga *Prototheca wickerhamii* (Chlorococcales): Ultrastructural and chemical studies, *J. Phycol.* **23:**649–656.

Raynaud, J-F., Lugardon, B., and Lacrampe-Couloume, G., 1989, Lamellar structures and bacteria as main components of the amorphous matter of source rocks, *Bull. C.R. Explor. Prod. Elf-Aquitaine* **13**:1–21.

Sabelle, S., Oliver, E., Metzger, P., Derenne, S., and Largeau, C., 1993, Variability in phenol moieties in the resistant biomacromolecules of the A and B races of *Botryococcus braunii*. Geochemical implications, in: *Organic Geochemistry* (K. Øygard, ed.) Falch Hurtigtrykk, Oslo, pp. 558–562.

Schouten, S., Ahmed, M., Moerkerken, P., and de Leeuw, J. W., 1995, RuO$_4$ oxidation of algaenans, cutans and kerogens, in preparation.

Serruya, C., Edelstein, M., Pollingher, U., and Serruya, S., 1974, Lake kinneret sediments: Nutrient composition of the pore water and mud water exchanges, *Limnol. Oceanogr.* **19**:489–508.

Shilo, M., 1970, Lysis of blue-green algae by Myxobacter, *J. Bacteriol.* **104**:453–461.

Strohl, W. R., Larkin, J. M., Good, B. H., and Chapman, R. L., 1977, Isolation of sporopollenin from four myxobacteria, *Can. J. Microbiol.* **23**:1080–1083.

Syrett, R. J., and Thomas, E. M., 1973, The assay of nitrate reductase in whole cells of *Chlorella:* Strain differences and the effect of cell walls, *New Phytol.* **72**:1307–1313.

Tegelaar, E. W., de Leeuw, J. W., and Holloway, P. J., 1989a, Some mechanisms of flash pyrolysis of naturally occurring higher plant polyesters, *J. Anal. Appl. Pyrol.* **15**:289–295.

Tegelaar, E. W., Derenne, S., Largeau, C., and de Leeuw, J. W., 1989b, A reappraisal of kerogen formation, *Geochim. Cosmochim. Acta* **53**:3103–3107.

Tegelaar, E. W., Matthezing, R. M., Jansen, B. H., Horsfield, B., and de Leeuw, J. W., 1989c, Possible origin of *n*-alkanes in high-wax crude oils, *Nature* **342**:529–531.

Templier, J., Diesendorf, C., Largeau, C., and Casadevall, E., 1992a, Metabolism of *n*-alkadienes in the A race of *Botryococcus braunii, Phytochemistry* **31**:113–120.

Templier, J., Largeau, C., Casadevall, E., and Berkaloff, C., 1992b, Chemical inhibition of resistant biopolymers in outer walls of the A and B races of *Botryococcus braunii, Phytochemistry* **31**:4097–4104.

Templier, J., Largeau, C., and Casadevall, E., 1993, Variations in external and internal lipids associated with inhibition of the resistant biopolymer from the A race of *Botryococcus braunii, Phytochemistry* **33**:1079–1086.

Tissot, B. P., and Welte, D. H., 1984, *Petroleum Formation and Occurrence,* 2nd ed., Springer, Heidelberg.

Woese, C. R., Magrum, L. J., and Fox, G. E., 1978, Archaebacteria, *J. Mol. Evol.* **11**:245–252.

Yamamoto, Y., and Suzuki, K., 1990, Distribution and algalysing activity of fruiting myxobacteria in lake Suwa, *J. Phycol.* **26**:457–462.

Zarsky, V., Kalina, T., and Sulek, J., 1985, Notes on the sexual reproduction of *Chlamydomonas geitleri, Arch. Protistenk,* **130**:343–353.

3

Interaction between Arbuscular Mycorrhizal Fungi and Plants

Their Importance in Sustainable Agriculture in Arid and Semiarid Tropics

D. J. BAGYARAJ and AJIT VARMA

1. Introduction

Sustainability refers to productive performance of a system over time. It implies use of natural resources to meet the present needs without jeopardizing the future potential. The concept has an undefined time dimension. The magnitude of the time dimension depends on one's objectives, being shorter for economic factors and longer for concerns pertaining to environment, soil productivity, and land degradation. The shorter time dimension is generally less than a decade, while the longer time span may be up to five decades or more. The time dimension is also clearly addressed in the definition of sustainability adopted by the Technical Advisory Committee of the Consultative Group of International Agricultural Research: "Successful management of resources for agriculture to satisfy changing human needs while maintaining or enhancing the quality of the environment and conserving natural resources" (TAC, 1989).

A sustainable system must be sustainable both ecologically and economically. From an agroecological perspective, sustainability may be defined as a measure of productivity over time per unit input of nonrenewable or a limiting resource.

Most soils of the tropics are of low inherent fertility. Crop yields are expectedly low, unless the nutrient status is enhanced by a regular or substantial addi-

D. J. BAGYARAJ • Department of Agricultural Microbiology, University of Agricultural Sciences, Bangalore 560 065, India.   AJIT VARMA • School of Life Sciences, Jawaharlal Nehru University, New Delhi 110 067, India.
Advances in Microbial Ecology, Volume 14, edited by J. Gwynfryn Jones. Plenum Press, New York, 1995.

tion of nutrients (fertilizers, manures, etc.). Soils of arid and semiarid regions are prone to production constraints imposed by physical and climatic processes, e.g., crusting, compaction, hard-setting, drought, erosion by wind and water, and high soil temperature. Productivity depends on input. Presently, agriculture in arid and semiarid tropics has low rates of energy input and low-to-moderate levels of traditional/subsistence farming. Much of the low-input or traditional agriculture of the developing world is not sustainable. The population pressure on the ecosystem in which these low-input systems are used results in severe land degradation. In view of the need to increase food production immediately and substantially and to ameliorate soil degradation, there is presently an enthusiastic interest in sustainable farming systems in the tropics. Some of the endeavors to achieve these in arid or semiarid ecosystems are: conserving water, controlling erosion, use of improved cultivar and cropping systems, enhancing soil fertility, application of fertilizers as organic amendments, and increasing biological activity of soil flora and fauna. This approach reduces but does not eliminate the need for fertilizer and pesticides, and farm machinery and motorized equipment.

Currently there is considerable resistance against the use of chemical pesticides and fertilizers, because of their hazardous influence on the environment, and on soil, plant, animal, and human health. Hence, use of biofertilizers and biocontrol agents are recommended in practical agriculture. Nitrogen and phosphorus are two important plant nutrients. There are a large number of soil bacteria that are capable of fixing atmospheric nitrogen and making it available for plant growth. Bacteria like rhizobia, which live in association with roots of legumes, and the actinomycete *Frankia,* which lives in association with casuarina, alder, and so forth, are examples of symbiotic nitrogen fixation. Bacteria like *Azotobacter* and *Azospirillum* and cyanobacteria like *Nostoc* and *Anabaena* are examples of organisms living freely in soil but capable of fixing atmospheric nitrogen. Phosphorus, which is the other major plant nutrient, is less mobile in soil solution. In most of the tropical soils, phosphorus is fixed and is in a form not readily available for plant growth. There are a number of fungi and bacteria that solubilize unavailable forms of phosphate and make it available for plant growth. There are certain fungi that form symbiotic association with the roots of plants and help in the uptake of phosphate from the labile pool. These fungi, which form symbiotic association with roots of plants, are referred to as mycorrhizal fungi, and the association itself is referred to as mycorrhiza. Five broad groups of mycorrhizas are: the ectomycorrhizas, the arbuscular mycorrhizas, the ericaceous mycorrhizas, the ectendomycorrhizas, and the orchidaceous mycorrhizas (Harley, 1994; Marx et al., 1994). Ectomycorrhiza are commonly associated with temperate tree species; ericoid mycorrhiza commonly are found in plants belonging to the family Ericaceae; and orchid mycorrhizal associations are found in orchids. The most common mycorrhizal association found in tropics and cultivated crop plants throughout the world is the arbuscular mycorrhizal fungi

(AMF) type, and ectendomycorrhizas type. In the arid and semiarid regions, the last two mycorrhizal types are predominant (Varma *et al.*, 1994).

2. Arbuscular Mycorrhiza and Their Importance

Arbuscular mycorrhiza, probably evolved with the Devonian land flora, are formed by most angiosperms, gymnosperms, ferns, and bryophytes (Trappe, 1987). Among the different 2,600,000 plant known species, 240,000 have been estimated to have the potential to form mycorrhizal associations with 6000 fungal species, representatives of the Zygomycotina, Ascomycotina, Basidiomycotina, and Deuteromycotina (Wyss and Bonfante, 1993). According to Gerdemann (1975), it is easier to list plant families that do not form such mycorrhiza than to list those that do. Families not forming arbuscular mycorrhiza include Betulaceae, Commelinaceae, Ericaceae, Fumariaceae, Orchidaceae, Pinaceae, and Urticaceae. Families that rarely form mycorrhiza of this type include Brassicaceae, Polygonaceae, Cyperaceae, and Chenopodiaceae. Families that form both ectomycorrhiza and endomycorrhiza include Caesalpiniaceae, Fagaceae, Juglandiaceae Myrtaceae, Salicaceae, and Triliaceae (Gerdemann, 1975). In addition to the widespread distribution of mycorrhiza throughout the plant kingdom, the association is geographically ubiquitous, occurring in plants from the artic to the tropical regions over a broad ecological range, from aquatic to desert environments (Mosse *et al.*, 1981). These mycorrhiza are formed by nonseptate zygomycetous fungi belonging to the genera *Acaulospora, Gigaspora, Scutellospora, Entrophospora, Glomus,* and *Sclerocystis* in the family Endogonaceae of the order Endogonales (Table I). At present, over 160 species of fungi forming

**Table I. Endomycorrhizae:
New Ordinal and Family
Structure Concept[a]**

Glomales (order)
  Glominaeae (suborder)
    Glomaceae (family)
      Glomus (genera)
      Sclerocystis
    Acaulosporaceae
      Acaulospora
      Entrophospora
  Gigasporineae
    Gigasporaceae
      Gigaspora
      Scutellospora

[a]Compare Morton *et al.* (1995);
Walker (1992).

arbuscular mycorrhiza have been described (Kendrick and Berch, 1985). Many are cosmopolitan in distribution, but some may be strictly tropical like *Acaulospora foveata* and *A. tuberculata*. The arbuscular endophytes are not host-specific, although evidence is growing that certain endophytes may form preferential associations with certain host plants (Bagyaraj, 1992).

Simon *et al.* (1992a,b) have sequenced ribosomal DNA genes from 12 species of AMF, and their phylogenetic analysis confirms the existence of three families (Walker, 1992; Walker and Trappe, 1993): *Glomaceae, Acaulosporaceae,* and *Gigasporaceae*. They have estimated the origin of AMF of 383–462 million years ago (Simon *et al.*, 1993a), which is consistent with the hypothesis that endomycorrhizal fungi were instrumental in the colonization of land plants by ancient plants (Harley, 1989). *Glomaceae* appeared first, followed by *Acaulosporaceae* and *Gigasporaceae*, and diverged from each other in the late Paleozoic, 250 million years ago. The split between *Entrosphospora* and *Acaulospora* occurred during the Cretaceous (Simon *et al.*, 1993b).

The fungi forming mycorrhiza have the largest known resting spores of any fungi, their size ranging from 20 to 400 μm. They can be isolated from soil by the wet sieving and decantation method (Gerdemann and Nicolson, 1963). A root system colonized by these fungi does not show any morphological variations from the normal root system. Hence, the status of a root system in this respect can only be known after staining the roots with trypan blue (Phillips and Hayman, 1970). The presence of vesicles and arbuscules is the diagnostic criterion for identifying AMF in a root. The mycorrhizal fungus biotrophically colonizes the parenchymatous root tissue, forming intraradical aseptate hyphae, highly branched haustoriumlike intracellular structures termed arbuscules, and in some species lipid-rich intracellular ovoid bodies called vesicles. During this process, several hyphae proliferate in the soil and build up an extended extramatrical mycelium, which may extend distances up to 14 cm from the root system, but no longer extensions are expected, and forms a network. Internal and external hyphae are in contact with up to 10 entry points/cm root (Fortuna *et al.*, 1992). Mycorrhizal fungal mycelium appears to be more resistant than the root itself to abiotic stresses such as drought, toxicity of elements, and soil acidity. AMF elicit specific responses from plants and soil acidity.

It is a well-established fact that the mycorrhizae improve plant growth. The main effect of these fungi in improving plant growth is through improved uptake of nutrients, especially phosphorus, which is due to the exploration by the external hyphae of the soil beyond the root hair and phosphorus depletion zones (Gerdemann, 1975; Tinker *et al.*, 1994). Hyphae are also known to absorb phosphorus from lower concentrations compared to nonmycorrhizal roots. Experiments with ^{32}P-labeled phosphate indicate that the mycorrhizal hyphae obtain their extra phosphate from the labile pool rather than by dissolving insoluble phosphate (Raj *et al.*, 1981). The better utilization of sparingly soluble rock phosphate is explained by the hyphae making closer physical contact with the ions that dissociate at the particle surface (Hayman, 1983). The mycorrhizal fungi produce phosphatases that allow utilization of organic phosphorus, especially under humid tropical conditions where hyphae can be seen to be in close contact with finely divided litter, when the acid soil reaction reduces the absorption of organic phosphates on soil surfaces (Tinker, 1975; Tinker *et al.*, 1994). Arbuscular mycorrhiza have also been reported to improve the uptake of minor elements and water. They also produce plant hormones, increase the activity of nitrogen-fixing organisms in the root zone, and reduce the severity of disease caused by root pathogens (Bagyaraj, 1992; Garbaye, 1991; Pankow *et al.*, 1991). Thus, the benefits the plant would derive from mycorrhizal inoculation seem to be very many (Fitter, 1991).

The beneficial effect of mycorrhiza on plant growth is particularly spectacular in P-deficient soils and because of the generally low availability of P in tropical soils, the potential for the exploitation of mycorrhiza in agriculture seems to be much greater than in temperate soils. Carnet and Diem (unpublished data) found that seedlings of *Acacia raddiana* failed to grow in a sterilized soil from Senegal unless they were mycorrhizal or supplemented with a soluble phosphate source. In these conditions, they grew well even though they were not nodulated. This suggests that in some tropical soils, P deficiency may be even more important than N deficiency, as a factor limiting plant growth. Further, there are many other obligatory mycotropic plant species that will not survive if not colonized by mycorrhizal fungi. Obligate mycotrophs do not respond to fertilizer applications, suggesting that their obligate dependence on mycorrhiza is not only for nutrients but also for some other factors. As Black (1980) pointed out, mycorrhiza will be important factors for increasing plant productivity in developing tropical countries for several reasons: the intrinsically low availability of P or highly P-fixing ability of many tropical soils, the difficulties for locally manufacturing soluble phosphate fertilizers and the high cost of importation, and the mycotropic habits of many tropical plant species of economic importance, especially legumes.

Further, because of the greater fungal activity at high temperatures prevalent in tropical countries, mycorrhiza research may have particular relevance in the tropics. It has the added advantage that techniques are simple and require little specialized equipment. Because of the obligate symbiotic nature, AMF are now maintained and mass produced as "pot cultures" on suitable host plants, which is labor-intensive. Labor needed for maintenance of pot cultures, inoculum production, and field inoculation work will be available without much difficulty in tropical countries.

3. Multiplication of AMF

Techniques are available for the production of arbuscular mycorrhizal inoculum in an almost sterile environment through nutrient film techniques, circulation hydroponic culture systems (Jarstfer and Sylvia, 1995), aeroponic culture systems, root organ culture, and tissue culture (Nopamornbodi *et al.*, 1988; Becard and Piche, 1992). For large-scale trials, however, the only convenient method of producing large quantities of inoculum is by the traditional pot culture technique developed about 30 years ago by Mosse and Gerdemann (see Sieverding, 1991; Wood, 1985).

Pot cultures are complex systems comprising host plants, mycorrhizal fungi, soil microflora and microfauna, and supporting soil. Soil cultures can sometimes harbor root pathogens, and thus can also create transport problems, and many countries restrict soil importation despite assurances of sterility. The problem is therefore to manipulate the host, substrate, and environment in such a way as to produce pot cultures yielding mycorrhizal inoculum with high inoculum potential, few contaminants, longer shelf life, and easy transportation. Several host plants, including Sudan grass (*Sorghum bicolor* var. *sudanense*), bahia grass (*Paspalum notatum*), guinea grass (*Panicum maximum*), cenchrus grass (*Cenchrus ciliaris*), clover (*Trifolium subterraneum*), strawberry (*Fragaria* sp.), sorghum (*Sorghum vulgare*), maize (*Zea mays*), onion (*Allium cepa*), coprosma (*Coprosma robusta*), and *Coleus* sp., have been studied for their suitability in producing mycorrhizal inoculum. It was reported that Rhodes grass is the best host for mass production of *Glomus fasciculatum* in the tropics (Sreenivasa and Bagyaraj, 1988a). In view of the disadvantages of using soil as the substrate for producing mycorrhizal inocula, it is desirable, where possible, to produce mycorrhizal inoculum in a partially, if not completely, artificial substrate. We have found 1:1 perlite:Soilrite (Soilrite is a commercial preparation with perlite, vermiculite, and peat moss of 1:1:1 proportion by volume) to be the best substrate for mass production of *G. fasciculatum* (Sreenivasa and Bagyaraj, 1988b). Dehne and Backhaus (1986) suggested the use of expanded clay for multiplying mycorrhizal fungi. Moisture content and temperature of the substrate, as well as irradiance, mineral content, pot size, and presence of mycoparasites and associated organisms can all influence mycorrhizal inoculum production (Bagyaraj, 1984), and hence should be evaluated for each fungal isolate.

Pot cultures of arbuscular mycorrhizal inoculum are usually harvested after about 8 months. The inoculum, consisting of the substrate and the roots chopped into small bits, is air-dried and packed in polyethylene bags and stored. Ferguson and Woodhead (1982) observed that *G. fasciculatum* inoculum produced in this way was viable after 4 years storage. We found such inoculum viable after 20 months even at 40 °C (Harinikumar *et al.*, 1982). Attempts to freeze-dry mycorrhizal inoculum as a method of storage have not been successful (Crush and Pattison, 1975); however, it can be preserved for up to 6 years using L-drying or cryopreservation techniques (Tommerup and Kidby, 1979).

4. Important Factors Influencing Arbuscular Mycorrhizal Associations in Arid and Semiarid Regions

Even though several physical, chemical, and biological factors influence mycorrhizal associations, the most important ones in the arid and semiarid regions are temperature, light, water stress, and the biological factors. Hence, only these three factors are discussed.

4.1. Temperature

Temperature has been shown to have a significant influence on colonization and sporulation by mycorrhizal fungi under greenhouse conditions. Higher temperature generally results in greater root colonization and increased sporulation. Studying the effects of temperature on mycorrhizal establishment, Schenck and Schroder (1974) observed that the maximum arbuscule development occurred near 30 °C. Daniels and Trappe (1980) observed that optimum temperature for the germination of *Glomus* and *Acaulospora* spores is around 20–25 °C, while *Gigaspora* had a much higher optima. These studies suggest that increased soil temperature hastens the development of mycorrhiza. This may explain the slow development of infection in agricultural crops in temperate soils (Black and Tinker, 1979), where soil temperatures are low compared to tropical soils. Since most species of endomycorrhizal fungi are worldwide in their distribution, it is possible that strains and species may be temperature-adapted. This is suggested by the work of Schenck *et al.* (1975), who found that two isolates of *Glomus* from Florida germinated best at 34 °C, whereas one from Washington had an optimum of 20 °C.

4.2. Light

Arbuscular mycorrhizal fungi obtain their carbon source from the host plant, and thus rely on both the photosynthetic ability of the plant and the translocation of photosynthates to the root. Hence, light can strongly affect the development of mycorrhiza. The stimulatory effect of light on the development of mycorrhiza has been shown by Furlan and Fortin (1977). Shading not only reduces root colonization and spore production but also the plant response to mycorrhiza (Gerdemann, 1968). It was earlier postulated that day length may play an important role in mycorrhiza development, and this was confirmed by Daft and El-Giahmi (1978). However, the effect of light on mycorrhiza seems to depend on the photosensitivity of the species of host plant (Redhead, 1975). In fact, a photoperiod of 12 hr or more may be more important than light intensity in providing high levels of root colonization; but if suitable day length is provided, increased light intensity may still increase colonization. It would be particularly interesting to examine this aspect in plantation crops like coffee, cardamom, and so forth that are cultivated under shade trees in the tropics (Bagyaraj, 1990).

4.3. Water Stress

Arbuscular mycorrhizal fungi occur over a wide range of soil water regimens. Colonization has been found in arid regions in xerophytes (Khan, 1974; Singh and Varma, 1981). Mycorrhizal fungi can improve the drought resistance of plants (Puppi and Bras, 1990), although some reports indicate that drought resistance is unaffected or decreased by mycorrhiza (Simpson and Daft, 1990). In greenhouse studies, mycorrhiza have been shown to increase the drought resistance of cultivated crops such as wheat (*Triticum aestivum* L.), soybean [*Glycine max* (L.) Merr.], onion (*Allium cepa* L.), pepper (*Capsicum annuum* L.), red clover (Trifolium pratense L.), as well as several native plant species (Sylvia and Williams, 1992). Improved crop nutritional status during periods of soil water deficit reportedly enhances drought resistance (Pai *et al.*, 1993). Other factors associated with mycorrhizal colonization are changes in leaf elasticity (Auge *et al.*, 1987a), improved water and turgor potentials, and maintenance of stomatal opening and transpiration (Auge *et al.*, 1987b) and increased rooting length and depth (Kothari *et al.*, 1990), which may also influence drought resistance.

Most experiments with mycorrhiza have been conducted in controlled greenhouse or growth chamber environments. There is relatively sparse information on the function of mycorrhiza in field environments. Fitter's (1988) field studies suggested that mycorrhiza improve the drought resistance of plants. White *et al.* (1992) showed, in a study of the Red Desert of Wyoming, that irrigation schedule was more important than irrigation rate for enhancing establishment of functional mycorrhizal biomass. Sylvia *et al.* (1993) conducted field trials, over three seasons, to directly test the effect of mycorrhiza on water-stressed corn (*Zea mays* L.). Plots were fumigated with methyl bromide and fertilized before each growing season. Two inoculation treatments (inoculated or not with *Glomus etunicatum*) were applied to subplots of three water management regimens each year. Inoculum was placed in a furrow 10-cm deep at an average rate of 1500 propagules per meter of row. Inoculation increased both the number of leaves produced by 45 days and the number of the last leaf that had collared by 52 days. Inoculation also increased the concentrations of P and Cu in both shoots and grain on all measurement dates. Grain and total aboveground biomass yields increased linearly with irrigation, and a positive response to mycorrhizal inoculation was constant across irrigation levels of both grain yield and total biomass. Due to the smaller size of water-stressed plants but a consistent growth response to inoculation across water treatment, the proportional response of corn to inoculation with *G. etunicatum* increased with increasing water stress. This study provides the first direct evidence that mycorrhizal fungi improve resistance of plants to water stress in the field.

4.4. Host Plant

The presence or absence of a host plant obviously plays an important role in whether or not colonization and subsequent sporulation will occur. Nonhost plants such as Chenopodiaceae and Brassicaceae species can become minimally colonized by mycorrhizal fungi (Kruckelmann, 1975), particularly when grown in the presence of host plants (Ocampo et al., 1980). The influence of nonhost plants on the colonization of host plants has been studied with contradictory results. For example, the presence of nonmycorrhizal plants has resulted in reduced colonization of mycorrhizal host plants possibly because of toxic nonhost exudates (Iqbal and Qureshi, 1976). In contrast, Ocampo et al. (1980) detected no reduced colonization of mycorrhizal plants together with nonmycorrhizal hosts. In fact, onions became more colonized when grown with nonhost rutabagas than when grown alone; similar results were observed in barley cropped together with rape (Neeraj, 1992). These experiments were conducted in the greenhouse. If planting host and nonhost species together results in more rapid utilization of soil nutrients because there is more root per pot, an increase in colonization might not be unexpected. The presence of a so-called nonhost seems to be better for a mycorrhizal fungus than no plant at all (Harinikumar and Bagyaraj, 1988), since mycorrhizal hyphae have been observed colonizing the root surfaces of nonhosts (Ocampo et al., 1980), thereby procuring an ecological niche in which to survive in the absence of a host root. Not only does this indicate that these nonhost roots are not releasing toxic factors into the surrounding soil, it also raises the possibility of some saprophytic growth in rhizosphere soil by mycorrhizal fungi.

That the crop species itself can exert a selective effect on which mycorrhizal species in a mixed indigenous population become predominent is well illustrated in a study of Schenck and Kinloch (1980). They found marked difference in population of mycorrhizal fungi between different crops grown for 7 years in monoculture. Three species of *Gigaspora* were most prevalent in association with soybean roots, in contrast to cotton and peanut where *Acaulospora* species dominated and bahia grass where two species of *Glomus* were most numerous. The largest number of mycorrhizal species was found with sorghum.

Although AMF have extremely wide host ranges (Mosse, 1973), the existence of host preference has been suggested by many researchers (Bagyaraj et al., 1989; Neeraj et al., 1991). The preferential association between certain plant and fungal species can be evaluated with respect to combinations that provide the greatest plant growth stimulation. At present, we do not have much explanation for the variation in mycorrhizal dependency of different host plants. One possibility is that plants with coarse roots and relatively fewer root hairs are more dependent on mycorrhiza compared to those plants with fine roots and long root hairs (St. John, 1980; Allen, 1992; Sylvia and Williams, 1992). Recent investigations suggest that mycorrhizal colonization could be treated as a genetic trait (Krishna *et al.*, 1985; Mercy *et al.*, 1990). Like resistance or susceptibility of crop plants to infection by fungal pathogens, the extent of root colonization is a plant-heritable trait. There are no reports at present of exclusive colonizations between a mycorrhizal fungus and a host plant. The factors that determine host–symbiont affinities have not been studied and no doubt are of considerable importance. Reduction in plant size by pruning and defoliation of plants (Sreenivasa and Bagyaraj, 1988a) can decrease mycorrhizal root colonization and sporulation.

4.5. Arbuscular Mycorrhizal Fungal Efficiency

Mycorrhizal fungi are not always equally infective to any one plant species, and they certainly vary in their physiological interactions with different plants, and hence in their effects on plant growth. Species and strains of mycorrhizal fungi have been shown to differ in the extent to which they increase nutrient uptake and plant growth (Powell *et al.*, 1980; Powell and Bagyaraj, 1982; Sieverding, 1991; Varma, 1995). These observations have led to introduction of the term "efficient" or "effective" strains (Abbott and Robson, 1981). Generally, those fungi that infect and colonize the root system more rapidly are considered to be efficient strains (Munns and Mosse, 1980). The ability to form extensive external hyphae and the ability to absorb P from soil solution and improve plant growth are also important in determining the effectiveness of mycorrhizal endophytes (Bagyaraj, 1992; Smith, 1995). Comprehensive studies are needed with as many species and strains as possible to see whether better ones exist in nature or can perhaps be created by techniques like genetic manipulation.

4.6. Arbuscular Mycorrhizal Fungal Dormancy

Spore dormancy in mycorrhizal fungi was first reported by Tommerup (1987). *Acaulospora laevis* spores are dormant for about 6 months, *Gigaspora calospora* for 6–12 weeks, and *Gigaspora decipiens, Glomus caledonicum,* and *G. monosporum* for a few days. Dormancy periods were much shorter in dry soils than in moist soils. Dormancy can be broken by dessication or cold treatment. Tommerup (1983) suggests that short dormancy periods could prevent spores from germinating immediately after formation around the root; such germination would thwart their roles as infection agents for later crops. Long dormancy may protect spores against early false breaks in a season; selection pressure may have operated to select such a population of mycorrhizal fungi. More studies on dormancy of mycorrhizal fungi and its ecological implications are needed.

4.7. Other Soil Microorganisms

Of the various microorganisms colonizing the rhizosphere, mycorrhizal fungi occupy a unique ecological position, since they are partly inside and partly outside the host. The part of the fungus within the root does not encounter competition with other soil microorganisms. This advantage enables them to achieve a more functional biomass in intimate contact with the root, and thus increases their chances of exerting a greater effect on plants. The different aspects of biological interactions with mycorrhizal fungi have been reviewed (Linderman, 1988, 1992).

Mycorrhizal fungi markedly improve nodulation and nitrogen fixation by legume bacteria mainly by providing the high phosphorus requirement for the fixation process. Mycorrhizal colonization also allows the introduced populations of beneficial soil organisms like *Azotobacter* and phosphate-solubilizing bacteria to maintain higher numbers than around nonmycorrhizal plants and to exert synergistic effects on plant growth. In general, mycorrhiza decreases the severity of root diseases. The various mechanism of suppression of pathogens by mycorrhiza has also been reviewed (Bagyaraj, 1984; Pankow *et al.,* 1991). Secilia and Bagyaraj (1987) suggest one more possible mechanism of suppression of root pathogens by mycorrhizal fungi, that is, by stimulation of antagonistic actinomycetes in the rhizosphere.

Research on parasites and predators of mycorrhizal fungi should be intensified, since they are probably going to play an important role in the success or failure of mycorrhizal inoculum trials in the field and also on the survival and persistence of indigenous and introduced endophytes. Most of the biological interaction studies have been conducted in pots, and hence there is a need for exploiting this potential under field conditions. To begin with, experiments could be done with plants important in agriculture, horticulture, and forestry that are usually raised in the nursery and then planted out in the field. As only small quantities of the mycorrhizal inoculum would be required, the method could be applied immediately to practical farming without much difficulty. The literature published so far leads to conclusion that the main ecological significance of mycorrhizas concerns the protective rather than the productive phase of plant individual as well as of an ecosystem. This phase of the life cycle is all too often neglected, not only by those carrying out research on mycorrhizas but also by biologists in general. Scientists tend to focus on the early phase of life cycle—the productive phase of juvenile growth. Comparatively little is known about the protective phase of maturity. The criteria for measuring fitness change during the life cycles of individuals and ecosystems. During the productive phase of young plants and of early stages in a succession, the relative growth rate may be a useful measure of fitness, and the supply of abiotic resources may be the main factor limiting it. However, the limits to growth shift from the outside to the inside with time. Genetic constraints on shape and structure start to take effect, and the maintenance and cycling of resources already incorporated become more important than the acquisition of new ones (Atkinson, 1992; Garbaye, 1991). At this stage, competition for new resources among individuals in an ecosystem may be replaced by collaboration to maintain the resources acquired, and it is in this almost unexplored domain where the mycorrhizal symbiosis might have its key function.

5. Spread of Arbuscular Mycorrhizal Mycelia from the Host Root

The rate of mycorrhizal hyphal growth from plant roots has been studied by some workers. Powell (1979) determined that an efficient mycorrhizal fungus would advance 65 m in 150 years, or 0.43 m/year under greenhouse conditions. This slow extension limits the potential importance of this mode of dispersal. Powell further demonstrated that the introduced fungus spreads more readily through soils that already contain low populations of mycorrhizal fungi, particularly if the population has been depressed by cultivation to nonhost crops. More recently, it had been shown that plant species and root density may significantly influence the rate of mycorrhizal fungus spread (Warner and Mosse, 1982). In clover, the greatest rate of spread of *Glomus fasciculatum* was 1 cm/week, while in fescue, *G. fasciculatum* spread at only 0.7 cm/week. These experiments showed that root density is most critical when plants are young and root density is low. In fact, supraoptimal root density was achieved in fescue (a grass that develops an extensive root system) and the rate of fungal spread was reduced as plant size increased. A similar supraoptimal root density was not achieved in clover because of the less extensive root system. The previous experiments were conducted in the greenhouse in fumigated soils, and it is difficult to project the rate of mycelial growth and mycorrhizal fungus spread through field soils. Colonization by morphologically similar indigenous mycorrhizal fungi in nonsterile field soils make this type of study difficult. However, using sporulation by nonindigenous mycorrhizal fungi as proof of spread, Mosse *et al.* (1982) demonstrated that *Glomus caledonicum* was able to move 7 to 13 cm from an inoculation point after 13 weeks. No correlation was observed between rate of spread and plant size, but spread rate was greater in nonsterilized plots than in those receiving formalin treatments. In these experiments, host species also significantly affected spread rate.

Whether mycorrhizal fungi grow in a directed way, i.e., toward a root stimulus or randomly in soil, has been debated. Directed growth would most likely make optimum use of energy supplies in the spore and would increase the number of infective hyphal strands that reach a host (Wallace, 1978; Read, 1992). Powell (1976), using the buried slide technique in partially sterilized soil, demonstrated little or no attraction of mycorrhizal hyphae to root until random contact occurred, except with hyphae from honey-colored spores (*Acaulospora laevis*), which frequently grow toward the roots. Koske (1982) has demonstrated chemotactic attraction of hyphae of *Gigaspora margarita* to host roots suspended above germinating spores; the attractant is probably a volatile substance. Whether such chemotactic substances are produced under field conditions and can direct mycelial growth in the field have not been studied.

6. Arbuscular Mycorrhizal Fungal Association in Desert Environment

AMF are part of microbial community in the desert soil. They occupy an especially critical niche in the soil system because of their high metabolic rate and strategically diffused distribution, usually in the upper soil layers. In the Red Desert of southwest Wyoming, the Aridisols retained up to 81 spores per gram soil at a depth of 140 cm. The AMF biomass attenuation in relatively deep horizons of these arid soils may still allow adequate formation of propagules for use in reclamation of disturbed lands (White *et al.*, 1989). Plants growing with mycorrhiza in arid and semiarid areas and their significance in stressed environments have been studied (Hirsh, 1984; Mikola, 1980; Singh and Varma, 1980, 1981; Mathew *et al.*, 1991). These fungi play a significant role in plant growth (Fitter, 1991; Neeraj *et al.*, 1991) in sand dune stabilization (Koske and Polson, 1984) and are an important consideration in maximizing the range land productivity because mycorrhiza-dependent plants cannot succeed without their fungal associates (Trappe, 1981).

Many plants require AMF association in order to survive on disturbed lands, but most successful pioneers of harsh sites and new soil are probably nonmycorrhizal plants (Loree and Williams, 1984). Marx *et al.* (1994) hypothesized that many such plants are successful because they do not require infection and that succession of disturbed ecosystems might be influenced by the quality and quantity of mycorrhizal inoculum present in the soil.

In perennial native ecosystems, the soil disturbances alter the process of colonization, propagule population, and length of colonized root, leading to complete loss of infectivity (Johnson and Pfleger, 1992). The presence of grasses may result in high numbers of mycorrhizal roots in the soil, and combined with high numbers of spores at these sites could give a large reserve of mycorrhizal propagules, while the disturbance reduces infectivity. Jasper *et al.* (1989a,b) reported that in a forest soil and heathland soil, the percentage of root length colonization of test plants was almost halved if the soil were disturbed; but in pasture soils, many more propagules survived disturbance. Evans and Miller

(1990) have given evidence indicating that the undisturbed mycelial network may also increase the nutrient absorption independent of the degree of colonization, though the hyphae in soil are especially vulnerable to disturbances.

7. Selection of Efficient AMF

Mycorrhizal workers rarely provide the rationale for the selection of particular mycorrhizal endophytes in their experiments. There are a few favorite spore types for field inoculation, including *Glomus fasciculatum, G. mosseae, G. etunicatum, G. tenue,* and *Gigaspora margarita.* These fungi were probably chosen because of their ready availability. Ideally, mycorrhizal fungi selected for inoculation into agricultural soils must be able to enhance nutrient uptake by plants and persist in soils. Many of the characteristics required by inoculant fungi correspond to those essential for the success of inoculant strains of *Rhizobium* for legumes (Chatel *et al.,* 1968). These characteristics are infectiveness, effectiveness, extent of colonization, and survival in the rhizosphere and in soil. An excellent discussion on the selection of mycorrhizal fungi for possible use in agriculture has been published by Abbott and Robson (1982).

Mycorrhizal fungi are known to improve plant growth by increasing nutrient uptake, primarily by increasing the volume of soil explored (Bethlenfalvay, 1992; Harley, 1994). Therefore, characteristics that could be associated with differences between them in their effectiveness and increasing nutrient uptake are (1) the ability to form extensive well-distributed mycelium in soil, (2) the ability to form extensive colonization, (3) the efficiency of absorbing phosphorus from soil solution, and (4) the time that hyphae remain effective in transporting nutrients to the plant. It was suggested by Abbott and Robson (1982) that the initial selection of inoculant fungi may be conducted under controlled conditions; after that, selected mycorrhizal fungi can be evaluated in the field.

There have been a few attempts to select for efficient mycorrhizal fungi on a rational basis, such as the ability to stimulate plant growth. Govinda Rao *et al.* (1983) suggested that several fungi can be screened for symbiotic response using a test host through pot culture, followed by microplot and then field trials. Seedlings were first raised in unsterile nursery soils containing indigenous mycorrhizal fungi to which different test mycorrhizal fungi had been added. It was important to add the same number of infective propagules of different fungi based on most-probable-number counts (Porter, 1979). Once the infection had developed, the seedlings were transplanted to experimental field plots and plant growth was monitored. This technique followed precisely the procedure used by farmers in India and many Third-World countries. It is extremely good for selecting fungi for the preinoculation of transplanted crops raised in unsterile soils. The fungi thus selected will compete well with indigenous AMF. In this selection process it is essential to include mycorrhizal fungi isolated from the root zone of the test host. This procedure for selecting efficient fungi has led to the selection of inoculant mycorrhizal fungi for many economically important forest tree species, such as *Leucaena leucocephla, Tamarindus indica, Acadia nilotica, Calliandra calothyrsus,* and *Casuarina equisetifolia,* and root stocks of citrus and mango (Bagyaraj, 1992; Vasanthakrishna and Bagyaraj, 1993; Balakrishna Reddy and Bagyaraj, 1994). Inoculation of citrus root stocks with efficient mycorrhizal fungi made the plants ready for budding 4–6 months early, a feature of great importance in citriculture.

Schubert and Hayman (1986) have also outlined procedures for selecting mycorrhizal fungi for crops sown directly in the field and pastures. These experiments can be criticized because fungi were screened in greenhouse pot experiments, but they are still useful. Before laying down a field trial with onions, it was determined in a pot experiment that a mixture of *Glomus* spp. was the best inoculant for the onion cultivar Pukekohe Long Keeper. This *Glomus* inoculum was then shown to be the best inoculant in a field trial (Powell and Bagyaraj, 1982), amply vindicating the effort of prior fungal selection even for a directly field-sown crop.

8. Mixed Inoculation and Succession of Mycorrhizal Fungi

While screening and selecting efficient mycorrhizal fungi for inoculating a particular crop or tree, it is likely that more than one fungus could be identified as equally good for inoculation. For *Leucaena leucocephala, Glomus mosseae* was found to be the best mycorrhizal fungus for inoculation; the next best fungus was *Gigaspora margarita*. The two fungi, *G. mosseae* and *G. margarita,* were inoculated singly as well as together and the growth response of *L. leucocephala* was monitored (Mallesha and Bagyaraj, 1990). The result showed that the two fungi did not compete with each other for root colonization; in fact, the effect was additive as reflected in root colonization percentage as well as plant biomass. It was thus deduced that mixed inoculation with two mycorrhizal fungi could be advantageous, as soil/agroclimate if hostile to one fungus may be conducive to the other fungus in forming effective symbiosis.

Succession of ectomycorrhizal fungi in forest tree species like birches and pines has been reported (Mason *et al.,* 1982). In *Eucalyptus dumosa* seedlings, it was observed that early infection was entirely by arbuscular mycorrhizal fungi, which was later replaced by ectomycorrhizal fungi (Lapeyrie and Chilvers, 1985). A recent study clearly brought out the occurrence of mycorrhizal succession in *Casuarina equisetifolia*. The fungi colonizing the root varied with increasing age of the host. Though *Glomus* was the genus associated with *Casuarina,* the species associated with trees of different ages varied (M. Vasanthakrishna *et al.,* unpublished data). This study suggested the need for isolating mycorrhizal fungi from young seedlings and using them in screening trials to select efficient inoc-ulant fungi, as inoculation is done to the young seedlings in the nursery.

9. Management of Indigenous AMF

There are two principle ways of ensuring that benefits to crop production are obtained from mycorrhizal associations: (1) by inoculating the crop with selected efficient fungi, and (2) by promoting the activity of effective indigenous mycor-rhizal fungi by adequate cultural practices (Sieverding, 1986, 1991; Wood and Cummings, 1992).

Mycorrhiza workers have shown more interest in the process of inoculating plants with efficient mycorrhizal fungi. In order to promote mycorrhizal activity, information is required about the indigenous mycorrhizal fungi that must be managed. It may then be possible to alter the composition of mycorrhizal fungal species in the soil by certain agronomic practices, in such a way that effective fungi are enhanced and ineffective fungi are depressed. Such attempts have already been made by some workers. It is known that prolonged monoculture with certain agricultural crops, such as wheat, increased mycorrhizal population, while continuous cropping with mustard decreases mycorrhizal population. Heavy doses of phosphatic fertilizer depress the size of mycorrhizal fungal population, while organic manure increases their size. Cultivation of a non-mycorrhizal host like mustard or kale, or leaving the land fallow is known to reduce the number of native mycorrhizal fungi significantly (Bagyaraj, 1990).

Because of the vast amount of information required and the lack of an easy, rapid methodology to define the status of the mycorrhizal population in the field, the use of this management strategy must at present be considered questionable. However, it can be expected in the years to come that some general recommenda-tions for the management of indigenous mycorrhizal fungi to improve crop production may become available for each soil and crop.

10. Projections for Future Research and Conclusions

1. Differences in host response to morphologically indistinguishable isolates of mycorrhizal fungi have been reported (Stahl and Christensen, 1991), but little is known about geographic variation or physiological specialization in these fungi. Data are needed to support the hypothesis that there is significant genetic and physiological diversity in populations of mycorrhizal fungi from dissimilar environment.

2. The ability to distinguish between species and even strains of the same species has clear implications for taxonomy and will lead to identification and potential of exploitation of those fungi that maintain a favorable association with

particular groups of plants. Selection of AMF for use from natural populations by general, arbitrary criteria (copious spore production, aggressive, early colonization, etc.) has been recommended in the past. Such criteria will enhance mycorrhiza formation, but they do not guarantee specific host–plant responses. In order to enhance AMF effectiveness for specific applications, directional selection traits need to be exploited. The products of selection (inocula) may be used in solving specific problems in the plant–soil system or in improving plant products or processes. Plants inoculated with spores from the same population may be screened for several desirable traits concurrently. This is possible only through the employment of molecular approaches, namely serological and DNA probes for studying the mycorrhizal fungi (Varma, 1995; Varma and Hock, 1994; Varma and Schuepp, 1994).

3. The role of mycorrhizal fungi on soil aggregation needs concentrated research. The growth and proliferation of mycorrhizal hyphae in the soil matrix by the different strains of the fungi that play a role in the formation of soil macroaggregates has to be studied in greater detail. The importance of the extraradical hyphae phase of the mycorrhizal association that increases soil aggregates is now beginning to be recognized. While screening and selecting efficient mycorrhizal fungi for different crop plants, it is essential to quantify the extent of extramatrical hyphae in addition to determining the plant growth responses. Especially in arid and semiarid tropics, soil aggregation by extraradical hyphae plays an important role in soil conservation. Optimum fungal plant combinations that produce copious external hypha could be identified. Such plants and mycorrhizal isolates may be used in erosion control and accretion of soil organic matter (Varma et al., 1994).

4. Mycorrhizal fungi significantly change the host physiology. This results in altered root exudation and in turn a changed rhizosphere microflora. The synergistic interaction between mycorrhizal fungi and beneficial soil organisms like nitrogen fixers, phosphate solubilizers, and plant growth-promoting rhizobacteria are well documented (Bethlenfalvay and Linderman, 1992; Varma et al., 1994). Similarly, there are several reports on mycorrhiza-suppressing root pathogens. Most of the interaction studies of mycorrhizal fungi with beneficial soil organisms or root pathogens have been carried out in the laboratory or in the greenhouse (Mondal and Varma, 1994). There is an urgent need to extend these studies under field conditions. Such field studies will reveal the possibility of their use as a crop management strategy in sustainable agriculture.

5. Recent studies have shown that the degree to which cultivars are colonized by and benefit from mycorrhiza is a heritable trait selectable through plant breeding (Mercy et al., 1990). If mycorrhiza are to be managed in sustainable crop production systems, crop breeders should consider mycorrhizal responsiveness during selection. Toth et al. (1990) showed that lines of corn inbred for resistance to fungal pathogens were poorly colonized by mycorrhizal fungi,

matured more slowly, and had larger root systems when compared with disease-susceptible inbreds. Breeding for pathogen resistance may inadvertently reduce the mycorrhizal susceptibility of manipulated plants. What does breeding for various stresses and increased crop yield do to plant colonization by mycorrhizal fungi? If breeding does select against mycorrhizal fungi, it may have repercussions on soil structure, soil erosion, fertilizer use efficiency, and general soil health (Varma, 1994).

Colonization by AMF alters the physiology, morphology, and nutritional status of the host plant, and host–soil biota and structure. The over 160 known species of AMF colonize almost all crop plants. There is no host–plant or host–soil specificity, but some plant–fungus and some soil–fungus combinations are more effective than others. Effectiveness is not well-defined, but it can be measured by a wide range of host responses. Work during the last two decades shows that AMF have potential benefits to agriculture. Yet, this potential has so far not been realized, because research efforts are fragmented, data synthesis and modeling on the ecosystem level are lacking, and there is no information on the genetic potential of AMF to tolerate environmental or cultural conditions to modulate host plant and host soil responses. Thus, we are beginning to view the arbuscular mycorrhizal fungus as a soil symbiont as well as a plant symbiont, but more importantly, also as an interface between plant and soil communities.

References

Abbott, L. K., and Robson, A. D., 1981, Infectivity and effectiveness of vesicular arbuscular mycorrhizal fungi: Effect of inoculum type, *Aust. J. Agric. Res.* **32**:631–639.

Abbott, L. K., and Robson, A. D., 1982, The role of vesicular arbuscular mycorrhizal fungi in agriculture and the selection of fungi for inoculation, *Aust. J. Agric. Res.* **33**:389–408.

Allen, M. J., 1992, *Mycorrhizal Functioning. An Integrative Plant–Fungal Process,* Chapman and Hall, New York.

Atkinson, D., 1992, Tree root development: The role of models in understanding the consequences of arbuscular endomycorrhizal infection, *Agronomie* **12**:817–820.

Auge, R. M., Schekel, K. A., and Wample, R. L., 1987a, Rose leaf elasticity changes in response to mycorrhizal colonization and drought acclimation, *Physiol. Plant* **70**:175–182.

Auge, R. M., Schekel, K. A., and Wample, R. L., 1987b, Leaf water and carbohydrate status of VA mycorrhizal rose exposed to drought stress, *Plant Soil* **99**:291–302.

Bagyaraj, D. J., 1984, Biological interactions with VA mycorrhizal fungi, in: *VA Mycorrhiza* (C. L. Powell and D. J. Bagyaraj, eds.), CRC Press, Boca Raton, Fl., pp. 131–153.

Bagyaraj, D. J., 1990, Ecology of VA mycorrhizae, in: *Handbook of Applied Mycology,* Vol. 1 (D. K. Arora, Bharat Rai, K. G. Mukerji, and G. R. Knudsen, eds.), Marcel Dekker New York, pp. 3–34.

Bagyaraj, D. J., 1992, Mycorrhiza: Application in agriculture, in: *Methods in Microbiology,* Vol. 24 (J. R. Norris, D. J. Read and A. K. Varma, eds.), Academic Press, London, pp. 359–374.

Bagyaraj, D. J., Reddy, M. S. B., and Nalini, P. A., 1989, Selection of an efficient inoculant VA mycorrhizal fungus for *Leucaena, Forest Ecol. Manag.* **27**:81–85.

Balakrishna Reddy, and Bagyaraj, D. J., 1994, Selection of efficient VA mycorrhizal fungi for inoculating mango rootstock "Nekkare," *Scientia Hort.* **59**:69–73.

Becard, G., and Piche, Y., 1992, Establishment of vesicular–arbuscular mycorrhiza in root organ culture: Review and proposed methodology, in: *Methods in Microbiology*, Vol. 24 (J. R. Norris, D. J. Read and A. K. Varma, ed.), Academic Press, London, pp. 89–108.

Bethlenfalvay, G. J., 1992, Mycorrhizae and crop productivity, in: *Mycorrhizae in Sustainable Agriculture* (G. J. Bethlenfalvay and R. G. Linderman, eds.), ASA Special Publication No. 54, American Society of Agronomy, Madison, Wisconsin, pp. 1–27.

Bethlenfalvay, G. J., and Linderman, G. J., eds., 1992, *Mycorrhizae in Sustainable Agriculture*, ASA Special Publication No. 54, American Society of Agronomy, Madison, Wisconsin.

Black, R., 1980, The role of mycorrhizal symbiosis in the nutrition of tropical plants, in: *Tropical Mycorrhiza Research* (P. Mikola, ed.), Clarendon Press, Oxford, England, pp. 191–202.

Black, R., and Tinker, P. B., 1979, The development of endomycorrhizal root systems. II. Effect of agronomic factors and soil conditions on the development of vesicular arbuscular mycorrhizal infection in barley and on the endophyte spore density, *New Phytol.* **83**:401–413.

Chatel, D. L., Greenwood, R. M., and Parker, C. A., 1968, Saprophytic competence as an important character in the selection of *Rhizobium* for inoculation, *Proc. 9th Int. Congr. Soil Sci. Trans.* **11**:65–73.

Crush, J. R., and Pattison, A. C., 1975, Preliminary results on the production of vesicular arbuscular mycorrhizal inoculum by freeze drying, in: *Endomycorrhizas* (F. E. Sanders, B. Mosse, and P. B. Tinker, eds.), Academic Press, London, pp. 485–493.

Daft, M. J., and El-Giahmi, A. A., 1978, Effect of arbuscular mycorrhiza on plant growth. VIII. Effect of defoliation and light on selected hosts, *New Phytol.* **80**:365–372.

Daniels, B. A., and Trappe, J. M., 1980, Factors affecting spore germination of the vesicular arbuscular mycorrhizal fungus, *Glomus epigaeus, Mycologia* **72**:457–471.

Dehne, H. W., and Backhaus, G. F., 1986, The use of vesicular arbuscular mycorrhizal fungi in plant production. I. Inoculum production, *Z. Pflanzenkr. Pflanzenschutz.* **93**:415–424.

Evans, D. G., and Miller, M. H., 1990, The soil of the external hyphae mycelial network in the effect of soil disturbance–induced upon vesicular–arbuscular mycorrhizal colonization of young maize, *New Phytol.* **114**:65–71.

Ferguson, J. J. and Woodhead, S. H., 1982, Production of endomycorrhizal inoculum, in: *Methods and Principles of Mycorrhizal Research* (N. C. Schenck, ed.), American Phytopathological Society, St. Paul, Minn., pp. 47–54.

Fitter, A. H., 1988, Water relations of red clover *Trifolium pratense* L. as affected by VA mycorrhizal colonization of phosphorus supply before and during drought. *J. Exp. Bot.* **39**:595–603.

Fitter, A. H., 1991, Costs and benefits of mycorrhizas: Implications for functioning under natural conditions, *Experientia* **47**:350–354.

Fortuna, P., Citernesi, S., Morini, S., Giovannetti, M., and Loreti, F., 1992, Infectivity and effectiveness of different species of arbuscular mycorrhizal fungi in micropropagated plants of Mr S 2/5 plum rootstock, *Agronomie* **12**:825–830.

Furlan, V., and Fortin, J. A., 1977, Effect of light intensity on the formation of vesicular–arbuscular endomycorrhizas on *Allium cepa* by *Gigaspora calospora, New Phytol.* **79**:335–340.

Garbaye, J., 1991, Biological interactions in the mycorrhizosphere, *Experientia* **47**:370–375.

Gerdemann, J. W., 1968, Vesicular arbuscular mycorrhiza plant growth, *Annu. Rev. Phytopathol.* **6**:397–418.

Gerdemann, J. W., 1975, Vesicular arbuscular mycorrhizae, in: *The Development and Function of Roots* (J. G. Torrey and D. T. Clarkson, eds.), Academic Press, London, pp. 576–591.

Gerdemann, J. W., and Nicolson, T. H., 1963, Spores of mycorrhizal *Endogone* species extracted from soil by wet-sieving and decanting, *Trans. Br. Mycol. Soc.* **46**:235–244.

Govinda Rao, Y. S., Bagyaraj, D. J., and Rai, P. V., 1983, Selection of efficient VA mycorrhizal fungus for finger millet. 2. Screening under field conditions, *Zentralbl. Mikrobiol.* **138**:413–419.

Harinikumar, K. H., and Bagyaraj, D. J., 1988, The effect of season on VA mycorrhiza of leucaena and mango in a semi-arid tropic, *Arid Soil Res. Rehabil.* **2**:139–143.

Harinikumar, K. H., Bagyaraj, D. J., and Secilia, J., 1992, Survival of *Glomus fasciculatum* inocula at three storage temperatures, *J. Soil Biol. Ecol.* **12**:30–34.

Harley, J. L., 1989, The significance of mycorrhiza, *Mycol. Res.* **92**:92–129.

Harley, J. L., 1994, Introduction: The state of the art, in: *Techniques for Mycorrhizal Research* (J. R. Norris, D. J. Read, A. K. Varma, eds.), Academic Press, London, pp. 1–24.

Hayman, D. S., 1983, The physiology of vesicular arbuscular endomycorrhizal symbiosis, *Can. J. Bot.* **61**:944–963.

Iqbal, S. H., and Qureshi, K. S., 1976, The influence of mixed sowing (cereals and crucifers) and crop rotation on the development of mycorrhiza and subsequent growth of crops under field conditions, *Biologia* **22**:287–291.

Jarstfer, A. G., and Sylvia, D. M., 1995, Aeroponic culture, in: *Mycorrhizae: Structure, Function, Molecular Biology and Biotechnology* (A. Varma and B. Hock, eds.), Springer-Verlag, Germany, pp. 427–442.

Jasper, D. A., Abbott, L. K., and Robson, A. D., 1989a, Soil disturbance reduces the infectivity of external hyphae of vesicular–arbuscular mycorrhizal fungi, *New Phytol.* **112**:93–99.

Jasper, D. A., Abbot, L. K., and Robson, A. D., 1989b, Hyphae of a vesicular–arbuscular mycorrhizal fungus maintain infectivity in dry soil, except when the soil is disturbed, *New Phytol.* **112**:101–107.

Johnson, N. C., and Pfleger, F. L., 1992, Vesicular–arbuscular mycorrhizae and cultural stresses, in: *Mycorrhizae in Sustainable Agriculture* (G. J. Bethlenfalvay and R. G. Linderman (eds.), ASA Special Publication, No. 54, American Society of Agronomy, Madison, Wisconsin, pp. 71–99.

Kendrick, B., and Berch, S., 1985, Mycorrhizae: Applications in agriculture and forestry, in: *Comprehensive Biotechnology*, Vol. 4 (C. W. Robinson, ed.), Pergamon Press, Oxford, England, pp. 109–150.

Khan, A. G., 1974, The occurrence of mycorrhizas in halophytes, hydrophytes and xerophytes and of *Endogone* spores in adjacent soils, *J. Gen. Microbiol.* **81**:7–14.

Koske, R. E., 1982, Evidence for a volatile attractant from plant roots affecting germ tubes of a VA fungus, *Trans. Br. Mycol. Soc.* **79**:305–310.

Koske, R. E., and Polson, W. R., 1984, Are VA mycorrhizae required for sand dune stabilization? *Bioscience* **34**:420–424.

Kothari, S. K., Marschner, H., and George, E., 1990, Effect of VA mycorrhizal fungi and rhizosphere microorganisms on root and shoot morphology, growth and water relations in maize, *New Phytol.* **116**:303–311.

Krishna, K. R., Shetty, K. G., Dart, P. J., and Andrews, D. J., 1985, Genotype dependent variation in mycorrhizal colonization and response to inoculation of pearl millet, *Plant Soil* **86**:113–135.

Kruckelmann, H. W., 1975, Effects of fertilizers, soils, soil tillage and plant species on the frequency of *Endogone* chlamydospores and mycorrhizal infection in arable soils, in: *Endomycorrhizas* (F. E. Sanders, B. Mosse, and P. B. Tinker, eds.), Academic Press, London, pp. 511–525.

Lapeyrie, F. F., and Chilvers, G. A., 1985, An endomycorrhiza–ectomycorrhiza succession cited with enhanced growth of *Eucalyptus dumosa* seedlings planted in calcareous soil, *New Phytol.* **100**:93–104.

Linderman, R. G., 1988, Mycorrhizal interactions with the rhizosphere microflora: The mycorrhizosphere effect, *Phytopathology* **78**:366–371.

Linderman, R. G., 1992, Vesicular–arbuscular mycorrhizae and soil microbial interactions, in: *Mycorrhizae in Sustainable Agriculture* (G. J. Bethlenfalvay, and R. G. Linderman, eds.), ASA Special Publication No. 54, American Society of Agronomy, Madison, Wisconsin, pp. 45–70.

Loree, M. A. J., and Williams, S. E., 1984, Vesicular-arbuscular mycorrhizae and severe land disturbance, in: *VA Mycorrhizae and Reclaimation of Arid and Semi-arid Lands* (S. E. Williams and M. F. Allen, eds.), University of Wyoming Publications, Laramie, Wyoming, pp. 1–14.

Mallesha, B. C., and Bagyaraj, D. J., 1990, Reduction of leucaena to single inoculation versus dual inoculation with *Glomus mosseae* and/or *Gigaspora margarita*, *Leucaena Res. Rep.* **11:** 56–57.

Marx, D. H., Ruehle, J. L., and Cordell, C. E., 1994, Methods for studying nursery and field responses of trees to specific ectomycorrhizas, in: *Mycological Research* (J. R. Norris, D. J. Read, and A. K. Varma, eds.), Academic Press, London, pp. 383–412.

Mason, P. A., Last, T. F., Pelham, J., and Ingleby, K., 1982, Ecology of some fungi associated with an ageing stand of birches (*Betula pendula* and *B. pubescens*), *Forest Ecol. Manag.* **4:**19–39.

Mathew, J., Shankar, A., Neeraj, Varma, A., 1991, Glomaceous fungi associated with cacti, a fodder supplement in deserts. *Trans. Soc. Japan* **32:**225–233.

Mercy, M. A., Shivashankar, G., and Bagyaraj, D. J., 1990, Mycorrhizal colonization in cowpea is host dependent and heritable, *Plant Soil* **121:**292–294.

Mikola, P., 1980, *Tropical Mycorrhizal Research,* Clarendon Press, Oxford, England.

Mondal, N., and Varma, A., 1994, Mycorrhizosphere: Impact in sustaining soil fertility, in: *Perspectives in Life Science* (M. L. Naik, ed.), in press.

Morton, J. B., Franke, S. P., and Bentivenga, S. P., 1995, Systematic developmental foundations for morphological diversity among endomycorrhizal fungi in Glomales (Zygomycetes), in: *Mycorrhizae: Structure, Function, Molecular biology and Biotechnology* (A. Varma and B. Hock, eds.), Springer-Verlag, Germany, pp. 669–684.

Mosse, B., 1973, The role of mycorrhiza in phosphorus solubilization, in: *Proceedings of the 4th International Conference on Global Impacts of Applied Microbiology,* (J. S. Furtado, ed.), Sau Paulo, Brazil, pp. 543–561.

Mosse, B., Stribley, D. P., and Le Tacon, F., 1981, Ecology of mycorrhizae and mycorrhizal fungi, *Adv. Microbiol. Ecol.* **5:**137–210.

Mosse, B., Warner, A., and Clarke, C. A., 1982, Plant growth response to vesicular arbuscular mycorrhiza. XIII. Spread of an introduced VA endophyte in the field and residual growth effects of inoculation in the second year. *New Phytol.* **90:**521–528.

Munns, D. N., and Mosse, B., 1980, Mineral nutrition of legume corps, in: *Advances in Legume Science* (R. J. Summerfield, and A. H. Bunting, eds.), University of Reading Press, England, pp. 115–125.

Neeraj, 1992, *Better Uptake of Nutrients in Transplantation Crops by Inoculation of Endomycorrhizal Fungi and Potential Biological Nitrogen-fixing Microorganisms,* Kanpur University, India, Ph.D. thesis.

Neeraj, Shankar, A., and Varma, A., 1991, Occurrence of VA mycorrhizae within Indian semi-desert arid soils, *Biol. Fertil. Soils* **11:**140–144.

Nopamornbodi, O., Rojanasiriwong, W., and Thamsuakul, S., 1988, Production of VAM fungi, *Glomus intraradices* and *G. mosseae* in tissue culture, in: *Mycorrhiza for Green Asia* (A. Mahadevan, N. Raman, and K. Natarajan, eds.), University of Madras, India, pp. 315–316.

Ocampo, J. A., Martin, J., and Hayman, D. S., 1980, Influence of plant interactions on vesicular arbuscular mycorrhizal infections. I. Host and non-host plants grown together, *New Phytol.* **84:**27–35.

Pai, G., Bagyaraj, D. J., and Padmavathi Ravindra, T., 1993, Effect of moisture stress on growth and water relations of VA mycorrhizal and non-mycorrhizal cowpea grown at different levels of phosphorus, *J. Soil Biol. Ecol.* **13:**14–24.

Pankow, P., Boiler, T., and Weimken, A., 1991, The significance of mycorrhizas for protective ecosystems, *Experientia* **47:**391–394.

Phillips, J. H., and Hayman, D. S., 1970, Improved procedures for clearing roots and staining parasitic and vesicular arbuscular mycorrhizal fungi for rapid assessment of infection, *Trans. Br. Mycol. Soc.* **55**:158–161.

Porter, W. M., 1979, The "most probable number" method for enumerating infective propagules of vesicular arbuscular mycorrhizal fungi in soil, *Aust. J. Soil Res.* **17**:515–519.

Powell, C., 1976, Development of mycorrhizal infections from *Endogone* spores infected root segments, *Trans. Br. Mycol. Soc.* **66**:439–445.

Powell, C. L., 1979, Spread of mycorrhizal fungi through soil, *N. Z. J. Agric. Res.* **22**:335–339.

Powell, C. L., and Bagyaraj, D. J., 1982, VA mycorrhizal inoculation of field crops, *Proc. N. Z. J. Agron. Soc.* **12**:85–88.

Powell, C. L., Metcalfe, D. M., Buwalda, J. G., and Waller, J. E., 1980, Phosphate response curves of mycorrhizal and nonmycorrhizal plants. II. Response to rock phosphate, *N. Z. J. Agric. Res.* **23**:477–482.

Puppi, G., and Bras, A., 1990, Nutrient and water relations of mycorrhizal white clover, *Agric. Ecosyst. Environ.* **29**:317–322.

Raj, J., Bagyaraj, D. J., and Manjunath, A., 1981, Influence of soil inoculation with vesicular arbuscular mycorrhiza and a phosphate dissolving bacterium on plant growth and ^{32}P uptake, *Soil Biol. Biochem.* **13**:105–108.

Read, D. J., 1992, Mycorrhizal mycelium, in: *Mycorrhizal Functioning: An Intergrative Plant–Fungal Process* (M. J. Allen, ed.), Chapman & Hall, New York, pp. 102–133.

Redhead, J. F., 1975, Endotrophic mycorrhizas in Nigeria: Some aspects of the ecology of the endotrophic mycorrhizal association of *Khaya grandifolia* C. D. D., in: *Endomycorrhizas* (F. E. Sanders, B. Mosse, and P. B. Tinker, eds.), Academic Press, London, pp. 447–459.

Schenck, N. C., and Kinloch, R. A., 1980, Incidence of mycorrhizal fungi on six field crops in monoculture on a newly cleared woodland site, *Mycologia* **72**:445–455.

Schenck, N. C., and Schroder, V. N., 1974, Temperature response of *endogone* mycorrhiza on soybean roots, *Mycologia* **66**:600–605.

Schenck, N. C., Graham, S. O., and Green, N. E., 1975, Temperature and light effects on contamination and spore germination of vesicular arbuscular mycorrhizal fungi, *Mycologia* **67**:1189–1192.

Schubert, A., and Hayman, D. S., 1986, Plant growth responses to vesicular arbuscular mycorrhiza. XVI. Effectiveness of different endophytes at different levels of soil phosphate. *New Phytol.* **103**:79–80.

Secilia, J., and Bagyaraj, D. J., 1987, Bacteria and actinomycetes associated with pot culture of VA mycorrhizas, *Can. J. Microbiol.* **33**:1069–1073.

Sieverding, E., 1986, Research model towards practical application of VA mycorrhizal fungi in tropical agriculture, in: *Physiological and Genetical Aspects of Mycorrhizae* (V. Gianinazzi-Pearson and S. Gianinazzi, eds.), INRA Press, Dijon, France, pp. 475–478.

Sieverding, E., 1991, *Vesicular–Arbuscular Mycorrhiza Management in Tropical Argo Systems,* Deutsche Gesellschaft für Technische Zusammenarbeit (GTZ), Eschborn, Germany.

Simon, L., Lalonde, M., and Bruns, T. D., 1992a, Specific amplification of 18S fungal ribosomal genes from vesicular–arbuscular endomycorrhizal fungi colonizing roots, *Appl. Environ. Microbiol.* **58**:291–295.

Simon, L., Roger, C., Lévesque, R. C., and Lalonde, M., 1992b, Rapid quantification by PCR of endomycorrhizal fungi colonising roots, *PCR Meth. Application* **2**:76–80.

Simon, L., Bousquet, J., Lévesque, R. C., and Lalonde, M., 1993a, Origin and diversification of endomycorrhizal fungi and coincidence with vascular plants, *Nature* **363**:67–69.

Simon, L., Lévesque, R. C., and Lalonde, M., 1993b, Identification of endmycorrhizal fungi colonizing roots by fluorescent single-strand conformation polymorphism-polymerase chain reaction. *Appl. Environ. Microbiol* **59**:4211–4215.

Simpson, D., and Daft, M. J., 1990, Interactions between water-stress and different mycorrhizal inocula on plant growth and mycorrhizal development in maize and sorghum, *Plant Soil* **121:**179–186.

Singh, K., and Varma, A. J., 1980, Host specificity in Endogonaceae, *Trans. Mycol. Soc. Japan* **21:**477–482.

Singh, K., and Varma, A., 1981, Endogonaceous spores associated with xerophytic plants in northern India, *Trans. Br. Mycol. Soc.* **77:**655–658.

Smith, S. E., 1995, State of art, discoveries, discussion, and directions in mycorrhizal research, in: *Mycorrhiza: Structure, Function, Molecular Biology and Biotechnology*, Springer-Verlag, Germany, pp. 3–24.

Sreenivasa, M. N., and Bagyaraj, D. J., 1988a, *Chloris gayana* (Rhodes grass), a better host for mass production of *Glomus fasciculatum*, *Plant Soil* **106:**289–290.

Sreenivasa, M. N., and Bagyaraj, D. J., 1988b, Selection of suitable substrate for mass multiplication of *Glomus fasciculatum*, *Plant Soil* **109:**125–127.

Stahl, P. D., and Christensen, M., 1991, Population variation in the mycorrhizal fungus *Glomus mosseae:* Breadth of environmental tolerance, *Mycol. Res.* **95:**300–307.

St. John, T. V., 1980, Root size, root hairs and mycorrhizal infection. A re-examination of Baylis's hypothesis with tropical trees, *New Phytol.* **84:**483–487.

Sylvia, D. M., and Williams, S. E., 1992, Vesicular arbuscular mycorrhiza and environmental stress, in: *Mycorrhiza in Sustainable Agriculture* (G. J. Bethlenfalvay and R. G. Linderman, eds.), ASA Special Publication No. 54, American Society of Agromony, Madison,Wisconsin, pp. 101–124.

Sylvia, D. M., Hammond, L. C., Bennett, J. M., Haas, J. H., and Linda, S. B., 1993, Field response of maize to a VAM fungus and water management, *Agron. J.* **85:**193–198.

TAC Secretariat, 1989, *Sustainable Agricultural Production: Implication for International Agricultural Research*, Consultative Group on International Agriculture Research, Washington, D.C.

Tinker, P. B., 1975, The soil chemistry of phosphorus and mycorrhizal effects on plant growth, in: *Endomycorrhizas* (F. E. Sanders, B. Mosse, and P. B. Tinker, eds.), Academic Press, London, pp. 353–371.

Tinker, P. B., Jones, M. D., and Durall, D. M., 1994, Principles of use of radioisotopes in mycorrhizal studies, in: *Mycorrhizal Research* (J. R. Norris, D. J. Read, and A. K. Varma, eds.), Academic Press, London, pp. 295–308.

Tommerup, I C., 1983, Spore dormancy in vesicular arbuscular mycorrhizal fungi, *Trans. Br. Mycol. Soc.* **81:**37–45.

Tommerup, I. C., 1987, Physiology and ecology of VAM spore germination and dormancy in soil, in: *Mycorrhizae in the Next Decade: Practical Applications and Research Priorities* (D. M. Sylvia, L. L. Hung, and H. H. Graham, eds.), University of Florida Publication, Gainesville, pp. 175–177.

Tommerup, I. C., and Kidby, D. K., 1979, Preservation of spores of vesicular arbuscular endophytes by L-drying, *Appl. Environ. Microbiol.* **37:**831–835.

Toth, R., Toth, D., Starke, D., and Smith, D. R., 1990, Vesicular arbuscular mycorrhizal colonization in *Zea mays* affected by breeding for resistance to fungal pathogens, *Can. J. Bot.* **66:**1039–1044.

Trappe, J. M., 1981, Mycorrhiza and productivity of arid and semi-arid rangelands, in: *Advances in Food Producing Systems for Arid and Semi-arid Lands* (J. T. Manassah, and E. J. Briskrel, eds.), Academic Press, London, pp. 581–599.

Trappe, J. M., 1987, Phytogenetic and ecological aspects of mycotrophy in the angiosperms from an evolutionary stand point, in: *Ecophysiology of VA Mycorrhizal Plants* (G. R. Safir, ed.), CRC Press, Boca Raton, Fl., pp. 5–25.

Varma, A., 1994a, Molecular approaches: Identification of mycorrhizal fungi, in: *Plant Microbe*

Interactions (P. F. Kidwai and R. Singh, eds.), Department of Science and Technology Publications, New Delhi, India, pp. 20–22.

Varma, A., 1995, Ecophysiology of mycorrhizal fungi, in: *Mycorrhizae: Structure, Function, Molecular Biology and Biotechnology,* Springer-Verlag, Germany, pp. 561–592.

Varma, A., and Schuepp, H., 1994, Infectivity and effectiveness of *Glomus intraradices, Mycorrhiza* **5:**29–37.

Varma, A., Verma, S., and Schuepp, H., 1994, Mycorrhizae: What we know and what should we know? in: *New Approaches in Microbial Ecology* (J. P. Tewari, G. Saxena, N. Mittal, and I. Tewari, eds.), Aditya Books, New Delhi, India, pp. 143–154.

Vasanthakrishna, M., and Bagyaraj, D. J., 1993, Selection of efficient VA mycorrhizal fungi for inoculating *Casuarina equisetifolia, Arid Soil Res. Rehabil.* **7:**377–380.

Walker, C., 1992, Systematics and taxonomy of arbuscular endomycorrhizal fungi (Glomales) a possible way forward, *Agronomie* **12:**887–897.

Walker, C., and Trappe, J. M., 1993, Names and epithets in the Glomales and endogonales, *Mycol. Res.* **97:**329–344.

Wallace, H. R., 1978, Dispersal in time and space: Soil pathogens, in: *Plant Diseases: An Advanced Treatise,* Vol. 2 (J. G. Horsefall and E. B. Cowling, eds.), Academic Press, New York, pp. 181–202.

Warner, A., and Mosse, B., 1982, Factors affecting the spread of vesicular arbuscular mycorrhizal fungi in soil. I. Root density, *New Phytol.* **90:**529–536.

White, J. A., Munn, L. C., and Williams, S. E., 1989, Dispersal agents of vesicular arbuscular mycorrhizal fungi in a disturbed arid ecosystem, *Mycologia* **79:**721–730.

White, J. A., Depuit, E. J., Smith, J. L., and Williams, S. E., 1992, Vesicular arbuscular mycorrhizal fungi and irrigated mined land reclamation in South Western Wyoming, *Soil Sci. Soc. Am. J.* **56:**1466–1471.

Wood, T., 1985, Commercial pot culture inoculum production: Quality control and other headaches, in: *Proceedings of 6th North American Conferences on Mycorrhizae* (R. Molina, ed.), Forest Research Laboratory, Oregon State University, Corvallis, p. 84.

Wood, T. E., and Cummings, B., 1992, Biotechnology and the future of VAM commercializations, in: *Mycorrhizal Functioning, An Integrated Plant-Fungal Process* (M. J. Allen ed.), Chapman & Hall, New York, pp. 462–487.

Wyss, P., and Bonfante, P., 1993, Amplification of genomic DNA of arbuscular mycorrhizal (AM) fungi by PCR using short arbitrary primers, *Mycol. Res.* **97:**1351–1357.

4

The Microbial Logic and Environmental Significance of Reductive Dehalogenation

JAN DOLFING and JACOBUS E. M. BEURSKENS

1. Introduction

In the last 25 years, Western society has decisively changed its attitude toward halogenated compounds. Until the end of the 1960s, "chemicals" were applied indiscriminately in a wide variety of agricultural and industrial processes. Many of these chemicals were chlorinated compounds. They were used because they had many useful characteristics. One of these characteristics was that they were very stable and rather resistant to chemical and biological degradation. With hindsight it is thus not surprising that these halogenated compounds proved to be quite persistent in the environment. Many of these generally hydrophobic compounds have the tendency to accumulate in biota to such levels that they caused considerable damage or even death. The eloquent outcry of Rachel Carson (1962) and others in the 1960s resulted in a drastic reappraisal of the wisdom of using halogenated organic compounds indiscriminately.

Halogenated compounds are rare in nature, and this led to the perception that halogenated compounds are xenobiotics. In the last decade, however, scientists are beginning to appreciate that nature produces a wide variety of halogenated organic compounds. This leads to the conclusion that it is not correct to state that halogenated compounds are by definition xenobiotic. On the other hand, it is also not correct to conclude that halogenated compounds are not xenobiotic, since the possibility that nature consistently produces compounds like decachlorinated biphenyl or halons and chlorofluorocarbons seems negli-

JAN DOLFING • DLO–Research Institute for Agrobiology and Soil Fertility (AB-DLO), NL-9750 AC Haren, The Netherlands.   JACOBUS E. M. BEURSKENS • National Institute of Public Health and Environmental Protection (RIVM), NL-3720 BA Bilthoven, The Netherlands.
Advances in Microbial Ecology, Volume 14, edited by J. Gwynfryn Jones. Plenum Press, New York, 1995.

gible. However, for many organohalogens, even for unexpected ones like chlorinated dioxins and tetrachloroethylene, there are now indications that they are produced in nature at appreciable, that is measurable, rates. Thus there is a tendency toward a somewhat looser use of the term xenobiotic, backed up by the statement "sole dosis facit xenobioticum." Such discussions are distinctly beyond the scope of this chapter.

This chapter deals with the ecophysiology of reductive dechlorination in anoxic environments. In the last decade it has been found that anaerobic microorganisms are able to degrade a wide variety of halogenated organic compounds. Both aliphatic and aromatic organohalogens are susceptible to microbial degradation, and in anoxic environments their degradation is generally initiated by a reductive dechlorination step. This encompasses a two-electron transfer that results in the removal of a halogen atom as a halide ion and its replacement by a hydrogen atom. Polyhalogenated compounds may undergo stepwise dehalogenation. The loss of halide(s) results in fewer halogenated products and may lead to further biodegradation, either anaerobically or aerobically, under appropriate environmental conditions (Tiedje *et al.*, 1987; Zitomer and Speece, 1993).

1.1. Aim and Scope of This Chapter

We will attempt to analyze the logic behind microbial catalysis of reductive dehalogenation and to pinpoint the relation of this process to the other processes that occur in anoxic environments. To that end we will include information on the concentrations of organohalogens observed in the environment and give an overview of what is known about the actual rate and occurrence of reductive dechlorination in the environment. The objective is not to give an all encompassing overview of the present knowledge on microbial reductive dehalogenation, since this has been done recently and expertly by Mohn and Tiedje (1992).

1.2. Definition of Reductive Dehalogenation

Reductive dehalogenation has been defined as the removal of a halogen substituent from a molecule with concurrent addition of two electrons to the molecule (Mohn and Tiedje, 1992). It should be kept in mind, however, that in the majority of redox reactions the two electrons are transferred in sequential steps (Eberson, 1987), and there is no reason to believe that this is different for reductive dehalogenation reactions. Essentially, two types of reductive dehalogenation processes have been identified. The first process, hydrogenolysis, is the replacement of a halogen substituent of a molecule with a hydrogen atom (Fig. 1). The second process, vicinal reduction, also known as dihaloelimination, encompasses the removal of two halogen substituents from adjacent carbon atoms with the formation of an additional bond between the carbon atoms (Fig.1). The emphasis here will be on the hydrogenolysis process, since this is the type

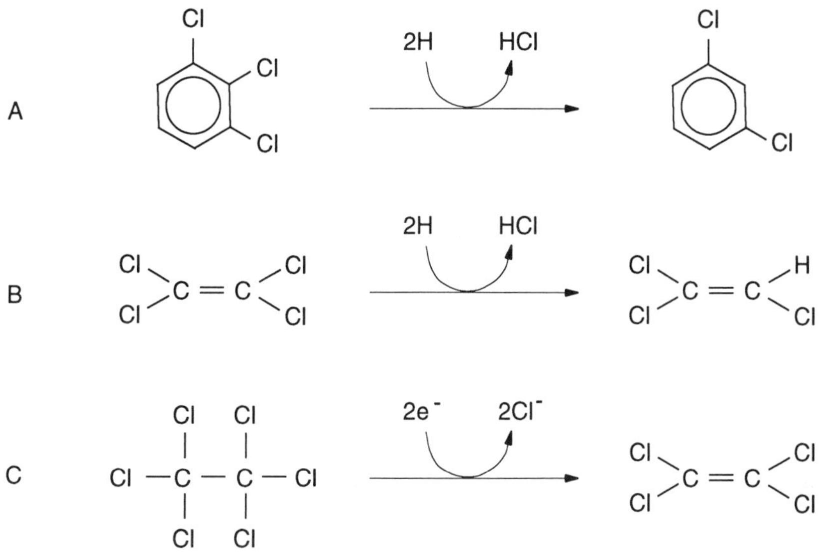

Figure 1. Examples of reductive dehalogenation. (A) Aryl hydrogenolysis of 1,2,3-trichlorobenzene to 1,3-dichlorobenzene; (B) alkyl hydrogenolysis of tetrachloroethylene to trichloroethylene; and (C) vicinal reduction of hexachloroethane to tetrachloroethylene.

that is most often observed. In this chapter, whenever we use the term *reductive dehalogenation* we actually mean hydrogenolysis. The features of reductive dehalogenation that need to be mentioned here are: (1) that in all reported examples the halogen atoms are released as halide ions, which makes reductive dehalogenation a proton-generating reaction, and, more importantly, (2) that the process requires an electron donor.

1.3. Industrial Production of Organohalogens

Data on the production of chemicals by the US chemical industry are published annually by the American Chemical Society (Anonymus, 1993). Two halogenated compounds, ethylene dichloride and vinyl chloride, are among the top 20 chemicals (by volume). Of the top 45 organic chemicals, 9 are halogenated. The list (Table I) includes carbon tetrachloride, chloroform, ethyl chloride, ethylene dichloride, methyl chloride, methylchloroform, methylene chloride, perchloroethylene, and vinyl chloride. Thus all the top halogenated organics are aliphatics, and in all cases the halogen is chlorine. There is no general trend over the last decade in the production figures of these compounds (Table I). For some organohalogens production has increased consistently, while for other compounds production has decreased considerably.

Table I. Production of Organohalogens by the US Chemical Industry

| | Production (10^6 kg/year) | | Percent annual change |
	1991	1992	1982–1992
1,2-Dichloroethane	6226	7235	7
Chloroethylene	5058	6005	11
Chloromethane	416	390	9
1,1,1-Trichloroethane	292	326	2
Trichloromethane	229	na	na
Dichloromethane	177	163	−4
Tetrachloromethane	143	na	na
Tetrachloroethylene	109	110	−3
Chloroethane	na[a]	na	na

[a]na, Not available; the latest data available for chloroethane was 68 million kg in 1990.

The production of nonhydrogenated chlorofluorocarbons (CFCs) has decreased sharply over the last few years. For CFC-11 (trichlorofluoromethane) and CFC-12 (dichlorodifluoromethane), for example, the annual US production dropped by about 75% between 1991 and 1992 (Anonymus, 1993). The latter trend is in agreement with the "Montreal protocol," prescribing the international action to halt depletion of the ozone layer (Benedick, 1991) and the new time table attached to this protocol as agreed in Copenhagen (Anonymus, 1992). An important milestone set in the Copenhagen agreement is 1996, when production of not only CFCs but also of carbon tetrachloride and 1,1,1-trichloroethane must be phased out.

Commercial production of polychlorinated biphenyls (PCBs) in the United States started in 1929. The world production reached a maximum in the early 1960s, and PCBs became widely used as dielectric fluids, heat transfer fluids, hydraulic fluids, plasticizers, and flame retardants. More than 2×10^9 kg have been produced worldwide (Tanabe, 1988). Because of elevated levels found in a range of environmental samples and their suspected toxicity, their industrial use in much of Europe and America has been restricted since the mid-1970s (Rapaport and Eisenreich, 1988). The toxicity of PCBs depends on the spatial configuration of the chlorines. Congeners with no or only one ortho, two para, and two or more meta chlorines, resemble 2,3,7,8-tetrachlorodibenzo-p-dioxin in their biological and toxic effects (Safe, 1990). These so-called nonortho and monoortho congeners are now considered to be mainly responsible for the toxicity associated with PCBs (de Voogt et al., 1990), even though they were only present at trace levels in the original PCB mixtures (Hong et al., 1993). Emissions of chlorinated aromatics had their maxima in the 1960s and 1970s. Contemporary emissions for PCBs and hexachlorobenzene reach levels close to those in the mid-1940s (Alcock *et al.*, 1993; Jones *et al.*, 1992; Rapaport and Eisenreich, 1988).

For most chlorinated compounds it is difficult to estimate what percentage of the annual production actually ends up in the environment. These percentages may vary considerably. Pearson (1982) has estimated that virtually all of the solvents tetra- and trichloroethylene eventually end up in the environment, versus only 2 and 10% of the industrially produced chloroethylene and 1,2-dichloroethane, since these are intermediates in the production of other industrial chemicals. Another factor that complicates the estimation of global budgets of chlorinated compounds is that some chlorinated compounds, e.g., chlorinated benzenes, are also produced as by-products in the manufacturing of other chlorinated compounds, or they end up in the environment as products of the degradation of other industrial products, as has been reported for chlorinated anilines (Howard, 1989). Chlorinated benzoates on the other hand occur in the environment as a result of their use as herbicide or as a product of the aerobic degradation of other chlorinated aromatics such as PCBs (Furukawa *et al.*, 1983; Flanagan and May, 1993).

1.4. Naturally Occurring Halogenated Organic Compounds

Man's contribution to the pool of halogenated compounds should be set against the background of natural abiotic (Symonds *et al.*, 1988) and biotic production (Petty, 1961; Lovelock, 1975; Pearson, 1982; Harper, 1993; Grimvall and de Leer, 1995; De Jong *et al.*, 1994). Natural sources such as forest fires and decomposition of seaweeds release 10 to 100 times more chloromethane than that manufactured by the chemical industry (Leisinger, 1983). The chemical diversity of naturally occurring halogenated compounds is impressive (Fig. 2). Some commercially important chlorinated antibiotics are produced by bacteria, for example, chlortetracycline and chloramphenicol (Neidleman and Geigert, 1986). Common wood and forest litter-degrading fungi produce chlorinated anisyl metabolites (Fig. 2). These compounds occur at high concentrations in the environment (De Jong *et al.*, 1994). Chlorinated anisyl metabolites can be detoxified by microbial mineralization and by incorporation into humus. Besides these detoxification reactions, microbial metabolism of chlorinated anisyl metabolites may result in potentially more toxic compounds, for example, chlorinated dioxins, as recently postulated by De Jong *et al.* (1994). The presence of dioxins in the environment may therefore not exclusively have an anthropogenic origin. The production of chlorinated dioxins from chlorophenols during wastewater treatment and composting suggest a similar mechanism (Öberg *et al.*, 1993).

The presence of natural halogenated compounds in the environment, in many instances produced as agents to inhibit the growth of competing or pathogenic species, provides a major selective pressure for the evolution of detoxifica-

chlortetracycline

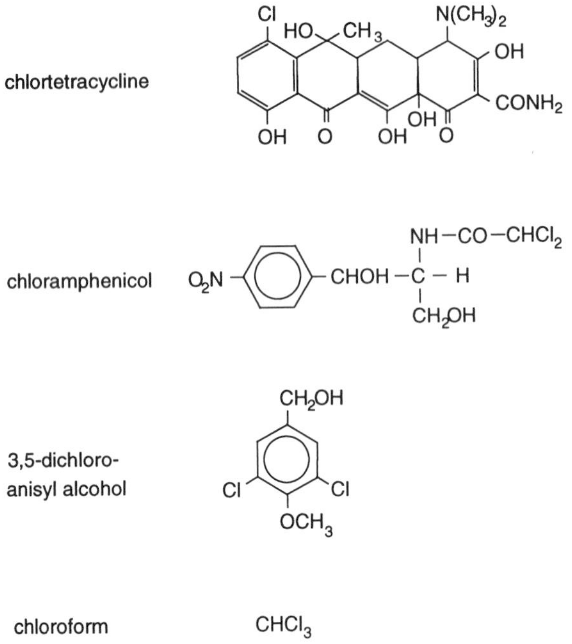

chloramphenicol O_2N —⟨ ⟩— CHOH —C̈ — H

3,5-dichloro-
anisyl alcohol

chloroform $CHCl_3$

Figure 2. Examples of naturally occurring halogenated compounds (Neidleman and Geigert, 1986).

tion mechanisms. This may be exemplified by the biodehalogenation of 2,4-dichlorophenol observed in marine sediments that contain natural sources of halophenols produced as bactericidal agents (King, 1986, 1988). Microorganisms capable of the metabolism of a wide range of haloaromatic and aliphatic substrates have been isolated (Häggblom, 1992; Mohn and Tiedje, 1992; Dolfing *et al.*, 1993), and there is little doubt that these organisms are important factors in determining the fate of both natural and anthropogenic halogenated substances in the environment.

1.5. Physicochemical Processes That Affect the Distribution of Organohalogens in the Environment

An excellent treatise on the factors that govern the behavior of organic chemicals in the environment was published recently by Schwarzenbach, Gschwend, and Imboden (1993). For this chapter, it suffices to state that the mobility and distribution of organohalogens in air, water, and soil depends on their physicochemical properties, such as volatility and hydrophobicity.

The atmosphere appears to be the major "sink" for many volatile halogenated compounds, especially the halogenated aliphatics. Field measurements indicated that, for example, tetrachloroethylene, but also 1,4-dichlorobenzene, were predominantly eliminated from Lake Zurich by evaporation (Schwarzenbach *et al.*, 1979). Compounds that are considered "semivolatile" such as PCBs and chlorinated dioxins were also found to be transported through the atmosphere. Such semivolatile compounds can be transported in the atmosphere either via the vapor phase or as particle-bound molecules (Eitzer and Hites, 1989). In fact, sorption is generally considered to be the single most important fate in the physicochemical behavior of most organohalogens in the environment. Most organohalogens are rather hydrophobic compounds. An environmentally meaningful way to quantify the hydrophobicity of a compound is by using the octanol/water partition coefficient ($K_{o/w}$) (Schwarzenbach *et al.*, 1993). For example, the mobility of halogenated compounds in aquifers with an organic carbon content $>0.1\%$ correlated well with the $K_{o/w}$ of these compounds (McCarty *et al.*, 1981; Schwarzenbach and Westall, 1981; Schwarzenbach *et al.*, 1983).

Because of their hydrophobic properties, many organohalogens that have been released into waterways partition into aquatic sediments. Sediments are often anoxic, and this makes reductive dechlorination potentially a very important transformation reaction for chlorinated compounds in the environment. A consequence of the hydrophobic properties of organohalogens, however, is also that these compounds are poorly soluble in water. This may limit the rate at which a compound can transfer from sediments, through (pore) water, to another phase (e.g., microorganisms). Therefore, the hydrophobic properties of PCBs may limit the bioavailability of organohalogens, especially highly chlorinated PCBs, to microorganisms in anoxic sediments.

1.6. Toxicity of Organohalogens in Anoxic Environments

Biological systems have been exposed to numerous naturally occurring organohalides for perhaps hundreds of millions of years. A substantial fraction of the marine biota produces a remarkable array of aliphatic and aromatic compounds containing chlorine, bromine, or iodine. Many of the compounds that have been isolated have antimicrobial activities, and it has been proposed that these compounds play an ecological role (Sheikh and Djerassi, 1975). A well-known example of such a proposed role is the high concentrations (up to several hundred micromolar) of 2,4-dibromophenol in the burrow microenvironment of the hemichordate *Saccoglossus kowalewskii* (Ashworth and Cornier, 1967; King, 1986, 1988). At the observed *in situ* concentrations of 2,4-dibromophenol, aerobic microbial metabolism was selectively inhibited, but only minimal inhibition of growth on nutrient broth occurred for anaerobes exposed to 1 mM 2,4-dibromophenol. It is thus likely that 2,4-dibromophenol was excreted to inhibit the growth of aerobic bacteria on the inner surface of *S. kowalewskii* burrows. One of the reasons why anaerobes may be less sensitive than aerobes is that ATP

production by anaerobes is generally not as dependent on oxidative phosphorylation, the process effectively inhibited by dibromophenol (Stockdale and Selwyn,
1971).

Data on the toxicity of chlorinated organic compounds toward anaerobic
microorganisms are scarce (Zhang and Wiegel, 1990; Sierra-Alvarez and Lettinga, 1991; Davies-Venn et al., 1992; Madsen and Aamand, 1992; Renard et
al., 1993). Highly chlorinated compounds are generally more hydrophobic than
their lower chlorinated analogues (Table II). Sierra-Alvarez and Lettinga (1991)
have shown that for homologous series of chlorinated benzenes and phenols a
positive correlation exists between the hydrophobicity of these compounds and
their inhibitory effect on the methanogenic degradation of acetate (Fig. 3). Such a
relationship is plausible when factors such as partitioning and transport control
the toxic effects of the compound. In such cases, when reductive dechlorination
reduces the hydrophobicity, it also results in a decrease of the toxicity of chlorinated compounds, at least in theory. Whether such a reduction also occurs in
practice remains to be seen. Less chlorinated dechlorination products sorp less
strongly to hydrophobic materials present in the natural environment, so it is
conceivable that the effect of dechlorination is that not only the actual concentration of toxicants in the aqueous phase increases but also the "toxicity" of the
aqueous phase, especially because lower chlorinated compounds have the ten-

Table II. Hydrophobicity[a] and Inhibitory
Concentration Towards Acetoclastic
Methanogenesis of Some Chlorinated Benzenes
and Phenols[b]

Compound	$\log K_{o/w}$	50% IC[c]
Benzenes		
Benzene	1.95	18.91
Chlorobenzene	2.84	3.38
1,2-Dichlorobenzene	3.53	1.22
1,2,4-Trichlorobenzene	4.26	0.52
Phenols		
Phenol	1.46	11.69
2-Chlorophenol	2.17	3.19
2,4-Dichlorophenol	3.15	0.49
2,4,6-Trichlorophenol	3.38	0.59
Pentachlorophenol	5.01	0.03

[a]The hydrophobicity of a compound is traditionally expressed as the
logarithm of the partition coefficient n-octanol/water ($\log K_{o/w}$).
[b]Data are form Sierra-Alvarez and Lettinga (1991).
[c]The compound concentration that caused 50% inhibition of the
acetoclastic methanogenic activity.

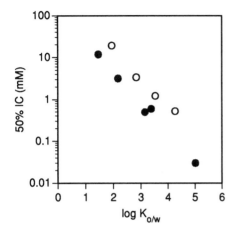

Figure 3. Correlation between the hydrophobicity of a series of chlorinated phenols and of a series of chlorinated benzenes, as quantified by their log $K_{o/w}$ values, and the concentrations at which these compounds caused a 50% inhibition of the conversion of acetate into methane (50% IC). (After Sierra-Alvarez and Lettinga, 1991.)

dency to accumulate transiently during phases where microbial assemblages have to adopt to the degradation of chlorinated compounds.

From the data that are available (Zhang and Wiegel, 1990; Sierra-Alvarez and Lettinga, 1991; Davies-Venn *et al.*, 1992; Madsen and Aamand, 1992; Renard *et al.*, 1993) some indication of the toxicity can be obtained, but the question of what the ecological meaning of this toxicity is immediately arises. It is well known that microbial ecosystems can adapt to stress in various ways, and chlorinated compounds are merely another source of stress. The most elegant way in which microbial ecosystems can deal with the stress exerted by halogenated compounds is by simply eliminating these compounds, i.e., by developing the appropriate enzymatic machinery to dehalogenate the stress factor. How the system may do this is discussed elsewhere in this chapter.

2. Potential Role of Organohalogens as Electron Acceptors

The potential of halogenated organic compounds to serve as electron acceptors in anoxic environments is now well established (Vogel *et al.*, 1987; Tiedje *et al.*, 1987; Mohn and Tiedje, 1992). A wide variety of both aliphatic and aromatic compounds are known to be dehalogenated under anoxic conditions. To evaluate to what extent microorganisms can potentially benefit from catalyzing reductive dehalogenation, it is useful to know the size of the associated change in Gibbs free energy.

2.1. Introduction

In recent years a number of papers have been published that list the Gibbs free energy of formation from the elements for halogenated compounds (Dean,

1985; Dolfing and Harrison, 1992; Holmes *et al.*, 1993; Dolfing and Janssen, 1994). When using such data, care should be taken that the values are in the appropriate format. The convention in microbiology is that Gibbs free energy of formation values are given for a temperature of 25 °C and a pressure of 1 atmosphere (101 kPa). In aqueous solutions, the standard condition of all solutes is 1 mole/liter (Thauer *et al.*, 1977). The main advantage of this way of presenting free energy of formation data is that estimation of the amount of Gibbs free energy that is available from a reaction ($\Delta G'$) for *in situ* conditions can easily be made by substituting the actual concentration for the standard concentration in Equation (1):

$$\Delta G' = \Delta G^{0\prime} + RT \ln [C]^c[D]^d/[A]^a[B]^b \qquad (1)$$

for a hypothetical reaction $aA + bB \rightarrow cC + dD$. $\Delta G^{0\prime}$ in Eq. (1) is calculated from the relationship $\Delta G^{0\prime} = \Sigma \Delta G_f^0$ (products) $- \Sigma \Delta G_f^0$ (substrates) at pH = 7.

The conversion of gas-phase free energy of formation values to aqueous-phase Gibbs free energy of formation values can be made via the formula $\Delta Gf^0_{aq} = \Delta G_f^0{}_{gas} + RT\ln H$, where $\Delta G_f^0{}_{gas}$ is the gas-phase free energy of formation of the compound, R is the universal gas constant (8.314 J/K per mole), T is the temperature (K), and H is the Henry constant in atm/liter per mole. An example of the outcome of such a conversion is given in Table III where the Henry constants and Gibbs free energy of formation values for chlorinated ethylenes are listed. Such data can then be used to quantify the amount of Gibbs free

Table III. Henry's Constants and Gibbs Free Energy of Formation Values for Chlorinated Ethylenes, Ethylene, and Ethane[a]

Compound	H (atm/liter per mole)	ΔGf^0 (kJ/mole)	
		Gas	Aqueous
Tetrachloroethylene	17.7	20.5	27.6
Trichloroethylene	9.58	19.9	25.5
1,1-Dichloroethylene	26.1	24.2	32.3
cis-1,2-Dichloroethylene	4.08	24.4	27.8
trans-1,2-Dichloroethylene	9.38	26.6	32.1
Vinylchloride	27.8	51.5	59.8
Ethylene	214	68.2	81.6
Ethane	501[b]	−32.8	−17.4

[a]All values are for 25 °C. Data for Henry's constant are from Gossett (1987); the Gibbs free energy of formation values for the gaseous state are from *Lange's Handbook of Chemistry* (Dean, 1985).
[b]After Mackay and Shiu (1981).

Table IV. Hydrogen-Consuming Reactions Involved in the Complete Reductive Dechlorination of Tetrachloroethylene [a]

Substrate	Product	$\Delta G^{0'}$ (kJ/mole)	E'_0 (mV)
Tetrachloroethylene	Trichloroethylene	−173.8	487
Trichloroethylene	trans-1,2-Dichloroethylene	−165.0	441
Trichloroethylene	cis-1,2-Dichloroethylene	−169.3	463
Trichloroethylene	1,1-Dichloroethylene	−164.9	440
trans-1,2-Dichloroethylene	Vinylchloride	−144.0	332
cis-1,2-Dichloroethylene	Vinylchloride	−139.7	310
1,1-Dichloroethylene	Vinylchloride	−144.2	333
Vinylchloride	Ethylene	−149.9	362
Ethylene	Ethane	−98.9	99

[a] The $\Delta G^{0'}$ values for the reaction Substrate + H_2 ⟶ Product + H+ + Cl− have been calculated using the free energies of formation listed in Table I. H_2 is in the gaseous state; all other substances are in the aqueous state at 1 mole/kg activity, with the exception of H+ and OH−, which are at 10^{-7} mole/kg activity, (i.e., pH 7 conditions). The redox potentials have been calculated from the $\Delta G^{0'}$ values for redox compound reductions with H_2 after Thauer et al. (1977); the free energy of formation value for Cl− was taken form Stumm and Morgan (1981).

energy that becomes available when the compound is reductively dechlorinated with, for example, H_2 as electron donor (Table IV). Also given in Table IV are the redox potentials of the corresponding redox couples, calculated after Thauer et al. (1977), by using the formula $E^{0'} = (\Delta G^{0'}/-0.193) - 414$. The background behind this formula is that the redox potential is directly related to the change in Gibbs free energy via the equation $\Delta E^{0'} = \Delta G^{0'}/-0.193$; the value of 414 stems from the redox potential of the redox couple $H_2/H^+ + 2e^-$, which is generally chosen as the reference state in biological systems (Thauer et al., 1977). It should be mentioned here that the redox potentials in Table IV are thus for $2e^-$ transfer.

The outcome of these calculations is that the amount of Gibbs free energy that is available from reductive dechlorination of chlorinated ethylenes with molecular H_2 is in the range of 140 to 175 kJ/mole of halogen removed (Table IV). This is essentially the same range as for other chlorinated aliphatic compounds like chlorinated methanes, ethanols, acetates, and propionates (Dolfing and Janssen, 1994), as well as for chlorinated aromatic compounds like chlorinated benzenes, phenols, benzoates (Dolfing and Harrison, 1992), and PCBs (Holmes et al., 1993), where these values range between 130 and 190 kJ/mole of halogen removed. These Gibbs free energy values indicate that organohalogens are indeed good electron acceptors.

The redox potential for reductive dehalogenation of halogenated compounds is in the range of 260 to 570 mV. This is slightly below the value for oxygen ($E^{0'}$

= 818 mV), and comparable to the value for the redox couple NO_3^-/NO_2^- ($E^{0'}$ = 433 mV). These values are much higher than the value for sulfate ($E^{0'} = -217$ mV) (Thauer et al., 1977).

Recently it was reported that microorganisms can dechlorinate tetrachloroethylene via trichloroethylene, cis-1,2-dichloroethylene, and vinyl chloride to ethylene, with the eventual formation of ethane (de Bruin et al., 1992). The data presented here indicate that all these steps can be used by microorganisms as a source of energy with hydrogen as the source of electrons. So far, one dehalogenating organism from this enrichment culture has been described, namely Dehalobacter restrictus (Holliger, 1992; Holliger et al., 1993). This bacterium obtains energy for growth from the reductive dechlorination of tetrachloroethylene to cis-1,2-dichloroethylene with hydrogen as electron donor, but cannot dechlorinate this compound any further. Thus it is likely that other organisms exist that can grow on the reductive dechlorination of dichloroethylene and vinyl chloride.

The amount of energy (expressed per mole of hydrogen consumed) available from the reductive dechlorination of chlorinated ethylenes is much higher than the amount of energy available to sulfate reducers (38 kJ/mole H_2) or methanogens (34 kJ/mole H_2), the most likely hydrogenotrophic competitors of dechlorinating organisms in anoxic environments.

The Gibbs free energy values given in Table IV are all for H_2 concentrations of 1 atmosphere (101 kPa; 600 μM). H_2 concentrations in the aqueous anaerobic environment are generally in the range of 0.2 to 200 nM (Goodwin et al., 1991). Under those conditions the amount of available free energy drops by 20 to 30 kJ per mole of hydrogen consumed (Dolfing, 1992). The advantage of using chlorinated compounds rather than sulfate or carbon dioxide as electron acceptors is thus under in situ conditions greater than under standard conditions. Under standard conditions the energy yield is 140 kJ per mole H_2 for dechlorinating organisms and 35 kJ per mole H_2 for sulfate-reducing or methanogenic bacteria, respectively, but under in situ conditions these yields are 115 and 10 kJ per mole H_2, i.e., a difference of a factor of 10 rather than a factor of 4.

2.2. The Energetic Consequences of the Presence of Halogen Substituents

The presence of a halogen instead of a hydrogen group makes a molecule more oxidized and therefore decreases the number of reducing equivalents that are released when the molecule is mineralized. The presence of a chlorine substituent on an aromatic ring causes a reduction of the amount of energy that is obtained upon complete mineralization of a halogenated compound by, on the average, 82 kJ per mole and substituent if O_2 would have been the alternative electron acceptor (Fig. 4).

If O_2 is the final electron acceptor, the cost of the presence of a halogen

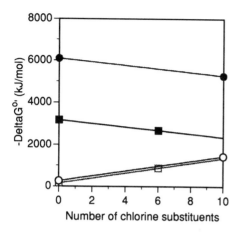

Figure 4. Effect of the presence of chlorine substituents and the type of electron acceptor (O_2 or CO_2) on the Gibbs free energy change for the complete mineralization of chlorinated aromatic compounds ($\Delta G^{0\prime}$) under standard conditions (pH = 7; 298 K; O_2, H_2 CO_2, and CH_4 at partial pressures of 1 atm; all other compounds at concentrations of 1 M). ■, benzenes (aerobic); ●, biphenyls (aerobic); □, benzenes (anaerobic); ○, biphenyls (anaerobic). The calculations were made after Thauer et al. (1977); Gibbs free energy of formation values were from Thauer et al. (1977), Dolfing and Harrison (1992), and Holmes et al. (1993).

substituent is rather small in relative terms. The total decrease in energy gain for chlorobenzenes is only 15% for six chlorine substituents, and for PCBs the decrease in energy gain is still lower: 15% for ten chlorine substituents (Fig. 5). The total energy gain with CO_2 as the alternative electron acceptor is for hexachlorobenzene six times the energy gain of benzene and for decachlorobiphenyl 5.5 times the energy gain of biphenyl. Thus the impetus for microorganisms to use chlorinated compounds, under anaerobic conditions, is very strong. This is true for all compounds with chlorine substituents on the aromatic (benzenoid) ring. For chlorinated alkanes this trend is less distinct (Dolfing and Janssen, 1994).

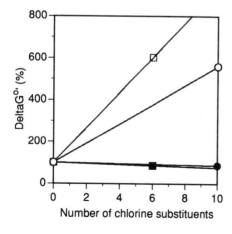

Figure 5. Effect of the presence of chlorine substituents and the type of electron acceptor (O_2 or CO_2) on the Gibbs free energy change for the complete mineralization of chlorinated aromatic compounds ($\Delta G^{0\prime}$) under standard conditions (see Fig. 4); the Gibbs free energy change for the mineralization of the nonchlorinated congeners was taken to be 100%. ■, benzenes (aerobic); ●, biphenyls (aerobic); □, benzenes (anaerobic); ○, biphenyls (anaerobic).

2.3. The Source of Reducing Equivalents for the Dehalogenation Reaction

In the first series of papers on reductive dechlorination, the need for reducing equivalents was not mentioned, and reducing equivalents were not added explicitly to potentially dechlorinating cultures. For those cultures in which the dechlorinated compound is subsequently mineralized, this is generally not a problem, since reducing equivalents will be generated and probably excreted during the mineralization process. This can be illustrated, for example, by the anaerobic mineralization of 3-chlorobenzoate in a three-tiered microbial mixed culture consisting of the dehalogenating bacterium *Desulfomonile tiedjei*, a syntrophic benzoate degrader, and a hydrogenotrophic methanogen, *Methanospirillum* sp. (Dolfing and Tiedje, 1986). All three organisms are needed to achieve mineralization of this compound: first, the dechlorinating organism, which converts 3-chlorobenzoate into benzoate; second, a syntrophic organism, which ferments benzoate to acetate and hydrogen; and third, a methanogen, which converts hydrogen with CO_2 into methane. Hydrogen balance studies indicate that in this food-chain-like assemblage, part of the hydrogen produced by the benzoate degrader is used by the dechlorinating organism as a source of reducing equivalents for the dechlorination reaction. 2,5-Dichlorobenzoate, on the other hand, is only dechlorinated to 2-chlorobenzoate in such a community (Dolfing and Tiedje, 1991a,c) and is therefore not available as a growth substrate for this mixed culture. This explains why it will be difficult if not impossible to obtain enrichments on highly chlorinated compounds such as hexachlorobenzene or pentachlorophenol without addition of additional sources of reducing equivalents.

2.4. Competition for Hydrogen as a Source of Reducing Equivalents

The role of H_2 as an electron donor for reductive dechlorination in the three-tiered mixed culture grown by methanogenic conversion of 3-chlorobenzoate (Dolfing and Tiedje, 1986, 1987) mentioned above suggests that under methanogenic conditions competition for hydrogen may occur between methanogenic and dechlorinating organisms. The actual situation, however, can be more complex (Dolfing and Tiedje, 1991b). Methanogens and dechlorinating organisms in the aforementioned mixed culture draw their hydrogen both from the same pool, while the rate at which this pool is filled depends on the level in the pool, and thus on the rate with which the pool is emptied. This is because the source of hydrogen in the mixed culture is the degradation of benzoate, and the rate at which benzoate is degraded is inversely correlated with the hydrogen partial pressure in the mixed culture. Thus the two groups of hydrogen-consuming organisms supplement each other's activities and create each their own flux of hydrogen.

Cord-Ruwisch *et al.* (1988) have provided evidence in support of the thesis that the threshold for hydrogen uptake decreases if the energy gain from a hydrogen-consuming reaction increases. Since the energy gain from reductive dechlorination is significantly higher than the energy gain from sulfate reduction or methanogenesis, this would imply that the threshold for hydrogen uptake in dechlorinating organisms is significantly lower than in sulfate-reducing bacteria. However, Madsen and Aamand (1991) and Wiegel *et al.* (1992) have presented evidence that competition for H_2 between sulfate reduction and dechlorination can explain the sulfate-related inhibition of the degradation of chlorophenols in their mixed cultures. The rule that the threshold for hydrogen uptake is inversely correlated with the energy gain of the hydrogen-consuming reaction (Cord-Ruwisch *et al.*, 1988) is so appealing that it is tempting to try to reason this violation of the rule away, for example, by suggesting that reductive dechlorination is a relatively new process, with the consequence that the metabolism of the organisms that perform the reaction is not yet optimally adapted. An alternative explanation could be that hydrogen is not always the source of reducing equivalents in mixed cultures (Mohn and Tiedje, 1992).

Coming back to the question of how important the role of reductive dehalogenation potentially is: It is presently possible only to state that chlorinated compounds are in principle very "powerful" electron acceptors. In theory, chlorinated compounds should be able to divert the flow of electrons from other electron acceptors like sulfate or bicarbonate. In most environments, however, there will probably be a large surplus of electrons; so the presence of chlorinated compounds in these environments will not result in a lowering of the electron pressure in such systems, not only because the amounts of organohalogens present will be much lower than the concentrations of sulfate and bicarbonate, but also because in many cases after their dechlorination these compounds will subsequently be further degraded, thereby serving as an electron source rather than as an electron sink.

2.5. Reductive versus Hydrolytic Dehalogenation

Thermodynamic calculations show that, under standard conditions, hydrolytic dehalogenation of chlorinated benzenes and benzoates yields between 65 and 80 kJ per reaction, while reductive dehalogenation of the same series of compounds yields between 140 and 170 kJ per reaction. This indicates that bacteria may exist that thrive on the hydrolytic dehalogenation of chlorinated aromatic compounds.

3. Chemical versus Biological Dechlorination in the Environment

3.1. Substrates for Microbial Reductive Dehalogenation

Both alkyl halides and aryl halides can be dechlorinated by microorganisms. The list of substrates for dechlorination is long (Mohn and Tiedje, 1992), and includes, for example, organochlorine pesticides like dichloro-diphenyl-trichloroethane (DDT), lindane, mirex, and toxaphene, and halogenated C_1 and C_2 solvents such as chloroform, tetrachloroethylene, and trichloroethylene, which will be discussed in more detail. Other examples of halogenated alkanes that are subject to microbial reductive dehalogenation are potentially ozone-destroying compounds like freons CFC-11 and CFC-113 (Semprini et al., 1991; Sonier et al., 1994).

Reductive aryl dehalogenation is also frequently observed. The compound classes that have been proved to be open to microbial reductive dechlorination are the halobenzoates, the halophenols, the haloanilins, the halobenzenes, the PCBs, and polybrominated biphenyls (PBBs) (Mohn and Tiedje, 1992; Morris et al., 1992; Ramanand et al., 1993a), and the chlorinated dibenzo-p-dioxins (Beurskens et al., 1995a). There are no reports on microbially catalyzed reductive dehalogenation of halogenated dibenzofurans. It was recently shown that a wide variety of halogenated heterocyclic compounds, including chlorinated pyridines, bromacil, and 5-bromouracil, can also be dehalogenated via a reductive mechanism (Adrian and Suflita, 1990; Ramanand et al.; 1993b; Adrian and Suflita, 1994; Liu, 1995). Hydrolytic dehalogenation has been reported for several other related compounds, including the triazine herbicides (Kuhn and Suflita, 1989). It is thought that the low electron density on the carbon atoms of the triazine ring facilitates nucleophilic replacement reactions and that these compounds are therefore more likely to undergo hydrolytic dehalogenations.

3.2. Microbial Reductive Dehalogenation

Many pure bacteria cultures are reported to catalyze reductive dechlorination of alkyl halides. The list is long and the organisms are phylogenetically diverse. Two such organisms have been isolated via selection for reductive alkyl dehalogenation activity, viz. *Dehalobacter restrictus* (Holliger, 1992; Holliger et al., 1993) and *Dehalospirillum multivorans* (Scholz-Muramatsu et al., 1995). An interesting characteristic of *D. restrictus*, which was obtained in pure culture recently (Holliger, personal communication, 1995), is that only H_2 and formate are used as electron donor, while the series of electron acceptors is restricted to tetra- and trichloroethylene. *D. multivorans*, on the other hand, can use a wide variety of energy substrates.

Only a few bacterial capable of reductive dehalogenation of aryl halides are currently available in pure culture. Of these organisms, most have been isolated for this specific feature. The first organism to be isolated for its ability to dehalogenate ayl halides was *Desulfomonile tiedjei* (DeWeerd et al., 1990), formerly known as DCB-1, for dechlorinating bacterium number 1 (Shelton and Tiedje, 1984). This organism was obtained from a 3-chlorobenzoate degrading methanogenic enrichment culture. More recently, a series of chlorophenol dechlorinating bacteria have been obtained in pure culture: strain DCB-2 (Madsen and Licht, 1992), strain 2CP-1 (Cole *et al.*, 1994), and *Desulfitobacterium dehalogenans* (Utkin *et al.*, 1994). Whether these chlorophenol degraders conserve energy from the dechlorination reaction, as *D. tiedjei, D. restrictus,* and *D. multivorans* do (Dolfing, 1990; Mohn and Tiedje, 1990; Holliger *et al.*, 1993; Scholz-Muramatsu *et al.*, 1995) still needs to be studied.

Evidence for microbial involvement in the reductive dehalogenation of other classes of aromatic compounds has been obtained with enrichment cultures, rather than with pure cultures. The fact that it has been possible to obtain enrichment cultures with these compounds suggests that energy conservation from the dehalogenation reaction is possible with these compounds as well, at least in those cases where dehalogenation was studied with the halogenated compound with which the enrichment was selected.

3.3. Chemical Dehalogenation

The observed presence of many chlorinated organic compounds in the environment suggests that they are not degraded, or only very slowly. The long half-lives that have been tabulated by Vogel *et al.* (1987) for abiotic degradation processes appear to confirm this. The mere fact that many chlorinated compounds are chemically degradable via reductive dehalogenation steps is interesting, however, not only because the instability of these compounds implies that microorganisms can potentially benefit, i.e., gain energy, from catalyzing such degradation reactions, but also because studies of the chemical mechanisms behind the degradation of chlorinated compounds can give clues about the mechanisms that microorganisms use to degrade these compounds.

In the literature there are few reports on the chemical degradation of chlorinated compounds in the environment. Kriegman-King and Reinhard (1992, 1994) have reported on the transformation of carbon tetrachloride in the presence of sulfide, pyrite, and clays such as biotite, vermiculite, and pyrite. This work clearly showed the significance of mineral surfaces on the transformation rate of a halogenated organic compound. In homogeneous solution with 1 mM HS^- present, the half-life of CCl_4 was 2600 days, but biotite (56 m^2/liter) reduced the half-life to 50 days. Reductive dechlorination contributed only 5–15% to the total CCl_4 transformation, the main transformation product being CS_2, which hydrolyzes to CO_2 at rates that may be appreciable relative to groundwater residence times (Kriegman-King and Reinhard, 1992). More recently, these authors showed that reductive dechlorination of carbon tetrachloride to $CHCl_3$ by pyrite is a surface-controlled reaction. Interestingly, sulfide appeared to inhibit the transformation of CCl_4, probably because sulfide blocked CCl_4 reaction sites at the pyrite surface (Kriegman-King and Reinhard, 1994).

The reactivity of hydrogen sulfide species with alkyl halides is of considerable significance because of the toxicity of many of the organic sulfide reaction products. Roberts *et al.* (1992) recently presented a study showing that abiotic reactions of dihalomethanes with bi- and polysulfide species results in the formation of toxic poly(thiomethylene) and dimercaptomethane at rates that exceed the hydrolysis rates of the halogenated substrates at sulfide concentrations that are commonly encountered in anoxic aquatic systems.

Reduction of carbon tetrachloride by humic acids in the presence of Fe^{2+} and HS^- has also been reported (Curtis, 1991). The humic acids increased the dechlorination about tenfold over the rate observed with the reducing agents alone. Hydroquinone groups present in humic acids may be responsible for at least some of the catalytic activity of this complex organic material.

Sulfide-mediated dehalogenation of methyl halides like methyl bromide in salt marsh sediments is a chemical rather than a biological reaction (Oremland *et al.*, 1994). The product of this nucleophilic substitution reaction is methanethiol, which undergoes further chemical and biological reactions to form dimethyl sulfide, which in turn is metabolized by methanogenic and sulfate-reducing bacteria. Hence, methyl bromide removal is ultimately under biological control (Table V).

A more biochemical approach was taken in studies by the groups of Thauer, Vogel, and Wackett. Transition-metal coenzymes like cobalamin (vitamin B_{12}) and factor F_{430} catalyze the complete reductive dechlorination of tetrachloroethylene via trichloroethylene, dichloroethylene, and monochloroethylene to ethylene (Gantzer and Wackett, 1991). With hematin as the catalyst, the reaction stops at *cis*-1,2-dichloroethylene.

Chlorinated methanes have been shown to undergo reductive dechlorination in a similar manner (Krone *et al.*, 1989a,b). When the chlorine atoms were present on one carbon only in chlorinated ethanes, reductive dechlorination was again observed. But when the chlorine substituents were present on both carbon atoms, the major route was a desaturative elimination. Hexachloroethane, for example, yielded tetrachloroethylene when exposed to hematin or vitamin B_{12} (Schanke and Wackett, 1992). These observations are consistent with observa-

Table V. Degradation of Methyl Bromide in Anoxic Sediments[a]

Reaction type	Substrates	Products
Chemical	$2CH_3Br + 2HS^-$	$\longrightarrow$ $2CH_3SH + 2Br^-$
Chemical	$2CH_3SH + 2CH_3Br$	$\longrightarrow$ $2(CH_3)_2S + 2Br^- + 2H^+$
Biological	$2(CH_3)_2S + 2OH^-$	$\longrightarrow$ $3CH_4 + CO_2 + 2HS^-$
Overall:		
Biogeochemical	$4CH_3Br + 2H_2O$	$\longrightarrow$ $3CH_4 + CO_2 + 4H^+ + 4Br^-$

[a]Based on the work of Oremland *et al.* (1994).

tions that hexachloroethane in waste water and anoxic sediments is converted into tetrachloroethylene. Reductive dechlorination of carbon tetrachloride by vitamin B_{12} proceeds by a one-electron reduction of vitamin B_{12}. Two vitamin B_{12} molecules are involved per reacting carbon tetrachloride molecule (Assaf-Anid *et al.*, 1994).

Cobalamine and hematin can also catalyze reductive dehalogenation of aromatic compounds. The work of Gantzer and Wackett (1991) has shown that hexachlorobenzene can be dechlorinated to pentachlorobenzene by hematin. Cobalamin goes even one step further and dechlorinates hexachlorobenzene via pentachlorobenzene to 1,2,3,5-tetrachlorobenzene (20%) and 1,2,4,5-tetrachlorobenzene (80%), while pentachlorophenol is reductively dechlorinated to 2,3,4,6-tetrachlorophenol (50%) and 2,3,5,6-tetrachlorophenol by cobalamine with titanium citrate as an electron donor (Gantzer and Wackett, 1991). This reduced form of vitamin B_{12}, vitamin B_{12s}, also dechlorinates all tetra- and trichlorophenols, but at much lower rates (Smith and Woods, 1994). Vitamin B_{12s} mainly removes chlorines in positions *meta* or *para* to the hydroxylgroup of chlorophenols. Interestingly, this results in a chemical dechlorination pattern that differs substantially from the pattern exhibited by anaerobic microbial consortia. The regiospecificity of chlorophenol reductive dechlorination by vitamin B_{12s} has been rationalized with the aid of thermodynamics (Dolfing, 1995) and correlates with the most electrophilic position on the ring (Woods and Smith, 1995), which is not unexpected since vitamin B_{12s} is one of the most powerful nucleophiles known (Schrauzer, 1968).

Assaf-Anid *et al.* (1992) have reported that vitamin B_{12} can also dechlorinate 2,3,4,5,6-pentachlorobiphenyl. Consistent with results of *in vivo* experiments with anoxic sediments (Nies and Vogel, 1991), the proton from water was shown to be the source of the hydrogen atom used for the replacement of chlorine on the biphenyl ring. Assaf-Anid *et al.* (1992) used 1,4-dioxane to dissolve the chlorinated model compounds, and diluted this solution with anoxic aqueous buffer to yield a completely miscible 1 : 1 mixture of aqueous buffer and solvent. Reductive dechlorination took place to a lesser extent when methanol was used instead of dioxane. No dechlorination was observed with vitamin B_{12} in a 100% aqueous medium, indicating that the reaction might be limited by the solubility of the chlorinated compounds. Nies (1993) used tetrahydrofuran as a water-miscible solvent to enhance the aqueous solubility of PCBs and chlorinated benzenes in water. In this "biomimetic" system, with vitamin B_{12} as an electron transfer mediator, a difference in redox potential of chlorinated benzenes of only 20 mV in the range of 350 to 450 mV corresponded to a change in dechlorination rate of one order of magnitude.

Recently there has been a series of papers that indicate that microorganisms are not necessarily the active agents in the reductive dechlorination in natural environments. Peijnenburg *et al.* (1992a) have collected data for a range of fresh water sediments with organic carbon contents ranging between 0.5 and 33%. The halogenated aromatic test compounds included benzenes, phenols, toluenes, and anilines. The first step in the degradation route appeared to be a reductive dechlorination. A pseudo-first-order rate of reduction was observed, which was after a time period of varying length followed by an increased rate of reaction, which could again be adequately described by a first-order rate constant. Control experiments with γ-irradiated sediments indicated that initially abiotic dehalogenation took place, and that after some time a faster biological process took over. A quantitative structure–activity relationship (QSAR) was developed to relate these initial and final pseudo-first-order rate constants to four molecular descriptors, namely (1) the bond strength of the carbon–halogen bond to be broken, (2) the summation of the Hammet sigma constants of the additional substituents to the benzene ring, (3) the summation of the inductive constants of the additional substituents to the benzene ring, and (4) the summation of the steric factors of additional substituents (Peijnenburg *et al.*, 1992b). Interestingly, a comparison of the initial and the final QSARs suggested that the same agent may have been involved as the reductant in both processes. The rate constants for reductive dechlorination of seven haloaromatics obtained in ten different sediment samples were correlated with the organic carbon content of the samples. This correlation was observed for both nonadapted and adapted sediments, once again pointing toward a close similarity between both processes. The half-lives in the experiments of Peijnenburg *et al.* (1992a) ranged between <0.5 to 320 days.

The results of the experiments with γ-irradiated sediments preclude a direct coupling between microbial activity and reductive dechlorination, but it is difficult to evaluate to what extent the observed "abiotic" reduction processes are coupled to the activity of microbes, since microbes have created the anoxic, reducing conditions in sediments in the first place and are likely to have been involved in the formation of the hydrogen sulfide and reduced iron that are probably responsible for the observed initial reduction reactions.

It should be stressed that the results are only indicative for the potential of anoxic sediments to catalyze abiotic reductive dechlorination. It is not known to what extent these processes really occur *in situ*. The observed long adaptation times (between 12 and 78 days) suggest that microorganisms are not directly involved in reductive dechlorination of the test compounds *in situ,* hopefully simple because these pollutants were not present *in situ*. It is not known what will happen *in situ* when sediments are exposed to these compounds over a longer period of time. It is conceivable that these compounds are degraded abiotically, but it is also possible that the microflora develops the ability to degrade these compounds. Our knowledge on the processes that govern microbial adaptation to the presence of new chemicals in the environment is still in its infancy (Nishino and Spain, 1993; Pries *et al.,* 1994; van der Meer *et al.,* 1992).

Peijnenburg and co-workers (1991) have presented preliminary information that indicates that reductive dechlorination of aliphatic compounds also occurred in these sediments and could also be described adequately by a similar QSAR. It is not clear from their scant data, however, to what extent the described processes were chemical or biological by nature.

The general picture that emerges from the above discussion is that the abiotic dechlorination of halogenated aliphatic compounds in the environment is possible and can indeed occur at significant rates, e.g., in sulfide-containing environments like salt marsh sediments. For halogenated aromatic compounds, on the other hand, the picture is less clear. Here, the tentative conclusion is that abiotic dehalogenation only occurs in the presence of "biogenic" cofactors.

4. Kinetics of Microbial Dehalogenation Reactions

The kinetic parameters of the dehalogenating organisms determine their efficacy in the removal of halogenated compounds. Since reductive dehalogenation is a redox reaction, it is important to have information on the affinity toward both the electron donor and the electron acceptor. The affinity for the electron donor will determine to what extent the dehalogenating organisms can compete with other organisms for their source of reducing equivalents. This parameter plus the affinity toward the electron acceptor determines at what compound concentrations the organisms can function efficiently in the environment.

Published data on kinetic parameters for reductive dechlorination reactions are scarce. Data from two studies directed toward the development of bioreactor systems for the cleaning of waste streams show that in engineered systems perchloroethylene can be converted at rates of 3 to 25 μmole/liter per hr. Carter and Jewell (1993) reported conversion of tetrachloroethylene to mainly vinyl chloride at a rate of 32 μmole/g volatile solids per day (1.25 μmole/g per hr). At an apparent biomass concentration of 20 g volatile solids per liter this activity resulted in a volumetric conversion rate of 25 μmole/liter per hr. In their attached-film system, which was operated at the fairly low temperature of 15 °C with sucrose as a growth substrate, the apparent K_s value for tetra- and trichloroethylene were 0.054 μM and about 0.3 μM, respectively. These low values suggest that tetra- and trichloroethylene can be degraded when present in groundwater at low concentrations. The low K_s value especially for tetrachloroethylene (9 μg/liter) implies that tetrachloroethylene can be degraded to below the effluent limit of the Safe Drinking Water Act (5 μg/liter) in the anaerobic bioreactor system used by these authors. de Bruin et al. (1992) have reported a volumetric activity of 3.7 μmole/liter per hr for a fixed-bed reactor at temperatures between 10 and 30 °C. In their enrichment, lactate served as source of reducing equivalents, and reductive dechlorination was complete, with ethene and ethane being the final dechlorination products. There are too many differences between the two systems to make a kinetic comparison fruitful, but these two independent studies show that reductive dechlorination can be applied in engineered systems at considerable rates. A recent study by DiStefano et al. (1992) suggests that there is still room for improvement. These authors used methanol rather than sucrose as a source of reducing equivalents and found that dechlorination rates (to ethene as main and vinyl chloride as minor product) in their batch experiments were 200 times higher than those reported for the sucrose-fed systems of Carter and Jewell (1993). It remains to be seen, however, whether this advantage still exists at the low tetrachloroethylene concentrations at which the reactors are to be operated in practice.

The dechlorination rate for tetrachloroethylene by *D. restrictus* with hydrogen as the electron donor was 1 mmole/g cells per hr (Holliger, 1992). This rate is in the same order of magnitude as the rate reported by DiStefano et al. (1991) for their methanol-fed culture, in which the main part of the electrons obtained from methanol degradation was used for reductive dechlorination. The hydrogen consumption rate measured with *D. restrictus* is 10 to 20 times lower than the hydrogen consumption rates reported for hydrogenotrophic methanogens (Dolfing and Bloemen, 1985). The maximal growth rate was $0.024/\text{hr}$ ($t_d = 29$ hr) and the growth yield was at least 2.1 g protein/mole Cl^-. This growth rate was in the same order of magnitude as the growth rate of hydrogenotrophic methanogens, while the growth yield expressed per mole of hydrogen consumed is significantly higher than the yield of methanogens (Vogels et al., 1988). The half saturation constant for hydrogen was not reported by Holliger (1992), so a comparison with the affinity of methanogens and acetogens for hydrogen is not yet possible. In his enrichment procedure, Holliger got rid of hydrogenotrophic methanogens by using 2-bromoethanesulfonate. Omission of selenium and tungsten from the medium resulted in the loss of the homoacetogens from the enrichment culture. The specific activity of *D. restrictus* was five to six times higher than the specific activities of organisms that cometabolically degrade tetrachloroethylene, like *Methanosarcina* and *D. tiedjei* (Fathepure et al., 1987). The molar growth yields of *D. restrictus* and *D. multivorans* on tetrachloroethylene were at 2.1 and 1.4 g protein/mole Cl^-, respectively (Holliger, 1992; Scholz-Muramatsu et al., 1995), comparable to, though somewhat lower than, the molar growth yield of *D. tiedjei* on 3-chlorobenzoate (6.6 g protein/mole Cl^-) (Dolfing, 1990). The growth yield of strain 2CP-1 on 2-chlorophenol was 3 g protein/mole Cl^- (Cole et al., 1994).

Microorganisms are able to adapt to their environment, and dechlorinating organisms are no exception to this. Adaptation can be at various levels, and kinetics are one level where this can be readily observed. For example, the 3-chlorobenzoate-degrading enrichment culture containing *D. tiedjei* was originally reported to exhibit Michaelis–Menten kinetics with an apparent K_s value for chlorobenzoate of 67 μM. After having been cultured in the laboratory for 5 years, however, the organism in pure culture exhibited zero-order kinetics and the K_s had dropped to well below 1 μM (DeWeerd et al., 1986).

Another chlorinated compound that has recently been studied in bioreactor systems is pentachlorophenol. Hendriksen and Ahring (1992) report that glucose-grown granular sludge is able to dechlorinate pentachlorophenol at volumetric activities of 153 μg pentachlorophenol/liter per hr with an apparent K_s value of 2.2 μM. The main dechlorination pathway in this biomass proceeded from pentachlorophenol via 2,3,5,6-tetrachlorophenol and 2,3,5-trichlorophenol to 3,5-dichlorophenol. The maximum activities on these substrates were 0.5, 0.8, 1.3, and 0.4 mg/g per day respectively, a result that was consistent with the observation that dichlorophenols were the major products in the effluent from which these granules were taken. On a molar basis these activities, of about 0.15 to 0.45 umole/g protein per hr, are about 1000-fold lower than the conversion potential for acetate in most granular sludges (Dolfing and Bloemen, 1985). This indicates that there should be possibilities to improve the dechlorinating activity in granular biomass. The pentachlorophenol degrading biomass had been cultivated with glucose as an additional carbon and energy source, and it seems likely that most of the biomass of the sludge granules consisted of organisms that were involved in glucose degradation rather than pentachlorophenol dechlorination.

The chlorophenol-transforming isolate DCB-2 has specific dechlorinating activities toward a number of chlorinated phenols in the range of 25 to 500 μmole/g protein per hr (Madsen and Licht, 1992). The activity of D. tiedjei toward pentachlorophenol falls at 54 μmole/g protein per hr, well within this range (Mohn and Kennedy, 1992). These values, too, suggest that chlorophenol-degrading bacteria formed less than 1% of the biomass in the experiments of Hendriksen et al. (1992).

The kinetic data that are currently available on reductive dehalogenation in laboratory cultures indicate that the specific activities and growth yields of organisms that can grow with reductive dechlorination as energy-generating reaction are at least as "good" as those of methanogenic bacteria. This explains the observations that such dechlorinating organisms are able to degrade halogenated compounds at substantial rates in anaerobic reactor systems.

5. Actual Role of Organohalogens as Electron Acceptors in the Environment

5.1. Man-Made Environments

A solid discussion of the impact of reductive dechlorination on the electron flux in anaerobic environments requires information on two items: the flux of electrons through the systems, and the flux of chlorinated compounds through the system. Currently, there are not many ecosystems for which such information is available. The best-studied example is the original 3-chlorobenzoate-degrading methanogenic enrichment culture with which microbial reductive dechlorination was first described and the three-tiered 3-chlorobenzoate-degrading mixed culture that was set up to mimic the conditions in the original enrichment culture (Dolfing and Tiedje, 1986). In this mixed culture, one third of the electrons produced during degradation of benzoate, the dechlorination product of 3-chlorobenzoate, are, in the form of H_2, used for the dechlorination reaction. In this example, 3-chlorobenzoate is not a net electron acceptor, since the subsequent mineralization of the dechlorination product benzoate results in the generation of more electrons than originally used in the dechlorination reaction.

This work has been followed up by Ahring et al., (1992), who incorporated D. tiedjei into the microbial biomass known as granular methanogenic sludge (Dolfing, 1986) that is widely used in anaerobic wastewater treatment systems of the upflow ananerobic sludge blanket (UASB) reactor type. In these incorporation studies, 6 mM formate was added to the medium plus 0.75 mM 3-chlorobenzoate. Thus, about 10% of the electron flux was channeled into the dechlorination reaction.

Another series of bioreactor studies where chlorinated compounds played a role as electron acceptors was carried out by Hendriksen and co-workers (Hendriksen and Ahring, 1992, 1993; Hendriksen et al., 1992), who studied the degradation of pentachlorophenol in the presence of phenol and glucose. Here, the ratio between glucose and pentachlorophenol was such that less than 0.1% of the electrons liberated from glucose degradation was involved in reductive dechlorination. In a parallel experiment, Hendriksen and co-workers operated a reactor in which only phenol was added as a cosubstrate at a molar ratio (phenol : pentachlorophenol) of 160 : 17. This may seem a large surplus of reducing equivalents, but close scrutiny (Table VI) shows that pentachlorophenol is a net electron acceptor in anaerobic environments. Complete dechlorination of 1 mole of pentachlorophenol requires 5 moles of H_2. The degradation pathway of phenol proceeds via benzoate to acetate and H_2. If we assume that acetate cannot act as

Table VI. The Requirement for an Additional Source of Reducing Equivalents for the Degradation of Pentachlorophenol in Methanogenic Habitats

Pentachlorophenol + 5 H_2	$\longrightarrow$ Phenol + 5HCl
Phenol + H_2 + CO_2	$\longrightarrow$ Benzoic acid + H_2O
Benzoic acid + 6H_2O	$\longrightarrow$ 3Acetic acid + 3H_2 + CO_2
Sum:	
Pentachlorophenol + 3H_2 + 5H_2O	$\longrightarrow$ 3Acetic acid + 5HCl

an electron donor, the mineralization of pentachlorophenol requires 3 moles of H_2 per mole of pentachlorophenol mineralized. A molar ratio of 160 : 17 between phenol and pentachlorophenol in the feed solution therefore implies that probably (that is if the pathway outlined in Table VI is correct) more than 20% of the electrons that were available in the form of H_2 flowed via the dechlorination reaction.

Anaerobic degradation of phenol generally yields only 2 moles of reducing equivalents in the form of H_2 per mole of phenol degraded. Within the group of the chlorinated phenols only the monochlorinated congeners are therefore net electron donors in anaerobic environments; tri-, tetra-, and pentachlorophenol are net electron acceptors. A similar line of reasoning for halobenzoates indicates that here the watershed is at trichlorobenzoate. This discussion can serve to illustrate that it is sometimes necessary to have rather detailed insight into biochemical degradation pathways if one has to evaluate the importance of a compound in the flow of electrons in microbial habitats. Without such knowledge, it would have been logical to base our evaluation on the assumption that degradation of phenol results in the generation of 14 moles of H_2 per mole of phenol mineralized, according to the equation phenol $+ 11 \ H_2O \rightarrow 6CO_2 + 14H_2$. This would have led to the erroneous conclusion that pentachlorophenol is a net electron donor in methanogenic environments, while it is actually a net electron acceptor under environmentally realistic conditions.

For polyhalogenated benzenes and PCBs, analogous evaluations cannot be made as long as the major degradation steps of benzene and biphenyl in anaerobic environments are not known. If we assume that these compounds are not biodegradable in the absence of O_2, the evaluation becomes straightforward that halogenated benzenes and biphenyls are potential electron acceptors in methanogenic ecosystems. For *D. tiedjei* is has been shown that this organism can also use acetate as source of reducing equivalents for reductive dechlorination, but the organism does this only if hydrogen is not available (Dolfing and Tiedje, 1991c).

The currently available information suggests that ethane and ethene are not biodegradable in the absence of O_2. This implies that chlorinated compounds like tetrachloroethylene are net electron acceptors in methanogenic environments.

5.2. Removal Rates of Aryl Halides in the Environment

In environmental chemistry, transformation kinetics are generally described as pseudo-first-order reactions in models that simulate the behavior and fate of pollutants in the environment (Schwarzenbach *et al.*, 1993). The transformation rate is assumed to be solely proportional to the actual concentration of the compound that is studied. The prefix "pseudo" indicates that in reality more variables may govern the reaction rate, for example, the number of actively involved microorganisms, but these variables are assumed to remain constant. Whether this is the most appropriate way to describe the kinetics of microbially catalyzed dehalogenation reactions in the environment may be subject to discussion, but the fact is that generally additional information for deriving more variables is lacking in the few studies that describe *in situ* reactions. In fact, even the calculation of first-order rate constants is frequently hampered. Comparisons of concentrations in the past to present concentrations yield only two time points, and the resulting rate "constants" are minimum values. Environmental rate constants are frequently reported as half-lives, which are inversely proportional to these first-order rate constants. Consequently, the estimated half-lives are maximum values.

The use of pseudo-first-order rate constants may be outlandish to microbiologists who are more used to Monod- and Michaelis–Menten-type kinetics, but the approach certainly has its merits in environmental microbiology. In a recent study, Peijnenburg *et al.* (1992a), for example, showed that after adaptation periods of various lengths the dechlorination rates for a wide range of halogenated aromatic compounds could be described adequately by first-order rate constants. Furthermore, it should be kept in mind that at concentrations that are well below the half-saturation constants of the enzymatic machinery of the cells degradation rates also are pseudo-first-order.

5.3. Natural Environments

Presently, there is no information on the flux of halogenated compounds through anaerobic sediments, let alone on the amount of electrons that is channeled through reductive dechlorination, so the best we can do is evaluate how large the flows of electrons through anaerobic sediments are and compare these flows with some rough estimates on the size of the dechlorination process.

The size of the electron flux, measured as hydrogen, through anaerobic lake sediments varies between 0.02 and 1 μmole/liter per hr (Goodwin *et al.*, 1988). Compared to these rates the rates for reductive dechlorination of, for example, perchloroethylene in bioreactors is at 40 μM/hr (Chu and Jewell, 1994) orders of magnitude higher. This implies that reductive dechlorination has the potential to divert a significant part of the electrons that normally flow to methane, provided that halogenated compounds are available in adequate amounts. To what extent this is the case in sediments can potentially be better approximated by evaluating the potential for reductive dechlorination in actual sediments. The work of Peijnenburg *et al.* (1992a,b) indicates that half-life times of 10 to 30 days are potentially realistic for many easily degradable halogenated compounds in dutch sediments. Taking into account that these values were obtained at sediment water ratios of on the average 1 to 4, this corresponds to dehalogenation rates of approximately 0.1 to 0.3 μmole/liter per day at concentrations of 1 μM. If the actual concentrations are lower, the dehalogenation rates decrease accordingly. A comparison of these values with the flow of electrons to methane in anaerobic sediments indicates that reductive dehalogenation does not play a major role as electron sink reaction in anoxic environments.

Recent preliminary results on the generation of chlorinated aliphatic compounds in the subsurface suggest that under certain conditions chlorinated compounds are produced at rates of about 2 μmole/m^2 per year (Hoekstra and de Leer, 1993, and personal communication). If we assume that this amount would be dechlorinated in a sediment layer of 1 cm, the activity should be 0.2 μmole/liter per year. Clearly, this activity will go unnoticed in eutrophic sediments where the hydrogen flux is 1,000 to 10,000 times this value. This back-of-the-envelope calculation also indicates that reductive dehalogenation does not play a major role in the flow of electrons through natural environments.

5.4. Dehalogenation of Halogenated Aromatics in Polluted Sediments

The determination of *in situ* biodegradation is plagued by major methodological limitations. In order to demonstrate *in situ* biodegradation, information should include (1) laboratory studies that demonstrate the biodegradation potential of the native microbial population; (2) field studies that show concentration profiles at the site suggesting contaminant losses that exceed those expected for abiotic processes and changes in reactants or products indicative of microbial metabolism; and (3) an unequivocal distinction between biotic and abiotic processes (Madsen, 1991). The possibilities to prove *in situ* biodegradation are determined by the circumstances at each field site as well as by our knowledge of the microbial metabolism of the compound under investigation.

In the last decades there have been several reports on the occurrence of halogenated aromatic compounds in sediments, especially in harbors and estuaries. As stated in the introduction, organohalogens are hydrophobic compounds and therefore have the propensity to sorb to suspended solids present in lakes and rivers. These particles tend to settle in estuaries and harbors when streaming velocities of rivers decrease. This phenomenon has contributed to the generation of some highly polluted sediments in, for example, northwestern Europe. Data on the presence of chlorinated benzenes indicate that the sediment levels for the dichlorinated isomers can reach values of several hundred micrograms per kilogram sediment (Table VII), while hexachlorobenzene levels are generally well below 100 μg/kg.

A comparison of recent and old sediment core data indicated a disappearance of hexachlorobenzene in anoxic lake sediment and an increase of tri- and dichlorobenzenes (Beurskens *et al.*, 1993a,b). The maximum half-life for hexachlorobenzene in lake sediment was estimated to be 7 years. Sediment

Table VII. Mean Concentrations of Chlorinated Benzenes in Surface Sediments[a]

	Location				
Chlorinated benzene	Hamburg Harbor[b] ($n = 32$)	Elbe River[c] ($n = 8$)	Lake Ketelmeer[d] ($n = 5$)	Scheldt River[e] ($n = 4$)	Lake Ontario[f] ($n = 11$)
Mono		355			
1,2-Di-	111	249	220	21	11
1,3-Di-	132	181	110	18	74
1,4-Di-	539	536	210	62	94
1,2,3-Tri-	5	5	2	6	7
1,2,4-Tri-	84	54	70	43	94
1,3,5-Tri-	54	13	50	1	60
1,2,3,4-Tetra-	6	5	5	6	33
1,2,3,5-Tetra-	3		2	<dl	6
1,2,4,5-Tetra-	8	11[g]	20	<dl	52
Penta-	5	6	10	3	32
Hexa-	92	50	40	23	97

[a]Concentrations in μg/kg dry weight.
[b]Hamburg Harbor (Germany) is located in the downstream area of the Elbe River. Data from Götz et al. (1990).
[c]Sampling points were mainly upstream Harmburg Harbor. Samples consisted of bed sediment and sediment chamber material. Data from Götz et al. (1993).
[d]Lake Ketelmeer (The Netherlands) is a downstream sedimentation area of the Rhine River; sediment deposited around 1985. Data from Beurskens et al. (1994b).
[e]Data from Oliver and Nicol (1982). dl, detection limit.
[f]Suspended solids, sampled in 1987–1989. Data from van Zoest and van Eck (1991).
[g]Sum of 1,2,3,5- and 1,2,4,5-tetrachlorobenzene.

collected from this area and incubated with freshly spiked hexachlorobenzene demonstrated that the microbial population in this sediment dechlorinated the spiked hexachlorobenzene with a half-life of 11 weeks (Beurskens et al., 1993a). Fathepure et al. (1988) reported a similar half-life for hexachlorobenzene in sewage sludge incubations in the laboratory. In another sedimentation area of the Rhine river, Hollands Diep, sediment core data were related to long-term trends of hexachlorobenzene in the water column; mathematical modeling of sedimentation and transformation processes suggested a half-life for hexachlorobenzene of 2 years in the anoxic sediment (Zwolsman et al., 1993).

    One of the factors that cause the discrepancy between the half-lives in the field and in the laboratory is probably that the actual in situ concentrations of chlorinated benzenes in the pore water of the polluted sediments are relatively low. Sediment levels of several hundreds of micrograms of hexachlorobenzene per kilogram of sediment correspond to a few micromoles per kilogram. At K_d values of approximately 10,000, this translates into concentrations of less than 1 nM in the pore water. It is not surprising that microbial activity is slow at these

Table VIII. Mean Concentrations of Polychlorinated Biphenyls in Surface Sediments[a]

Polychlorinated biphenyl		Location							
IUPAC No.	Structure	Hamburg Harbor[b] (n = 32)	Elbe River[c] (n = 6)	Lippe River[d] (n = 16)	Lake Ketelmeer[e] (n = 18)	Scheldt River[f]	Lake Ontario[g] (n = 38)	Hudson River[h] (n = 97)	Sheboygan River[i] (n = 20)
28	2,4,4'-Tri-		15	47	52	4	17		
52	2,2'5,5'-Tetra-		29	35	44	7	25		
101	2,2',4,5,5'-Penta-		14	20	39	10	27		
138	2,2',3,4,4',5'-Hexa-		18	16	34	18	15		
153	2,2',4,4',5,5'-Hexa-		14	36	34	18	15		
180	2,2',3,4,4',5,5'-Hepta-		3	12	21	10	13		
ΣPCB		507	93	166	224	64	122		
Total PCB (=5 × ΣPCB)		2,535	465	830	1,120	320	570	10,000	150,000

[a] Concentrations in µg/kg dry weight.
[b] Data from Götz et al. (1990).
[c] Sample locations mainly downstream of Hamburg Harbor. Samples consisted of suspended solids. Data from Sturm and Gandrass (1988).
[d] The Lippe River (Germany) is a tributary of the Rhine River. Data from Friege et al. (1989).
[e] Lake Ketelmeer (The Netherlands) is a downstream sedimentation area of the Rhine River. Data form Winkels et al. (1993).
[f] Concentration in suspended solids, collected in the period 1984–1991, and expressed standardized to 5% organic carbon content. Personal communication, B. van Eck, Tidal Waters Division, Rijkswaterstaat, Middelburg, The Netherlands.
[g] Data from Oliver and Niimi (1988). Total PCB concentration is the sum of 67 identified congeners.
[h] Total PCB measured as Aroclor 1242. Data from Bopp et al. (1981).
[i] Total PCB measured as Aroclor 1248. Data form Blasland and Bouck Engineers (1992).

low concentrations. Another factor that may limit the rate at which halogenated compounds are dechlorinated in these sediments is the low rate at which these compounds desorb from sediment particles. Recently, Allard *et al.* (1994) demonstrated that endogenous chlorocatechols in contaminated sediment were not accessible to microorganisms with dechlorinating activity.

For PCBs, essentially the same story can be told. The *in situ* concentrations in northwestern Europe are below or around 1000 μg/kg sediment (total PCBs) (Table VII) and are a reason for public concern, while on the other hand the corresponding concentrations in the pore water are extremely low and may limit dechlorination. There are nevertheless clear indications that slow but distinct dehalogenation of biphenyls occurs *in situ*. The first reports were based on the severely polluted Hudson sediment (Table VIII). Four field studies provide information that enable the calculation of half-lives for individual PCB congeners (Table IX). Congener distribution patterns for PCBs in sediments of the Hudson River (Brown *et al.*, 1987a,b) and Acushnet Estuary (Brown and Wagner, 1990) suggested that reductive dechlorination was a microbial process in these sediments. For the Hudson River, this was confirmed in the laboratory when microorganisms eluted from these PCB-contaminated sediments were shown to dechlorinate commercial PCB mixtures (Quensen *et al.*, 1990). The observations made in the extensive field study in Acushnet Estuary strongly suggest that here too anaerobic microbial populations were involved in changing the congener composition of PCBs in the sediments, but this has not been verified in laboratory experiments. For two other locations, New Bedford Harbor and Lake Ketelmeer, laboratory experiments did confirm the role of native anaerobic microorganisms in PCB dechlorination (Alder *et al.*, 1993; Beurskens *et al.*, 1995a).

PCB half-lives for individual congeners range from a few to more than 100 years; there is no clear relationship between the observed half-lives and the chlorination level of the congeners (Table IX). PCB dechlorination rates can differ considerably between sites. For example, the half-life for 2,3',4',6-tetrachlorobiphenyl was 15 years in Hudson River sediment versus more than 200 years in Acushnet Estuary sediment. For certain other PCB congeners, on the other hand, half-lives were very similar at different sites (Table IX).

Recently, a limited *in situ* debromination of PBBs was demonstrated by Morris *et al.* (1993) by combining field observations with laboratory incubations. In Pine River sediment, an average concentration of approximately 40 mg/kg of PBBs was found. The data indicated a 10% loss of parent compounds during 16 years of "environmental incubation"; maximum half-lives of 100 years for the penta-, hexa-, and heptabromobiphenyls were estimated from these results.

The influence of environmental factors on the dechlorination rates of dichlorophenols to monochlorophenols in pond sediments was studied by Hale *et al.* (1991) by repetitive field sampling. Half-lives of the various dichlorophenols varied between 6 and 215 days. Sediment pH, redox potential, and concentration of sulfate and nitrate accounted for 83% of the variation in half-lives for 2,5- and 3,4-dichlorophenol. The number of 2,4-dichlorophenol-dechlorinating microorganisms was not correlated with the half-lives of this isomer. These results suggest that environmental factors may influence dechlorination of dichlorophenol to a greater extent than chlorine substitution pattern on the aromatic ring or densities of the microbial population.

Complete microbial degradation of halogenated aromatics can be achieved in a two-step reaction, anaerobic dechlorination of the higher chlorinated isomers followed by mineralization of the dechlorination products under air (Fig. 6) (Fathepure and Vogel, 1991; Anid et al., 1991). The first step may take place spontaneously in the deeper anoxic sediment layers and result in an accumulation of di- and trichlorinated isomers, as for PCBs in the Hudson River (Brown et al., 1987a,b). Biological cleanup of polluted sediments demands detailed information with regard to (1) capabilities of the native microbial population; (2) influence of environmental conditions on the specific reactions, e.g., temperature, pH, nutrients, carbon sources, etc.; and (3) the availability of the chlorinated aromatics for the microbial population. If one of these factors causes a serious limitation for a successful biological decontamination, specific manipulations may be helpful. Complete degradation of hexachlorobenzene and PCBs in a two-step reaction has been demonstrated on laboratory scale (Fathepure and Vogel, 1991; Anid et al., 1991), full-scale application will be complex and generally time consuming but not unrealistic as recently demonstrated by Harkness et al. (1993). Several attempts to stimulate in situ microbial dehalogenation have been reported in the literature, sometimes with contradictory results. The addition of organic substrates to stimulate the overall anaerobic microbial activity is a frequently applied concept. Addition of fatty acids enhanced PCB-dechlorinating activity in Hudson River sediment, but did not stimulate the PCB dechlorination in the New Bedford Harbor and Silver Lake sediments (Alder et al., 1993). Addition of butyrate to aquifer slurries capable of dehalogenating chlorinated anilines resulted in the use of an additional dechlorination pathway besides stimulation of the overall reaction rate (Kuhn et al., 1990). Addition of various nutrients resulted in only minor rate accelerations of PCB dechlorination in Hudson sediment (Abramowicz et al., 1993).

The addition of halogenated analogues in order to stimulate the dehalogenation reaction has been reported only once. In laboratory experiments 2,6-dibromobiphenyl (2,6-DBB) was shown to stimulate the dechlorination of Aroclor 1260 (Bedard and Van Dork, 1992). 2,6-DBB was readily dehalogenated to biphenyl by the native microbial activity in pond sediment; however, application in a field trial did not result in the expected dichlorobiphenyl formation (D. L. Bedard, personal communication, 1993).

Table IX. Estimated Half-lives (years) for PCB Congeners in Sediments from Several Sites

IUPAC No.	Structure	Hudson River[a]	Acushnet Estuary[b]	New Bedford Harbor[c]	Lake Ketelmeer[d]
6	2,3′	>150*			
8	2,4′	>150*			
13	3,4′	15*			
16	2,2′,3		<20		
17	2,2′,4	<15*			
18	2,2′,5	<15*			
20	2,3,3′		8		
22	2,3,4′		8		
28	2,4,4′	10			
31	2,4′,5	10		13, 465	
33	2′,3,4	3	7		
35	3,3′,4		7		
37	3,4,4′	3	7		
42	2,2′,3,4′	4	10, >40		
44	2,2′,3,5′	4	10, >40		
45	2,2′3,6	21	20, >200		
46	2,2′,3,6′	15	>200		
47	2,2′,4,4′	150*			
49	2,2′4,5′	150*			
52	2,2′,5,5′	150*			
57	2,3,3′,5		7, 20		
60	2,3,4,4′	3			
63	2,3,4′,5		7, 20		
66	2,3′,4,4′	3	10		
67	2,3′,4,5		10		
70	2,3′,4′,5	3	10		
71	2,3′,4′6	15	>200		
74	2,4,4′,5	3	10		
82	2,2′,3,3′,4	6			
84	2,2′3,3′,6	50			
85	2,2′,3,4,4′	8	6		
87	2,2′,3,4,5′	8	6		
91	2,2′,3,4′,6	30	>200		
95	2,2′,3,5′,6	30	>200		
96	2,2′,3,6,6′		>200		
97	2,2′,3′,4,5	8			
99	2.2′,4,4′,5	10	20		
101	2,2′,4,5,5′	10	20		
105	2,3,3′,4,4′	4		4, 8	9
110	2,3,3′,4′,6	15	>200		
114	2,3,4,4′,5	4			
118	2,3′,4,4′,5			7	
123	2′,3,4,4′,5		10		

(continued)

Table IX. (*Continued*)

IUPAC No.	Structure	Hudson River[a]	Acushnet Estuary[b]	New Bedford Harbor[c]	Lake Ketelmeer[d]
124	2′,3,4,5,5′		10		
126	3,3′,4,4′,5				6
128	2,2′,3,3′,4,4′	10			
133	2,2′,3,3′,4,6′		10		
135	2,2′,3,3′,5,6′		>200		
136	2,2′,3,3′,6,6′		>200		
137	2,2′,3,4,4′,4	8	7		
138	2,2′,3,4,4′,5′	10			
139	2,2′,3,4,4′,6		7		
141	2,2′,3,4,5,5′	8	7		
144	2,2′,3,4,5′,6		7		
149	2,2′,3,4′,5′,6	>30	>200		
153	2,2′,4,4′,5,5′	10		19	
156	2,3,3′,4,4′,5				10
162	2,3,3′,4′,5,5′		20		
163	2,3,3′,4′,5,6		>200		
167	2,3′,4,4′,5,5′		20		
169	3,3′,4,4′,5,5′				6
170	2,2′,3,3′,4,4′,5	21	10		
172	2,2′,3,3′,4,5,5′		10		
174	2,2′,3,3′,4,5,6′		10		
175	2,2′,3,3′,4,5′,6		7		
176	2,2′,3,3′,4,6,6′		10		
177	2,2′,3,3′,4′,5,6		10		
180	2,2′,3,4,4′,5,5′	21	10		
181	2,2′,3,4,4′,5,6	15	10		
183	2,2′,3,4,4′,5′,6		7		
185	2,2′,3,4,5,5′,6	15	10		

[a]Half-lives based on data from Brown *et al.* (1987b). *Dechlorination half-lives may be underestimated by simultaneous formation of these congeners via dechlorination of higher congeners.

[b]Half-lives based on data from Brown and Wagner (1990); two half-lives for a single congener refer to different dechlorination patterns.

[c]From Lake *et al.* (1992); two half-lives for a single congener refer to two separate locations.

[d]From Beurskens *et al.* (1993a).

Taking into account the slow release of chlorinated aromatics from sediments, microbial dehalogenation generally takes years to decades. One should realize the limited impact of stimulation measures on this time scale. Probably measures have to be repeated frequently to be effective, and this will be costly.

The sparse information compiled above indicates that *in situ* microbial dehalogenation of chlorinated benzenes and biphenyls in a slow process that proceeds with half-lives of at least several years. Dichlorophenols, on the other

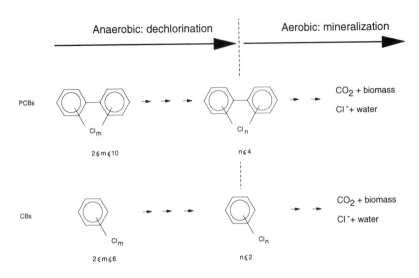

Figure 6. Scheme for the complete microbial degradation of halogenated aromatic compounds in a two-step reaction: anaerobic dechlorination of the higher chlorinated isomers followed by aerobic mineralization of the dechlorination products.

hand, appear to be dechlorinated in the environment with half-lives of less than 1 year.

5.5. Microbial Dehalogenation in Marine Sediments

Dehalogenation of brominated and chlorinated phenols in marine sediments was described by King (1988). Dehalogenation with the consequent production of phenol appeared to initiate anaerobic degradation of 2,4-dibromophenol. Sulfate-reducing bacteria did not dehalogenate dibromophenol, but appeared to degrade phenol or end products of phenol fermentation. The information that bromophenols, which are naturally produced in intertidal marine mudflats, are reductively dehalogenated by the native anaerobic microflora (King, 1986) may give the impression that the potential for reductive dehalogenation is ubiquitous in marine environments. This is not necessarily correct, however, as shown by results of Kohring et al. (1989), who could not detect dechlorinating activity in anaerobic samples obtained from a sea grass bed at the Bahamas. Abrahamsson and Klick (1991) recently described that they observed anaerobic degradation of halogenated phenols in marine sediments that had been contaminated with effluent from a paper and pulp mill, but these authors could not detect dechlorination in nonpolluted anaerobic sediments.

In a study by Häggblom and Young (1990), chlorophenol dechlorination and subsequent substrate oxidation in marine sediment was shown to be coupled to sulfate reduction. This observation is different from the results obtained by King (1988), and underscores that a single observation should not be taken as representative for "the" marine environment.

Microbial dechlorination of hexachlorobenzene in the presence of 27 mM sulfate by microorganisms from estuarine environments is described by Beurskens *et al.* (1995b). Brown and Wagner (1990) describe the dechlorination of PCBs in an estuarine sediment, but unfortunately detailed information about salinity and sulfate concentrations are lacking. Under sulfate-reducing conditions, PCB-dechlorinating activity was demonstrated in sediment that originated from an estuarine area, New Bedford Harbor; however, two freshwater sediments showed no PCB dechlorination when incubated under sulfate-reducing conditions (Alder *et al.*, 1993). The influence of sulfate on dehalogenation reactions has been studied frequently and with contradictory results. Several reports conclude that reductive dehalogenation by anaerobic microbial communities is inhibited by sulfate (Kuhn *et al.*, 1990; Genthner *et al.*, 1989; Allard *et al.*, 1992; Mohn and Tiedje, 1992). However, exceptions to the above findings exist (Bosma *et al.*, 1988; Kohring *et al.*, 1989; Häggblom and Young, 1990; Häggblom *et al.*, 1993). Inhibitory effects were generally observed with freshwater sediments or inocula obtained from freshwater sediments that generally are methanogenic. Introduction of high-sulfate concentrations in these incubations means a drastic change in environmental conditions. One of the changes will involve the concentrations, the nature, and the flux of the reducing equivalents, which consequently may become less available to the dechlorinating organisms. Adding sulfate to methanogenic sediments may provide insight into the diversity of the dehalogenating microbial population. Unfortunately, these experiments have led to the impression that microbial dehalogenation in marine environments might be rare. In fact, demonstration of microbial dehalogenation in marine sediments requires different research efforts, namely incubations with originally sulfidogenic, marine sediments. Additional subjects such as the influence of salinity on microbial dehalogenation can then be addressed, too. Presently, only a few studies with marine or estuarine sediments are available; this type of research clearly needs more attention in the future.

5.6. Dehalogenation of Chlorofluorocarbons

The threat of volatile organohalogens toward the ozone layer (Molina and Rowland, 1974) does not need much introduction. Recent work with anoxic sediments and soils has shown that anaerobic microorganisms can degrade at least some of the CFCs, namely CFC-11 and CFC-12 (Table X), even at the low concentrations that are actually present in the atmosphere (Lovley and Woodward, 1992). In this study, no dehalogenation products could be detected; so it

Table X. Nomenclature and Degradation of Fluoroalkanes

		Anaerobic		Aerobic
Abbreviation	Formula	Sediment[a]	Corrinoids[b]	methanotrophic[c]
CFC-11	CFCl$_3$	+	+	−
CFC-12	CF$_2$Cl$_2$	+	+	
CFC-13	CF$_3$Cl		+	
HFC-134	CHF$_2$CHF$_2$			−
HFC-134a	CF$_3$CH$_2$F			−
HFC-143	CHF$_2$CH$_2$F			+
HCFC-21	CHCl$_2$F			+
HCFC-123	CF$_3$CHCl$_2$			−
HCFC-131	CHCl$_2$CHClF			+
HCFC-141b	CCl$_2$FCH$_3$			+
HCFC-142b	CClF$_2$CH$_3$			−

[a]Data are from Lovley and Woodward (1992).
[b]Data are from Krone et al. (1991).
[c]Data are from DeFlaun et al. (1992).

still has to be proved that the process that occurred was a reductive dechlorination reaction, but the observation that corrinoids can catalyze the reductive dechlorination of CFCs (Krone et al., 1991) makes this a plausible assumption. Unfortunately, uptake rates of CFCs in the different anoxic environments varied considerably—a well-known phenomenon in microbial ecology—and it is therefore premature to attempt to extrapolate global removal rates from the currently available data. Furthermore, anoxic environments other than sediments and soils that have been screened so far, such as termite mounds and the rumen of various herbivores, may also affect the global consumption of CFCs. The observed accumulation of CFCs in the troposphere, however, indicates that CFC uptake in anoxic habitats is minor compared to the current anthropogenic release of these compounds.

Other interesting observations in this context are (1) that the redox potential does not have to be at the level of methanogenesis for CFC consumption to proceed, since actively sulfate- or iron-reducing sediments can also consume CFCs; and (2) that there is no direct relationship between the rates of CFC-11 and CFC-12 consumption. It is not known which organisms are involved in anaerobic degradation of CFCs. Experiments with *Clostridium pasteurianum* have shown that anaerobic growth of this organism with glucose as a carbon and energy source resulted in CFC uptake. Similar studies with *Escherichia coli* did not result in CFC consumption (Lovley and Woodward, 1992). Recently, Sonier *et al.* (1994) have presented evidence for CFC dechlorination by sulfate-reducing bacteria.

In this context it is also worth mentioning that *Nitrosomonas europaea* is able to catalyze the reductive dehalogenation of the trichloromethyl group of nitrapyrin (Vannelli and Hooper, 1993). So far, this is the only demonstration of catalysis of a reductive dechlorination by ammonia mono-oxygenase, but the observation opens the possibility that catalysis of other reductive dechlorination reactions by this organism can occur. It will be especially interesting to see whether, and if so how, ammonia-oxidizing bacteria degrade CFCs. The activity of ammonia-oxidizing bacteria toward halogenated aliphatics shows many similarities with the activity of methanotrophic bacteria toward this class of compounds, and it was recently reported that methanotrophic bacteria can degrade some fluorocarbons (DeFlaun *et al.*, 1992). No metabolites were detected in these studies.

6. Dehalogenation Pathways in Anoxic Habitats

The observation that reductive dechlorination of polychlorinated compounds is a sequential process gives rise to the question of whether this results in the development of distinct dechlorination patterns, and if so, whether these patterns can be rationalized. Two approaches have been taken to predict or explain (microbial) degradation patterns in anoxic environments. One makes use of the redox potential of the various redox couples. The other one uses the bond charge of the carbon–halogen bond to be broken. The latter model is based on the assumption that the halogen in the region (represented by the carbon–chlorine bond charge) with the largest negative charge is most likely to be reductively dechlorinated (Cozza and Woods, 1992).

The rationale for the hypothesis that redox potentials can be used to predict the dehalogenation pathway of polychlorinated compounds is based on a presumed analogy to the preferential use of inorganic electron acceptors in microbial habitats. Here, the electron acceptor with the highest redox potential, generally oxygen, is used first, then nitrate, after that sulfate, and finally carbon dioxide (Fenchel and Blackburn, 1979; Zehnder and Stumm, 1988). The hypothesis is therefore that the microbially catalyzed dehalogenation pathway will follow the redox potential of the various redox couples, i.e., that the couples with the highest energy yield will be used preferentially (Dolfing and Harrison, 1993).

6.1. Chlorinated Benzenes

The first test of the hypothesis that the microbially catalyzed dehalogenation follows the redox potential of the various redox couples in such a way that the couples with the highest redox potential are used preferentially has been done by evaluating the observed dechlorination patterns for chlorinated benzenes in a hexachlorobenzene-adapted enrichment culture (Dolfing and Harrison, 1993). It was found that in this enrichment culture (Fathepure *et al.*, 1988) the major dechlorination pathway indeed followed the pathway "predicted" by thermodynamics: hexachlorobenzene was preferentially dechlorinated via pentachlorobenzene and 1,2,3,5-tetrachlorobenzene to 1,3,5-trichlorobenzene. The latter compound was not degraded any further in this enrichment. When challenged with hexachlorobenzene, a mixed culture enriched on 1,2,3-trichlorobenzene also followed this dechlorination pathway (Holliger *et al.*, 1992), thus lending further credit to the hypothesis that the main dechlorination pathway of hexachlorobenzene dechlorination proceeds in accordance to the redox potential of the various redox couples, with a step with a higher redox potential being preferred over a step with a lower redox potential.

The values for the redox potential used in the above evaluation had been obtained via thermodynamic calculations (Dolfing and Harrison, 1992). In the literature there are also measured reduction potentials for chlorinated benzenes, but these reduction potentials have been determined in dimethylsulfoxide rather than in water (Farwell *et al.*, 1975). the dechlorination pattern that would be predicted with these measured values did not match the observed microbial dechlorination pattern. The calculated redox potentials on the other hand were apparently accurate enough to make the correct prediction of the dechlorination pattern.

It should be noted that the above evaluation was based on only a few dechlorination steps with a few enrichment cultures, so there is clearly a need for more experimental data (also because the presumed analogy between inorganic compounds and chlorinated compounds as electron acceptors is merely an analogy). Inorganic compounds serve as final electron acceptors, and the organisms that use the electron acceptor with the highest redox potential have a clear energetic advantage over organisms that use an electron acceptor with a lower redox potential (Zehnder and Stumm, 1988; Fenchel and Blackburn, 1979). With a compound like hexachlorobenzene the situation is slightly different in that this compound is stepwise dechlorinated and the amount of potential energy available from dechlorination is independent of the pathway taken. Thus, to a microbial system as a whole, it will make no difference which pathway will be followed; the potential amount of energy available to the organisms involved is always the same and cannot be a criterion for the selection of the most efficient pathway in an ecosystem that is fully adapted to the use of these compounds as a source of energy (Dolfing and Harrison, 1993). This reservation is not valid in those cases where polyhalogenated compounds are dehalogenated only partially.

The dechlorination pathway for hexachlorobenzene predicted by the QSAR developed by Peijnenburg *et al.* (1992a,b) also predicts the observed dechlorination pathway correctly. This is interesting because the redox potential as such does not come up in the QSAR. There was only a poor correlation between the redox potential for the various dechlorination steps and the kinetic constants for these steps as determined by the QSAR.

As alluded to above, it is not uncommon to observe the simultaneous occurrence of more than one pathway in hexachlorobenzene-adapted anaerobic enrichment cultures. We recently described a hexachlorobenzene-adapted enrichment culture obtained from a sedimentation area of the Rhine with proven *in situ* dechlorination of hexachlorobenzene (Beurskens *et al.*, 1994a). With this culture, we observed a striking correlation between the selectivity of the culture toward the individual chlorobenzene congeners that could be dechlorinated and the thermodynamics of the various dechlorination steps. Of the 19 dechlorination reactions possible with benzenes that contain at least two chlorines, only the seven with the highest energy release (highest redox potential) took place.

6.2. PCBs

The apparent applicability of the idea that thermodynamics can be used to rationalize the dechlorination pattern of chlorinated benzenes makes it tempting to apply this concept also to the degradation pattern of PCBs. The paucity of single-congener studies, however, makes this a somewhat risky undertaking. The search for patterns in the dechlorination pathway started as soon as the first reports on environmental dechlorination of these compounds were published. Brown *et al.* (1987a,b) described a series of patterns, but there was no obvious way to rationalize these patterns. The only generalization that could be made at that time was that ortho-dechlorination was seldom observed, and that dechlorination of higher chlorinated congeners appeared easier than dechlorination of lower chlorinated PCBs (Quensen *et al.*, 1988). The observation that ortho-dechlorination is rare has been made by several groups. In fact, the observation of biologically mediated ortho-dechlorination was reported only recently (Van Dort and Bedard, 1991). If we accept that ortho-dechlorination does not generally occur, it appears possible again to use thermodynamics to predict or rationalize the dechlorination pattern observed in the single-congener studies that have been reported so far (Abramowicz *et al.*, 1993; Rhee *et al.*, 1993a,b,c; Williams, 1994). An example of such an evaluation is given in Table XI. The observed dechlorination steps indeed obey the rule that the dechlorination step that is taken follows from the redox potential of the corresponding redox couple. However, removal of the ortho-chlorine would have yielded the highest energy of the dechlorination steps possible at the different chlorination levels (Table XI). The frequently observed stability of ortho-chlorines indicates that factors other than thermodynamics are also involved in determining which dechlorination pathway is followed in the environment. Steric hindrance could be one of those factors.

The cultures that were used for the above-mentioned dechlorination studies

Table XI. Gibbs Free Energy Values for the Reductive Dechlorination ($\Delta G^{0\prime}$) of PCBs with Hydrogen as Electron Donor in Laboratory Incubations from Three Laboratories

Substrate	Product	$\Delta G^{0\prime}$ (kJ/reaction)	Position	Adjacent Cl	Dechlorination reaction[a]
Abramowicz *et al.* (1993)					
2,3,4,3',4'	3,4,3',4'	−164.1	*o*	1	−
	2,4,3',4'	−162.3	*m*	2	+
	2,3,3',4'	−154.0	*p*	1	−
	2,3,4,3'	−152.9	*p*	1	−
	2,3,4,4'	−152.1	*m*	1	−
2,4,3',4'	3,4,4'	−156.0	*o*	1	−
	2,4,3'	−152.5	*p*	1	+
	2,4,4'	−151.1	*m*	1	−
	2',3,4	−146.4	*p*	0	−
2,4,3'	3,4'	−157.3	*o*	0	−
	2,3'	−147.7	*p*	0	+
	2,4	−145.0	*m*	0	−
2,3'	3	−154.9	*o*	0	−
	2	−144.3	*m*	0	+
Rhee *et al.* (1993c)					
2,3,4	3,4	−167.6	*o*	1	−
	2,4	−162.1	*m*	2	+
	2,3	−155.8	*p*	1	−
2,4,5	3,4	−159.3	*o*	0	−
	2,5	−153.8	*p*	1	+
	2,4	−153.8	*m*	1	+
Williams (1994)					
2,3,4	3,4	−167.6	*o*	1	−
	2,4	−162.1	*m*	2	+
	2,3	−155.8	*p*	1	−
2,3,5	3,5	−164.1	*o*	1	−
	2,5	−153.9	*m*	1	+
	2,3	−147.6	*p*	0	−
2,3,6	2,5	−156.3	*o*	1	−
	2,6	−152.3	*m*	1	+
	2,3	−150.0	*o*	0	−
2,4,5	3,4	−159.3	*o*	0	−
	2,5	−153.8	*p*	1	+
	2,4	−153.8	*m*	1	−
2,4,6	2,4	−151.7	*o*	0	−
	2,6	−147.7	*p*	0	+
3,4,5	3,5	−163.1	*p*	2	+
	3,4	−158.4	*m*	1	−

[a]Presence (+) or absence (−) of reaction.

182

were taken directly from the environment. We recently did cross-acclimatization studies in which we analyzed hexachlorobenzene-adapted enrichments for cross-acclimatization toward PCBs (Beurskens *et al.,* 1995c). These studies were done with individual PCB congeners, and here, too, the dechlorination pattern obeyed the rule that the dechlorination step that is taken follows from the redox potential of the corresponding redox couple (J. Dolfing and J. E. M. Beurskens, unpublished data). For this culture, as well as for the experiments reported by Williams (1994), the rule can also be formulated in a more accessible way; namely, by stating that those chlorines that are flanked at both sites by other chlorines are removed preferentially. Originally we speculated that the pattern observed with the hexachlorobenzene-adapted enrichment culture had to be explained by the history of the culture that was used, namely the preadaptation toward hexachlorobenzene, since in the original hexachlorobenzene dechlorinating culture, too, those chlorines that were removed preferentially were the chlorines that were flanked at both sides by other chlorines (Beurskens *et al.,* 1994a). The observation of a similar pattern in cultures that were taken directly from the environment, however, implies that this is a general phenomenon.

Recent work with the above-mentioned hexachlorobenzene-adapted enrichment culture has revealed that this culture is not only cross-adapted to PCBs but also to chlorinated dioxins (Beurskens *et al.,* 1995a). With regard to dechlorination pathways of PCBs, several additional remarks can be made. One final point that needs to be stressed is that with respect to chlorines located at the meta or para position, it is the chlorination pattern rather than the position that determines which chlorine is removed preferentially (Rhee *et al.,* 1993a–c).

6.3. Chlorinated Phenols

Reductive dechlorination of chlorinated phenols has been observed for unacclimated and acclimated anoxic sewage sludges, sediments, soils amended with sewage sludge, and aquifers. Biotransformation pathways vary with the characteristics of the microorganisms present. Unacclimated cultures preferentially remove chlorines from the position ortho to the hydroxyl group. For example, unacclimated cultures transformed pentachlorophenol to produce 2,3,4,5-tetrachlorophenol and 3,4,5-trichlorophenol (Boyd *et al.,* 1983; Boyd and Shelton, 1984). Acclimated enrichment cultures may yield biotransformation pathways different from those of unacclimated cultures. Mikesell and Boyd (1986) observed that sludges acclimated to individual monochlorophenols produce different initial PCP products. Sludge acclimated to 2-chlorophenol produced 2,3,4,5-tetrachlorophenol from pentachlorophenol by reductive dechlorination of the ortho-chlorine. Sludge acclimated to 3-chlorophenol dechlorinates pentachlorophenol as the meta position to yield 2,3,4,6-tetrachlorophenol, and sludge acclimated to 4-chlorophenol yields 2,3,5,6-tetrachlorophenol by para-dechlorination. The presently available data suggest that environments exposed to pentachlorophenol over a long period may yield a large number of metabolites and a complex pentachlorophenol degradation pathway (Nicholson *et al.*, 1992). Our hypothesis is that during the development of the degradation pathway, organisms come to the fore that have a different preference for the various isomers. As stated above, the ortho-chlorines are removed first, but once these chlorines are removed, other enzymes/organisms come to the fore that dehalogenate the meta and para positions. Once these systems are present, they will also act on pentachlorophenol, and a mixed dechlorination pattern develops.

For completeness it should be mentioned that pentachlorophenol degradation in unadapted cultures does not necessarily proceed via the thermodynamically predicted pathway pentachlorophenol → 2,3,4,6-tetrachlorophenol → 2,4,6-trichlorophenol → 2,6-dichlorophenol → 2-chlorophenol → phenol (Hendriksen *et al.*, 1992; Hendriksen and Ahring, 1993).

Cozza and Woods (1992) have proposed using the bond charge of the carbon–halogen bond to predict the degradation pathway of halogenated aromatic compounds. Unfortunately, these authors did not provide data on the bond charge for chlorinated benzenes and PCBs, so for these compounds, where the use of the redox potential led to correct predictions, it is not possible to compare both approaches. For chlorinated phenols, however, it is possible to make this comparison. The data on the bond charge predict a dechlorination pattern for pentachlorophenol that starts with ortho removal followed by removal of the chlorosubstituent in para position, namely pentachlorophenol → 2,3,4,5-tetrachlorophenol → 3,4,5-trichlorophenol → 3,5-dichlorophenol → 3-chlorophenol → chlorophenol. The major pathway in studies with upflow anaerobic sludge blanket reactors observed by Hendriksen and co-workers, however, proceeded via 2,3,5,6-tetrachlorophenol. In parallel experiments with fixed-film reactors, the major intermediates were 2,3,4,5-tetrachlorophenol and 3,4,5-trichlorophenol (Hendriksen *et al.*, 1992). Results like these indicate that it is not very meaningful to try to predict a priori the pattern of pentachlorophenol degradation in unacclimated mixed cultures, and it is thus presently not possible to evaluate the relative merits of the use of the redox potential versus the use of the carbon halogen bond charge to predict the dechlorination pattern in anoxic microbial habitats.

It is tempting to speculate that a correct prediction of the dechlorination pathway of chlorinated aromatic compounds, if obtained by evaluating the bond strength of the various chlorosubstituents, would provide hints on a possible dechlorination mechanism. The rate of reductive dechlorination is probably related to the reduction potential of the transfer of the first electron from or to the compound of interest, if we assume that the two electrons that are involved in reductive dechlorination are transferred sequentially. There are currently no data available on one-electron reduction potentials for halogenated compounds in aqueous solution, but it seems reasonable to limit our evaluation to a qualitative discussion of the factors that determine the free-energy of activation for adding an electron to a polyhalogenated compound (Schwarzenbach *et al.*, 1993). The addition of an electron to a carbon–halogen bond causes a partial dissociation of this bond, which is subsequently cleaved to yield a carbon radical and a halide ion. Thus, the factors that will have the largest effect on the free energy of the activated complex are probably the strength of the carbon–halogen bond and the stability of the carbon radical formed, and a good correlation between the predicted and observed dechlorination pathways can be taken to suggest that the electrons in reductive dechlorination are transferred sequentially.

7. Consequences of Dehalogenation in the Environment

The lower- and nonchlorinated compounds formed during reductive dechlorination are less bioaccumulative, generally less toxic, and more suitable substrates for aerobic biomineralization. These environmentally beneficial consequences of microbial dehalogenation are generally emphasized in the literature. Formation of lower-chlorinated compounds, however, means also an increase in water solubility relative to the water solubility of the parent compound. Consequently, reductive dehalogenation in sediments may result in the formation of mobile metabolites. Another phenomenon that has not received much attention is that organohalogens like methyl iodide are known to mobilize heavy metals (from anoxic environments). Degradation of organohalogens deletes this phenomenon, and therefore may affect the presence and toxic effects of heavy metals in other environmental compartments.

7.1. Generation of Mobile Metabolites

At many sites in Western Europe, river water infiltrates through sediments into aquifers and transport of mobile metabolites of reductive dechlorination may have adverse effects on groundwater quality (Zoeteman *et al.*, 1980; McCarty *et al.*, 1981; Schwarzenbach *et al.*, 1983). Transport of di- and trichlorobenzenes from polluted sediments to deeper, originally unpolluted layers was demonstrated by Beurskens *et al.*, 1993a,b). Long-distance transport of dichlorobenzenes over 3.5 km in groundwater illustrates the mobility of these compounds when present in groundwater (Barber, 1988). One of the factors that determines the rate of downward seepage of lower chlorinated compounds is the sediment–water partition coefficient of these compounds. The octanol–water partition coefficient ($K_{o/w}$) can be used as an indication for the sediment–water partition coefficient and the related mobility (Schwarzenbach *et al.*, 1993). The plot of log $K_{o/w}$ versus the number of chlorines in chlorinated benzenes, biphenyls, and dioxins, as shown in Fig. 7, illustrates that the potential mobility of

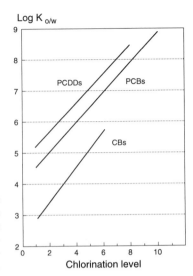

Figure 7. The relationship between the average octanol–water partitioning coefficient ($K_{o/w}$) and the number of chlorines present in chlorinated benzenes (CBs), polychlorinated biphenyls (PCBs), and polychlorinated dibenzo-*p*-dioxins (PCDDs). Log $K_{o/w}$ values from de Bruijn *et al.* (1989) and Shiu *et al.* (1988).

chlorinated compounds increases with decreasing chlorination levels. From Fig. 7 it is clear that the microbial dechlorination of PCBs and chlorinated dioxins results in metabolites that are more hydrophobic than trichlorobenzenes. Since trichlorobenzenes are virtually immobile in the environment, it can be concluded that dechlorination of PCBs and chlorinated dioxins will not result in the formation of metabolites that are transported easily in the environment. Dechlorination of chlorinated benzenes in anoxic sediments, however, may cause release of chlorinated compounds into previously noncontaminated environments.

Tri- and dichlorobenzenes are not necessarily persistent in anoxic environments. Bosma *et al.* (1988) have described the dechlorination of these compounds to monochlorobenzene in column studies with methanogenic Rhine sediment columns. Biomineralization of dichlorobenzenes and monochlorobenzene may occur if, after long-distance transport, these compounds arrive in environments where molecular oxygen or nitrate are present as potential electron acceptors (Bosma, 1994). In this way, the sequential redox conditions that mobile contaminants encounter may eventually result in complete elimination of toxicants (Zitomer and Speece, 1993).

7.2. Mobilization of Heavy Metals

Anoxic sediments frequently act as a sink for heavy metals, which precipitate as metalsulfides. About 10 years ago it was discovered that methyl iodide can mobilize metals such as iron, manganese, and lead out of anoxic sediments into water (Thayer *et al.*, 1984): methyl iodide attacks the sulfur atoms of metal sulfides, breaks bonds, destroys the lattice, and enables the metals to escape into the surrounding water:

$$FeS(s) + 2CH_3I(aq) \rightarrow (CH_3)_2S(g) + Fe^{2+}(aq) + 2I^-(aq)$$

Presently, it is not clear whether other methyl halides have the same potential. It is clear, however, that reductive dechlorination and methylation of heavy metals are two competing reaction pathways for methyl iodide, and that both reactions may occur in the same anoxic environment. The occurrence of reductive dechlorination can theoretically influence the remobilization of heavy metals from anoxic environments, but the surplus of sulfide in most anoxic environments suggests that the relevance of this process is negligible.

Another reaction that may also be of environmental importance is the enhancement of the dissolution of metals by methyl iodide (Thayer *et al.*, 1987). This reaction can be described by the equation

$$M(s) + CH_3I(aq) \rightarrow CH_3MI(aq) + H_2O(l) \rightarrow CH_4(g) + M(OH)^+(aq) + I^-(aq)$$

Not only other alkyl bromides and alkyl iodides like allyl iodide (= 3-iodopropene), but also iodoethanol and iodo- and bromoacetate gave dissolution patterns similar to those obtained with methyl iodide. The proposed reaction mechanism involves an initial adsorption of the halide to the surface of the metal, followed by reaction and subsequent dissolution.

The consequence of the above-described reactions of alkyl halides with metals is that the carbon–halogen bond is broken (Thayer, 1990). This implies that zero-valence metals, zinc and iron in particular, can be used as a means to degrade chlorinated aliphatic compounds abiotically (Matheson and Tratnyek, 1994). An *in situ* field test with a permeable reaction wall consisting of iron and sand showed that substantial amounts of, for example, tetra- and trichloroethylene can be degraded by this technology (Gillham *et al.*, 1993; Gillham and O'Hannesin, 1994). Microbial activities at the reaction wall and its proximity will have important practical consequences for the efficiency of this process. The success of this technology will depend on an accurate characterization of the biogeochemistry of this interesting process.

8. Selection Pressure and Evolutionary Aspects

8.1. Acquisition of the Dehalogenating Ability

To anaerobic microorganisms, halogenated compounds are a potential source of energy in two ways: as a compound that can be completely mineralized and as an electron acceptor. Thus anaerobes have at least three options to benefit from the conversion of these compounds: by harnessing the energy from the dehalogenation reaction, by obtaining energy from the degradation of the dehalogenated product, or from both parts of the degradation pathway.

The dechlorinating isolates *D. tiedjei* and *D. restrictus* use only the energy from the dechlorination reaction itself. In fact, this is counterintuitive. It would seem easier for an organism that already degrades, for example, benzoate, to extend its enzymatic capabilities toward also using halobenzoates via the acquisition of the extra dehalogenation step. The next ability to acquire would then be to couple dehalogenation to energy conservation. But apparently this is not what happens in nature. The organisms that dehalogenate do not necessarily have the enzymatic machinery to also metabolize the dehalogenated product. This was observed with DCB-1 (now known as *D. tiedjei*), with *D. restrictus*, and with the thus far obtained enrichment cultures that degrade polychlorinated benzenes but do not degrade monochlorobenzene. The development of specialists is a general phenomenon in anaerobic microbiology, where organic compounds are broken down by the concerted actions of a variety of microbes. The answer to the question of why anaerobic microbial ecosystems depend on specialists to perform certain metabolic steps, e.g., the removal of hydrogen by methanogens, is still open. The "evolution" of DCB-1 in the laboratory, however, is not consistent with this trend. When first obtained in pure culture, this organism was not able to metabolize benzoate, but currently *D. tiedjei* is able to do this (J. M. Tiedje, personal communication, 1992).

8.2. Phylogeny of Dehalogenating Anaerobes

The original discovery that *D. tiedjei* is a sulfate-reducing bacterium, most closely related to *Desulfuromonas acetoxidans* (DeWeerd *et al.*, 1990), gave hope that reductive dehalogenation would be widespread among sulfate-reducing bacteria. The *Desulfovibrio* species tested thus far, however, did not show dechlorinating activity. It is hard to draw conclusions, however, based on screening studies with laboratory strains since we can only guess how to properly induce these organisms for dechlorination. For example, the potential of *D. tiedjei* to dehalogenate chlorophenol was discovered only recently (Mohn and Kennedy, 1992). In previous studies this potential was missed since the dehalogenating enzyme(s) are not induced by chlorophenol.

Analysis of the partial 16S rRNA sequence of strain 2CP-1 indicates that this organism is a member of the delta subdivision of the proteobacteria, just like *D. tiedjei*. The 16S sequence of the 2-chlorophenol isolate, however, does not place it among the sulfate-reducing bacteria but instead maps it to a deep branch in the myxobacteria (Cole *et al.*, 1994). Evidence suggesting that the dehalogenating potential may be widespread in the delta subdivision of the purple bacteria stems from the observation that a recently described dehalogenating culture like the monofluorobenzoate-degrading diculture of Blotevogel (Drzyzga *et al.*, 1994) consists of a *Desulfotomaculum-* and a *Desulfovibrio*-like organism. *Desulfitobacterium dehalogenans* (Utkin *et al.*, 1994) uses sulfite and thiosulfate rather than sulfate as electron acceptors. This organism is more closely related to *Desulfotomaculum orientis* than to *D. tiedjei*. So the potential for reductive dechlorination among anaerobes is not restricted to sulfate-reducing bacteria. The chlorophenol-transforming bacterium strain DCB-2 (Madsen and Licht, 1992) cannot use sulfate as electron acceptor either. This organism was described as a gram-positive endospore-forming bacterium, so chances are that this is also a *Desulfotomaculum*-like organism.

Phylogenetic and physiological studies with *D. restrictus* have not given any indication that this organism also belongs to the delta subdivision of the purple bacteria. 16S rRNA analysis suggests that this organism is distantly related to anaerobes like *Selenomonas, Sporomusa,* and *Megasphaera*, strains that share a gram-negative cell wall on the one hand and a distinct though remote phylogenetic relationship to the 16S rRNA cluster of the gram-positive eubacteria on the other as common characteristics (Holliger, 1992).

8.3. Anaerobic Growth on Chlorinated Methanes

Recently, two other organisms have been described that can grow at the expense of halogenated aliphatic compounds. Strain MC, for example, is able to grow at the expense of methyl chloride as sole energy source (Traunecker *et al.*, 1991). The organism converts CH_3Cl to acetate in its energy metabolism. The first step in this route, however, is not reductive dechlorination of methyl chloride to methane, but the conversion of methylchloride to methyl tetrahydrofolate and HCl. This inducible dehalogenase activity is not a side reaction of an enzyme involved in the homoacetogenic pathway, but must rather be considered to be a novel enzyme activity specific for this organism (Messmer *et al.*, 1993).

Dichloromethane is another chlorinated substrate that can serve as a growth substrate for anaerobic bacteria, but is not metabolized via reductive dechlorination (Braus-Stromeyer *et al.*, 1993b). Culture DM grows on dichloromethane. Chloromethane, however, is not utilized as a carbon and energy source by this culture, but is transformed with H_2S to methanethiol and dimethylsulfide (Braus-Stromeyer *et al.*, 1993a). Leisinger and co-workers suggest that dehalogenation proceeds via a hydrolysis of dichloromethane to formaldehyde, and that the dehalogenating organism obtains its energy from the conversion of formaldehyde to formate. The energy available from the hydrolysis of dichloromethane is apparently not utilized. Interestingly, in this pathway reducing equivalents (in the form of formate) are produced rather than consumed by the dehalogenating organism. The formation of formate from dichloromethane is thermodynamically favorable. Growth of the dehalogenating organism in pure culture with dichloromethane as an energy source has not been successful so far. The problem may lie in the requirement for growth factors or for a carbon source secreted by a partner organism in mixed cultures. It is interesting to note that halomethanes are not reductively dechlorinated by these anaerobes that use these compounds as energy source, in spite of the facts that such a dechlorination reaction would be highly exergonic (Dolfing and Janssen, 1994), and reductive dechlorination is used in the microbial degradation of chlorinated ethylenes (de Bruin et al., 1992; Holliger et al., 1993). A similar observation has also been made for the anaerobic degradation of chloroacetic acids. Egli et al. (1989) have reported that mono- and dichloroacetic acid can serve as carbon and energy sources for stable methanogenic mixed cultures; these authors found no evidence that the degradation pathway of chloroacetic acids involved a reductive dehalogenation step, even though such a step would be highly exergonic (Dolfing and Janssen, 1994).

8.4. Molecular Biology of Dehalogenating Bacteria

The recent reports on the discovery of dehalogenating enzymes in aerobic microorganisms will probably foster research into the molecular biology behind the acquisition of the enzymatic machinery for reductive dehalogenation, and a further speculation on this subject are beyond the scope of this chapter. Presently, a number of new dechlorinating isolates are being characterized in much (molecular) detail at the NSF Center for Microbial Ecology in East Lansing, Michigan (J. M. Tiedje, personal communications, 1992, 1993). The results of these studies will probably begin to give answers on some of the questions regarding the molecular adaption to chlorinated compounds.

8.5. Selection Pressure and Reductive Dehalogenation

Two observations regarding selection pressure and reductive dechlorination should be mentioned here briefly. The first is that adaption is an ongoing process, even in the laboratory. The original D. tiedjei, strain DCB-1, was difficult to grow in the laboratory and had an apparent K_s value for 3-chlorobenzoate of about 60 µM. After a couple of years in the lab, the organism had become much less fastidious and had a K_m for 3-chlorobenzoate well below 5 µM.

Second, it is frequently observed that it is difficult to subculture primary enrichments obtained with chlorinated compounds. This observation has been made with various classes of halogenated compounds. Sometimes organisms even seem to lose the ability to degrade certain chlorinated compounds (Drzyzga et al., 1994), so selection pressure is not always powerful enough to ascertain the development of new organisms fast enough to suit the time schedule of the average graduate student. On the other hand, tetrachloroethylene is considered to be a xenobiotic compound. The organism obtained from Rhine sediment, however, appears to be able to degrade only this xenobiotic compound and nothing else. Thus here the selection pressure was apparently so strong that it gave rise to the development of a novel highly specialized mode of respiration within a rather short time, if we accept that tetrachloroethylene is a xenobiotic compound that has been used on a large scale only since the 1940s. Abrahamsson *et al.* (1995) have recently shown, however, that these compounds are also produced naturally in the presence of marine algae. The Rhine is not the only place where tetrachloroethylene dechlorinating organisms can be found. The United States harbors similar bacteria (DiStefano *et al.*, 1991, 1992).

8.6. The Origin of the Dechlorinating Enzyme(s)

One of the most frequent questions regarding the evolution of dechlorinating organisms and their dechlorinating machinery is where these abilities originate from. The knowledge that many halogenated compounds are naturally occurring in marine environments makes it likely that the dechlorinating machinery was developed first in marine environments and has spread from there to nonmarine environments. The recent discovery of the biogenesis of chlorinated aromatics by fungi in soil environments (De Jong *et al.*, 1994) implies, however, that this hypothesis is not necessarily correct.

8.7. Microbial Uptake of Halogenated Compounds

For an efficient use of halogenated substrates, these compounds should be taken up readily, the conversion should proceed at significant rates, and the energy should be conserved. In nature it could also be advantageous if the organisms can regulate the enzymatic machinery in such a way that the enzymes are only induced when the substrate is available; but this depends on the supply side of the chlorinated substrates, especially if the dechlorinating organism is able to use other compounds in addition to halogenated ones. If the supply is (almost) continuous, it is more efficient to have a constitutive enzymatic machinery; but if the supply is irregular, it may be more advantageous to have an inducible enzymatic machinery. The somewhat awkward term *enzymatic machinery* is used here on purpose because recent evidence indicates that uptake of 4-chlorobenzoate in the coryneform bacterium strain NTB-1 requires an active energy-dependent uptake system (Groenewegen *et al.*, 1990), and this may well be the rule for chlorinated aromatic compounds. Whether halogenated aliphatics are taken up by means of an active uptake system is not known (Parsons *et al.*, 1987), and it is also not known to what extent other classes of halogenated aromatic compounds, especially the apolar ones like benzenes and biphenyls, are taken up actively if at all. The pK values for chlorobenzoates range between about 2 and 4, i.e., they are well below the values at which the average microorganism thrives. Thus virtually all of the chlorobenzoate is in the dissociated form at physiological pH. The situation is slightly different for chlorinated phenols where the pK values range between 4.5 and 9.5. It is conceivable that for many of the chlorinated phenols with pK values close to the environmental pH, uptake is strongly affected by the environmental pH.

8.8. Dehalogenation at Growth-Sustaining Rates

The second condition that needs to be fulfilled if a compound is to be used as a source of energy is that its conversion should proceed at significant rates. This can be illustrated with the specific activity of D. restrictus toward tetra-chloroethene (Holliger, 1992). The bacterium converts this compound at rates that are 10^5 to 10^6 times higher than the activity observed in methanogens, where this conversion is fortuitous or cometabolic, or in D. tiedjei (Fathepure et al., 1987). In D. tiedjei, it is likely that tetrachloroethene conversion is cometabolic also, since the observed activity was about 0.01% (0.1 μmole of Cl^- h^{-1} g of protein^{-1}) of the dehalogenating activity of the organism with 3-chlorobenzoate (1.1 nmole of Cl^- h^{-1} g of protein^{-1}). This activity is even lower than the reductive biotransformation of tetrachloromethane by Shewanella putrefaciens, a fortuitous reduction catalyzed by c-type cytochromes (Picardal et al., 1993). The dechlorinating activity of D. tiedjei with pentachlorophenol was about 5% of the 3-chlorobenzoate dechlorination rate (Mohn and Kennedy, 1992). It will be inter-esting to see whether D. tiedjei can be adapted toward growth with chlorinated phenols as an energy source. The rates at which D. dehalogenans dehalogenates 2,3- and 2,6-dichlorophenol are, at 2.8 and 5.8 mmole h^{-1} (g of biomass dry weight)$^{-1}$ (Utkin et al., 1995), in principle high enough to be growth sustaining. Currently, there is rapid progress toward a better understanding of aerobic bacte-ria that obtain energy for growth from the conversion of halogenated compounds. Some of the striking observations there are that during growth on a halogenated compound like 1,2-dichloroethane dehalogenase can make up 30 to 40% of the total protein in the cells (van den Wijngaard et al., 1992). This hydrolytic dehalogenase, as found in Xanthobacter autotrophicus GJ10, is able to mutate at a relatively high rate toward utilization of structurally related compounds (Pries et al., 1994).

8.9. Reductive Dehalogenation by Aerobic Bacteria

So far, we have restricted our discussion of reductive dechlorination to anoxic systems. The reason for this is that reductive dechlorination is generally an anaerobic process. In many cases where the effect of oxygen was studied, oxygen inhibited or blocked the reductive dechlorination process (e.g., Horowitz et al., 1983). The observation that reductive dechlorination is generally restricted to anaerobic systems also makes sense. Oxygen is a better electron acceptor than most chlorinated compounds. Nevertheless, there are some examples of organ-isms that can catalyze reductive dechlorination aerobically. The coryneform bac-terium strain NTB-1, originally incorrectly classified as Alcaligenes denitrificans NTB-1 (Groenewegen et al., 1992), uses a reductive dechlorination step during the aerobic mineralization of 2,4-dichlorobenzoate (van den Tweel et al., 1987). Rhodococcus chlorophenolicus (Apajalahati and Salkinoja-Salonen, 1987) and a

Flavobacterium sp. (Steiert and Crawford, 1986) both put reductive dechlorination to use in the degradation of pentachlorophenol, a compound that is invulnerable to ring-cleaving oxygenases. Recently, Xun *et al.* (1992) reported the purification and characterization of a reductive dehalogenase from a *Flavobacterium* sp. This is the first report of the purification of a reductive dehalogenase from a bacterium. The enzyme uses glutathione but not NADPH or NADH as the reducing agent. The enzyme uses glutathione to reductively dechlorinate tetrachloro-*p*-hydroquinone to trichloro-*p*-hydroquinone with the production of oxidized glutathione, probably via a glutathione–hydroquinone conjugate as an intermediate. The enzyme did not stop at trichloro-*p*-hydroquinone, but reductively dechlorinated this compound further to dichloro- and monochloro-*p*-hydroquinone. Interestingly, the enzyme seems to be sensitive toward oxygen, because a loss of enzyme activity occurred when dithithreitol was not included in elution buffers during purification. At about the same time, Nayler *et al.* (1992) reported the isolation of a reductively dechlorinating enzyme from the slime mold *Dictyostelium discoideum*. This enzyme, which probably plays a role in the induction of amoeboid cells to differentiate into stalk cells, also uses glutathione as its cofactor. Molecular studies will have to show how related these enzymes really are. Nayler *et al.* (1992) have pointed out that the *Dictyostelium* enzyme may be related to the type I thyroxine deiodinase from rat liver (Berry *et al.*, 1991), which would make it a selenocysteine enzyme. The *in vivo* source of reducing agents for reductive dechlorination in *D. tiedjei* is not yet known. A variety of possibilities, including NADH and NADPH, have been tested, but glutathione was not (DeWeerd and Suflita, 1990).

8.10. Energy Conservation from Reductive Dehalogenation

Energy conservation from reductive dechlorination is by definition essential if an organism is to grow on the dechlorination of a halogenated compound with H_2 as an electron donor. What mechanism(s) such organisms use is currently not known. All we know is that this activity in *D. tiedjei* is membrane-bound, while there is some evidence to suggest that energy conservation proceeds via a chemiosmotic (respiratory) mechanism (Mohn and Tiedje, 1991). The apparent inability of chlorophenols to induce dehalogenation activity in *D. tiedjei* has been taken to suggest that dechlorination of chlorophenols is a fortuitous activity (Mohn and Kennedy, 1992); but it will nevertheless be interesting to see whether dechlorination of chlorophenols in this organism is, or after exposure to the proper selection conditions can be, coupled to ATP generation. In this context, it is tempting to mention that ATP generation in dehalogenating organisms may be coupled to the generation and excretion of the halide ion. The possession of such a respiratory mechanism would make it relatively easy for microorganisms to use this "module" for the generation of energy from the reductive dehalogenation of, in principle, all halogenated compounds if the proper enzymes are present.

9. Summary and Outlook

The first rigorous reports on aryl reductive dehalogenation as a microbial process (Suflita *et al.*, 1982, 1983; Horowitz *et al.*, 1983) are now about 10 years old. In these 10 years, microbiological research has focused on the isolation of dehalogenating organisms, which turned out to be more difficult than originally hoped. As a result of this unforeseen difficulty, considerable attention has also been given to the factors that influence dehalogenation in the laboratory. The main conclusion from this work is that there are no strict rules. Reductive dehalogenation is possible under a wide variety of environmental conditions in various habitats, i.e., both in salt water and in fresh water, both in the presence and absence of sulfate, and for some compounds even in the presence of oxygen. With many enrichments, the process proceeds best at moderate temperatures; but there are reports that dechlorination proceeds also at fairly low (Beurskens *et al.*, 1994a) or high temperatures (Larsen *et al.*, 1991), and there is no reason to believe that the temperature limits for dechlorination are different from those for other microbial reactions. The same is true for the pH.

Field studies on halogenated compounds have yielded a large data base on the presence of various organohalogens in nature. Unfortunately, this data base mainly contains information on the concentrations of organohalogens in different environmental compartments. Information on turnover times and fluxes is only recently becoming available. This lack of data makes it currently impossible to properly evaluate the impact of reductive dehalogenation on the flow of electrons in anoxic environments, although there is no reason to assume that this impact is large. This does, on the other hand, not exclude that reductive dehalogenation is of great importance for the pool of organohalogens in anoxic environments.

The availability of a number of dechlorinating organisms in pure culture will make it possible to develop specific and less specific probes for the detection of these and related organisms in the environment. Information on the distribution and activity of these organisms may give further insight into the importance of dechlorination in and for the environment.

The importance, or better, the logic, of reductive dechlorination for the microorganisms that perform this reaction is in principle crystal clear. The dechlorinating organisms can potentially use dechlorination as a source of energy for growth. How the dechlorinating organisms harness this energy is still not clear, but the first steps toward elucidating this problem have been made. It is also clear that the energy available from the dechlorination reaction is not always used by the dechlorinating organism. Sometimes the process is fortuitous, a phenomenon that is not unusual in microbiology.

One of the main messages of this chapter is therefore that reductive dechlorination should not be regarded as a freak event: This type of reaction is not intrinsically different from other microbially catalyzed redox reactions. The ecological aspects of the process are subject to the same rules as other microbial degradation processes. This conclusion should of course not be abused by claiming that microorganisms in nature will be able to deal with all halogenated compounds. Halogenated compounds can no longer be considered xenobiotics merely because they contain a carbon–halogen bond, but this does not automatically make them biodegradable in the environment. The elevated levels of halogenated compounds in sediments underscore that reductive dehalogenation is a very slow process, if it takes place at all, in the environment. There is clearly a need to verify the occurrence of dechlorination reactions in such environments and to identify possible limiting factors.

References

Abrahamsson, K., and Klick, S., 1991, Degradation of halogenated phenols in anoxic natural marine sediments, *Mar. Pollut. Bull.* **22**:227–233.

Abrahamsson, K., Ekdahl, A., Collen, J., Fahlström, E., and Pedersen, M., 1995, The natural formation of trichloroethylene and perchloroethylene in seawater, in: *Naturally-Produced Organohalogens* (A. Grimvall and E. W. B. de Leer, eds.), Kluwer Academic Publishers, Netherlands, pp. 327–331.

Abramowicz, D. A., Brennan, M. J., van Dort, H. M., and Gallager, E. L., 1993, Factors influencing the rate of polychlorinated biphenyl dechlorination in Hudson River sediments, *Environ. Sci. Technol.* **27**:1125–1131.

Adrian, N. R., and Suflita, J. M., 1990, Reductive dehalogenation of a nitrogen heterocyclic herbicide in anoxic aquifer slurries, *Appl. Environ. Microbiol.* **56**:292–294.

Adrian, N. R., and Suflita, J. M., 1994, Anaerobic biodegradation of halogenated and nonhalogenated *N*-, *S*-, and *O*-heterocyclic compounds in aquifer slurries, *Environ. Toxicol. Chem.* **13**:1551–1557.

Ahring, B. K., Christiansen, N., Mathrani, I., Hendriksen, H. V., Macario, A. J. L., and Conway de Macario, E., 1992, Introduction of a *de novo* bioremediation ability, aryl reductive dechlorination, into anaerobic granular sludge by inoculation of sludge with *Desulfomonile tiedjei*, *Appl. Environ. Microbiol.* **58**: 3677–3682.

Alcock, R. E., Johnston, A. E., McGrath, S. P., Berrow, M. L., and Jones, K. C., 1993, Long-term changes in the polychlorinated biphenyl content of United Kingdom soils, *Environ. Sci. Technol.* **27**:1918–1923.

Alder, A. C., Häggblom, M. M., Oppenheimer, S. R., and Young, L. Y., 1993, Reductive dechlorination of polychlorinated biphenyls in anaerobic sediments, *Environ. Sci. Technol.* **27**:530–538.

Allard, A.-S., Hynning, P.-A., Remberger, M., and Neilson, A. H., 1992, Role of sulfate concentration in dechlorination of 3,4,5-trichlorocatechol by stable enrichment cultures grown with coumarin and flavone glycones and aglycones, *Appl. Environ. Microbiol.* **58**:961–968.

Allard, A.-S., Hynning, P.-A., Remberger, M., and Neilson, A. H., 1994, Bioavailability of chlorocatechols in naturally contaminated sediment samples and of chloroguaiacols covalently bound to C2-guaiacyl residues, *Appl. Environ. Microbiol.* **60**:777–784.

Anid, P. J., Nies, L., and Vogel, T. M., 1991, Sequential anaerobic–aerobic biodegradation of PCBs in the river model, in: *Proceedings: On-site Reclamation Processes for Xenobiotic Treatment and Hydrocarbon Treatment* (R. E. Honchee and R. F. Olfenbuttel, eds.), Butterworth-Heinemann, Boston, pp. 428–436.

Anonymus, 1992, Montreal protocol: Faster cuts agreed in Copenhagen, *Chem. Indust.* 7 December, p. 887.

Anonymus, 1993, Facts and figures for the chemical industry, *Chem. Eng. News* 40–45.

Apajalahti, J. H. A., and Sakinoja-Salonen, M. S., 1987, Complete dechlorination of tetrachlorohydroquinone by cell extracts of pentachlorophenol-induced *Rhodococcus chlorophenolicus*, *J. Bacteriol.* **169**:5125–5130.

Ashworth, R. B., and Cornier, M. J., 1967, Isolation of 2,6-dibromophenol from the marine hemichordate, *Balanoglossus biminiensis, Science* **156**:158–159.

Assaf-Anid, N., Nies, L., and Vogel, T. M., 1992, Reductive dechlorination of a polychlorinated biphenyl congener and hexachlorobenzene by vitamin B$_{12}$, *Appl. Environ. Microbiol.* **58**:1057–1060.

Assaf-Anid, N., Hayes, K. F., and Vogel, T. M., 1994, Reductive dechlorination of carbon tetrachloride by cobalamin(II) in the presence of dithiothreitol: Mechanistic study, effect of redox potential and pH, *Environ. Sci. Technol.* **28**:246–252.

Barber, II, L. B., 1988, Dichlorobenzene in ground water: Evidence for long-term persistence, *Ground Water.* **26**:696–702.

Bedard, D. L., and Van Dort, H. M., 1992, Brominated biphenyls can stimulate reductive dechlorination of endogenous Aroclor 1260 in methanogenic sediment slurries, *Abstr. 92nd Gen. Mtg. Amer. Soc. Microbiol.,* p. 339, Q-26.

Benedick, R. E., 1991, *Ozone Diplomacy: New Directions in Safeguarding the Planet,* Harvard University Press, Cambridge.

Berry, M. J., Banu, L., and Larsen, R., 1991, Type I iodothyronine deiodinase is a selenocysteine-containing enzyme, *Nature* **349**:438–440.

Beurskens, J. E. M., Dekker, C. G. C., Jonkhoff, J., and Pompstra, L., 1993a, Microbial dechlorination of hexachlorobenzene in a sedimentation area of the Rhine River, *Biochemistry* **19**: 61–81.

Beurskens, J. E. M., Mol, G. A. J., Barreveld, H. L., van Munster, B., and Winkels, H. J., 1993b, Geochronology of priority pollutants in a sedimentation area of the Rhine River, *Environ. Toxicol. Chem.* **12**:1549–1566.

Beurskens, J. E. M., Dekker, C. G. C., van den Heuvel, H., Swart, M., De Wolf, J., and Dolfing, J., 1994a, Dechlorination of chlorinated benzenes by an anaerobic microbial consortium that selectively mediates the thermodynamic most favorable reactions, *Environ. Sci. Technol.* **28**: 701–706.

Beurskens, J. E. M., Winkels, H. J., de Wolf, J., and Dekker, C. G. C., 1994b, Trends of priority pollutants in the Rhine during the last 50 years, *Wat. Sci. Technol.* **29**:77–85.

Beurskens, J. E. M., Toussaint, M., de Wolf, J., van der Steen, J., Slot, P., Commandeur, L. C. M., and Parsons, J. R., 1995a, Dehalogenation of chlorinated dioxins by an anaerobic microbial consortium from sediment, *Environ. Toxicol. Chem.* **14**: 939–943.

Beurskens, J. E. M., de Wolf, J., and Dolfing, J., 1995b, Enrichment of a hexachlorobenzene dechlorinating consortium from marine sediment, submitted.

Beurskens, J. E. M., de Wolf, J., and van den Heuvel, H., 1995c, Reductive dechlorination of some polychlorinated biphenyls by enrichment culture from a sedimentation area of the Rhine River, submitted.

Blasland and Bouck Engineers, 1992, Sheboygan River and harbor biodegradation pilot study work plan, internal report.

Bopp, R. F., Simpson, H. J., Olsen, C. R., and Kostyk, N., 1981, Polychlorinated biphenyls in sediments of the tidal Hudson River, New York, *Environ. Sci. Technol.* **15**:210–216.

Bosma, T. N. P., 1994, *Simulation of Subsurface Biotransformation,* Agricultural University, Wageningen, The Netherlands, Ph.D. thesis.

Bosma, T. N. P., van der Meer, J. R., Schraa, G., Tros, M. E., and Zehnder, A. J. B., 1988,

Reductive dechlorination of all trichloro- and dichlorobenzene isomers, *FEMS Microbiol. Ecol.* **53**:223–229.

Boyd, S. A., and Shelton, D. R., 1984, Anaerobic biodegradation of chlorophenols in fresh and acclimated sludge, *Appl. Environ. Microbiol.* **47**:272–277.

Boyd, S. A., Shelton, D. R., Berry, D., and Tiedje, J. M., 1983, Anaerobic biodegradation of phenolic compounds in digested sludge, *Appl. Environ. Microbiol.* **46**:50–54.

Braus-Stromeyer, S. A., Cook, A. M., and Leisinger, T., 1993a, Biotransformation of chloromethane to methanethiol, *Environ. Sci. Technol.* **27**:1577–1579.

Braus-Stromeyer, S. A., Hermann, R., Cook, A. M., and Leisinger, T., 1993b, Dichlormethane as the sole carbon source for an acetogenic mixed culture and isolation of a fermentative, dichloromethane-degrading bacterium, *Appl. Environ. Microbiol.* **59**:3790–3797.

Brown, J. F., and Wagner, R. E., 1990, PCB movement, dechlorination, and detoxification in the Acushnet Estuary, *Environ. Toxicol. Chem.* **9**:1215–1233.

Brown, J. F., Jr., Bedard, D. L., Brennan, M. J., Carnahan, J. C., Feng, H., and Wagner, R. E., 1987a, Polychlorinated biphenyl dechlorination in aquatic sediments, *Science* **236**:709–712.

Brown, J. F., Jr., Wagner, R. E., Feng, H., Bedard, D. L., Brennan, M. J., Carnahan, J. C., and May, R. J., 1987b, Environmental dechlorination of PCBs, *Environ. Toxicol. Chem.* **6**:579–593.

Carson, R., 1962, *Silent Spring,* Fawcett Crest, New York.

Carter, S. R., and Jewell, W. J., 1993, Biotransformation of tetrachloroethylene by anaerobic attached-films at low temperatures, *Water Res.* **27**:607–615.

Chu, K. H., and Jewell, W. J., 1994, Treatment of tetrachloroethylene with anaerobic attached film process, *J. Environ. Eng.* **120**:58–71.

Cole, J. R., Cascarelli, A. L., Mohn, W. W., and Tiedje, J. M., 1994, Isolation and characterization of a novel bacterium growing via reductive dehalogenation of 2-chlorophenol, *Appl. Environ. Microbiol.* **60**:3536–3542.

Cord-Ruwisch, R. H., Seitz, H.-J., and Conrad, R., 1988, The capacity of hydrogenotrophic anaerobic bacteria to complete for traces of hydrogen depends on the redox potential of the terminal electron acceptor, *Arch. Microbiol.* **149**:350–357.

Cozza, C. L., and Woods, S. L., 1992, Reductive dechlorination pathways for substituted benzenes: A correlation with electronic properties, *Biodegradation* **2**:265–278.

Curtis, G. P., 1991, Ph.D. thesis; cited by: Semprini, L., Hopkins, G. D., McCarty, P. L., and Roberts, P. V., 1992, *In situ* transformation of carbon tetrachloride and other halogenated compounds resulting from biostimulation under anoxic conditions, *Environ. Sci. Technol.* **26**:2454–2461.

Davies-Venn, C., Young, J. C., and Tabak, H. H., 1992, Impact of chlorophenols and chloroanilines on the kinetics of acetoclastic methanogenesis, *Environ. Sci. Technol.* **26**:1627–1635.

Dean, J. A., 1985, *Lange's Handbook of Chemistry,* 13th ed., McGraw-Hill, New York.

de Bruijn, J., Busser, F., Seinen, W., and Hermens, J., 1989, Determination of octanol/water partition coefficients for hydrophobic organic chemicals with the "slow-stirring" method, *Environ. Toxicol. Chem.* **8**:499–512.

de Bruin, W. P., Kotterman, M. J. J., Posthumus, M. A., Schraa, G., and Zehnder, A. J. B., 1992, Complete reductive transformation of tetrachloroethene to ethane, *Appl. Environ. Microbiol.* **58**: 1996–2000.

DeFlaun, M. F., Ensley, B. D., and Steffan, R. J., 1992, Biological oxidation of hydrochlorofluorocarbons (HCFCs) by a methanotrophic bacterium, *Bio/technology* **10**:1576–1578.

De Jong, E., Field, J. A., Spinnler, H.-E., Wijnberg, J. P. B. A,. and de Bont, J. A. M., 1994, Significant biogenesis of chlorinated aromatics by fungi in natural environments, *Appl. Environ. Microbiol.* **60**:264–270.

de Voogt, P., Wells, D. E., Reutergardh, L., and Brinkman, U. A. T., 1990, Biological activity, determination and occurrence of planar, mono- and di-ortho PCBs, *Int. J. Environ. Anal. Chem.* **40**:1–46.

DeWeerd, K. A., and Suflita, J. M., 1990, Anaerobic aryl reductive dehalogenation of halobenzoates by cell extracts of "*Desulfomonile Tiedjei*," *Appl. Environ. Microbiol.* **56**:2999–3005.

DeWeerd, K. A., Suflita, J. M., Linkfield, T. G., Tiedje, J. M., and Pritchard, P. H., 1986, The relationship between reductive dehalogenation and other aryl substituent removal reactions catalyzed by anaerobes, *FEMS Microbiol. Ecol.* **38**:331–339.

DeWeerd, K. A., Mandelco, L., Tanner, R. S., Woese, C. R., and Suflita, J. M., 1990, *Desulfomonile tiedjei* gen. nov. and sp. nov., a novel anaerobic, dehalogenating, sulfate-reducing bacterium, *Arch. Microbiol.* **154**:23–30.

DiStefano, T. D., Gossett, J. M., and Zinder, S. H., 1991, Reductive dechlorination of high concentrations of tetrachloroethene to ethene by an anaerobic enrichment culture in the absence of methanogenesis, *Appl. Environ. Microbiol.* **57**:2287–2292.

DiStefano, T. D., Gossett, J. M., and Zinder, S. H., 1992, Hydrogen as an electron donor for dechlorination of tetrachloroethene by an anaerobic mixed culture, *Appl. Environ. Microbiol.* **58**:3622–3629.

Dolfing, J., 1986, Granulation in UASB reactors, *Water Sci. Technol.* **18**(12):15–25.

Dolfing, J., 1990, Reductive dechlorination of 3-chlorobenzoate is coupled to ATP production and growth in an anaerobic bacterium, strain DCB-1, *Arch. Microbiol.* **153**:264–266.

Dolfing, J., 1992, The energetic consequences of hydrogen gradients in methanogenic ecosystems, *FEMS Microbiol. Ecol.* **101**:183–187.

Dolfing, J., 1995, Regiospecificity of chlorophenol reductive dechlorination by vitamin B_{12s} [letter to the editor], *Appl. Environ. Microbiol.* **61**: 2450–2451.

Dolfing, J., and Bloemen, W. G. B. M., 1985, Activity measurements as a tool to characterize the microbial composition of methanogenic environments, *J. Microbiol. Meth.* **4**:1–12.

Dolfing, J., and Harrison, B. K., 1992, Gibbs free energy of formation of halogenated aromatic compounds and their potential role as electron acceptors in anaerobic environments, *Environ. Sci. Technol.* **26**:2213–2218.

Dolfing, J., and Harrison, B. K., 1993, Redox and reduction potentials as parameters to predict the degradation pathway of chlorinated benzenes in anaerobic environments, *FEMS Microbiol. Ecol.* **13**:23–30.

Dolfing, J., and Janssen, D. B., 1994, Estimates of Gibbs free energies of formation of chlorinated aliphatic compounds, *Biodegradation* **5**:21–28.

Dolfing, J., and Tiedje, J. M., 1986, Hydrogen cycling in a three-tiered food web growing on the methanogenic conversion of 3-chlorobenzoate, *FEMS Microbiol. Ecol.* **38**:293–298.

Dolfing, J., and Tiedje, J. M., 1991a, Influence of substituents on reductive dehalogenation of 3-chlorobenzoate analogs, *Appl. Environ. Microbiol.* **57**:820–824.

Dolfing, J., and Tiedje, J. M., 1991b, Kinetics of two complementary hydrogen sink reactions in a defined 3-chlorobenzoate degrading methanogenic co-culture, *FEMS Microbiol. Ecol.* **86**:25–32.

Dolfing, J., and Tiedje, J. M., 1991c, Acetate as a source of reducing equivalents in the reductive dechlorination of 2,5-dichlorobenzoate, *Arch. Microbiol.* **156**:356–361.

Dolfing, J., van den Wijngaard, A. J., and Janssen, D. B., 1993, Microbiology aspects of the removal of chlorinated hydrocarbons from air, *Biodegradation* **4**:261–282.

Drzyzga, O., Jansen, S., and Blotevogel, K.-H., 1994, Mineralization of monofluorobenzoate by a diculture under sulfate-reducing conditions, *FEMS Microbiol. Lett.* **116**:215–220.

Eberson, L., 1987, *Electron Transfer Reactions in Organic Chemistry,* Springer, Berlin.

Egli, C., Thüer, D., Cook, A. M., and Leisinger, T., 1989, Monochloro- and dichloroacetic acids as carbon and energy sources for a stable, methanogenic mixed culture, *Arch. Microbiol.* **152**:218–223.

Eitzer, B. D., and Hites, R. A., 1989, Atmospheric transport and deposition of polychlorinated dibenzo-*p*-dioxins and dibenzofurans, *Environ. Sci. Technol.* **23**:1396–1401.

Farwell, S. O., Beland, F. A., and Geer, R. D., 1975, Reduction pathways of organohalogen compounds part 1. Chlorinated benzenes, *J. Electroanal. Chem.* **61**:303–313.

Fathepure, B. Z., and Vogel, T. M., 1991, Complete degradation of polychlorinated hydrocarbons by a two-stage biofilm reactor, *Appl. Environ. Microbiol.* **57**:3418–3422.

Fathepure, B. Z., Tiedje, J. M., and Boyd, S. A., 1987, Anaerobic bacteria that dechlorinate perchloroethylene, *Appl. Environ. Microbiol.* **53**:2671–2674.

Fathepure, B. Z., Tiedje, J. M., and Boyd, S. A., 1988, Reductive dechlorination of hexachlorobenzene to tri- and dichlorobenzenes in anaerobic sewage sludge, *Appl. Environ., Microbiol.* **54**:327–330.

Fenchel, T., and Blackburn, T. H., 1979, *Bacteria and Mineral Cycling*, Academic Press, London.

Flanagan, W. P., and May, R. J., 1993, Metabolite detection as evidence for naturally occurring aerobic PCB biodegradation in Hudson River sediments, *Environ. Sci. Technol.* **27**:2207–2212.

Friege, H., Stock, W., Alberti, J., Poppe, A., Juhnke, I., Knie, J., and Schiller, W., 1989, Environmental behaviour of polychlorinated mono- methyl-substituted diphenyl-methanes (Me-PCDMs) in comparison with polychlorinated biphenyls (PCBs). II Environmental residues and aquatic toxicity, *Chemosphere* **18**:1367–1378.

Furukawa, K., Tomizuka, N., and Kamibayashi, A., 1983, Metabolic breakdown of Kaneclors (polychlorobiphenyls) and their products by *Acinetobacter* sp., *Appl. Environ. Microbiol.* **46**:140–145.

Gantzer, C. J., and Wackett, L. P., 1991, Reductive dechlorination catalyzed by bacterial transition-metal coenzymes, *Environ. Sci. Technol.* **25**:715–722.

Genthner, S. B. R., Price, W. A., and Pritchard, P. H., 1989, Characterization of anaerobic dechlorinating consortia derived from aquatic sediments, *Appl. Environ. Microbiol.* **55**:1466–1471.

Gillham, R. W., and O'Hannesin, S. F., 1994, Enhanced degradation of halogenated aliphatics by zero-valent iron, *Ground Wat.* **32**:958–967.

Gillham, R. W., O'Hannesin, S. F., and Orth, W. S., 1993, Metal enhanced abiotic degradation of halogenated aliphatics: Laboratory tests and field trials, paper presented at the 1993 HazMat Central Conference, Chicago, Illinois, March 9–11.

Goodwin, S., Conrad, R., and Zeikus, J. G., 1988, Influence of pH on microbial hydrogen metabolism in diverse sedimentary ecosystems, *Appl. Environ. Microbiol.* **54**:590–593.

Goodwin, S., Giralo-Gomez, E., Mobarry, B., and Switzenbaum, M. S., 1991, Comparison of diffusion and reaction rates in anaerobic microbial aggregates, *Microb. Ecol.* **22**:161–174.

Gossett, J. M., 1987, Measurement of Henry's law constants for C_1 and C_2 chlorinated hydrocarbons, *Environ. Sci. Technol.* **21**:202–208.

Götz, R., Schumacher, E., Roch, K., Specht, W., and Weeren, R. D., 1990, Chlorierte kohlenwasserstoffe (CKWs) in Hamburger hafensedimenten, *Vom Wasser* **75**:393–415.

Götz, R., Friesel, P., Roch, K., Papke, O., Ball, M., and Lis, A., 1993, Polychlorinated-*p*-dibenzodioxins (PCDDs), dibenzofurans (PCDFs), and other chlorinated compounds in the river Elbe: Results on bottom sediments and fresh sediments collected in sedimentation chambers, *Chemosphere* **27**:105–111.

Grimvall, A., and de Leer, E. W. B., 1995, *Naturally-Produced Organohalogens*, Kluwer Academic Publishers, Dordrecht, The Netherlands.

Groenewegen, P. E. J., Driessen, A. J. M., Konings, W. N., and de Bont, J. A. M., 1990, Energy-dependent uptake of 4-chlorobenzoate in the coryneform bacterium NTB-1, *J. Bacteriol.* **172**:419–423.

Groenewegen, P. E. J., van den Tweel, W. J. J., and de Bont, J. A. M., 1992, Anaerobic bioformation of 4-hydroxybenzoate from 4-chlorobenzoate by the coryneform bacterium NTB-1, *Appl. Microbiol. Biotechnol.* **36**:541–547.

Häggblom, M. M., 1992, Microbial breakdown of halogenated aromatic pesticides and related compounds, *FEMS Microbiol. Rev.* **103**:29–72.

Häggblom, M. M., and Young, L. Y., 1990, Chlorophenol degradation coupled to sulfate reduction, *Appl. Environ. Microbiol.* **56**:3255–3260.

Häggblom, M. M., Rivera, M. D., and Young, L. Y., 1993, Influence of alternative electron acceptors on the anaerobic biodegradability of chlorinated phenols and benzoic acids, *Appl. Environ. Microbiol.* **59**:1162–1167.

Hale, D. D., Rogers, J. E., and Wiegel, J., 1991, Environmental factors correlated to dichlorophenol dechlorination in anoxic freshwater sediments, *Environ. Toxicol. Chem.* **10**:1255–1265.

Harkness, M. R., McDermott, J. B., Abramowicz, D. A., Salvo, J. J., Flanagan, W. P., Stephens, M. L., Mondello, F. J., May, R. J., Lobos, J. H., Carroll, K. M., Brennan, M. J., Bracco, A. A., Fish, K. M., Warner, G. L., Wilson, P. R., Dietrich, D. K., Lin, D. T., Morgan, C. B., and Gately, W. L., 1993, *In situ* stimulation of aerobic PCB biodegradation in Hudson River sediments, *Science* **259**:503–507.

Harper, D. B., 1993, Biogenesis and metabolic role of halomethanes in fungi and plants, in: *Metal Ions in Biological Systems*, Vol. 29 (H. Sigel and A. Sigel, eds.), Marcel Dekker, New York, pp. 346–388.

Hendriksen, H. V., and Ahring, B. K., 1992, Metabolism and kinetics of pentachlorophenol transformation in anaerobic granular sludge, *Appl. Microbiol. Biotechnol.* **37**:662–666.

Hendriksen, H. V., and Ahring, B. K., 1993, Anaerobic dechlorination of pentachlorophenol in fixed-film and upflow anaerobic sludge blanket reactors using different inocula, *Biodegradation* **3**:399–408.

Hendriksen, H. V., Larsen, S., and Ahring, B. K., 1992, Influence of a supplemental carbon source on anaerobic dechlorination of pentachlorophenol in granular sludge, *Appl. Environ. Microbiol.* **58**:365–370.

Hoekstra, E. J., and de Leer, E. W. B., 1993, Natural production of chlorinated aromatic compounds in soil, in: *Contaminated Soil 93* (F. Arendt, G. J. Annokkee, R. Bosman, and W. J. van den Brink, eds.), Kluwer Academic Publishers, Dordrecht, The Netherlands, pp. 215–224.

Holliger, H. C., 1992, *Reductive Dehalogenation by Anaerobic Bacteria*, Agricultural University, Wageningen, The Netherlands, Ph.D. thesis.

Holliger, C., Schraa, G., Stams, A. J. M., and Zehnder, A. J. B., 1992, Enrichment and properties of an anaerobic mixed culture reductively dechlorinating 1,2,3-trichlorobenzene to 1,3-dichlorobenzene, *Appl. Environ. Microbiol.* **58**:1636–1644.

Holliger, C., Schraa, G., Stams, A. J. M., and Zehnder, A. J. B., 1993, A highly purified enrichment culture couples the reductive dechlorination of tetrachloroethene to growth, *Appl. Environ. Microbiol.* **59**:2991–2997.

Holmes, D. A., Harrison, B. K., and Dolfing, J., 1993, Estimation of Gibbs free energies of formation for polychlorinated biphenyls, *Environ. Sci. Technol.* **27**:725–731.

Hong, C.-S., Bush, B., Xiao, J., and Qiao, H., 1993, Toxic potential of non-ortho and mono-ortho coplanar polychlorinated biphenyls in Aroclors, seals, and humans, *Arch. Environ. Contam. Toxicol.* **25**:118–123.

Horowitz, A., Suflita, J. M., and Tiedje, J. M., 1983, Reductive dehalogenations of halobenzoates by anaerobic lake sediment microorganisms, *Appl. Environ. Microbiol.* **45**:1459–1465.

Howard, P. H., 1989, *Fate and Exposure Data for Organic Chemicals*, Lewis Publishers, Chelsea, England.

Jones, K. C., Sanders, G., Wild, S. R., Burnett, V., and Johnston, A. E., 1992, Evidence for a decline of PCBs and PAHs in rural vegetation and air in the United Kingdom, *Nature* **356**:137–140.

King, G. M., 1986, Inhibition of microbial activity in marine sediments by a bromophenol from a hemichordate, *Nature* **323**:257–259.

King, G. M., 1988, Dehalogenation in marine sediments containing natural sources of halophenols, *Appl. Environ. Microbiol.* **54**:3079–3085.

Kohring, G.-W., Zhang, X., and Wiegel, J., 1989, Anaerobic dechlorination of 2,4-dichlorophenol in freshwater sediments in the presence of sulfate, *Appl. Environ. Microbiol.* **55**:2735–2737.

Kriegman-King, M. R., and Reinhard, M., 1992, Transformation of carbon tetrachloride in the presence of sulfide, biotite, and vermiculite, *Environ. Sci. Technol.* **26**:2198–2208.

Kriegman-King, M. R., and Reinhard, M., 1994, Transformation of carbon tetrachloride by pyrite in aqueous solution, *Environ. Sci. Technol.* **28**:692–700.

Krone, U. E., Thauer, R. K., and Hogenkamp, H. P. C., 1989a, Reductive dehalogenation of chlorinated C_1-hydrocarbons mediated by corrinoids, *Biochemistry* **28**:4908–4914.

Krone, U. E., Laufer, K., Thauer, R. K., and Hogenkamp, H. P. C., 1989b, Coenzyme F_{430} as a possible catalyst for the reductive dehalogenation of chlorinated C_1 hydrocarbons in methanogenic bacteria, *Biochemistry* **28**:10061–10065.

Krone, U. E., Thauer, R. K., Hogenkamp, H. P. C., and Steinbach, K., 1991, Reductive formation of carbon monoxide from CCl_4 and FREONs 11, 12, and 13 catalyzed by corrinoids, *Biochemistry* **30**:2713–2719.

Kuhn, E. P., and Suflita, J. M., 1989, Dehalogenation of pesticides by anaerobic microorganisms in soils and groundwater—a review, in: *Reactions and Movements of Organic Chemicals in Soils* (B. L. Sawhney and K. Brown, eds.), Special Publication 22, Soil Science Society of America and American Society of Agronomy, Madison, Wisconsin, pp. 111–180.

Kuhn, E. P., Townsend, G. T., and Suflita, J. M., 1990, Effect of sulfate and organic carbon supplements on reductive dehalogenation of chloroanilines in anaerobic aquifer slurries, *Appl. Environ. Microbiol.* **56**:2630–2637.

Lake, J. L., Pruell, R. J., and Osterman, F. A., 1992, An examination of dechlorination processes and pathways in New Bedford Harbor sediments, *Marine Environ. Res.* **33**:31–47.

Larsen, S., Hendriksen, H. V., and Ahring, B. K., 1991, Potential for thermophilic (50°C) anaerobic dechlorination of pentachlorophenol in different ecosystems, *Appl. Environ. Microbiol.* **57**:2085–2090.

Leisinger, T., 1983, Microorganisms and xenobiotic compounds, *Experientia* **39**:1183–1191.

Liu, S. M., 1995, Anaerobic dechlorination of chlorinated pyridines in anoxic freshwater sediment slurries, *J. Environ. Sci. Health Part A-Environ. Sci. Eng.* **30**:485–503.

Lovelock, J. E., 1975, Natural halocarbons in the air and in the sea, *Nature* **256**:193–194.

Lovley, D. R., and Woodward, J. C., 1992, Consumption of freons CFC-11 and CFC-12 by anaerobic sediments and soils, *Environ. Sci. Technol.* **26**:925–929.

Mackay, D., and Shiu, W. Y., 1981, A critical review of Henry's law constants for chemicals of environmental interest, *J. Phys. Chem. Ref. Data* **10**:1175–1199.

Madsen, E. L, 1991, Determining *in situ* biodegradation, *Environ. Sci. Technol.* **25**:1663–1673.

Madsen, T., and Aamand, J., 1991, Effects of sulfuroxy anions on degradation of pentachlorophenol by a methanogenic enrichment culture, *Appl. Environ. Microbiol.* **57**:2453–2458.

Madsen, T., and Aamand, J., 1992, Anaerobic transformation and toxicity of trichlorophenols in a stable enrichment culture, *Appl. Environ. Microbiol.* **58**:557–561.

Madsen, T., and Licht, D., 1992, Isolation and characterization of an anaerobic chlorophenol-transforming bacterium, *Appl. Environ. Microbiol.* **58**:2874–2878.

Matheson, L. J., and Tratnyek, P. G., 1994, Reductive dehalogenation of chlorinated methanes by iron metal, *Environ. Sci. Technol.* **28**:2045–2053.

McCarty, P. L., Reinhard, M., and Rittmann, B. E., 1981, Trace organics in groundwater, *Environ. Sci. Technol.* **15**:40–51.

Messmer, M., Wohlfarth, G., and Diekert, G., 1993, Methyl chloride metabolism of the strictly anaerobic, methyl chloride-utilizing homoacetogen strain MC, *Arch. Microbiol.* **160**:383–387.

Mikesell, M. D., and Boyd, S. A., 1986, Complete reductive dechlorination and mineralization of pentachlorophenol by anaerobic microorganisms, *Appl. Environ. Microbiol.* **52**:861–865.

Mohn, W. W., and Kennedy, K. J., 1992, Reductive dehalogenation of chlorophenols by *Desulfomonile tiedjei* DCB-1, *Appl. Environ. Microbiol.* **58**:1367–1370.

Mohn, W. W., and Tiedje, J. M., 1990, Strain DCB-1 conserves energy for growth from reductive dechlorination coupled to formate oxidation, *Arch. Microbiol.* **153**:267–271.

Mohn, W. W., and Tiedje, J. M., 1991, Evidence for chemiosmotic coupling of reductive dechlorination and ATP synthesis in *Desulfomonile tiedjei*, *Arch. Microbiol.* **157**:1–6.

Mohn, W. W., and Tiedje, J. M., 1992, Microbial reductive dechlorination, *Microbiol. Rev.* **56**:482–507.

Molina, M. J., and Rowland, F. S., 1974, Stratospheric sink for chlorofluoromethanes: Chlorine atom-catalysed destruction of ozone, *Nature* **249**:810–812.

Morris, P. J., Quensen, J. F., Tiedje, J. M., and Boyd, S. A., 1992, Reductive debromination of the commercial polybrominated biphenyl mixture firemaster BP6 by anaerobic microorganisms from sediments, *Appl. Environ. Microbiol.* **58**:3249–3256.

Morris, P. J., Quensen, III, J. F., Tiedje, J. M., and Boyd, S. A., 1993, An assessment of the reductive debromination of polybrominated biphenyls in the Pine River Reservoir, *Environ. Sci. Technol.* **27**:1580–1586.

Nayler, O., Insall, R., and Kay, R. R., 1992, Differentiation-inducing-factor dechlorinase, a novel cytosolic dechlorinating enzyme from *Dictyostelium discoideum*, *Eur. J. Biochem.* **208**:531–536.

Nicholson, D. K., Woods, S. L., Istok, J. D., and Peek, D. C., 1992, Reductive dechlorination of chlorophenols by a pentachlorophenol-acclimated methanogenic consortium, *Appl. Environ. Microbiol.* **58**:2280–2286.

Neidelman, S. L., and Geigert, J., 1986, *Biohalogenation: Principles, Basic Roles and Applications,* Ellis Horwood, Chichester, England.

Nies, L. F., 1993, *Microbial and Chemical Reductive Dechlorination of Polychlorinated Biphenyls and Chlorinated Benzenes,* The University of Michigan, Ann Arbor, Ph.D. thesis.

Nies, L. F., and Vogel, T. M., 1991, Identification of the proton source for the microbial reductive dechlorination of 2,3,4,5,6-pentachlorobiphenyl, *Appl. Environ. Microbiol.* **57**:2771–2774.

Nishino, S. F., and Spain, J. C., 1993, Cell-density dependent adaptation of *Pseudomonas putida* to biodegradation of *p*-nitrophenol, *Environ. Sci. Technol.* **27**:489–494.

Öberg, L. G., Andersson, R., Wågman, N., and Rappe, C., 1993, Formation of polychlorinated dibenzo-*p*-dioxins and dibenzofurans from chloroorganic precursors in activated sewage sludge and garden compost, paper presented at the International Conference on Naturally Produced Organohalogens, 19–24 September, Delft, The Netherlands.

Oliver, B. G., and Nicol, K. D., 1982, Chlorobenzenes in sediments, water, and selected fish from lakes Superior, Huron, Erie, and Ontario, *Environ. Sci. Technol.* **16**:532–536.

Oliver, B. G., and Niimi, A. J., 1988, Tropodynamic analysis of polychlorinated biphenyl congeners and other chlorinated hydrocarbons in Lake Ontario ecosystem, *Environ. Sci. Technol.* **22**:388–397.

Oremland, R. S., Miller, L. G., and Strohmaier, F. E., 1994, Degradation of methyl bromide in anaerobic sediments, *Environ. Sci. Technol.* **28**:514–520.

Parsons, J. R., Opperhuizen, A., and Hutzinger, O., 1987, Influence of membrane permeation on biodegradation kinetics of hydrophobic compounds, *Chemosphere* **16**:1361–1370.

Pearson, C. R., 1982, C_1 and C_2 halocarbons, in: *The Handbook of Environmental Chemistry,* Vol. 3B (O. Hutzinger, ed.), Springer-Verlag, Berlin, pp. 69–88.

Peijnenburg, W. J. G. M., 't Hart, M. J., den Hollander, H. A., van de Meent, D., Verboom, H. H., and Wolfe, N. L., 1991, QSARs for predicting biotic and abiotic reductive transformation rate constants of halogenated hydrocarbons in anoxic sediment systems, *Sci. Total Environ.* **109/110**:283–300.

Peijnenburg, W. J. G. M., 't Hart, M. J., den Hollander, H. A., van de Meent, D., Verboom, H. H., and Wolfe, N. L., 1992a, Reductive transformations of halogenated aromatic hydrocarbons in anaerobic water-sediment systems: Kinetics, mechanisms and products, *Environ. Toxicol. Chem.* **11**:289–300.

Peijnenburg, W. J. G. M., 't Hart, M. J., den Hollander, H. A., van de Meent, D., Verboom, H. H., and Wolfe, N. L., 1992b, QSARs for predicting reductive transformation rate constants of halogenated aromatic hydrocarbons in anoxic sediment systems, *Environ. Toxicol. Chem.* **11**:301–314.

Petty, M. A., 1961, An introduction to the origin and biochemistry of microbial halometabolites, *Bacteriol. Rev.* **25**:111–160.

Picardal, F. W., Arnold, R. G., Cough, H., Little, A. M., and Smith, M. E., 1993, Involvement of cytochromes in the anaerobic biotransformation of tetrachloromethane by *Shewanella putrefaciens* 200, *Appl. Environ. Microbiol.* **59**:3763–3770.

Pries, F., van der Ploeg, J. R., Dolfing, J., and Janssen, D. B., 1994, Degradation of halogenated aliphatic compounds: The role of adaptation, *FEMS Microbiol. Rev.* **15**:279–295.

Quensen, III, J. F., Tiedje, J. M., and Boyd, S. A., 1988, Reductive dechlorination of polychlorinated biphenyls by anaerobic microorganisms from sediments, *Science* **242**:752–754.

Quensen, III, J. F., Boyd, S. A., and Tiedje, J. M., 1990, Dechlorination of four commercial polychlorinated biphenyl mixtures (Aroclors) by anaerobic microorganisms from sediments, *Appl. Environ. Microbiol.* **56**:2360–2369.

Ramanand, K., Balba, M. T., and Duffy, J., 1993a, Reductive dehalogenation of chlorinated benzenes and toluenes under methanogenic conditions, *Appl. Environ. Microbiol.* **59**:3266–3272.

Ramanand, K., Nagarajan, A., and Suflita, J. M., 1993b, Reductive dechlorination of the nitrogen heterocyclic herbicide picloram, *Appl. Environ. Microbiol.* **59**:2251–2256.

Rapaport, R. A., and Eisenreich, S. J., 1988, Historical atmospheric inputs of high molecular weight chlorinated hydrocarbons to eastern North America, *Environ. Sci. Technol.* **22**:931–941.

Renard, P., Bouillon, C., Naveau, H., and Nyns, E.-J., 1993, Toxicity of a mixture of polychlorinated organic compounds towards an unacclimated methanogenic consortium, *Biotechnol. Lett.* **15**:195–200.

Rhee, G.-Y., Sokol, R. C., Bush, B., and Bethoney, C. M., 1993a, Long-term study of anaerobic dechlorination of Aroclor 1254 with and without biphenyl enrichment, *Environ. Sci. Technol* **27**:714–719.

Rhee, G.-Y., Bush, Bethoney, C. M., DeNucci, A., Oh, H.-M., and Sokol, R. C., 1993b, Reductive dechlorination of Aroclor 1242 in anaerobic sediments: Pattern, rate and concentration dependence, *Environ. Toxicol. Chem.* **12**:1025–1032.

Rhee, G.-Y., Sokol, R. C., Bethoney, C. M., and Bush, B., 1993c, Dechlorination of polychlorinated biphenyls by Hudson river sediment organisms: Specificity to the chlorination pattern of congeners, *Environ. Sci. Technol.* **27**:1190–1192.

Roberts, A. L., Sanborn, P. N., and Gschwend, P. M., 1992, Nucleophilic substitution reactions of dihalomethanes with hydrogen sulfide species, *Environ. Sci. Technol.* **26**:2263–2274.

Safe, S., 1990, Polychlorinated biphenyls (PCBs), dibenzo-*p*-dioxins (PCDDs), dibenzofurans (PCDFs), and related compounds: Environmental and mechanistic considerations which support the development of toxic equivalency factors (TEFs), *Crit. Rev. Toxicol.* **21**:51–88.

Schanke, C. A., and Wackett, L. P., 1992, Environmental reductive elimination reactions of polychlorinated ethanes mimicked by transition-metal coenzymes, *Environ. Sci. Technol.* **26**:830–833.

Scholz-Muramatsu, H., Neumann, A., Meßmer, M., Moore, E., and Diekert, G., 1995, Isolation and characterization of *Dehalospirillum multivorans* gen. nov., sp. nov., a tetrachloroethene-utilizing, strict anaerobic bacterium, *Arch. Microbiol.* **163**:48–56.

Schrauzer, G. N., 1968, Organocobalt chemistry of vitamin B$_{12}$ model compounds (cobaloximes), *Acc. Chem. Res.* **1**:97–103.

Schwarzenbach, R. P., and Westall, J., 1981, Transport of nonpolar organic compounds from surface water to groundwater. Laboratory studies, *Environ. Sci. Technol* **15**:1360–1367.

Schwarzenbach, R. P., Molnar-Kubica, E., Giger, W., and Wakeham, S. G., 1979, Distribution, residence time, and fluxes of tetrachloroethylene and 1,4-dichlorobenzene in Lake Zürich, Switzerland, *Environ. Sci. Technol.* **13**:1367–1373.

Schwarzenbach, R. P., Giger, W., Hoehn, E., and Schneider, J. K., 1983, Behavior of organic compounds during infiltration of river water to groundwater. Field studies, *Environ. Sci. Technol.* **17**:472–479.

Schwarzenbach, R. P., Gschwend, P. M., and Imboden, D. M., 1993, *Environmental Organic Chemistry,* John Wiley & Sons, New York.

Semprini, L., Hopkins, G. D., Roberts, P. V., and McCarty, P. L., 1991, *In situ* biotransformation of carbon tetrachloride, Freon-113, Freon-11 and 1,1,1-TCA under anoxic conditions, in: *On-Site Bioreclamation Processes for Xenobiotic and Hydrocarbon Treatment* (R. E. Hinchee, and R. F. Olfenbuttel, eds.), U.S.A. Reed Publishers, Newton, MA, pp. 41–58.

Sheikh, Y. M., and Djerassi, C., 1975, 2,6-Dibromophenol and 2,4,6-tribromophenols—antiseptic secondary metabolites of *Phoronopsis viridis, Experientia* **31**:265–266.

Shelton, D. R., and Tiedje, J. M., 1984, Isolation and partial characterization of bacteria in an anaerobic consortium that mineralizes 3-chlorobenzoic acid, *Appl. Environ. Microbiol.* **48**:840–848.

Shiu, W. Y., Doucette, W., Gobas, F. A. P. C., Andren, A., and Mackay, D., 1988, Physical–chemical properties of chlorinated dibenzo-*p*-dioxins, *Environ. Sci. Technol.* **22**:651–658.

Sierra-Alvarez, R., and Lettinga, G., 1991, The effect of structure on the inhibition of acetoclastic methanogenesis in granular sludge, *Appl. Environ. Microbiol.* **34**:544–550.

Smith, M. H., and Woods, S. L., 1994, Regiospecificity of chlorophenol reductive dechlorination by vitamin B$_{12s}$, *Appl. Environ. Microbiol.* **60**:4111–4115.

Sonier, D. N., Duran, N. L., and Smith, G. B., 1994, Dechlorination of trichlorofluoromethane (CFC-11) by sulfate-reducing bacteria from an aquifer contaminated with halogenated aliphatic compounds, *Appl. Environ. Microbiol.* **60**:4567–4572.

Steiert, J. G., and Crawford, R. L., 1986, Catabolism of pentachlorophenol by a *Flavobacterium* sp., *Biochem. Biophys. Res. Commun.* **141**:825–830.

Stockdale, M., and Selwyn, M. J., 1971a, Influence of ring substituents on the action of phenols on some dehydrogenases, phosphokinases and the soluble ATPase from mitochondria, *Eur. J. Biochem.* **21**:416–423.

Stockdale, M., and Selwyn, M. J., 1971b, Influence of ring substituents on the activity of phenols as inhibitors and uncouplers of mitochondrial respiration, *Eur. J. Biochem.* **21**:565–574.

Stumm, W., and Morgan, J. J., 1981, *Aquatic Chemistry,* 2nd ed. Wiley-Interscience, New York.

Sturm, R., and Gandrass, J., 1988, Verhalten von schwerflüchtigen chlorkohlenwasserstoffen an schwebstoffen des Elbe-ästuars, *Vom Wasser* **70**:265–280.

Suflita, J. M., Horowitz, A., Shelton, D. R., and Tiedje, J. M., 1982, Dehalogenation: a novel pathway for the anaerobic biodegradation of haloaromatic compounds, *Science* **218**:1115–1117.

Symonds, R. B., Rose, W. I., and Reed, M. H., 1988, Contribution of Cl- and F bearing gases to the atmosphere by volcanoes, *Nature* **334**:415–417.

Tanabe, S., 1988, PCB problems in the future: Foresight from current knowledge, *Environ. Poll.* **50**:5–28.

Thauer, R. K., Jungermann, K., and Decker, K., 1977, Energy conservation in chemotrophic anaerobic bacteria, *Bacteriol. Rev.* **41**:100–180.

Thayer, J. S., 1990, React metals, organic halides, and—water!?! *Chemtech* **20**:188–191.

Thayer, J. S., Olson, G. J., and Brinckman, F. E., 1984, Iodomethane as a potential metal mobilizing agent in nature, *Environ. Sci. Technol.* **18**:726–729.

Thayer, J. S., Olson, G. J., and Brinckman, F. E., 1987, A novel flow process for metal and ore solubilization by aqueous methyl iodide, *Appl. Organometal. Chem.* **1**:73–79.

Tiedje, J. M., Boyd, S. A., and Fathepure, B. Z., 1987, Anaerobic degradation of chlorinated aromatic hydrocarbons, *Dev. Ind. Microbiol.* **27**:117–127.

Traunecker, J., Preuss, A., and Diekert, G., 1991, Isolation and characterization of a methyl chloride utilizing, strictly anaerobic bacterium, *Arch. Microbiol.* **15**:416–421.

Utkin, I., Woese, C., and Wiegel, J., 1994, Isolation and characterization of *Desulfitobacter dehalogenans* gen. nov., sp. nov., an anaerobic bacterium which reductively dechlorinates chlorophenolic compounds, *Int. J. Syst. Bact.* **44**:612–619.

Utkin, I., Dalton, D. D., and Wiegel, J., 1995, Specificity of reductive dehalogenation of substituted *ortho*-chlorophenols by *Desulfitobacterium dehalogenans* JW/IU-DC-1, *Appl. Environ. Microbiol.* **61**:346–351.

van den Tweel, W. J. J., Kok, J. B., and de Bont, J. A. M., 1987, Reductive dechlorination of 2,4-dichlorobenzoate to 4-chlorobenzoate and hydrolytic dehalogenation of 4-chloro-, 4-bromo-, and 4-iododbenzoate by *Alcaligenes denitrificans* NTB-1, *Appl. Environ. Microbiol.* **53**:810–815.

van den Wijngaard, A. J., van der Kamp, K. W. H. J., van der Ploeg, J., Kazemier, B., Pries, F., and Janssen, D. B., 1992, Degradation of 1,2-dichloroethane by *Ancylobacter aquaticus* and other facultative methylotrophs, *Appl. Environ. Microbiol.* **58**:976–983.

van der Meer, J. R., de Vos, W. M., Harayama, S., and Zehnder, A. J. B., 1992, Molecular mechanisms of genetic adaptation to xenobiotic compounds, *Microbiol. Rev.* **56**:677–694.

VanDort, H. M., and Bedard, D. L., 1991, Reductive *ortho* and *meta* dechlorination of a polychlorinated biphenyl congener by anaerobic microorganisms, *Appl. Environ. Microbiol.* **57**:1576–1578.

Vannelli, T., and Hooper, A. B., 1993, Reductive dehalogenation of the trichloromethyl group of nitrapyrin by the ammonia-oxidizing bacterium *Nitrosomonas europaea*, *Appl. Environ. Microbiol.* **59**:3597–3601.

van Zoest, R., and van Eck, G. T. M., 1991, Occurrence and behaviour of several groups of organic micropollutants in the Scheldt estuary, *Sci. Total Environ.* **103**:57–71.

Vogel, T. M., Criddle, C. S., and McCarty, P. L., 1987, Transformations of halogenated aliphatic compounds, *Environ. Sci. Technol.* **21**:722–736.

Vogels, G. D., Keltjens, J. T., and van der Drift, C., 1988, Biochemistry of methane production, in: *Biology of Anaerobic Microorganisms* (A. J. B. Zehnder, ed.), John Wiley & Sons, New York, pp. 707–770.

Wiegel, J., Kohring, G.-W., Zhang, X., Utkin, I., Dalton, D., He, Z., Wu, Q., and Bedard, D., 1992, Temperature an important factor in the anaerobic transformation and degradation of chlorophenols and PCBs, in: *Soil Decontamination Using Biological Processes*, DECHEMA, Germany, pp. 101–108.

Williams, W. A., 1994, Microbial reductive dechlorination of trichlorobiphenyls in anaerobic sediment slurries, *Environ. Sci. Technol.* **28**:630–635.

Winkels, H. J., Vink, J. P. M., Beurskens, J. E. M., and Kroonenberg, S. B., 1993, Distribution and geochronology of priority pollutants in a large sedimentation area, River Rhine, The Netherlands, *Appl. Geochem. (Supp)* **2**:95–101.

Woods, S. L., and Smith, M. H., 1995, Regiospecificity of chlorophenol reductive dechlorination by vitamin B$_{12s}$ [letter to the editor], *Appl. Environ. Microbiol.* **61**: 2450–2451.

Xun, L., Topp, E., and Orser, C. Y., 1992, Purification and characterization of a tetrachloro-*p*-hydroquinone reductive dehalogenase from a *Flavobacterium* sp., *J. Bacteriol.* **174**:8003–8007.

Zehnder, A. J. B., and Stumm, W., 1988, Geochemistry and biogeochemistry of anaerobic habitats, in: *Biology of Anaerobic Microorganisms* (A. J. B. Zehnder, ed.), John Wiley & Sons, New York, pp. 1–38.

Zhang, X., and Wiegel, J., 1990, Sequential anaerobic degradation of 2,4-dichlorophenol in freshwater sediments, *Appl. Environ. Microbiol.* **56:**1119–1127.

Zitomer, D. H., and Speece, R. E., 1993, Sequential environments for enhanced biotransformation of aqueous contaminants, *Environ. Sci. Technol.* **27:**227–244.

Zoeteman, B. C. J., Harmsen, K., Linders, J. B. H. J., Morra, C. F. H., and Slooff, W., 1980, Persistent organic pollutants in river water and ground water of The Netherlands, *Chemosphere* **9:**231–249.

Zwolsman, G. J., Sonneveldt, H. L. A., and Ruijgh, E. F. W., 1993, Onderzoek Noordelijk Deltabekken, Zuidrand. Toepassing accumulatiemodel waterbodem Hollands Diep, onzekerheidsanalyse en calibratie. Delft Hydraulics Report No. T 262 (in Dutch).

5

Soil Microbial Processes Involved in Production and Consumption of Atmospheric Trace Gases

RALF CONRAD

1. Introduction

World climate is not only a function of atmospheric physics but of atmospheric chemistry as well. In fact, the science community investigating questions on atmospheric chemistry developed rapidly after publication of the pioneering monograph by Christian Junge in 1963. Since then, it has become clear that the chemistry of the atmosphere is highly dynamic. In fact, the composition of the atmosphere is presently changing in a direction that ultimately may alter the global and regional climate of Earth (Houghton *et al.*, 1990). Interesting to the microbiologists is the fact that the chemistry of the atmosphere is to a large extent the result of the Earth's biosphere, and with that of microbial processes. For example, it was the evolution of oxygenic photosynthesis by cyanobacteria that allowed the production of significant amounts of O_2 (Schidlowski, 1983; Holland, 1984). This microbial O_2 production finally resulted in the accumulation of O_2 in a so-far anoxic atmosphere, and thus resulted in one of the most dramatic shifts in the habitability of our planet, i.e., shift in the dominance of anaerobic to aerobic life.

Now, microbial metabolism is also of great importance for the composition of our atmosphere. Some scientists even argue that the atmospheric composition is kept at close homeostasis by the activity of the biosphere (Lovelock and Margulis, 1974). Indeed, there are only a few atmospheric constituents (e.g., Rn, O_3) that exhibit a cycle that is purely due to physicochemical reactions. Most

RALF CONRAD • Max Planck Institut für Terrestrische Mikrobiologie, D-35043 Marburg, Germany.

Advances in Microbial Ecology, Volume 14, edited by J. Gwynfryn Jones. Plenum Press, New York, 1995.

Table I. Contribution of Soil to the Global Cycles of Atmospheric Trace Gases

Trace gas	Lifetime (d)	Mixing ratio (ppbv)	Total budget (Tg yr^{-1})	Annual increase (%)	Contribution (%) of soils as		Impact	References[a]
					Source	Sink		
N_2O	60,000	310	15	0.2–0.3	70	?	Stratospheric chemistry; greenhouse effect	1,2,3,4
CH_4	4,000	1700	540	<0.8	60	5	Greenhouse effect; tropospheric and stratospheric chemistry	4,5,6,7
H_2	1,000	550	90	0.6	5	95	Insignificant	8,9
OCS	>350	0.5	1.2	?	25	?	Aerosol formation	10,11,12
CO	100	100	2,600	1.0[b]	1	15	Tropospheric chemistry	8,13
NO	1	<0.1	60	?	20	?	Tropospheric chemistry	3,14

[a]1. Khalil and Rasmussen (1992); 2. Bouwman (1992); 3. Davidson (1991); 4. Prinn (1994); 5. Cicerone and Oremland (1988); 6. Khalil and Rasmussen (1990a); 7. Steele et al. (1992); 8. Conrad (1988); 9. Khalil and Rasmussen (1990b); 10. Andreae and Jaeschke (1992); 11. Chin and Daviss (1993); 12. Möller (1984); 13. Khalil and Rasmussen (1990c); 14. Conrad (1990).

[b]The long-term recently reversed; CO is now decreasing by about 6% per year (Novelli et al., 1994).

of the atmospheric gases undergo cycles that are more or less dominated by the biosphere. Within the biosphere, soils probably show the highest abundance and diversity of microorganisms in nature, and thus it is not surprising that soils are involved in and even dominate the cycling of many atmospheric trace gases. Table I lists a selection of atmospheric trace gases that all involve soils as important sources and/or sinks and that are all (maybe with the exception of H_2) of great importance for atmospheric chemistry and climate. In order to understand and predict potential changes in the source and sink strengths of these trace gases, it is necessary to know the processes involved in generating the net flux at the soil–atmosphere interface. In general, this net flux is a function of production processes, consumption processes, and gas transport.

Unfortunately, not many microbiologists are familiar with the metabolism of atmospheric trace gases by the soil microflora, and few are aware of the many scientific problems that still exist and must be solved with the help of microbiologists to obtain a comprehensive understanding of the cycling of atmospheric trace gases. In the following, I will review our knowledge of the soil microbial processes involved in the cycling of some selected trace gases, i.e., methane (CH_4), hydrogen (H_2), carbon monoxide (CO), carbonyl sulfide (OCS), nitrous oxide (N_2O), and nitric oxide (NO). In doing so, I will emphasize, on the one hand, the diversity of microorganisms producing or consuming the same trace gas and, on the other hand, the versatility of a single group of organisms being involved in different biogeochemical reactions. I will also distinguish between upland and wetland soils, which offer basically different conditions to soil microorganisms. I define *upland soils* as being non-water-saturated, well-aerated soils that are generally oxic, and *wetland soils* as being water-saturated soils that are generally anoxic. A glossary of the different microorganisms and metabolic reactions that I will discuss with respect to the cycling of atmospheric trace gases is given in Table II.

2. Methane

Methane absorbs long-wave radiation, and thus is a greenhouse gas that leads to the warming of the Earth's surface and the lower atmosphere. In this respect, a CH_4 molecule is about 30 times more effective than a CO_2 molecule (Dickinson and Cicerone, 1986; Ramanathan *et al.*, 1987). It also plays a great role in the production of tropospheric ozone (Crutzen, 1979), is a significant source of water vapor in the stratosphere (Crutzen and Schmailzl, 1983), and contributes to the scavenging of chlorine atoms involved in stratospheric O_3 depletion (Cicerone, 1987).

Soils are the most important source in the atmospheric CH_4 budget, contributing about 60% (Cicerone and Oremland, 1988). The active CH_4-emitting soils include all kinds of wetlands (Crill *et al.*, 1991; Bartlett and Harriss, 1993),

Table II. Glossary of Microorganisms and Metabolic Reactions Potentially Involved in the Cycling of Different Trace Gases

Microorganisms, enzymes	Relevant reaction	Relation to O_2	Growth by reaction	Other characteristics
CH_4				
Methanotrophs	$CH_4 + 2O_2 \rightarrow CO_2 + 2H_2O$	Oxic	+	
Common				K_m, V_{max}, threshold high
Unknown				K_m, V_{max}, threshold low
Nitrifiers (NH_4^+)	$CH_4 + O_2 + 2H^+ + 2e^- \rightarrow CH_3OH + H_2O$	Oxic	–	K_m high, V_{max} low
Methanogens	$4H_2 + CO_2 \rightarrow CH_4 + 2H_2O$	Anoxic	+	
	e.g., acetate $\rightarrow CH_4 + CO_2$			
H_2				
Abiontic soil hydrogenase	$2H_2 + O_2 \rightarrow 2H_2O$	Oxic	–	K_m, V_{max}, threshold low
Knallgas bacteria	$2H_2 + O_2 \rightarrow 2H_2O$	Oxic	+	K_m, V_{max}, threshold high
Methanogens	$4H_2 + O_2 \rightarrow CH_4 + 2H_2O$	Anoxic	+	K_m, threshold decreasing
Anaerobic respiratory bacteria	$4H_2 + SO_4^{2-} + H^+ \rightarrow HS^- + 4H_2O$	Anoxic	+	K_m, threshold decreasing
	$2.5 H_2 + NO_3^- + H^+ \rightarrow 0.5 N_2 + 3H_2O$	Anoxic	+	K_m, threshold decresing
Nitrogenase	$2H^+ + 2e^- \rightarrow H_2$	Anoxic	–	
Fermenting bacteria	e.g., pyruvate $\rightarrow$ acetate + CO_2 + H_2	Anoxic	+	
CO				
Nitrifiers (NH_4^+)	$CO + O_2 + 2e^- + 2H^+ \rightarrow CO_2 + H_2O$	Oxic	–	K_m, V_{max} low
Oligotrophs	$CO + 0.5 O_2 \rightarrow CO_2$	Oxic	+	K_m, V_{max} low
Carboxydotrophs	$CO + 0.5 O_2 \rightarrow CO_2$	Oxic	+	K_m, V_{max} high
Anaerobes, e.g., homoacetogens	$4CO + 2H_2O \rightarrow$ acetate + $2CO_2$	Anoxic	+	K_m, V_{max} high

OCS				
Thiobacilli	$OCS + H_2O \rightarrow CO_2 + H_2S$	Oxic/anoxic	–	Carbonic anhydrase
Thiobacilli	$H_2S + 2O_2 \rightarrow H_2SO_4$	Oxic	+	Subsequent catabolism of H_2S
	$SCN^- + H_2O \rightarrow OCS + H_2S$	Oxic/anoxic	–	
N$_2$O				
Nitrifiers, autotrophic	$NH_4^+ \rightarrow NO_2^- \rightarrow N_2O$	Oxic	+	Cell carbon from CO_2
Nitrifiers, heterotrophic	$NH_4^+ \rightarrow NO_2^- \rightarrow N_2O\ (\rightarrow N_2)$	Oxic	+	Cell carbon from org. C
Methanotrophs	$NH_4^+ \rightarrow\ \rightarrow N_2O$	Oxic	–	
Denitrifiers	$NO_3^- \rightarrow NO_2^- \rightarrow N_2O \rightarrow N_2$	Anoxic	+	
DNRA, etc.	$NO_3^- \rightarrow NO_2^- \rightarrow NH_4^+$	Anoxic	+	
	$\rightarrow N_2O \rightarrow N_2$			
NO				
Nitrifiers, autotrophic	$NH_4^+ \rightarrow NO_2^- \rightarrow NO$	Oxic	+	Cell carbon from CO_2
Nitrifiers, heterotrophic	$NH_4^+ \rightarrow NO_2^- \rightarrow NO\ (\rightarrow N_2)$	Oxic	+	Cell carbon from org. C
Methanotrophs	$NH_4^+ \rightarrow NO \rightarrow ?$	Oxic	–	
Nitrifiers (NO_2^-)	$NO_2^- \rightarrow NO$	Anoxic	+	Org. C as electron donor
	$NO \rightarrow NO_3^-$			
Pseudomonas	$NO \rightarrow NO_3^-$	Oxic	–	
Denitrifiers	$NO_3^- \rightarrow NO_2^- \rightarrow NO \rightarrow N_2O \rightarrow N_2$	Anoxic	+	
DNRA	$NO_3^- \rightarrow NO_2^- \rightarrow NH_4^+$	Anoxic	+	
	$\rightarrow NO \rightarrow ?$			

flooded rice fields (Schütz et al., 1990; Matthews et al., 1991), and termite nests (Seiler et al., 1984; Khalil et al., 1990). Common to all these sites is the production of CH_4 in anoxic zones by the strictly anaerobic methanogenic bacteria. In wetland soils, anoxic conditions establish in most of the soil because O_2 diffusion from the atmosphere is limiting and thus favors the activities of fermenting and finally of methanogenic bacteria (Conrad, 1989, 1993). Upland soils, by contrast, are generally oxic, and thus usually do not emit CH_4, but instead act as a sink for atmospheric CH_4 (Conrad, 1989). However, CH_4 production also occurs in upland soils if they contain "hot spots" that provide anoxic conditions for the operation of methanogens. Vice versa, CH_4 oxidation also occurs in wetland soils if they contain sites where O_2 is available. Furthermore, in marine sediments, CH_4 may also be consumed under anoxic conditions (Alperin and Reeburgh, 1984).

2.1. Upland Soils

Upland soils are now recognized as an important sink in the atmospheric CH_4 budget (Reeburgh et al., 1993). The CH_4 consumption activity in soil is abolished by autoclaving and shows a temperature optimum, demonstrating that it is a biological process (King and Adamsen, 1992). Its requirement for O_2 and its susceptibility to inhibition by acetylene or methylfluoride indicate that CH_4 oxidation is due to methanotrophic bacteria (Bender and Conrad, 1993, 1994a; Oremland and Culbertson, 1992). The kinetic properties of the CH_4 consumption activity are unusual and indicate the existence of two different populations of methanotrophs.

Bender and Conrad (1992, 1993) have shown that soils exhibit a much lower K_m and threshold for CH_4 than predicted from pure-culture studies with known methanotrophic bacteria. Soils typically exhibit a high affinity for CH_4, with K_m values for CH_4 being in the range of 30–60 nM (equivalent to a partial pressure of 20–45 ppmv), whereas known methanotrophic bacteria all exhibit a relatively low affinity for CH_4, with K_m values generally higher than 1000 nM (Carlsen et al., 1991). In addition, the threshold for CH_4 consumption in soils is usually less than 0.2 ppmv, i.e., significantly lower than ambient CH_4 (Bender and Conrad, 1993), whereas pure cultures of methanotrophic bacteria often do not consume CH_4 at atmospheric concentrations at all and never maintain long-term activity with such low CH_4 concentrations (King, 1993). These results substantiate the existence of two different populations of methanotrophs. One population consists of the commonly known methanotrophic bacteria (further on called "common methanotrophs"), which have a low affinity for CH_4 and grow and metabolize CH_4 only at elevated concentrations. The second population consists of so-far unknown methanotrophic bacteria (further on called "unknown methanotrophs"), which have a high affinity for CH_4 and are able to metabolize

CH_4 even at atmospheric concentrations (1.8 ppmv). The effect of the two methanotrophic populations on the kinetics of CH_4 oxidation by soil is shown in Fig. 1.

The nitrifying bacteria (ammonium oxidizers) are also potentially involved in the oxidation of methane. The nitrifiers are able to oxidize CH_4 fortuitously, without using it for growth (Hyman and Wood, 1983; Jones and Morita, 1983a). Although in pure cultures the maximum rates of CH_4 consumption per bacterium are about two to three orders of magnitude lower in the nitrifiers than in the common methanotrophs, the nitrifiers nevertheless could contribute to CH_4 oxidation if present in sufficient numbers. However, this is probably not the case, as numbers of nitrifiers are usually similar to numbers of methanotrophs (only the common methanotrophs can be counted) (e.g., Bender and Conrad, 1994b).

Unfortunately, a direct approach to distinguish the CH_4 oxidation activity of the common nitrifiers and methanotrophs in soil is not easily achieved, since they

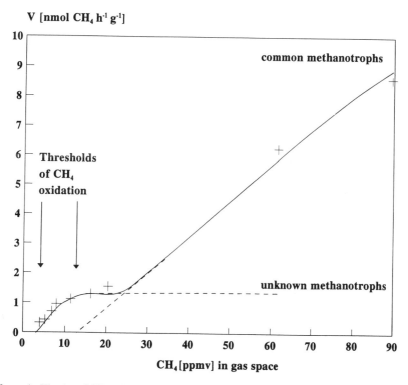

Figure 1. Kinetics of CH_4 oxidation in a forest soil exhibiting activity of both the "common" (high K_m, high threshold) and the "unknown" (low K_m, low threshold) methanotrophic bacteria. (Adapted from Bender and Conrad, 1992.)

Table III. Determination of the Role of Methanotrophic versus Nitrifying Bacteria for CH$_4$ Oxidation in Surface (0–2 cm) Soil of Fertilized (N-P-K) and Unfertilized Fields[a]

| Treatment | Nitrapyrin sensitive oxidation [nmole g^{-1}hr^{-1}] of | | | Dominant CH$_4$ oxidizers |
	^{14}CO	^{14}CH$_4$	Ratio	
Control				
1	199 ± 47	5.7 ± 0.1	0.03	Methanotrophs
2	90 ± 97	6.0 ± 0.8	0.07	Methanotrophs
Fertilized				
1	198 ± 21	1.5 ± 0.1	0.008	Nitrifiers
2	303 ± 32	1.2 ± 0.1	0.004	Nitrifiers

[a]Adapted from Castro et al. (1994).

have a similar K_m for CH$_4$ (in the micromolar range) and share the same inhibitors (Bedard and Knowles, 1989). Jones et al., (1984) used an indirect approach to distinguish between CH$_4$ oxidation by nitrifiers versus common methanotrophs. This approach exploits the capacity of both types of bacteria to fortuitously oxidize CO at similar rates per cell, while CH$_4$ is oxidized preferentially by the methanotrophs. Using nitrapyrin as an inhibitor of both nitrifiers and methanotrophs, the rates of CH$_4$ oxidation were normalized to those of nitrapyrin-sensitive CO oxidation to indicate whether the observed CH$_4$ oxidation might be due to nitrifiers (CH$_4$/CO <0.01) or methanotrophs (CH$_4$/CO > 0.01). This approach was recently used by Castro et al. (1994), and indicated that N-fertilized soils exhibit an increased contribution of nitrifiers to CH$_4$ oxidation (Table III). The possible problem with this approach is that it has been developed using cultures of the common nitrifiers and methanotrophs. The common methanotrophs, however, seem to be unable to oxidize CH$_4$ at atmospheric concentrations. The common nitrifiers have to my knowledge not yet been tested for this ability. Both common nitrifiers and methanotrophs have the same low affinity (high K_m) for CH$_4$. The unknown methanotrophs with the high affinity for CH$_4$ have not yet been isolated, and it is therefore not known whether they would also oxidize CO and whether this activity would be inhibited by nitrapyrin. The approach of Jones et al. (1984) is therefore meaningful only if the unknown methanotrophs can oxidize CO at rates per cell similar to those of the nitrifiers. In this case, it would also be interesting to check whether soils dominated by CH$_4$-oxidizing nitrifiers exhibit kinetics with the same low K_m for CH$_4$ as soils dominated by CH$_4$-oxidizing methanotrophs. In this case, we have to expect an as yet unknown population of nitrifiers with a low K_m for CH$_4$. I will not discuss this problem and will deal only with the role of the methanotrophic bacteria. The activity of the common methanotrophs was only observed in soils that had previously been exposed to artificially elevated CH_4 concentrations (Bender and Conrad, 1992). In these soils, the population size of the common methanotrophs was increased. These results are reasonable in light of previously reported calculations (Conrad, 1984) that show that the common methanotrophs should be able to grow and maintain cell integrity only at elevated but not at atmospheric CH_4 concentrations because of their low substrate affinity. However, the unknown methanotrophs with a K_m for CH_4 in the range of tens of nanomolar, as found in untreated soils, should theoretically be able to grow at atmospheric CH_4 concentrations (Conrad, 1984). In addition, they should also be able to grow at elevated CH_4 and this even better than at atmospheric CH_4, since their K_m value—though it is relatively low—is still more than tenfold higher than atmospheric CH_4 concentrations. In fact, soils exposed to elevated CH_4 for 3 weeks indeed showed an increased activity of the unknown methanotrophs (Bender and Conrad, 1992). This increased activity was probably due to their increased populations size, although this cannot be stated with certainty, since the numbers of the unknown methanotrophs cannot be determined with currently available techniques.

The common methanotrophs utilize only C-1 compounds, i.e., CH_4, methanol, or methylamines (Anthony, 1986). Nevertheless, the common methanotrophs are found in significant numbers in upland soils, although atmospheric CH_4 is the only obvious source of available energy (Heyer et al., 1984). They seem to be generally present in upland soils even if an apparent CH_4 consumption activity is lacking. As common methanotrophs are able to form cysts or exospores, they may just be present in a resting stage to become active when exposed to CH_4 concentrations that are sufficiently high to allow growth. Such an activation was observed in several studies (Megraw and Knowles, 1987; Bender and Conrad, 1992; Nesbit and Breitenbeck, 1992).

This conclusion is in contrast to most studies in which vertical profiles of CH_4 concentrations were measured in upland soils (Born et al., 1990; Striegl et al., 1992; Whalen et al., 1992; Adamsen and King, 1993; Dörr et al., 1993; Koschorreck and Conrad, 1993). These studies show that CH_4 concentrations decrease exponentially with soil depth, which is consistent with net consumption of atmospheric CH_4 within the soil profile. Methane concentrations higher than ambient are usually not observed in upland soils, and exceptions have rarely been reported (Yavitt et al., 1990).

However, even oxic upland soils have the potential to microbially produce CH_4 if they are incubated under anoxic conditions (Sexstone and Mains, 1990; Koschorreck and Conrad, 1993; Bender and Conrad, 1994a). Even desert soils contain small numbers of methanogenic bacteria that become active and start to multiply some time after submergence of the soil (Peters and Conrad, 1995). Hence, we may hypothesize that upland soils contain hot spots in which methanogenic bacteria are temporarily active (Fig. 2). This should only be possible in microniches where O_2 is depleted. Oxygen-deficient microniches do exist in

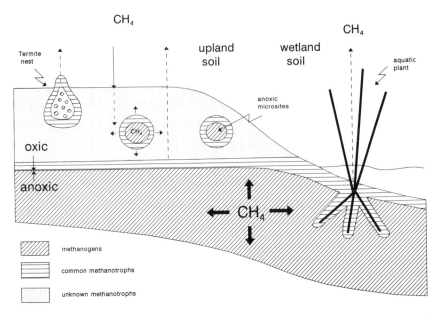

Figure 2. Scheme of distribution of microorganisms involved in CH_4 cycling in upland and wetland soils.

generally oxic soils (Sexstone *et al.*, 1985; Tiedje *et al.*, 1984; Zausig *et al.*, 1993). Methane production in hot spots could provide sufficiently high concentrations of CH_4 to allow growth of the common methanotrophs. The common methanotrophs would consume the CH_4 at the periphery of the hot spot in which CH_4 is produced, and elevated CH_4 concentrations would not become evident in the bulk soil (Fig. 2). This concept implies, however, that the common methanotrophs develop activity in close spatial and temporal vicinity to the methanogens. The same concept has been put forward for CH_4 cycling in oxic surface water of the ocean, i.e., production and consumption of CH_4 in detritus particles (Sieburth, 1987). Vecherskaya *et al.* (1993) observed methanotrophic microcolonies covering the surface of organic particles in tundra soils.

There is circumstantial evidence of CH_4 production in generally oxic soils (Keller *et al.*, 1986, 1993). However, the CH_4 production processes have so far not been studied. A clue to the study of CH_4 production in these oxic soils could be the so-called lower threshold concentrations for CH_4 that are generally observed in upland soils (Bender and Conrad, 1993). It cannot completely be ruled out that these thresholds are at least sometimes compensation concentrations between production and consumption of CH_4 rather than the ultimate concentration to which the methanotrophs can consume CH_4. A similar clue may be the vertical profiles of CH_4 measured in soil (Whalen *et al.*, 1992; Dörr *et al.*, 1993;

Koschorreck and Conrad, 1993), which in some cases may be not only due to diffusion and consumption of CH_4 but may be influenced by CH_4 production. Landfill soils may be regarded as a special type of upland soil. These soils have such a high content of degradable organic matter and obviously have so many CH_4-producing hot spots that they usually exhibit within their soil profiles CH_4 concentrations that are substantially higher than ambient (Jones and Nedwell, 1990, 1993). These soils consequently act as sources rather than as sinks for atmospheric CH_4 (Bingemer and Crutzen, 1987). Methane oxidation in these soils seems to be dominated by the common methanotrophic bacteria, exhibiting a relatively high K_m for CH_4 (Whalen et al., 1990).

A special kind of hot spot in upland soils may be provided by termites (Fig. 2). It is well established that several arthropod families, especially termites, produce CH_4, since their guts provide sites for methanogenic bacteria (Breznak, 1975; Zimmermann et al., 1982; Rasmussen and Khalil, 1983; Collins and Wood, 1984; Rouland et al., 1993; Hackstein and Stumm, 1994). However, termites are usually not dispersed in the soil but are concentrated in mounds that act as significant sources of atmospheric CH_4. Termites thus provide huge hot spots on an areal basis. Field measurements indicate that some of the CH_4 produced may be consumed within the termite mound before it can escape into the atmosphere (Seiler et al., 1984; Khalil et al., 1990). This consumption is probably accomplished by the common methanotrophs, as the CH_4 concentrations in termite mounds should be sufficiently high.

2.2. Wetland Soils

Wetland soils are generally anoxic, and thus provide an excellent habitat for methanogenic microbial communities. The factors controlling CH_4 emission from flooded rice fields and other wetland soils have been reviewed in some detail (Conrad, 1989, 1993; Schütz et al., 1991; Schimel et al., 1993; Neue and Roger, 1993; Neue and Sass, 1994). From the viewpoint of CH_4 cycling, the most striking difference between wetland and upland soils is that the former contain CH_4-consuming hot spots in a CH_4-producing environment, whereas the latter contain CH_4-producing hot spots in a CH_4-consuming environment (Fig. 2). The CH_4-consuming hot spots are largely provided by the roots of macrophytes, which supply the adjacent soils with O_2 that is transported from the atmosphere to allow root respiration (Schütz et al., 1991; Chanton and Dacey, 1991). In addition to the root surface, O_2 is usually also available at the soil–water interface. At both sites, CH_4 is oxidized so efficiently that sometimes less than 10% of the CH_4 produced in the wetland soil escapes to the atmosphere (Conrad, 1993; Reeburgh et al., 1993).

Methane oxidation is thus a major regulator of the CH_4 source strength of wetland soils. Methane oxidation in wetland soils and sediments seems to be accomplished by the common methanotrophs, which typically exhibit K_m values for CH_4 in the micromolar range (Lidstrom and Somers, 1984; Bender and Conrad, 1992; King, 1990; Bosse et al., 1993). This is reasonable since the oxic hot spots are supplied with CH_4 at relatively high concentrations. The exact concentrations available to the methanotrophs are presently not well known, since CH_4 oxidation occurs at the intercept of two steep gradients, CH_4 and O_2, which are difficult to analyze. Measurements with a recently developed microprobe for CH_4 indicate CH_4 concentrations in the boundary layer in the micromolar range (Conrad and Rothfuss, 1991; Rothfuss and Conrad, 1994).

Paddy soil that is kept below a moisture content equivalent to field capacity exhibits, in addition to the activity of the common methanotrophs, the high-affinity and low-threshold activity typical of the unknown methanotrophs (Bender and Conrad, 1992). It is not known whether this activity can persist in a flooded rice field that is permanently producing CH_4, and thus should be more favorable for the development of the common rather than the unknown methanotrophs.

There is presently no evidence that wetland soils exhibit anaerobic CH_4 oxidation, which is frequently observed in marine and hypersaline environments (Alperin and Reeburgh, 1984; Iversen and Joergensen, 1985; Iversen et al., 1987). In these environments, part of the produced CH_4 is consumed before it reaches the oxic zone at the sediment surface. The microorganisms responsible for anaerobic CH_4 oxidation are still completely enigmatic and have so far escaped isolation. Obviously, however, anaerobic CH_4 oxidation plays an important role in the control of CH_4 emission from marine sediments (Henrichs and Reeburgh, 1987; Reeburgh et al., 1993).

3. Hydrogen

With CH_4, hydrogen constitutes a source of water vapor in the stratosphere. However, the concentration gradient of H_2 from the troposphere into the stratosphere is so small that the flux of H_2 into the stratosphere is negligible and H_2 is cycled almost exclusively within the troposphere (Schmidt, 1974). In recent times, the H_2 cycle has apparently been disturbed, with H_2 concentrations increasing. The increase in H_2 is probably due to photochemical decomposition of CH_4 and other hydrocarbons (Khalil and Rasmussen, 1990b). To date, there is no indication that the global H_2 cycle has a major impact on atmospheric chemistry or climate.

The H_2 cycle is nevertheless of scientific interest, especially since H_2 is the only trace gas that is almost exclusively decomposed by biological processes in upland soils (Conrad, 1988). Wetland soils are potential sources of atmospheric H_2, but virtually all the produced H_2 is converted to CH_4 within the soil before it can escape into the atmosphere (Conrad, 1988). However, H_2 is produced during N_2 fixation, constituting a small source of atmospheric H_2 (Conrad, 1988).

3.1. Upland Soils

The net flux of H_2 between soil and atmosphere is the result of simultaneously operating production and utilization reactions, which both are located in the uppermost soil layers (Liebl and Seiler, 1976; Seiler, 1978). This has been shown by the existence of a compensation concentration at which H_2 production equals H_2 destruction. In most upland soils, the compensation concentration is much lower than the atmospheric H_2 concentration (500 ppbv) and close to the detection limit (about 10 ppbv H_2); thus, the direction of H_2 flux is into the soil. However, if the soil is covered by N_2-fixing legumes, the compensation concentration may increase to values >550 ppbv to direct the net flux into the atmosphere (Conrad and Seiler, 1979).

The processes involved in H_2 cycling in upland soils have been reviewed by Conrad (1988). The major production process is the formation of H_2 by the nitrogenase, the key enzyme of N_2 fixation. Furthermore, H_2 may be produced by fermenting bacteria within anoxic microniches. Hence, the situation of H_2 in upland soils is in principle similar to that of CH_4 with H_2 being produced in hot

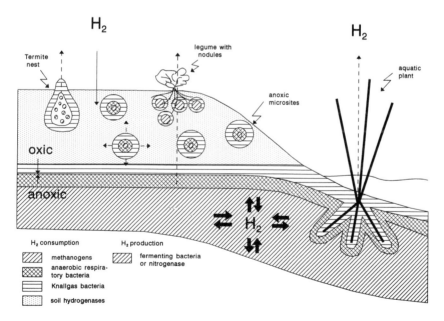

Figure 3. Scheme of distribution of microorganisms involved in H_2 cycling in upland and wetland soils.

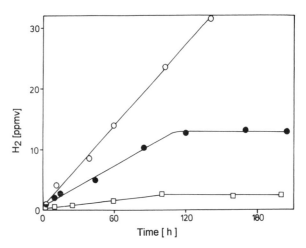

Figure 4. H_2 release by *Vicia faba* (Hup^-) planted in sterile soil (o), in nonsterile soil (□), and in sterile soil amended with Knallgas bacteria (•). (Adapted from Schuler and Conrad, 1991.)

spots that show either fermentation or N_2 fixation (Fig. 3). The situation is also similar with respect to H_2 consumption, since there are two distinctly different activities involved, i.e., aerobic H_2-oxidizing bacteria for the oxidation of H_2 at elevated concentrations, and soil hydrogenases that oxidize H_2 at atmospheric concentrations.

The existence of two H_2-oxidizing activities in upland soils has been shown by kinetic experiments in which a high-affinity system with K_m values in the range of tens of nanomolar H_2 and a low-affinity system with K_m values in the micromolar range have been demonstrated (Schuler and Conrad, 1990; Häring and Conrad, 1994). The low-affinity system is typical of aerobic Knallgas bacteria, which grow with H_2 as energy substrate, but have a threshold for H_2 that is too high to allow the oxidation of atmospheric H_2 (Conrad, 1988; Schuler and Conrad, 1990; Häring and Conrad, 1991; Klüber and Conrad, 1993).

Knallgas bacteria apparently contribute to the oxidation of H_2 at the relatively high concentrations found in the vicinity of H_2-producing root nodules of N_2-fixing legumes (Fig. 3) (LaFavre and Focht, 1983; Popelier *et al.*, 1985; Cunningham *et al.*, 1986; Schuler and Conrad, 1991). Thus, Schuler and Conrad (1991) demonstrated the existence of a relatively high H_2 compensation concentration, when sterile soil with legumes was inoculated with Knallgas bacteria. These Knallgas bacteria consumed the H_2 produced by the legumes only to concentrations of about 13 ppmv H_2 (Fig. 4). The consumption of H_2 to still lower concentrations was possible only if the soil hydrogenases were active in addition, i.e., in nonsterile soil. However, it is likely that the activity of the soil hydrogenases operates in line with the H_2-oxidizing bacteria, since the soil hydrogenases are reversibly inhibited by exposure to high H_2 concentrations

(Conrad and Seiler, 1981). Thus, the Knallgas bacteria probably achieve a decrease of H_2 to lower concentrations, which then allow further oxidation of H_2 by the soil hydrogenases.

Besides legumes, high H_2 concentrations may emerge in hot spots analogous to those assumed for production of CH_4 in upland soils, e.g., anoxic microsites (especially in landfills) and termite nests (Fig. 3). In fact, the H_2 produced by termites seems to be completely oxidized within the termite mounds, which thus apparently do not act as a source for atmospheric H_2 (Khalil et al., 1990). Also, cores of solid waste that were flushed with high H_2 concentrations developed increased numbers of H_2-oxidizing bacteria that effectively oxidized the H_2 (Dugnani et al., 1986).

The production of H_2 in hot spots of upland soils rarely develops to a net source of atmospheric H_2. So far, this potential has been shown only for legume fields (Conrad and Seiler, 1980). This is because the emerging H_2 is consumed not only by Knallgas bacteria, but is further consumed by soil hydrogenases that seem to be ubiquitous in upland soils (Conrad, 1988). Soil enzymes (Skujins, 1978) include not only free extracellular enzymes and enzymes bound to inert soil components but also active enzymes within dead or nonproliferating cells and others associated with the dead cell fragments. Nobody has so far succeeded in isolating a soil hydrogenase. Therefore, the characterization of this activity as a soil enzyme is preliminary (Conrad, 1988). The activity catalyzes the oxidation of H_2 as well as the exchange of tritium with water, is mainly localized in the upper soil layers, and is attached to soil grains (Häring and Conrad, 1994, Häring et al., 1994). Its distribution among the different sized fractions of soil grains is different from that of the bulk microbial biomass (Häring et al., 1994). Most notable is its high affinity for H_2, which distinguishes this hydrogenase activity from all hydrogenases known from the literature. Furthermore, it is the only H_2-consuming activity known so far that is able to consume H_2 at concentrations down to at least 10 ppbv H_2, equivalent to an aqueous concentration of about 10 pM H_2. If this activity did not exist, H_2 would certainly accumulate in the atmosphere to significantly higher concentrations than observed today.

3.2. Wetland Soils

The anoxic parts of wetland soils show an intensive cycling of H_2 (Conrad, 1995). Hydrogen is an important intermediate in the degradation of organic matter under anoxic conditions. Hydrogen is produced by various fermenting bacteria and is consumed mainly by the methanogenic bacteria that convert H_2 plus CO_2 to CH_4. The H_2 turnover in anoxic wetland soils is thus an important component of the production and emission of CH_4 from these environments (Conrad, 1989). The turnover of H_2 in the anoxic parts of wetland soils results in steady-state H_2 partial pressures in the order of 10–100 ppmv H_2, i.e., significant supersaturations with respect to atmospheric H_2. Therefore, wetland soils, sediments, and similar environments should still act as a source of atmospheric H_2. This happens indeed if gas bubbles are formed and released but these events contribute only marginally to the H_2 budget (Schütz et al., 1988). Hydrogen could also escape by diffusion or vascular transport, similar to CH_4. However, the diffusional loss is marginal (Schütz et al., 1988), mainly because H_2 is further consumed by respiring bacteria on its way to the atmosphere.

These respiring bacteria include species that reduce sulfate, ferric iron, nitrate, or O_2. They are active in zones where their respective oxidants are available (Fig. 3). The availability of O_2, nitrate, ferric iron, and sulfate decreases with increasing depth from a few millimeters down to a few centimeters of wetland soils or sediments (Zehnder and Stumm, 1988) or similarly around the roots of aquatic macrophytes (Conrad, 1989, 1993). The respiring bacteria consume H_2 down to concentrations that are lower than those left by the methanogens. These threshold concentrations typically decrease with increasing redox potential of the oxidant (Cord-Ruwisch et al., 1988; Lovley and Goodwin, 1988). Therefore, H_2 is finally consumed to partial pressures in the order of a few ppmv (parts per million by volume) H_2, which is only little more than atmospheric H_2. At the soil–water interface and in the water column, H_2 concentrations sometimes decrease even further, down to concentrations below 1 ppmv partial pressure (Conrad, 1995). This decrease may be due to the activity of abiontic hydrogenases present in surface sediment and water. However, the existence of these enzymes in aquatic environments has not been proved yet. We obtained strong indications of their presence in one lake (Conrad et al., 1983), but were unable to find indications of them in another one (Schmidt and Conrad, 1993). In summary, different microorganisms consume H_2 within the wetland soil so efficiently that almost no H_2 is released into the atmosphere.

4. Carbon Monoxide

Carbon monoxide plays a central role in the chemistry of the troposphere where it contributes significantly to the control of OH and the production of ozone (Logan et al., 1981; Crutzen and Gidel, 1983; Lu and Khalil, 1993). Carbon monoxide has increased in the atmosphere at about 1% per year, probably due to increasing emissions from anthropogenic activities (Khalil and Rasmussen, 1988, 1990c). Recently, however, the trend has reversed (Novelli et al., 1994).

Soils constitute only a small sink relative to the photochemical destruction of CO in the atmosphere (Conrad, 1988). However, the total sink strength of soil for atmospheric CO is still relatively uncertain, since soils can also act as sources of atmospheric CO (Conrad, 1988). Only two field measurements on CO flux have been published since the studies of Seiler's group (reviewed in Conrad, 1988), i.e., in rice fields (Conrad, et al., 1988) and in Venezuelan savanna and forest (Scharffe et al., 1990). It appears that wetland soils play only a marginal role in the atmospheric cycle of CO (Conrad *et al.*, 1988) and that most of the CO exchange takes place in upland soils.

4.1. Upland Soils

The net flux of CO between soil and atmosphere is the result of simultaneously operating production and consumption reactions, both of which are located in the uppermost soil layers (Liebl and Seiler, 1976; Seiler, 1978). This has been shown by the existence of a compensation concentration at which CO production equals CO consumption. Carbon monoxide production increases dynamically with soil temperature and shows a pronounced diurnal cycle, as does the CO compensation concentration that may change between <10 and >800 ppbv (Conrad and Seiler, 1985a). Therefore, soils may act as a source for atmospheric CO during the day and as a sink during the night. As a net effect, some soils may even, over a season, constitute a net source rather than a net sink for atmospheric CO (Conrad, 1988; Scharffe *et al.*, 1990).

The CO production process in soil seems to be exclusively abiological, i.e., thermochemical decomposition of soil organic matter (Conrad and Seiler, 1985b). Most of the soil organic matter is found in the upper soil layers, and these layers also exhibit the highest amplitudes of diel temperature changes. This fact may explain the highly dynamic nature of CO production observed in the field.

The CO consumption processes have been reviewed by Conrad (1988). They are due to microbial activities since they show a temperature optimum and are sensitive to antibiotics. Again, there seem to be two different CO-oxidizing activities, one that uses CO at atmospheric concentrations and another that uses CO at elevated concentrations. The first activity has a high affinity for CO (K_m in the range of tens of nanomolar) and seems to be due to nitrifiers (ammonium oxidizers) (Jones and Morita, 1983b; Jones *et al.*, 1984) and unknown oligotrophic bacteria (Conrad and Seiler, 1982). These bacteria oxidize CO fortuitously and do not use it for growth.

The second activity has a low affinity for CO and is due to carboxydotrophic bacteria that use CO as substrate for growth (Conrad *et al.*, 1981). In contrast to the common methanotrophs and the Knallgas bacteria, carboxydotrophic bacteria are able to oxidize CO also at atmospheric concentrations. A recent report (Mörsdorf *et al.*, 1992) even indicates that some carboxydotrophic bacteria may have higher affinities for CO than found on soil by Conrad *et al.* (1981). Therefore, the question as to which type of bacteria is involved in oxidation of atmospheric CO has not yet been resolved. Recent results from our group (Bender and Conrad, 1994b) indicate, however, that (at least in some soils) carboxydotrophic bacteria may primarily be involved in the oxidation of CO at elevated concentrations, whereas nitrifiers may primarily be responsible for oxidation of CO at atmospheric concentrations (Fig. 5). The question is whether there are any processes in soil that supply CO at a sufficiently high concentration to allow growth

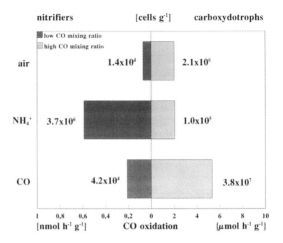

Figure 5. Oxidation of CO at low atmospheric (0.13 ppmv) and at elevated (2300 ppmv) CO concentrations by meadow soil. Soil samples were preincubated either under air or with ammonium or with CO to increase the population size of ammonium-oxidizing nitrifying bacteria or carboxydotrophic bacteria, respectively. Bars give CO oxidation rates and numbers give the population sizes of nitrifiers and carboxydotrophs. (Adapted from Bender and Conrad, 1994b.)

of carboxydotrophic bacteria. This CO concentration should be higher than the CO compensation concentration that develops due to CO production from soil organic matter (>1 ppmv). It is possible that concentrations can locally be much higher and have so far escaped detection by the sampling techniques used.

Another interesting feature is the marked difference of the sensitivity of CO production and CO consumption processes to temperature changes in the field. While CO production increases strongly with increasing temperature at the soil surface, CO consumption shows only a very modest increase (Conrad and Seiler, 1985a). The easiest explanation is that most of the CO oxidizers are localized in somewhat deeper soil layers where temperature amplitudes are lower than are the maxima of abiological CO production activities. Such an inhomogenous distribution would require that CO diffuses over some distance to reach the CO-consuming layer, and should result in CO concentration maxima within the upper soil layers. Alternatively, the CO-oxidizing bacteria may have a relatively low activation energy. Both explanations have to be tested.

4.2. Wetland Soils

Almost nothing is known about a possible cycling of CO within the anoxic parts of wetland soils. Carbon monoxide is known to be an intermediate in the metabolism of anaerobic bacteria. For example, it is produced and consumed by methanogenic, homoacetogenic, and sulfate-reducing bacteria (Zeikus, 1983; Fuchs, 1986; Conrad, 1988). Carbon monoxide is found in concentrations of about 5–30 nM (partial pressures of 6–40 ppmv) within wetland soils (Conrad *et al.*, 1988; Krämer and Conrad, 1993). However, the turnover of CO in anoxic environments has so far not been studied. Some of the CO dissolved in anoxic wetland soils is released into the atmosphere by pathways similar to those of CH_4 and H_2, i.e., mainly by plant-mediated transport and to a minor extent also by ebullition (Conrad *et al.*, 1988). A problem of all studies using soil systems that include the vegetation is the fact that plants produce relatively large amounts of CO (Seiler *et al.*, 1978; Bauer *et al.*, 1979), which may mask the CO turnover in the soil. It is possible that some CO is oxidized on its way through the oxic zones around the plant roots or through the sediment–water interface (in analogy to the oxidation of CH_4 and H_2), but so far this has not been studied. Therefore, we do not know which types of microorganisms may be involved and also do not know whether they possibly provide a filter against CO emission. This question may become important if the activity of the microbial CO filter changes due to future environmental changes.

5. Carbonyl Sulfide

Carbonyl sulfide (OCS) is the most abundant and most stable sulfur species in the troposphere, exhibiting a lifetime of at least 1 year (Möller, 1984; Warneck, 1988; Andreae and Jaeschke, 1992). Because of its long atmospheric lifetime, it is the only sulfur compound that can enter the stratosphere, and thus contributes to the maintenance of the stratospheric sulfate aerosol layer (Crutzen, 1976). The tropospheric lifetime of OCS seems to be controlled largely by uptake into the vegetation (Chin and Davis, 1993). Natural sources include upland and wetland soils, which represent about 25% of the total sources (Andreae and Jaeschke, 1992; Chin and Davies, 1993). However, our knowledge of the global OCS cycle is still poor. This is also true for our knowledge of the processes involved in the OCS cycle.

The presently available field data show that upland as well as wetland soils act as a source of atmospheric OCS. One problem in field studies is the differentiation between effects of the soil and effects of the vegetation. Plants have been shown to act as one of the major sinks for atmospheric OCS (Goldan *et al.*, 1988; Kesselmeier, 1992). Recent measurements show low OCS compensation concentrations of <0.2 ppbv in rape seed and corn plants (Kesselmeier and Merk, 1993). The OCS is apparently metabolized to CO_2 and H_2S by fortuitous reactions of enzymes involved in CO_2 fixation, especially of carbonic anhydrase (Protoschill-Krebs and Kesselmeier, 1992). Despite the uptake by plants, OCS release from soil is so strong that flushed chambers usually register a net emission of OCS even from vegetated soils (Goldan *et al.*, 1987; Hines and Morrison, 1992).

Another problem in assessing the flux of OCS between soil and atmosphere is the technique used for measurement. In most of the studies, the chambers are flushed with sulfur-free air or even with nitrogen (Aneja *et al.*, 1979; Adams *et al.*, 1981; Joergensen and Okholm-Hansen, 1985; Carroll *et al.*, 1986; Lamb *et al.*, 1987; Goldan *et al.*, 1987; Hines and Morrison, 1992; Hines *et al.*, 1993). This approach will result in a net release of OCS into the atmosphere even if soils normally consume OCS, simply because the OCS concentration will probably be below the compensation concentration. Soils do indeed have the capacity to consume OCS (Kluczewski *et al.*, 1985). Furthermore, experiments by Castro and Galloway (1991) demonstrated that chambers that were flushed with ambient air resulted in the soil acting as a sink for OCS, whereas chambers that were flushed with sulfur-free air resulted in soil acting as a source for OCS. Obviously, the flux of OCS between soil and atmosphere is the result of simultaneously operating OCS production and consumption processes, and the direction of the flux may change from emission to deposition when the atmospheric OCS concentration is sufficiently high.

There are only a few field measurements that have used gases with ambient sulfur concentrations for measuring OCS fluxes, and are thus unbiased by artificially low atmospheric sulfur concentrations (Steudler and Peterson, 1984, 1985; Castro and Galloway, 1991; Hofmann *et al.*, 1992; Bartell *et al.*, 1993; Huber, 1994). These studies show that OCS is at least occasionally taken up by the soil environments. It is unclear, however, how much plants versus soil contribute to this uptake. Both upland and wetland soils contribute to the budget of atmospheric OCS.

5.1. Upland Soils

We recently studied the production and consumption of OCS by soil in the laboratory, using incubation vessels that were flushed with air containing OCS at increasing concentrations. When the OCS concentration became higher than a particular value (the compensation concentration), the OCS release turned into an OCS uptake (Lehmann and Conrad, in press). The compensation concentrations were different for different soils and ranged between 1 and 200 ppbv OCS. Since the ambient OCS concentration is about 0.5 ppbv (Warneck, 1988), all soils would have acted as a net source of atmospheric OCS. However, one of the soils (an acidic forest cambisol) with the lowest OCS compensation mixing ratio (0.56 ppbv) was close to acting as a sink.

The experimental setup also allowed us to quantify OCS production rates separately from OCS uptake rates and to study the effect of incubation conditions. Carbonyl sulfide production was not abolished by autoclaving; rather it was stimulated, indicating that OCS is produced by abiological processes in soil. Carbonyl sulfide production in autoclaved soil was so high that it completely masked the simultaneously operating OCS consumption. When a soil sample was incubated at different temperatures, the rate constants of OCS consumption increased with temperature, showed a broad optimum between 10 and 40 °C, and decreased at higher temperatures, indicating that the OCS-consuming activity in soil was due to biological processes. Another study (Derikx *et al.,* 1991) with composting organic material found that OCS production was drastically reduced by autoclaving. In this composting material, OCS production was obviously due to biological processes.

The biological and abiological production processes of OCS in upland soils are still enigmatic. In fact, our knowledge has not much improved since the review of Bremner and Steele (1978), who showed that OCS production in soil is stimulated by addition of various sulfur compounds (pesticides, lanthionine, djenkolic acid, plant proteins, isothiocyanates, thiocyanate). Cystine also results in stimulated OCS production (Minami and Fukushi, 1981a). However, the mechanisms by which OCS is produced from these sulfur compounds are unknown. Only the mechanism of OCS production from thiocyanate has been elucidated in the meantime. This activity is due to the action of microbial thiocyanate hydrolase, which hydrolyzes SCN^- to NH_3 and OCS (Kelly and Smith, 1990; Kelly *et al.,* 1993). The enzyme has been characterized in *Thiobacillus thioparus,* an obligately aerobic chemolithoautotrophic bacterium (Katayama *et al.,* 1992). Thiocyanate is a reasonable precursor for OCS in soils since it is produced by hydrolysis of several plant constituents (Kelly *et al.,* 1993). The question remains whether thiocyanate hydrolysis is really the dominant OCS-producing mechanism in soil and whether it is due only to thiobacilli or also to other microorganisms.

The microorganisms responsible for consumption of OCS in soil are not definitely known either. Possible candidates include thiobacilli, which are able to hydrolyze OCS to CO_2 and H_2S and use the H_2S thus formed for autotrophic growth (Kelly and Smith, 1990). It is not unlikely that OCS is used by these bacteria in a way similar to that used by plants. Both groups of organisms are able to fix CO_2 and possibly metabolize OCS by similar enzymes, e.g., carbonic anhydrase (Kesselmeier, 1992):

$$CO_2 + H_2O <==> HCO_3^- + H^+$$

$$CO_2 + H_2S <==> OCS + H_2O$$

Studies of Protoschill-Krebs and Kesselmeier (personal communication) indicate that carbonic anhydrase has a 1000-fold higher affinity for OCS than for CO_2. Therefore, it is feasible that this enzyme is also responsible for OCS consumption at ambient concentrations.

5.2. Wetland Soils

Coastal marsh soils have often been found to be stronger sources for OCS than upland soils (Chin and Davis, 1993). Wetland rice fields, on the other hand, were only a relatively small source of atmospheric OCS (Kanda et al., 1992). This may be due to the observation that OCS is transported through the plant's gas vascular system, as discussed for CH_4, H_2, and CO. On its passage through the vascular system, OCS is probably taken up by the rice plants (Kanda et al., 1992). The coastal marsh soils, on the other hand, are vegetated with salt grass (Spartina species). In contrast to freshwater plants such as rice, salt grass has a thicker root epidermis, and thus, for example, does not transport CH_4 (Sebacher et al., 1985). Therefore, in coastal marsh soils, OCS probably cannot be emitted by passage through the grass plants, and thus cannot be consumed by the plant. This may be one reason why salt marsh soils show higher OCS emission rates than wetland rice soils. Another reason may be that rates of sulfur turnover are higher in sulfate-rich coastal marsh soils than in wetland rice fields, and that the increased sulfur turnover results in higher rates of OCS production.

Soils themselves apparently produce more OCS when they are flooded and anoxic than when they are unflooded and oxic (Minami and Fukushi, 1981a,b). The production mechanisms may be the same as for upland soils, i.e., abiological reactions or hydrolysis of thiocyanate, but this remains to be investigated. It is unclear whether thiocyanate hydrolase also exists in anaerobic bacteria. Similarly, it is not known whether OCS is also consumed within anoxic wetland soils. Carbonyl sulfide may be hydrolyzed, for example, by the carbonic anhydrase found in anaerobic bacteria, as occurs in plants and thiobacilli. However, not all bacteria contain carbonic anhydrase (Karrasch et al., 1989). For example, OCS is consumed by an anaerobic homoacetogenic bacterium that utilizes this substrate simultaneously with CO (Smith et al., 1991). This species probably does not possess carbonic anhydrase (Karrasch et al., 1989). The threshold for OCS of the homoacetogenic bacterium tested seems to be much higher than that of a phototrophic bacterium that probably does contain carbonic anhydrase (Smith et al., 1991). This example is in agreement with the suggestion of Kesselmeier (1992) that carbonic anhydrase may be the most effective OCS-degrading enzyme. However, this has to be investigated further.

6. Nitrous Oxide

Nitrous oxide is a greenhouse gas that is about 150 times more effective than CO_2 (Dickinson and Cicerone, 1986; Ramanathan et al., 1987). It also plays a significant role in the destruction of stratospheric ozone layer (Crutzen, 1970; Cicerone, 1987). The atmospheric abundance of N_2O is presently increasing at about 0.3% annually (Khalil and Rasmussen, 1992). The budget of N_2O is believed to be controlled by the soil sources that constitute about 70% of the total known sources (Davidson, 1991). Data of N_2O emission rates from various sites have recently been compiled by Bouwman (1990). It is likely that the present increase of N_2O is due to intensified agriculture (Matson and Vitousek, 1990; Robertson, 1993), but a number of other minor anthropogenic sources may have increased as well (Khalil and Rasmussen, 1992). From our present knowledge, it is not possible to balance the N_2O budget with known sources that account for only 7.9 Tg (teragram) of the 14.1 Tg sink for N_2O-N per year (Robertson, 1993). Obviously, major sources are missing in the N_2O budget.

The N_2O emission from upland soils, especially in the tropics, seems to be much more important for the global N_2O budget than N_2O emission from wetland soil (Sahrawat and Keeney, 1986; Freney and Denmead, 1992). Both soil categories emit N_2O. Uptake of N_2O by soil from the atmosphere has only occasionally been observed, e.g., in grass swards (Ryden, 1981; Slemr *et al.*, 1984).

6.1. Upland Soils

Although N_2O is generally released into the atmosphere, soils have also the potential to consume N_2O (Blackmer and Bremner, 1976). Seiler and Conrad (1981) demonstrated in field experiments the existence of a compensation concentration for N_2O and concluded that the flux of N_2O is the result of simultaneous production and consumption processes. The N_2O compensation concentration in soils appears to be significantly higher than the ambient atmospheric N_2O concentration, thus resulting in a flux out of the soil. It is evident, however, that the N_2O flux is not identical with the N_2O production rate, but is controlled in addition by diffusion and consumption processes within the soil column (Seiler and Conrad, 1981). Thus, Seiler and Conrad (1981) showed that N_2O may occur in deeper soil layers at high concentrations, but may be consumed before reaching the atmosphere.

The N_2O production processes in soils are due mainly to denitrifying and nitrifying (ammonium-oxidizing) microorganisms (Fig. 6). Nitrification as source of N_2O was at first neglected, but in upland soils it appears to be of similar importance to denitrification as a N_2O-producing mechanism (Bremner and Blackmer, 1978; Hutchinson and Davidson, 1993). The biochemistry and relative importance of these two processes have been reviewed in detail (Knowles, 1982; Tiedje, 1988; Firestone and Davidson, 1989; Bock *et al.*, 1991; Davidson, 1991). Chemical reactions, e.g., decomposition of nitrite, are apparently not significant for N_2O production. However, production of N_2O may also be caused by other microorganisms, e.g., during dissimilatory reduction of nitrate to ammonium, nitrate reduction to nitrite, and nitrate assimilation (Smith and Zimmerman, 1981; Bleakley and Tiedje, 1982). The mechanism of N_2O

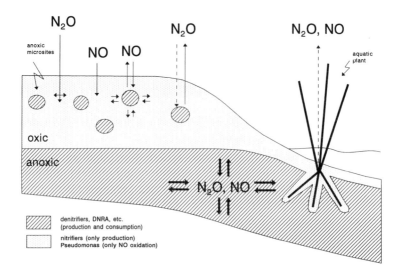

Figure 6. Scheme of distribution of microorganisms involved in cycling of N_2O and NO in upland and wetland soils.

production by these microorganisms is still not fully understood (Smith, 1983; Samuelsson, 1985; Costa *et al.*, 1990). It is also unknown which role these bacteria play in N_2O production in soil, since they cannot be differentiated from the denitrifiers by the usual acetylene inhibition techniques (Klemedtsson *et al.*, 1990).

The concept of the acetylene inhibition technique is illustrated in Fig. 7. It is based on the observation that ammonium oxidation by nitrifiers is inhibited by low acetylene partial pressures (1–10 Pa), thus allowing determination of the contribution of nitrification to the total N_2O production that is measured in the absence of acetylene. High acetylene partial pressures (10 kPa) also inhibit the N_2O reductase of denitrifying bacteria, thus allowing determination of the amount of N_2 that is formed by denitrification by measuring the additional production of N_2O.

While N_2O production processes have been studied relatively well, N_2O consumption processes have rarely been studied explicitly. This neglect is probably due to the fact that N_2O consumption by denitrifiers is trivial, since N_2O is an intermediate in the metabolism of denitrifiers, i.e., in the sequential reduction of nitrate $\rightarrow$ nitrite $\rightarrow$ NO $\rightarrow$ N_2O $\rightarrow$ N_2 (Knowles, 1982; Tiedje, 1988). This means that both production and consumption of N_2O are achieved by the same organism under anoxic conditions. Steady-state concentrations of N_2O in cultures of denitrifying bacteria are explained by modeling the kinetics of the N_2O-producing enzyme (NO reductase) and the N_2O-consuming enzyme (N_2O reductase) (Betlach and Tiedje, 1981; Zafiriou *et al.*, 1989). The rates of N_2O turnover in soil would thus depend primarily on the concentrations of these enzymes (i.e., NO reductase and N_2O reductase), the substrates (i.e., NO and N_2O), and the diffusion of the substrates between the site where they are generated, the bulk soil, and the site where they are consumed (Firestone and Davidson, 1989). The kinetic constants of the various denitrifiers have not yet been well studied. For example, N_2O reductase in bacterial cultures shows K_m values in a range of about 0.5 to 60 μM N_2O, and seems to be one to two orders of magnitude higher in soil (Betlach and Tiedje, 1981; Yoshinari *et al.*, 1977; Zumft and Kroneck, 1990). The higher K_m in soil is probably due to diffusion limitation biasing the kinetic measurements (Tiedje, 1988), but the data base of N_2O consumption is very poor, and the existence of bacterial populations with different affinities for N_2O cannot be excluded. It appears, however, that denitrifiers have the potential to consume N_2O to undetectable concentrations (<3 ppbv N_2O), i.e., there is no detectable threshold.

Besides denitrifiers, N_2O is also reduced by anaerobic bacteria that exhibit a dissimilatory reduction of nitrate to ammonium (DNRA), e.g., by *Wollinella succinogenes* (Teraguchi and Hollocher, 1989). The affinity of these bacteria for N_2O ($K_m = 7.5$ μM) is similar to that of denitrifiers. However, it is unclear which physiological role N_2O reduction plays in bacteria with DNRA, since

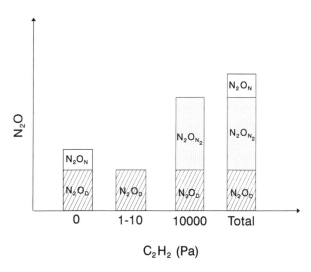

Figure 7. Concept of the acetylene inhibition technique to determine the contribution of N_2O production by denitrification (N_2O_D), N_2O production by nitrification (N_2O_N), and N_2 production by denitrification (N_2O_{N2}) to the total production N_2 plus N_2O in soil. (Adapted from Klemedtsson *et al.*, 1990.)

textbooks typically show the DNRA reaction sequence nitrate → nitrite → ammonium, in which N_2O is not an intermediate.

It should be noted that N_2O reduction by denitrifiers is an anaerobic process and that the N_2O reductase is repressed by O_2 (Tiedje, 1988). The other enzymes involved in the production and consumption of N_2O, i.e., nitrate reductase, nitrite reductase, NO reductase, and N_2O reductase, each have a different sensitivity to O_2, which furthermore differs between different denitrifier species (Körner et al., 1987, Körner et al., 1987; Coyne and Tiedje, 1990; Körner, 1993). Still, too little is known of the regulation pattern of N_2O turnover in denitrifiers to draw reliable conclusions for the situation in upland soils, where sites with different O_2 availabilities exist in various microniches side by side. It may be generalized, however, that N_2O turnover by denitrifiers depends on anoxic sites or on sites with drastically reduced O_2 concentrations in the soil (Fig. 6).

The production of N_2O by autotrophic nitrifiers (ammonium oxidizers, e.g., *Nitrosomonas europaea*) is basically a denitrification process as well in which the oxidation product of ammonium (i.e., nitrite) is reduced to N_2O (Poth and Focht, 1985). In analogy to denitrification, also N_2O production by nitrification is apparently inhibited by high O_2 concentrations, but remains active at reduced O_2 concentrations, since some O_2 is required for the oxidation of ammonium to nitrite. Little is known about the biochemistry of the nitrite reductase process in nitrifiers (Bock et al., 1991). Nothing is known about the existence of a N_2O reductase, although autotrophic nitrifiers (ammonium oxidizers) have been isolated that are able to reduce nitrite to N_2 (Poth, 1986). Thus, it is largely unknown whether autotrophic nitrifiers may contribute to consumption of N_2O in soil.

Besides the autotrophic nitrifiers (synthesizing cell-C from CO_2), heterotrophic nitrifiers (synthesizing cell-C from organic C) may contribute to the oxidation of ammonium, especially in acidic forest soils (Papen et al., 1993; Papen, personal communication). Some heterotrophic nitrifiers (e.g., *Alcaligenes faecalis*) are able to produce and consume N_2O as an intermediate in the denitrification of nitrate under anoxic conditions (Castignetti and Hollocher, 1981, 1984), and are obviously also able to do so during heterotrophic nitrification of ammonium to nitrite under oxic conditions (Papen et al., 1989; Anderson et al., 1993). However, the N_2O consumption process by heterotrophic nitrifiers under oxic conditions is not well defined, and its contribution to N_2O consumption in soil is only speculative at present.

With respect to aerobic N_2O consumption, two reports (Vedenina and Zavarzin, 1977; Vedenina et al., 1980) indicate the possibility of N_2O consumption by soil enzyme activity (i.e., catalase). This hypothesis, however, has since then not been investigated.

At the moment, we may assume that N_2O is mainly consumed by denitrifiers within anoxic microniches of upland soil. This consumption process (i.e., the N_2O reductase) has probably a dual function: it consumes N_2O present in soil air, and (more importantly) it allows only limited production by N_2O-producing denitrifiers (Fig. 6).

6.2. Wetland Soils

In wetland soils, N_2O cannot be produced by nitrification in most of the soil profile since O_2 is lacking. In these soil sections, N_2O can only be produced by denitrification. Other nitrate-reducing processes such as DNRA may also take place and produce N_2O (Tiedje, 1988). It has been believed that DNRA may be increasingly important relative to denitrification when the ratio of carbon substrate to nitrate increases in an environment, e.g., in sediments or submerged wetland soils (Tiedje, 1988). Recent studies indicate, however, that earlier measurements may have been biased by errors in methodology, and that denitrification is actually the predominant nitrate reduction process in sediments (Binnerup *et al.*, 1992). In sediments and unvegetated wetland soils, nitrate reduction was found to be the predominant source of N_2O, and is especially stimulated by high inputs of nitrogen (Joergensen *et al.*, 1984; Seitzinger, 1990). In rice fields, production of N_2O is also stimulated by nitrogen fertilization (Freney and Denmead, 1992). Nitrous oxide is further reduced by denitrifiers with K_m values of about 2 μM and 15 μM for N_2O in freshwater and marine sediments, respectively (Miller *et al.*, 1986). Miller *et al.* (1986) also observed that N_2O consumption was inhibited in the presence of nitrate at high concentrations. This observation is probably of importance for field conditions, since in rice fields emission of N_2O is stimulated especially if nitrate is present at high concentrations in the soils before flooding (Freney and Denmead, 1992).

In wetland rice fields, N_2O is emitted into the atmosphere via the plant gas vascular system (Mosier *et al.*, 1990). Nitrification is possible only in the shallow oxic zones at the soil–water interface and around the roots of plants, similar to the situation discussed previously for the oxidation of CH_4, H_2, and CO. These oxic zones are of great importance to the potential for N_2O emission from vegetated wetland soils for two reasons. First, these are the sites where N_2O is potentially produced during nitrification. The N_2O that is produced there may then rapidly be lost to the atmosphere via the plant's gas vascular system. Second, these are sites where ammonium is oxidized to nitrate, which then can diffuse into the adjacent anoxic zones to be reduced and thus to allow N_2O production. It is not known which portion of the N_2O flux is due to N_2O produced by the nitrifying bacteria directly or by denitrifying bacteria in nitrification–denitrification coupling. The efficiency of the plant root–sediment interface for coupling of nitrification–denitrification has been clearly demonstrated (Reddy and Patrick, 1986; Buresh and Austin, 1988; Reddy *et al.*, 1989). It is especially important in rice fields that are fertilized with ammonium or urea fertilizers. These fertilizers contain nitrogen in reduced form and first have to be oxidized to nitrate by nitrifiers before N_2O can be produced by denitrifiers. The presently available data indicate, however, that ammonium or urea, compared to nitrate fertilizers, results only in a minor stimulation of N_2O emission (Freney *et al.*, 1981; Smith *et al.*, 1982; Lindau *et al.*, 1990). I assume that nitrification–denitrification coupling results in a nitrate concentration that is so low that any N_2O produced is further reduced to N_2 to overcome the limitation of electron acceptors.

To summarize: The much lower potential of wetland soils versus upland soils for N_2O emission may be due to the fact that wetland soils are anoxic and contain only few oxic microniches, whereas upland soils are oxic and contain only few anoxic microniches (compare discussion for CH_4 in Section 2). Production of N_2O is enhanced when conditions are partially oxic, as in upland soils, since N_2O is produced by nitrifiers and since the production by denitrification requires nitrate, the oxidation product of ammonium. Consumption of N_2O, on the other hand, is enhanced only under anoxic conditions which are abundant in wetland soil but are restricted to anoxic microsites in the upland soils.

7. Nitric Oxide

Nitric oxide is a reactive gas that plays a major role in the photochemistry of the troposphere and contributes to the generation of acid rain (Crutzen, 1979; Logan, 1983; Singh, 1987). The photochemistry of NO is important for the generation of tropospheric O_3 (Crutzen, 1979), and it regulates the oxidizing capacity of the troposphere (Isaksen, 1988). Especially upland soils constitute a major source in the budget of tropospheric NO (Conrad, 1990; Davidson, 1991; Williams *et al.*, 1992). However, soils may also act as a sink of atmospheric NO, at least temporarily (review by Conrad, 1990).

7.1. Upland Soils

Upland soils may change rapidly between acting as a net source and acting as a net sink for atmospheric NO, depending on the NO concentration in the atmosphere (Slemr and Seiler, 1984, 1991). This capacity for changing the flux direction is due to the simultaneous operation of NO production and NO consumption processes in soil, which result in NO compensation concentrations being in the order of up to 600 ppbv NO (Johansson and Galbally, 1984; Remde *et al.*, 1989; Slemr and Seiler, 1991). Whenever the atmospheric NO concentration increases above the NO compensation concentration, soil will act as a sink for atmospheric NO. This event is likely to happen in areas with relatively high ambient NO concentrations, e.g., in urban areas and in soils where NO consumption activities are relatively higher than NO production activities.

There is a significant deficit in our knowledge concerning the quality and the quantity of NO turnover processes in soil (Conrad, 1990; Davidson, 1991). Studies have shown, however, that the flux of NO at the soil–atmosphere interface can indeed be modeled using rates of production, consumption, and transport of NO within the soil column (Galbally and Johansson, 1989; Remde et al., 1993). NO concentrations in deeper soil layers were found to be identical to the NO compensation concentration, showing that the NO concentrations in soil are the result of an intensive turnover process (Rudolph and Conrad, submitted).

The microbial processes with the capacity for NO production are basically the same as those for production of N_2O (Conrad, 1990), i.e., denitrification (Firestone et al., 1979), denitrification by nitrifiers (Remde and Conrad, 1990), heterotrophic nitrification (Papen et al., 1989; Krämer et al., 1990; Anderson et al., 1993), and reduction of nitrate by nondenitrifiers (Anderson and Levine, 1986; Kalkowski and Conrad, 1991). The biological role of NO in bacteria has recently been reviewed by Zumft (1993). Both nitrification and denitrification seem to be the major contributors to NO production in soil (Remde and Conrad, 1991a). The contribution of the two processes to NO production can be assessed by inhibition of the ammonium oxidation using, for example, nitrapyrin or acetylene (Fig. 7). The contribution of nitrification and denitrification to NO production (as well as N_2O production) theoretically depends on O_2 availability in soil and thus on soil moisture (Davidson, 1991, 1993). Most studies indicate that nitrification is the more important process at moistures below field capacity (Schuster and Conrad, 1992; Hutchinson et al., 1993; Williams et al., 1992), but some soils may also show a significant contribution by denitrification (Remde et al., 1993; Bollmann and Conrad, in preparation). With respect to NO production by nitrification, it is presently unclear whether autotrophic or heterotrophic nitrification is the more important source. As a working hypothesis, autotrophic nitrification may be more important in neutral agriculturally used soils and heterotrophic nitrification in acidic forest soils.

In contrast to N_2O, significant amounts of NO may also be produced by chemical decomposition of nitrite, especially under acidic conditions (Van Cleemput and Baert, 1984; Conrad, 1990). Although NO production depends in any case on the biological production of nitrite, Davidson (1991) emphasized the possibility that nitrification of ammonium can be coupled to the chemical decomposition of the produced nitrous acid. In cultures of nitrifiers, this process is negligible (Remde and Conrad, 1990). In soil, on the other hand, drying and wetting may cause enrichment of H^+ and nitrite ions at microsites, resulting in chemical conversion to NO (Davidson, 1991).

In all studies, consumption of NO in soil was sensitive to heating, and thus clearly due to microbial processes (e.g., Remde et al., 1989). Similar to N_2O, NO is an intermediate in denitrification and is produced according to a steady state between NO production by nitrite reductase and NO consumption by NO reductase (Zafiriou *et al.*, 1989). Denitrifiers have a remarkably high affinity for NO (K_m = 0.5–6.0 nM) and reduce it to N_2O (Remde and Conrad, 1991b; Kalkowski and Conrad, 1991; Schäfer and Conrad, 1993). In denitrifying soils, NO is consumed with a similarly high affinity (K_m < 6 nM NO) (Remde and Conrad, 1991b; Baumgärtner and Conrad, 1992; Schuster and Conrad, 1992). In these soils, NO consumption is stimulated when the soil is incubated under anoxic rather than oxic conditions (Remde and Conrad, 1991a; Baumgärtner and Conrad, 1992). Biosynthesis of NO reductase is very sensitive to O_2 inhibition (Körner, 1993). Once synthesized, however, NO reductase activity seems to be relatively insensitive against inactivation by O_2 and also operates to a relatively large extent in oxic soils. Thus, the resident denitrifier populations could account for the NO consumption activity that is observed there (Remde and Conrad, 1991b; Schäfer and Conrad, 1993). Typical kinetics of NO turnover in soil dominated by denitrifying bacteria are shown in Fig. 8 for oxic and anoxic incubation conditions.

In some soils, however, NO consumption is stimulated by oxic rather than anoxic incubation conditions (Baumgärtner *et al.*, to be submitted) and then exhibits a relatively low affinity with K_m values higher than the upper limit of our NO analysis system (> 30 nM NO). A threshold for NO uptake has so far not been observed, i.e., NO is consumed to concentrations < 0.5 ppbv NO. Apparently, an aerobic NO-consuming microbial population exists in addition to the denitrifiers. The typical kinetics of NO turnover in these soils are also shown in Fig. 8. However, the mechanism of NO consumption has so far not been elucidated. As NO is produced by autotrophic nitrifiers (ammonium oxidizers) during nitrification–denitrification, it has been hypothesized that NO may, in analogy to denitrification, be further reduced to N_2O (Conrad, 1990). However, this capacity of autotrophic nitrifiers (ammonium oxidizers) could so far not be demonstrated (Remde *et al.*, unpublished data).

Potential candidates for aerobic NO consumption are the following bacteria which all have been shown to consume NO at atmospheric concentrations. Nitrite-oxidizing autotrophic nitrifiers (e.g., *Nitrobacter winogradskii*) are able to consume NO under oxic conditions (Remde *et al.*, unpublished data). Experiments indicate that NO is both produced and consumed in *Nitrobacter* and that NO oxidation may be important for the generation of reducing equivalents (NADH) used for CO_2 fixation (Bock *et al.*, 1991). Methanotrophic bacteria have also been shown to consume NO (Krämer *et al.*, 1990). As methanotrophic bacteria have the ability for heterotrophic nitrification (oxidation of ammonium), the mechanism of NO consumption may be similar as in other heterotrophic nitrifiers that were recently shown to consume NO under oxic conditions (Anderson *et al.*, 1993). Finally, NO is also consumed by a heterotrophic *Pseudomonas* species, which seems to oxidize NO to nitrate in a fortuitous reaction (Baumgärtner *et al.*, to be submitted). All these NO-consuming reactions have not yet been characterized biochemically.

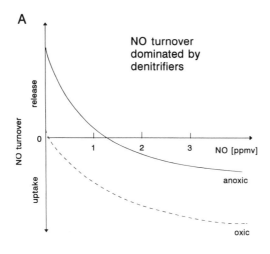

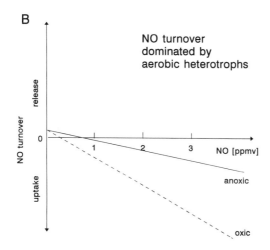

Figure 8. Typical kinetics of NO turnover in soils that are dominated by either (A) reductive or (B) oxidative NO consumption under oxic and anoxic incubation conditions. (Adapted from Remde and Conrad, 1991b; Baumgärtner and Conrad, 1992; Schuster and Conrad, 1992; and from unpublished data.)

It appears that NO turnover in upland soils is very dynamic with NO production occurring via nitrification and denitrification in more or less anoxic microniches, similar to N_2O production. In contrast to N_2O, however, NO is consumed as well as produced, involving both denitrification in anoxic microniches and other NO-consuming processes at oxic soil sites. Obviously, this high potential for NO consumption may cause dynamic changes in direction and magnitude of fluxes of NO between soil and atmosphere.

7.2. Wetland Soils

To my knowledge, there is only one study of NO flux in wetland soils that demonstrates a relatively small emission of NO from rice fields immediately after fertilization with urea (Galbally *et al.*, 1987). The processes involved in NO turnover in wetland soils have not been studied so far. However, at the moment we have no reason to believe that the principal processes are different from those involved in N_2O turnover in wetland soils.

8. Versatility and Diversity of Trace Gas Metabolism

This review of the microbial cycling of the different trace gases shows that some microorganisms have a striking versatility and are involved in the production and/or consumption of more than one trace gas (Table II). Methanotrophs are involved in the consumption of both CH_4 and NO, as well as in the production of N_2O. Ammonium-oxidizing nitrifiers produce NO and N_2O, but are also active in oxidation of CH_4 and CO. As CO_2-fixing autotrophs with carbonic anhydrase activity, they also should be able to hydrolyze OCS, although this has not yet been tested. Among anaerobic bacteria, for example, methanogens are involved in the cycling of CH_4, H_2, and CO, and perhaps also OCS. These examples indicate that a high versatility is especially found among the lithotrophic and methylotrophic bacteria. However, these trace gases are not important metabolites of these bacteria in every case, but they are cycled by fortuitous reactions that do not support growth (Table II). The hydrolytic reactions of carbonic anhydrase are one example, and the oxidative reactions of methane and ammonium monooxygenases are another. The great versatility of the different bacterial groups makes it difficult to predict rates of trace gas transformations in soil on a theoretical basis, even if the numbers, growth rates, and other major metabolic rates of the relevant microorganisms were known.

The situation becomes even more complicated by the fact that trace gases are cycled by a great diversity of microbial species that may differ in kinetic properties and regulation. Most striking is the observation that the consumption of CH_4, H_2, and CO in soil involves at least two activities with completely different kinetics. One activity operates at relatively high gas concentrations and consists of bacteria that use the gases for growth. These bacteria are relatively well known. The other activity operates at trace gas concentrations that are typical of the ambient atmosphere and are due to bacteria or soil enzymes that either have not been isolated (CH_4, H_2) or have been characterized only poorly (CO). Isolation of these activities that are important for atmospheric chemistry is a challenge for microbiologists in the future. The consumption of NO also involves two processes with different kinetic properties, of which one operates by reduction (involving dentrifiers) and the other one by oxidation (involving other bacteria) of NO. However, too little is known about the role of the latter bacteria in the cycling of NO in soil.

ACKNOWLEDGMENTS. I thank Dr. M. Bender and A. Meyer for the artwork of the figures and Drs. P. Frenzel and P. Janssen for critical reading of the manuscript.

References

Adams, D. F., Farwell, S. O., Robinson, E., Pack, M. R., and Bamesberger, W. L., 1981, Biogenic sulfur source strengths, *Environ. Sci. Technol.* **15**:1493–1498.

Adamsen, A. P. S., and King, G. M., 1993, Methane consumption in temperate and subarctic forest soils—rates, vertical zonation, and responses to water and nitrogen, *Appl. Environ. Microbiol.* **59**:485–490.

Alperin, M. J., and Reeburgh, W. S., (1984, Geochemical observations supporting anaerobic methane oxidation, in: *Microbial Growth on C-1 Compounds* (R. L. Crawford and R. S. Hanson, eds.), American Society for Microbiology, Washington, D.C., pp. 282–289.

Anderson, I. C., and Levine, J. S., 1986, Relative rates of nitric oxide and nitrous oxide production by nitrifiers, denitrifiers, and nitrate respirers, *Appl. Environ. Microbiol.* **51**:938–945.

Anderson, I. C., Poth, M., Homstead, J., and Burdige, D., 1993, A comparison of NO and N$_2$O production by the autotrophic nitrifier *Nitrosomonas europaea* and the heterotrophic nitrifier *Alcaligenes faecalis*, *Appl. Environ. Microbiol.* **59**:3525–3533.

Andreae, M. O., and Jaeschke, W. A., 1992, Exchange of sulfur between biosphere and atmosphere over temperate and tropical regions, in: *Sulphur Cycling on the Continents*, SCOPE Report 48 (R. W. Howarth, J. W. B. Stewart, and M. V. Ivanov, eds.), Wiley, Chichester, England, pp. 27–61.

Aneja, V. P., Overton, J. H., Cupitt, L. T., Durham, J. L., and Wilson, W. E., 1979, Carbon disulphide and carbonyl sulphide from biogenic sources and their contributions to the global sulphur cycle, *Nature* **282**:493–496.

Anthony, C., 1986, Bacterial oxidation of methane and methanol, *Adv. Microb. Physiol.* **27**:113–210.

Bartell, U., Hofmann, U., Hofmann, R., Kreuzburg, B., Andreae, M. O., and Kesselmeier, J., 1993, COS and H$_2$S fluxes over a wet meadow in relation to photosynthetic activity—an analysis of measurements made on 6 September 1990, *Atmos. Environ.* **27A**:1851–1864.

Bartlett, K. B., and Harriss, R. C., 1993, Review and assessment of methane emission from wetlands, *Chemosphere* **26**:261–320.

Bauer, K., Seiler, W., and Giehl, H., 1979, CO-Produktion höherer Pflanzen an natürlichen Standorten, *Z. Pflanzenphysiol.* **94**:219–230.

Baumgärtner, M., and Conrad, R., 1992, Role of nitrate and nitrite for production and consumption of nitric oxide during denitrification in soil, *FEMS Microbiol. Ecol.* **101**:59–65.

Baumgärtner, M., Koschorreck, M., and Conrad, R., Oxidative consumption of nitric oxide by heterotrophic bacteria in soil, to be submitted.

Bedard, C., and Knowles, R., 1989, Physiology, biochemistry, and specific inhibitors of CH$_4$, NH$_4$$^+$, and CO oxidation by methanotrophs and nitrifiers, *Microbiol. Rev.* **53**:68–84.

Bender, M., and Conrad, R., 1992, Kinetics of CH_4 oxidation in oxic soils exposed to ambient air or high CH_4 mixing ratios, *FEMS Microbiol. Ecol.* **101**:261–270.

Bender, M., and Conrad, R., 1993, Kinetics of methane oxidation in oxic soils, *Chemosphere* **26**:687–696.

Bender, M., and Conrad, R., 1994a, Methane oxidation activity in various soils and sediments: Occurrence, characteristics, vertical profiles and distribution on grain size fractions, *J. Geophys. Res.* **99**:16531–16540.

Bender, M., and Conrad, R., 1994b, Microbial oxidation of methane, ammonium and carbon monoxide, and turnover of nitrous oxide and nitric oxide in soils, *Biogeochem.* **27**:97–112.

Betlach, M. R., and Tiedje, J. M., 1981, Kinetic explanation for accumulation of nitrite, nitric oxide, and nitrous oxide during bacterial denitrification, *Appl. Environ. Microbiol.* **42**:1074–1084.

Bingemer, H. G., and Crutzen, P. J., 1987, The production of methane from solid wastes, *J. Geophys. Res.* **92**:2181–1187.

Binnerup, S. J., Jensen, K., Revsbech, N. P., Jensen, M. H., and Soerensen, J., 1992, Denitrification, dissimilatory reduction of nitrate to ammonium, and nitrification in a bioturbated estuarine sediment as measured with ^{15}N and microsensor techniques, *Appl. Environ. Microbiol.* **58**:303–313.

Blackmer, A. M., and Bremner, J. M., 1976, Potential of soil as a sink for atmospheric nitrous oxide, *Geophys Res. Lett.* **3**:739–742.

Bleakley, B. H., and Tiedje, J. M., 1982, Nitrous oxide production by organisms other than nitrifiers or denitrifiers, *Appl. Environ. Microbiol.* **44**:1342–1348.

Bock, E., Koops, H. P., Harms, H., and Ahlers, B., 1991, The biochemistry of nitrifying organisms, in: *Variations in Autotrophic Life* (J. M. Shively and L. L. Barton, eds.), Academic Press, London, pp. 171–200.

Born, M., Dörr, H., and Levin, I., 1990, Methane consumption in aerated soils of the temperate zone, *Tellus* **42B**:2–8.

Bosse, U., Frenzel, P., and Conrad, R., 1993, Inhibition of methane oxidation by ammonium in the surface layer of a littoral sediment, *FEMS Microbiol. Ecol.* **13**:123–134.

Bouwman, A. F., 1990, Exchange of greenhouse gases between terrestrial ecosystems and the atmosphere, in: *Soils and the Greenhouse Effect* (A. F. Bouwman, ed.), Wiley, Chichester, England, pp. 61–127.

Bremner, J. M., and Blackmer, A. M., 1978, Nitrous oxide: Emission from soils during nitrification of fertilizer nitrogen, *Science* **199**:295–296.

Bremner, J. M., and Steele, C. G., 1978, Role of microorganisms in the atmospheric sulfur cycle, *Adv. Microb. Ecol.* **2**:155–201.

Breznak, J. A., 1975, Symbiotic relationships between termites and their intestinal microbiota, in: *Symbiosis* (D. H. Tennings and D. L. Lee, eds.), Cambridge University Press, Cambridge, England, pp. 559–580.

Buresh, R. J., and Austin, E. R., 1988, Direct measurement of dinitrogen and nitrous oxide flux in flooded rice fields, *Soil Sci. Soc. Am. J.* **52**:681–688.

Carlsen, H. N., Joergensen, L., and Degn, H., 1991, Inhibition by ammonia of methane utilization in *Methylococcus capsulatus* (Bath), *Appl. Microbiol. Biotechnol.* **35**:124–127.

Carroll, M. A., Heidt, L. E., Cicerone, R. J., and Prinn, R. G., 1986, OCS, H_2S, and CS_2 fluxes from a salt water marsh, *J. Atmos. Chem.* **4**:375–395.

Castignetti, D., and Hollocher, T. C., 1981, Vigorous denitrification by a heterotrophic nitrifier of the genus Alcaligenes, *Curr. Microbiol.* **6**:229–231.

Castignetti, D., and Hollocher, T. C., 1984, Heterotrophic nitrification among denitrifiers, *Appl. Environ. Microbiol.* **47**:620–623.

Castro, M. S., and Galloway, J. N., 1991, A comparison of sulfur-free and ambient air enclosure techniques for measuring the exchange of reduced sulfur gases between soils and the atmosphere, *J. Geophys. Res.* **96:**15427–15437.

Castro, M. S., Peterjohn, W. T., Melillo, J. M., Steudler, P. A., Gholz, H. L., and Lewis, D., 1994, Effects of nitrogen fertilization on the fluxes of N_2O, CH_4, and CO_2 from soils in a Florida slash pine plantation, *Can. J. Forest Res.* **24:**9–13.

Chanton, J. P., and Dacey, J. W., 1991, Effects of vegetation on methane flux, reservoirs, and carbon isotopic composition, in: *Trace Gas Emissions by Plants* (J. E. Rogers and W. B. Whitman, eds.), Academic Press, New York, pp. 65–92.

Chin, M., and Davis, D. D., 1993, Global sources and sinks of OCS and CS_2 and their distributions, *Global Biogeochem. Cycles* **7:**321–337.

Cicerone, R. J., 1987, Changes in stratospheric ozone, *Science* **237:**35–42.

Cicerone, R. J., and Oremland, R. S., 1988, Biogeochemical aspects of atmospheric methane, *Global Biogeochem. Cycles* **2:**299–327.

Collins, N. M., and Wood, T. G., 1984, Termites and atmospheric gas production, *Science* **224:**84–86.

Conrad, R., 1984, Capacity of aerobic microorganisms to utilize and grow on atmospheric trace gases, in: *Current Perspectives in Microbial Ecology* (M. G. Klug and C. A. Reddy, eds.), American Society for Microbiology, Washington, D.C., pp. 461–467.

Conrad, R., 1988, Biogeochemistry and ecophysiology of atmospheric CO and H_2, *Adv. Microb. Ecol.* **10:**231–283.

Conrad, R., 1989, Control of methane production in terrestrial ecosystems, in: *Exchange of Trace Gases between Terrestrial Ecosystems and the Atmosphere. Dahlem Konferenzen* (M. O. Andreae and D. S. Schimel, eds.), Wiley, Chichester, England, pp. 39–58.

Conrad, R., 1990, Flux of NO_x between soil and atmosphere: Importance and soil microbial metabolism, in: *Denitrification in Soil and Sediment* (N. P. Revsbech and J. Soerensen, eds.), Plenum Press, New York, pp. 105–128.

Conrad, R., 1993, Mechanisms controlling methane emission from wetland rice fields, in: *Biogeochemistry of Global Change: Radiative Trace Gases* (R. S. Oremland, ed.), Chapman & Hall, New York, pp. 317–335.

Conrad, R., 1995, Anaerobic hydrogen metabolism in aquatic sediments, *Mitt. Internat. Ver. Limnol.* **25:**in press.

Conrad, R., and Rothfuss, F., 1991, Methane oxidation in the soil surface layer of a flooded rice field and the effect of ammonium, *Biol. Fertil. Soils* **12:**28–32.

Conrad, R., and Seiler, W., 1979, Field measurements of hydrogen evolution by nitrogen-fixing legumes, *Soil Biol. Biochem.* **11:**689–690.

Conrad, R., and Seiler, W., 1980, Contribution of hydrogen production by biological nitrogen fixation to the global hydrogen budget, *J. Geophys. Res.* **85:**5493–5498.

Conrad, R., and Seiler, W., 1981, Decomposition of atmospheric hydrogen by soil microorganisms and soil enzymes, *Soil Biol. Biochem.* **13:**43–49.

Conrad, R., and Seiler, W., 1982, Utilization of traces of carbon monoxide by aerobic oligotrophic microorganisms in ocean, lake and soil, *Arch. Microbiol.* **132:**41–46.

Conrad, R., and Seiler, W., 1985a, Influence of temperature, moisture and organic carbon on the flux of H_2 and CO between soil and atmosphere. Field studies in subtropical regions, *J. Geophys. Res.* **90:**5699–5709.

Conrad, R., and Seiler, W., 1985b, Characteristics of abiological CO formation from soil organic matter, humic acids, and phenolic compounds, *Environ. Sci. Technol.* **19:**1165–1169.

Conrad, R., Meyer, O., and Seiler, W., 1981, Role of carboxydobacteria in consumption of atmospheric carbon monoxide by soil, *Appl. Environ. Microbiol.* **42:**211–215.

Conrad, R., Aragno, M., and Seiler, W., 1983, Production and consumption of hydrogen in a eutrophic lake, *Appl. Environ. Microbiol.* **45:**502–510.

Conrad, R., Schütz, H., and Seiler, W., 1988, Emission of carbon monoxide from submerged rice fields into the atmosphere, *Atmos. Environ.* **22:**821–823.

Cord-Ruwisch, R., Seitz, H. J., and Conrad, R., 1988, The capacity of hydrogenotrophic anaerobic bacteria to compete for traces of hydrogen depends on the redox potential of the terminal electron acceptor, *Arch. Microbiol.* **149:**350–357.

Costa, C., Macedo, A., Moura, I., Moura, J. J. G., LeGall, J., Berlier, Y., Liu, M. Y., and Payne, W. J., 1990, Regulation of the hexaheme nitrite/nitric oxide reductase of *Desulfovibrio desulfuricans, Wolinella succinogenes* and *Escherichia coli*—A mass spectrometric study, *FEBS Lett.* **276:**67–70.

Coyne, M. S., and Tiedje, J. M., 1990, Induction of denitrifying enzymes in oxygen-limited *Achromobacter cycloclastes* continuous culture, *FEMS Microbiol. Ecol.* **73:**263–270.

Crill, P. M., Harriss, R. C., and Bartlett, K. B., 1991, Methane fluxes from terrestrial wetland environments, in: *Microbial Production and Consumption of Greenhouse Gases: Methane, Nitrogen Oxides, and Halomethanes* (J. E. Rogers and W. B. Whitman, eds.), American Society for Microbiology, Washington, D.C., pp. 91–109.

Crutzen, P. J., 1970, The influence of nitrogen oxides on the atmospheric ozone content, *Q. J. R. Meteor. Soc.* **96:**320–325.

Crutzen, P. J., 1976, The possible importance of CSO for the sulfate layer of the stratosphere, *Geophys. Res. Lett.* **3:**73–76.

Crutzen, P. J., 1979, The role of NO and NO_2 in the chemistry of the troposphere and stratosphere, *Annu. Rev. Earth Planet. Sci.* **7:**443–472.

Crutzen, P. J., and Gidel, L. T., 1983, A two-dimensional photochemical model of the atmosphere 2. The tropospheric budgets of the anthropogenic chlorocarbons CO, CH_4, CH_3Cl and the effect of various NO_x sources on tropospheric ozone, *J. Geophys. Res.* **88:**6641–6661.

Crutzen, P. J., and Schmailzl, U., 1983, Chemical budgets of the stratosphere, *Planet. Space Sci.* **31:**1009–1032.

Cunningham, S. D., Kapulnik, Y., and Phillips, D. A., 1986, Distribution of hydrogen-metabolizing bacteria in alfalfa field soil, *Appl. Environ. Microbiol.* **52:**1091–1095.

Davidson, E. A., 1991, Fluxes of nitrous oxide and nitric oxide from terrestrial ecosystems, in: *Microbial Production and Consumption of Greenhouse Gases: Methane, Nitrogen Oxides, and Halomethanes* (J. E. Rogers and W. B. Whitman, eds.), American Society for Microbiology, Washington, D.C., pp. 219–235.

Davidson, E. A., 1993, Soil water content and the ratio of nitrous oxide to nitric oxide emitted from soil, in: *Biogeochemistry of Global Change* (R. S. Oremland, ed.), Chapman & Hall, New York, pp. 369–386.

Derikx, P. J. L., Simons, F. H. M., Dencamp, H. J. M. O., VanderDrift, C., VanGriensven, L. J. L. D., and Vogels, G. D., 1991, Evolution of volatile sulfur compounds during laboratory-scale incubations and indoor preparation of compost used as a substrate in mushroom cultivation, *Appl. Environ. Microbiol.* **57:**563–567.

Dickinson, R. E., and Cicerone, R. J., 1986, Future global warming from atmospheric trace gases, *Nature* **319:**109–115.

Dörr, H., Katruff, L., and Levin, I., 1993, Soil texture parameterization of the methane uptake in aerated soils, *Chemosphere* **26:**697–713.

Dugnani, L., Wyrsch, I., Gandolla, M., and Aragno, M., 1986, Biological oxidation of hydrogen in soils flushed with a mixture of H_2, CO_2, O_2 and N_2, *FEMS Microbiol. Ecol.* **38:**347–351.

Firestone, M. K., and Davidson, E. A., 1989, Microbiological basis of NO and N_2O production and consumption in soil, in: *Exchange of Trace Gases between Terrestrial Ecosystems and the*

Atmosphere. Dahlem Konferenzen (M. O. Andreae and D. S. Schimel, eds.), Wiley, Chichester, England, pp. 7–21.

Firestone, M. K., Firestone, R. B., and Tiedje, J. M., 1979, Nitric oxide as intermediate in denitrification: Evidence from nitrogen-13 isotope exchange, *Biochem. Biophys. Res. Commun.* **91**:10–16.

Freney, J. R., and Denmead, O. T., 1992, Factors controlling ammonia and nitrous oxide emissions from flooded rice fields, *Ecol. Bull. (Copenhagen)* **42**:188–194.

Freney, J. R., Denmead, O. T., Watanabe, I., and Craswell, E. T., 1981, Ammonia and nitrous oxide losses following applications of ammonium sulfate to flooded rice, *Aust. J. Agric. Res.* **32**:37–45.

Fuchs, G., 1986, CO_2 fixation in acetogenic bacteria: Variations on a theme, *FEMS Microbiol. Rev.* **39**:181–213.

Galbally, I. E., and Johansson, C., 1989, A model relating laboratory measurements of rates of nitric oxide production and field measurements of nitric oxide emission from soils, *J. Geophys. Res.* **94**:6473–6480.

Galbally, I. E., Freney, J. R., Muirhead, W. A., Simpson, J. R., Trevitt, A. C. F., and Chalk, P. M., 1987, Emission of nitrogen oxides (NO_x) from a flooded soil fertilized with urea: Relation to other nitrogen loss processes, *J. Atmos. Chem.* **5**:343–365.

Goldan, P. D., Kuster, W. C., Albritton, D. L., and Fehsenfeld, F. C., 1987, The measurement of natural sulfur emissions from soils and vegetation: Three sites in the Eastern United States revisited, *J. Atmos. Chem.* **5**:439–467.

Goldan, P. D., Fall, R., Kuster, W. C., and Fehsenfeld, F. C., 1988, Uptake of COS by growing vegetation: A major tropospheric sink, *J. Geophys. Res.* **93**:14186–14192.

Hackstein, J. H. P., and Stumm, C. K., 1994, Methane production in terrestrial arthropods, *Proc. Natl. Acad. Sci. USA* **91**:5441–5445.

Häring, V., and Conrad, R., 1991, Kinetics of H_2 oxidation in respiring and denitrifying Paracoccus denitrificans, *FEMS Microbiol. Lett.* **78**:259–264.

Häring, V., and Conrad, R., 1994, Demonstration of 2 different H_2-oxidizing activities in soil using an H_2 consumption and a tritium exchange assay, *Biol. Fertil. Soils* **17**:125–128.

Häring, V., Klüber, H. D., and Conrad, R., 1994, Localization of atmospheric H_2-oxidizing soil hydrogenases in different particle fractions of soil, *Biol. Fertil. Soils* **18**:109–114.

Henrichs, S. M., and Reeburgh, W. S., 1987, Anaerobic mineralization of marine sediment organic matter: Rates and the role of anaerobic processes in the oceanic carbon economy, *Geomicrobiol. J.* **5**:191–237.

Heyer, J., Malashenko, Y., Berger, U., and Budkova, E., 1984, Verbreitung methanotropher Bakterien, *Z. Allg. Mikrobiol.* **24**:725–744.

Hines, M. E., and Morrison, M. C., 1992, Emissions of biogenic sulfur gases from Alaskan tundra, *J. Geophys. Res.* **97**:16703–16707.

Hines, M. E., Pelletier, R. E., and Crill, P. M., 1993, Emissions of sulfur gases from marine and freshwater wetlands of the Florida Everglades: Rates and extrapolation using remote sensing, *J. Geophys. Res.* **98**:8991–8999.

Hofmann, U., Hofmann, R., and Kesselmeier, J., 1992, Field measurements of reduced sulfur compounds over wheat during a growing season, in: *Precipitation Scavenging and Atmosphere–Surface Exchange,* Vol. 2 (S. E. Schwartz and W. G. N. Slinn, eds.), Hemisphere Publishing Corporation, Washington, DC, pp. 967–977.

Holland, H. D., 1984, *The Chemical Evolution of the Atmosphere and Oceans,* Princeton University Press, Princeton, N.J.

Houghton, J. T., Jenkins, G. J., and Ephraums, J. J. (eds.), 1990, *Climate Change. The IPCC Scientific Assessment,* Cambridge University Press, Cambridge, England.

Huber, B., 1994, Austausch flüchtiger Schwefelverbindungen in land- und forstwirtschaftlichen Ö kosystemen, Ph.D. thesis, Wissenschafts-Verlag Dr. W. Maraun, Frankfurt, Germany.

Hutchinson, G. L., and Davidson, E. A., 1993, Processes for production and consumption of gaseous nitrogen oxides in soil, in: *Agricultural Ecosystem Effects on Trace Gases and Global Climate Change* (L. A. Harper *et al.*, eds.), American Society of Agronomy, Madison, Wis., pp. 79–93.

Hutchinson, G. L., Guenzi, W. D., and Livingston, G. P., 1993, Soil water controls on aerobic soil emission of gaseous nitrogen oxides, *Soil Biol. Biochem.* **25:**1–9.

Hyman, M. R., and Wood, P. M., 1983, Methane oxidation by *Nitrosomonas europaea, Biochem. J.* **212:**31–37.

Isaksen, I. S. A., 1988, Is the oxidizing capacity of the atmosphere changing? in: *The Changing Atmosphere. Dahlem Konferenzen* (F. S. Rowland and I. S. A. Isaksen, eds.), Wiley, Chichester, England, pp. 141–157.

Iversen, N., and Joergensen, B. B., 1985, Anaerobic methane oxidation rates at the sulfate-methane transition in marine sediments from Kattegat and Skagerrak (Denmark), *Limnol. Oceanogr.* **30:**944–955.

Iversen, N., Oremland, R. S., and Klug, M. J., 1987, Big Soda Lake (Nevada). 3. Pelagic methanogenesis and anaerobic methane oxidation, *Limnol. Oceanogr.* **32:**815–824.

Joergensen, B. B., and Okholm-Hansen, B., 1985, Emissions of biogenic sulfur gases from a Danish estuary, *Atmos. Environ.* **19:**1737–1749.

Joergensen, K. S., Jensen, H. B., and Soerensen, J., 1984, Nitrous oxide production from nitrification and denitrification in marine sediment at low oxygen concentrations, *Can. J. Microbiol.* **30:**1073–1078.

Johansson, C., and Galbally, I. E., 1984, Production of nitric oxide in loam under aerobic and anaerobic conditions, *Appl. Environ. Microbiol.* **47:**1284–1289.

Jones, H. A., and Nedwell, D. B., 1990, Soil atmosphere concentrations profiles and methane emission rates in the restoration covers above landfill sites: Equipment and preliminary results, *Waste Manage Res.* **8:**21–31.

Jones, H. A., and Nedwell, D. B., 1993, Methane emission and methane oxidation in land-fill cover soil, *FEMS Microbiol. Ecol.* **102:**185–195.

Jones, R. D., and Morita, R. Y., 1983a, Methane oxidation by *Nitrosococcus oceanus* and *Nitrosomonas europaea, Appl. Environ. Microbiol.* **45:**401–410.

Jones, R. D., and Morita, R. Y., 1983b, Carbon monoxide oxidation by chemolithotrophic ammonium oxidizers, *Can. J. Microbiol.* **29:**1545–1551.

Jones, R. D., Morita, R. Y., and Griffiths, R. P., 1984, Method for estimating *in situ* chemolithotrophic ammonium oxidation using carbon monoxide oxidation, *Mar. Ecol. Progr. Ser.* **17:** 259–269.

Junge, C. E., 1963, *Air Chemistry and Radioactivity,* Academic Press, New York.

Kalkowski, I., and Conrad, R., 1991, Metabolism of nitric oxide in denitrifying *Pseudomonas aeruginosa* and nitrate-respiring *Bacillus cereus, FEMS Microbiol. Lett.* **82:**107–112.

Kanda, K., Tsuruta, H., and Minami, K., 1992, Emission of dimethyl sulfide, carbonyl sulfide, and carbon disulfide from paddy fields, *Soil Sci. Plant Nutr.* **38:**709–716.

Karrasch, M., Bott, M., and Thauer, R. K., 1989, Carbonic anhydrase activity in acetate grown *Methanosarcina barkeri, Arch. Microbiol.* **151:**137–142.

Katayama, Y., Narahara, Y., Inoue, Y., Amano, F., Kanagawa, T., and Kuraishi, H., 1992, A thiocyanate hydrolase of *Thiobacillus thioparus*—A novel enzyme catalyzing the formation of carbonyl sulfide from thiocyanate, *J. Biol. Chem.* **267:**9170–9175.

Keller, M., Kaplan, W. A., and Wofsy, S. C., 1986, Emissions of N_2O, CH_4 and CO_2 from tropical forest soils, *J. Geophys. Res.* **91:**11791–11802.

Keller, M., Veldkamp, E., Weitz, A. M., and Reiners, W. A., 1993, Effect of pasture age on soil trace-gas emissions from a deforested area of Costa Rica, *Nature* **365**:244–246.

Kelly, D. P., and Smith, N. A., 1990, Organic sulfur compounds in the environment. Biogeochemistry, microbiology, and ecological aspects, *Adv. Microb. Ecol.* **11**:345–385.

Kelly, D. P., Malin, G., and Wood, A. P., 1993, Microbial transformations and biogeochemical cycling of one-carbon substrates containing sulphur, nitrogen or halogens, in: *Microbial Growth on C1 Compounds* (J. C. Murrell and D. P. Kelly, eds.), Intercept, Andover, Mass., pp. 47–63.

Kesselmeier, J., 1992, Plant physiology and the exchange of trace gases between vegetation and the atmosphere, in: *Precipitation Scavenging and Atmosphere–Surface Exchange*, Vol. 2 (S. E. Schwartz and W. G. N. Slinn, eds.), Hemisphere Publishing Corporation, Washington, DC, pp. 949–966.

Kesselmeier, J., and Merk, L., 1993, Exchange of carbonyl sulfide (COS) between agricultural plants and the atmosphere—studies on the deposition of COS to peas, corn and rapeseed, *Biogeochemistry* **23**:47–59.

Khalil, M. A. K., and Rasmussen, R. A., 1988, Carbon monoxide in the earth's atmosphere: Indications of a global increase, *Nature* **332**:242–245.

Khalil, M. A. K., and Rasmussen, R. A., 1990a, Atmospheric methane: Recent global trends, *Environ. Sci. Technol.* **24**:549–553.

Khalil, M. A. K., and Rasmussen, R. A., 1990b, Global increase of atmospheric molecular hydrogen, *Nature* **347**:743–745.

Khalil, M. A. K., and Rasmussen, R. A., 1990c, The global cycle of carbon monoxide: Trends and mass balance, *Chemosphere* **20**:227–242.

Khalil, M. A. K., and Rasmussen, R A., 1992, The global sources of nitrous oxide, *J. Geophys. Res.* **97**:14651–14660.

Khalil, M. A. K., Rasmussen, R. A., French, J. R. J., and Holt, J. A., 1990, The influence of termites on atmospheric trace gases: CH_4, CO_2, $CHCl_3$, N_2O, CO, H_2, and light hydrocarbons, *J. Geophys. Res.* **95**:3619–3634.

King, G. M., 1990, Dynamics and controls of methane oxidation in a Danish wetland sediment, *FEMS Microbiol. Ecol.* **74**:309–323.

King, G. M., 1993, Ecophysiological characteristics of obligate methanotrophic bacteria and methane oxidation in situ, in: *Microbial Growth on C1 Compounds* (J. C. Murrell and D. P. Kelly, eds.), Intercept, Andover, Mass., pp. 303–313.

King, G. M., and Adamsen, A. P. S., 1992, Effects of temperature on methane consumption in a forest soil and in pure cultures of the methanotroph *Methylomonas rubra*, *Appl. Environ. Microbiol.* **58**:2758–2763.

Klemedtsson, L., Hanson, G., and Mosier, A., 1990, The use of acetylene for the quantification of N_2 and N_2O production from biological processes in soil, in: *Denitrification in Soil and Sediment* (N. P. Revsbech and J. Soerensen, eds.), Plenum Press, New York, pp. 167–180.

Klüber, H. D., and Conrad, R., 1993, Ferric iron-reducing *Shewanella putrefaciens* and N_2-fixing *Bradyrhizobium japonicum* with uptake hydrogenase are unable to oxidize atmospheric H_2, *FEMS Microbiol. Lett.* **111**:337–341.

Kluczewski, S. M., Brown, K. A., and Bell, J. N. B., 1985, Deposition of carbonyl sulphide to soils, *Atmos. Environ.* **19**:1295–1299.

Knowles, R., 1982, Denitrification, *Microbiol. Rev.* **46**:43–70.

Körner, H., 1993, Anaerobic expression of nitric oxide reductase from denitrifying *Pseudomonas stutzeri*, *Arch. Microbiol* **159**:410–416.

Körner, H., and Zumft, W. G., 1989, Expression of denitrification enzymes in response to the dissolved oxygen level and respiratory substrate in continuous culture of *Pseudomonas stutzeri*, *Appl. Environ. Microbiol.* **55**:1670–1676.

Körner, H., Frunzke, K., Döhler, K., and Zumft, W. G., 1987, Immunochemical patterns of distribution of nitrous oxide reductase and nitrite reductase (cytochrome *cd1*) among denitrifying pseudomonas, *Arch. Microbiol.* **148**:20–24.

Koschorreck, M., and Conrad, R., 1993, Oxidation of atmospheric methane in soil: Measurements in the field, in soil cores and in soil samples, *Global Biogeochem. Cycles* **7**:109–121.

Krämer, H., and Conrad, R., 1993, Measurement of dissolved H_2 concentrations in methanogenic environments with a gas diffusion probe, *FEMS Microbiol. Ecol.* **12**:149–158.

Krämer, M., Baumgärtner, M., Bender, M., and Conrad, R., 1990, Consumption of NO by methanotrophic bacteria in pure culture and in soil, *FEMS Microbiol. Ecol.* **73**:345–350.

LaFavre, J. S., and Focht, D. D., 1983, Conservation in soil of H_2 liberated from N_2 fixation by *hup*⁻ nodules, *Appl. Environ. Microbiol.* **46**:304–311.

Lamb, B., Westberg, H., Allwine, G., Bamesberger, L., and Guenther, A., 1987, Measurement of biogenic sulfur emissions from soils and vegetation: Application of dynamic enclosure methods with Natusch filter and GC/FPD analysis, *J. Atmos. Chem.* **5**:469–491.

Lehmann, S., and Conrad, R., Characteristics of turnover of carbonyl sulfide in four different soils, *J. Atmospher. Chem.*, in press.

Lidstrom, M. E., and Somers, L., 1984, Seasonal study of methane oxidation in Lake Washington, *Appl. Environ. Microbiol.* **47**:1255–1260.

Liebl, K. H., and Seiler, W., 1976, CO and H_2 destruction at the soil surface, in: *Microbial Production and Utilization of Gases* (H. G. Schlegel, G. Gottschalk, and N. Pfennig, eds.), E. Goltze, Göttingen, Germany, pp. 215–229.

Lindau, C. W., Patrick, W. H., DeLaune, R. D., and Reddy, K. R., 1990, Rate of accumulation and emission of N_2, N_2O and CH_4 from a flooded rice soil, *Plant Soil* **129**:269–276.

Logan, J. A., 1983, Nitrogen oxides in the troposphere: Global and regional budgets, *J. Geophys. Res.* **88**:10785–10807.

Logan, J. A., Prather, M. J., Wofsy, S. C., and McElroy, M. B., 1981, Tropospheric chemistry: A global perspective, *J. Geophys. Res.* **86**:7210–7254.

Lovelock, J. E., and Margulis, L., 1974, Atmospheric homeostasis by and for the biosphere: The GAIA hypothesis, *Tellus* **26**:2–10.

Lovley, D. R., and Goodwin, S., 1988, Hydrogen concentrations as an indicator of the predominant terminal electron-accepting reactions in aquatic sediments, *Geochim. Cosmochim. Acta* **52**:2993–3003.

Lu, Y., and Khalil, M. A. K., 1993, Methane and carbon monoxide in OH chemistry—the effects of feedbacks and reservoirs generated by the reactive products, *Chemosphere* **26**:641–655.

Matson, P. A., and Vitousek, P. M., 1990, Ecosystem approach to a global nitrous oxide budget, *BioScience* **40**:667–672.

Matthews, E., Fung, I., and Lerner, J., 1991, Methane emission from rice cultivation: Geographic and seasonal distribution of cultivated areas and emissions, *Global Biogeochem. Cycles* **5**:3–24.

Megraw, S. R., and Knowles, R., 1987, Methane production and consumption in a cultivated humisol, *Biol. Fertil. Soils* **5**:56–60.

Miller, L. G., Oremland, R. S., and Paulsen, S., 1986, Measurement of nitrous oxide reductase activity in aquatic sediments, *Appl. Environ. Microbiol.* **51**:18–24.

Minami, K., and Fukushi, S., 1981a, Detection of carbonyl sulfide among gases produced by the decomposition of cystine in paddy soils, *Soil Sci. Plant Nutr.* **27**:105–109.

Minami, K., and Fukushi, S., 1981b, Volatilization of carbonyl sulfide from paddy soils treated with sulfur-containing substances, *Soil Sci. Plant Nutr.* **27**:339–345.

Möller, D., 1984, On the global natural sulphur emission, *Atmos. Environ.* **18**:29–39.

Mörsdorf, G., Frunzke, K., Gadkari, D., and Meyer, O., 1992, Microbial growth on carbon monoxide, *Biodegradation* **3**:61–82.

Mosier, A. R., Mohanty, S. K., Bhadrachalam, A., and Chakravorti, S. P., 1990, Evolution of dinitrogen and nitrous oxide from soil to the atmosphere through rice plants, *Biol. Fertil. Soils* **9**:61–67.

Nesbit, S. P., and Breitenbeck, G. A., 1992, A laboratory study of factors influencing methane uptake by soils, *Agric. Ecosyst. Environ.* **41**:39–54.

Neue, H. U., and Roger, P. A., 1993, Rice agriculture: Factors controlling emissions, in: *Atmospheric Methane: Sources, Sinks, and Role in Global Change* (M. A. K. Khalil, ed.), Springer, Berlin, pp. 254–298.

Neue, H. U., and Sass, R. L., 1994, Trace gas emissions from rice fields, in: *Global Atmospheric– Biospheric Chemistry* (R. G. Prinn, ed.), Plenum Press, New York, pp. 119–147.

Novelli, P. C., Masarie, K. A., Tans, P. P., and Lang, P. M., 1994, Recent changes in atmospheric carbon monoxide, *Science* **263**:1587–1590.

Oremland, R. S., and Culbertson, C. W., 1992, Evaluation of methyl fluoride and dimethyl ether as inhibitors of aerobic methane oxidation, *Appl. Environ. Microbiol.* **58**:2983–2992.

Papen, H., VonBerg, R., Hinkel, I., Thoene, B., and Rennenberg, H., 1989, Heterotrophic nitrification by *Alcaligenes faecalis:* NO_2^-, NO_3^-, N_2O, and NO production in exponentially growing cultures, *Appl. Environ. Microbiol.* **55**:2068–2072.

Papen, H., Hellmann, B., Papke, H., and Rennenberg, H., 1993, Emission of N-oxides from acid irrigated and limed soils within a northern hardwood forest ecosystem, in: *Biogeochemistry of Global Change: Radiative Trace Gases* (R. S. Oremland, ed.), Chapman & Hall, New York, pp. 245–260.

Peters, V., and Conrad, R., 1995, Methanogenic and other strictly anaerobic bacteria in desert soil and other oxic soils, *Appl. Env. Microbiol.* **61**:1673–1676.

Popelier, F., Liessens, J., and Verstraete, W., 1985, Soil H_2-uptake in relation to soil properties and rhizobial H_2-production, *Plant Soil* **85**:85–96.

Poth, M., 1986, Dinitrogen production from nitrite by a *Nitrosomonas* isolate, *Appl. Environ. Microbiol.* **52**:957–959.

Poth, M., and Focht, D. D., 1985, ^{15}N kinetic analysis of N_2O production by *Nitrosomonas europaea:* An examination of nitrifier denitrification, *Appl. Environ. Microbiol.* **49**:1134–1141.

Prinn, R. G., 1994, Global atmospheric-biospheric chemistry, in: *Global Atmospheric–Biospheric Chemistry* (R. G. Prinn, ed.), Plenum Press, New York, pp. 1–18.

Protoschill-Krebs, G., and Kesselmeier, J., 1992, Enzymatic pathways for the consumption of carbonyl sulphide (COS) by higher plants, *Bot. Acta* **105**:206–212.

Ramanathan, V., Callis, L., Cess, R., Hansen, J., Isaksen, I., Kuhn, W., Lacis, A., Luther, F., Mahlman, J., Reck, R., and Schlesinger, M., 1987, Climate–chemical interactions and effects of changing atmospheric trace gases, *Rev. Geophys.* **25**:1441–1482.

Rasmussen, R. A., and Khalil, M. A. K., 1983, Global production of methane by termites, *Nature* **301**:700–702.

Reddy, K. R., and Patrick Jr., W. H., 1986, Fate of fertilizer nitrogen in the rice root zone, *Soil Sci. Soc. Am. J.* **50**:649–651.

Reddy, K. R., Patrick Jr., W. H., and Lindau, C. W., 1989, Nitrification–denitrification at the plant root-sediment interface in wetlands, *Limnol. Oceanogr.* **34**:1004–1013.

Reeburgh, W. S., Whalen, S. C., and Alperin, M. J., 1993, The role of methylotrophy in the global methane budget in: *Microbial Growth on C1 Compounds* (J. C. Murrell and D. P. Kelly, eds.), Intercept, Andover, Mass., pp. 1–14.

Remde, A., and Conrad, R., 1990, Production of nitric oxide in *Nitrosomonas europaea* by reduction of nitrite, *Arch. Microbiol.* **154**:187–191.

Remde, A., and Conrad, R., 1991a, Role of nitrification and denitrification for NO metabolism in soil, *Biogeochemistry* **12**:189–205.

Remde, A., and Conrad, R., 1991b, Metabolism of nitric oxide in soil and denitrifying bacteria, *FEMS Microbiol. Ecol.* **85**:81–93.

Remde, A., Slemr, F., and Conrad, R., 1989, Microbial production and uptake of nitric oxide in soil, *FEMS Microbiol. Ecol.* **62**:221–230.

Remde, A., Ludwig, J., Meixner, F. X., and Conrad, R., 1993, A study to explain the emission of nitric oxide from a marsh soil, *J. Atm. Chem.* **17**:249–275.

Robertson, G. P., 1993, Fluxes of nitrous oxide and other nitrogen trace gases from intensively managed landscapes: A global perspective, in: *Agricultural Ecosystem Effects on Trace Gases and Global Climate Change* (L. A. Harper *et al.*, eds.), American Society of Agronomy, Madison, Wis., pp. 95–108.

Rothfuss, F., and Conrad, R., 1994, Development of a gas diffusion probe for the determination of methane concentrations and diffusion characteristics in flooded paddy soil, *FEMS Microbiol. Ecol.* **14**:307–318.

Rouland, C., Brauman, A., Labat, M., and Lepage, M., 1993, Nutritional factors affecting methane emission from termites, *Chemosphere* **26**:617–622.

Rudolph, J., Rothfuss, F., and Conrad, R., Flux between soil and atmosphere, vertical concentration profiles in soil, and turnover of nitric oxide: 1. Measurements on a model soil core, *J. Atmospher. Chem.*, submitted.

Ryden, J. C., 1981, N_2O exchange between a grassland soil and the atmosphere, *Nature* **292**:235–237.

Sahrawat, K. L. and Keeney, D. R., 1986, Nitrous oxide emission from soils, *Adv. Soil Sci.* **4**:103–148.

Samuelsson, M.-O., 1985, Dissimilatory nitrate reduction to nitrite, nitrous oxide, and ammonium by *Pseudomonas putrefaciens*, *Appl. Environ. Microbiol.* **50**:812–815.

Schäfer, F., and Conrad, R., 1993, Metabolism of nitric oxide by *Pseudomonas stutzeri* in culture and in soil, *FEMS Microbiol. Ecol.* **102**:119–127.

Scharffe, D., Hao, W. M., Donoso, L., Crutzen, P. J., and Sanhueza, E., 1990, Soil fluxes and atmospheric concentration of CO and CH_4 in the northern part of the Guayana Shield, Venezuela, *J. Geophys. Res.* **95**:22475–22480.

Schidlowski, M., 1983, Evolution of photoautotrophy and early atmospheric oxygen levels, *Precambrian Res.* **20**:319–335.

Schimel, J. P., Holland, E. A., and Valentine, D., 1993, Controls on methane flux from terrestrial ecosystems, in: *Agricultural Ecosystem Effects on Trace Gases and Global Climate Change* (L. A. Harper *et al.*, ed.), American Society of Agronomy, Madison, Wis., pp. 167–182.

Schmidt, U., 1974, Molecular hydrogen in the atmosphere, *Tellus* **26**:78–90.

Schmidt, U., and Conrad, R., 1993, Hydrogen, carbon monoxide, and methane dynamics in Lake Constance, *Limnol. Oceanogr.* **38**:1214–1226.

Schuler, S., and Conrad, R., 1990, Soils contain two different activities for oxidation of hydrogen, *FEMS Microbiol. Ecol.* **73**:77–84.

Schuler, S., and Conrad, R., 1991, Hydrogen oxidation in soil following rhizobial H_2 production due to N_2 fixation by a *Vicia faba–Rhizobium leguminosarum* symbiosis, *Biol. Fertil. Soils* **11**:190–195.

Schuster, M., and Conrad, R., 1992, Metabolism of nitric oxide and nitrous oxide during nitrification and denitrification in soil at different incubation conditions, *FEMS Microbiol. Ecol.* **101**:133–143.

Schütz, H., Conrad, R., Goodwin, S., and Seiler, W., 1988, Emission of hydrogen from deep and shallow freshwater environments, *Biogeochemistry* **5**:295–311.

Schütz, H., Seiler, W., and Rennenberg, H., 1990, Soil and land use related sources and sinks of methane (CH_4) in the context of the global methane budget, in: *Soils and the Greenhouse Effect* (A. F. Bouwman, ed.), Wiley, Chichester, England, pp. 269–285.

Schütz, H., Schröder, P., and Rennenberg, H., 1991, Role of plants in regulating the methane flux to the atmosphere, in: *Trace Gas Emissions by Plants* (T. D. Sharkey, E. A. Holland, and H. A. Mooney, eds.), Academic Press, San Diego, Calif., pp. 29–63.

Sebacher, D. I., Harriss, R. C., and Bartlett, K. B., 1985, Methane emissions to the atmosphere through aquatic plants, *J. Environ. Qual.* **14**:40–46.

Seiler, W., 1978, The influence of the biosphere on the atmospheric CO and H$_2$ cycles, in: *Environmental Biogeochemistry and Geomicrobiology*, Vol. 3 (W. Krumbein, ed.), Ann Arbor Science Publishers, Ann Arbor, Mich., pp. 773–810.

Seiler, W., and Conrad, R., 1981, Field measurements of natural and fertilizer induced N$_2$O release rates from soils, *J. Air Poll. Contr. Assoc.* **31**:767–772.

Seiler, W., Giehl, H., and Bunse, G., 1978, The influence of plants on atmospheric carbon monoxide and dinitrogen oxide, *Pageoph* **116**:439–451.

Seiler, W., Conrad, R., and Scharffe, D., 1984, Field studies of methane emission from termite nests into the atmosphere and measurements of methane uptake by tropical soils, *J. Atmos. Chem.* **1**:171–186.

Seitzinger, S. P., 1990, Denitrification in aquatic sediments, in: *Denitrification in Soil and Sediment* (N. P. Revsbech and J. Soerensen, eds.), Plenum Press, New York, pp. 301–322.

Sexstone, A. J., and Mains, C. N., 1990, Production of methane and ethylene in organic horizons of spruce forest soils, *Soil Biol. Biochem.* **22**:135–139.

Sexstone, A. J., Revsbech, N. P., Parkin, T. B., and Tiedje, J. M., 1985, Direct measurement of oxygen profiles and denitrification rates in soil aggregates, *Soil Sci. Soc. Am J.* **49**:645–651.

Sieburth, J. M., 1987, Contrary habitats for redox-specific processes: Methanogenesis in oxic waters and oxidation in anoxic waters, in: *Microbes in the Sea* (M. A. Sleigh, ed.), Ellis Horwood, Chichester, England, pp. 11–38.

Singh, H. B., 1987, Reactive nitrogen in the troposphere. Chemistry and transport of NO$_x$ and PAN, *Environ. Sci. Technol.* **21**:320–327.

Skujins, J., 1978, History of abiontic soil enzyme research, in: *Soil Enzymes* (R. G. Burns, ed.), Academic Press, London, pp. 1–49.

Slemr, F., and Seiler, W., 1984, Field measurements of NO and NO$_2$ emissions from fertilized and unfertilized soils, *J. Atmos. Chem.* **2**:1–24.

Slemr, F., and Seiler, W., 1991, Field study of environmental variables controlling the NO emissions from soil, and of the NO compensation points, *J. Geophys. Res.* **96**:13017–13031.

Slemr, F., Conrad, R., and Seiler, W., 1984, Nitrous oxide emissions from fertilized and unfertilized soils in a subtropical region (Andalusia, Spain), *J. Atmos. Chem.* **1**:159–169.

Smith, C. J., Brandon, M., and Patrick Jr., W. H., 1982, Nitrous oxide emission following urea-N fertilization of wetland rice, *Soil Sci. Plant Nutr.* **28**:161–171.

Smith, K. D., Klasson, K. T., Ackerson, M. D., Clausen, E. C., and Gaddy, J. L., 1991, COS degradation by selected CO-utilizing bacteria, *Appl. Biochem. Biotechnol.* **28/29**:787–796.

Smith, M. S., 1983, Nitrous oxide production by *Escherichia coli* is correlated with nitrate reductase activity, *Appl. Environ. Microbiol.* **45**:1545–1547.

Smith, M. S., and Zimmerman, K., 1981, Nitrous oxide production by nondenitrifying soil nitrate reducers, *Soil Sci. Soc. Am. J.* **45**:865–871.

Steele, L. P., Dlugokencky, E. J., Lang, P. M., Tans, P. P., Martin, R. C., and Masarie, K. A., 1992, Slowing down of the global accumulation of atmospheric methane during the 1980s, *Nature* **358**:313–316.

Steudler, P. A., and Peterson, B. J., 1984, Contribution of gaseous sulphur from salt marshes to the global sulphur cycle, *Nature* **311**:455–457.

Steudler, P. A., and Peterson, B. J., 1985, Annual cycle of gaseous sulfur emissions from a New England *Spartina alterniflora* marsh, *Atmos. Environ.* **19**:1411–1416.

Striegl, R. G., McConnaughey, T. A., Thorstenson, D. C., Weeks, E. P., and Woodward, J. C., 1992, Consumption of atmospheric methane by desert soils, *Nature* **357**:145–147.

Teraguchi, S., and Hollocher, T. C., 1989, Purification and some characteristics of a cytochrome *c*-containing nitrous oxide reductase from *Wolinella succinogenes*, *J. Biol. Chem.* **264**:1972–1979.

Tiedje, J. M., 1988, Ecology of denitrification and dissimilatory nitrate reduction to ammonia, in: *Biology of Anaerobic Microorganisms* (A. J. B. Zehnder, ed.), Wiley, New York, pp. 179–244.

Tiedje, J. M., Sexstone, A. J., Parkin, T. B., Revsbech, N. P., and Shelton, D. R., 1984, Anaerobic processes in soil, *Plant Soil* **76:**197–212.

Van Cleemput, O., and Baert, L., 1984, Nitrite: A key compound in N loss processes under acid conditions? *Plant Soil* **76:**233–241.

Vecherskaya, M. S., Galchenko, V. F., Sokolova, E. N., and Samarkin, V. A., 1993, Activity and species composition of aerobic methanotrophic communities in tundra soils, *Curr. Microbiol.* **27:**181–184.

Vedenina, I. Y., and Zavarzin, G. A., 1977, Biological removal of nitrous oxide under oxidizing conditions, *Mikrobiologiya* **46:**898–903.

Vedenina, I. Y., Miller, Y. M., Kapustin, O. A., and Zavarzin, G. A., 1980, Oxidation of nitrous oxide during decomposition of hydrogen peroxide by catalase, *Mikrobiologiya* **49:**5–8.

Warneck, P., 1988, *Chemistry of the Natural Atmosphere*, Academic Press, San Diego, Calif.

Whalen, S. C., Reeburgh, W. S., and Sandbeck, K. A., 1990, Rapid methane oxidation in a landfill cover soil, *Appl. Environ. Microbiol.* **56:**3405–3411.

Whalen, S. C., Reeburgh, W. S., and Barber, V. A., 1992, Oxidation of methane in boreal forest soils—A comparison of 7 measures, *Biogeochemistry* **16:**181–211.

Williams, E. J., Hutchinson, G. L., and Fehsenfeld, F. C., 1992, NO_x and N_2O emissions from soil, *Global Biogeochem. Cycles* **6:**351–388.

Yavitt, J. B., Downey, D. M., Lang, G. E., and Sexstone, A. J., 1990, Methane consumption in two temperate forest soils, *Biogeochemistry* **9:**39–52.

Yoshinari, T., Hynes, R., and Knowles, R., 1977, Acetylene inhibition of nitrous oxide reduction and measurement of denitrification and nitrogen fixation in soil, *Soil Biol. Biochem.* **9:**177–183.

Zafiriou, O. C., Hanley, Q. S., and Snyder, G., 1989, Nitric oxide and nitrous oxide reduction and cycling during dissimilatory nitrite reduction by Pseudomonas perfectomarina, *J. Biol. Chem.* **264:**5694–5699.

Zausig, J., Stepniewski, W., and Horn, R., 1993, Oxygen concentration and redox potential gradients in unsaturated model soil aggregates, *Soil Sci. Soc. Am. J.* **57:**908–916.

Zehnder, A. J. B., and Stumm, W., 1988, Geochemistry and biogeochemistry of anaerobic habitats, in: *Biology of Anaerobic Microorganisms* (A. J. B. Zehnder, ed.), Wiley, New York, pp. 1–38.

Zeikus, J. G., 1983, Metabolism of one-carbon compounds by chemotrophic anaerobes, *Adv. Microb. Physiol.* **24:**215–299.

Zimmerman, P. R., Greenberg, J. P., Wandiga, S. O., and Crutzen, P. J., 1982, Termites: A potentially large source of atmospheric methane, carbon dioxide, and molecular hydrogen, *Science* **218:**563–565.

Zumft, W. G., 1993, The biological role of nitric oxide in bacteria, *Arch. Microbiol.* **160:**253–264.

Zumft, W. G., and Kroneck, P. M. H., 1990, Metabolism of nitrous oxide, in: *Denitrification in Soil and Sediment* (N. P. Revsbech and J. Soerensen, eds.), Plenum Press, New York, pp. 37–55.

6

The Biogeochemistry of Hypersaline Microbial Mats

DAVID J. DES MARAIS

1. Introduction

1.1. Significance of Microbial Mats

Microbial mats are structurally coherent macroscopic accumulations of microorganisms. Photosynthetic mats offer an opportunity to examine the dynamics of a complete microbial ecosystem. Microbial mats construct laminated "miniature reefs" called stromatolites, which occur typically as carbonate rocks and are among the oldest, most abundant fossil evidence of ancient life on Earth (Walter, 1976). Mats built by cyanobacteria created the most obvious and best-studied stromatolites in the fossil record (Walter, 1976; Walter *et al.*, 1992).

Although their distribution today is limited largely by predation and competition for space with other organisms, cyanobacterial mats still are remarkably cosmopolitan in their occurrence. They can be found in soils and rock pavements, freshwater lakes and streams, hot springs, and hypersaline lakes and lagoons. Benthic phototrophs growing on sediment surfaces or rocks can contribute significantly to the primary productivity of coastal marine environments (e.g., Grøntved, 1960; Cadée and Hegeman, 1974).

In order to interpret ancient marine cyanobacterial stromatolites, it is particularly important to understand the dynamic relationships between microorganisms that maintain the structure of the community. Accordingly, a study of marine hypersaline mats in Baja California, Mexico was initiated, and it has provided new insights. These subtidal cyanobacterial mats grow in the seawater evaporation ponds of the salt producer Exportadora de Sal, S.A. (ESSA). This

DAVID J. DES MARAIS • Ames Research Center, Moffett Field, California 94035-1000.
Advances in Microbial Ecology, Volume 14, edited by J. Gwynfryn Jones. Plenum Press, New York, 1995.

chapter cites and summarizes a set of published studies that, as a set, provide a broad overview of the ecology of a single mat ecosystem.

1.2. Guerrero Negro Field Site

The ESSA evaporation ponds are situated adjacent to Laguna Ojo de Liebre (Scammon's Lagoon) on the Pacific Coast at 28°N latitude, about 700 km south of the Mexican–US border. Annual precipitation ranges between 15 and 120 mm/year, although it can exceed 300 mm/year in exceptional years, which typically occur less than once per decade. Winds are west to west-northwesterly and average 5 m/sec. Daily air (water) summer temperatures range from 20 to 32°C and (22 to 29°C), winter temperatures vary between 8 and 24°C (14 and 22°C). Seawater pumped from the lagoon into pond 1 (Fig. 1) then flows to ponds 2, 3, 4, 5, 6, etc. These ponds receive no groundwater because they lie above the local hydrologic gradient. Runoff has minimal effect on the brines, except during very wet years. This saltern is the world's single largest producer of NaCl, yielding an annual harvest exceeding 6 million metric tons.

Along the salinity gradient, the evaporating brines display distinctive chem-

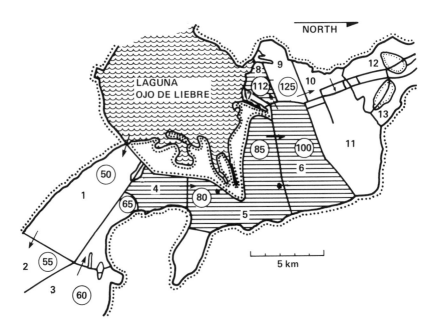

Figure 1. Map showing concentrator ponds of the company Exportadora de Sal, S.A., at Guerrero Negro, Baja California Sur, Mexico. Arrows indicate direction of flow of evaporating seawater. Circled numbers give average salinities (⁰/₀₀) for each pond. Areas in ponds 4 through 8 sustain permanently submerged ("subtidal") mats and are delineated by horizontal hatches. Mat study sites in pond 5 are indicated by black dots. (Modified from Des Marais *et al.*, 1992.)

ical trends (Des Marais *et al.*, 1989). For each of the concentrating ponds, long-term average salinities [expressed as parts per thousand (°/oo): g salts/kg brine; ESSA company records) are shown in Fig. 1. Although highly productive micro-bial mats occur in the salinity range 65 to 125°/oo, a strong net uptake of dis-solved inorganic carbon (DIC) by these mats is not indicated (Des Marais *et al.*, 1989). No strong trends were observed in the stable carbon isotopic composition of either DIC or mat organic matter (Des Marais *et al.*, 1989). Levels of phos-phate, nitrate, and ammonia are low (Javor, 1983a) due perhaps in part to negligible runoff. At these relatively low nutrient levels, benthic communities are favored over planktonic ones, perhaps because they retain nutrients more efficiently (Javor, 1983b).

In the salinity range 40 to 65°/oo (ponds 1, 2, 3, and part of 4), the ponds are dominated by a seagrass, *Ruppia* sp., and a green alga, *Enteromorpha* (Javor, 1983a). Also present are diatoms (*Navicula* sp., *Grammatophora* sp., *Striatella* sp., and *Licmophora* sp.) and cyanobacteria (*Entophysalis*). Well-developed per-manently submerged cyanobacterial mats occur in ponds 4 through 8, in the salinity range from 65 to 125°/oo. Filamentous cyanobacteria such as *Microcoleus* spp. and other *Oscillatoria* spp. are dominant between 65 and 100°/oo; unicellular cyanobacteria such as *Synechococcus* sp. become more abundant above 90°/oo (Jørgensen and Des Marais, 1986). The unicellular cyanobacterium *Aphanothece halophytica* is dominant in the range 90 to 200°/oo. *Dunaliella* sp., a halophilic green alga, is typically abundant in other hypersaline environments but is uncom-mon here (Javor, 1983a).

The best-developed, most coherent filamentous cyanobacterial mats are in pond 5. These mats were established about 1970 on a gypsum-encrusted evap-orite pan and have been remarkably stable since that time (ESSA personnel, personal communication). They are permanently submerged at 0.5 to 1 m water depth, where the salinity is maintained typically in the range 75 to 100°/oo (2½ to 3 times seawater salinity). The discussion that follows emphasizes these mats both because they have been most thoroughly studied and because they express most clearly all of the key components of a typical photosynthetic mat ecosystem.

2. Morphology

The mats in pond 5 have a cohesive, rubbery texture and range from 1 to 10 cm thick, with thicknesses usually in the range 4 to 7 cm (Des Marais *et al.*, 1992). The mat surface is typically smooth and olive-tan in color, although orange-tan-colored, blister-shaped communities of unicellular cyanobacteria and diatoms become more abundant at higher salinities. Internal laminations, fre-quently 0.3 to 3 mm thick, range in color from dark green to tan to orange-brown. These laminations represent variations over time in the relative contribu-tions to the mat by filamentous and unicellular cyanobacteria and organic de-tritusderived from disrupted mats. The top 1 to 2 mm of the mat is by far the most active zone metabolically (e.g., Canfield and Des Marais, 1993). At depths typically exceeding 5 to 7 cm, the mat color darkens and, at its base, the mat grades into a fine-grained, black, sulfidic gypsum mud.

The structure and microbiota of the pond 5 mat have been described in detail using light and transmission electron microscopy (D'Amelio *et al.*, 1989). The microbial community is embedded in a polysaccharide matrix excreted by cya-

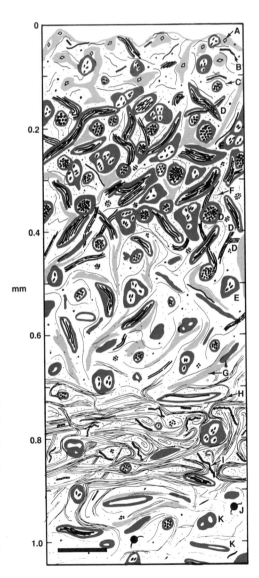

Figure 2. Schematic vertical section of the topmost 1 mm of subtidal mat in pond 5. Horizontal bar at lower left represents 100 μm. Letters along the right margin indicate the following: A, diatoms, B, *Spirulina* spp.; C, *Oscillatoria* spp.; D, *Microcoleus chthonoplastes;* E, nonphotosynthetic bacteria; F, unicellular cyanobacteria; G, fragments of bacterial mucilage; H, *Chloroflexus* spp., I, *Beggiatoa* spp.; J, meiofauna; K, abandoned cyanobacterial sheaths. (Modified from Des Marais *et al.*, 1992.)

nobacteria and other bacteria (Fig. 2). The relatively flat mat surface hosts a small diatom community (*Navicula* and *Nitzschia* spp.). Filamentous cyanobacteria, *Oscillatoria* spp. and *Spirulina labyrynthiformis*, and the unicellular *Synechococcus* spp. are abundant in the uppermost 300 μm of the mat. The cyanobacterium *Microcoleus chthonoplastes* dominates the photic zone from a depth of 50 μm to depths of 1 to 2 mm. *Microcoleus chthonoplastes* typically forms bundles having variable numbers (1 to 90) of trichomes enveloped in a common mucopolysaccharide sheath. Two populations can be distinguished by differences in their trichome diameters, which are 2.7 to 3 μm and 3.5 to 6 μm. Approximately 15% of the trichome bundles harbor a bacterium suggested to be a gliding, filamentous purple phototroph, with stacks of photosynthetic lamellae resembling those of *Ectothiorhodospira* sp. (D'Amelio *et al.*, 1987). Other cyanobacteria, including *Oscillatoria*, occupy the photic zone with *M. chthonoplastes*, and include *O. limnetica* and *O. salina*. The trichomes of these cyanobacteria are approximately 1 μm or larger in diameter. Other unidentified, nonphotosynthetic bacteria are also observed in the photic zone.

Near the base of the photic zone, a distinctive microbial community can be discerned at the "chemocline," where daytime ambient O_2 levels reach zero (D'Amelio *et al.*, 1987). Although some of the biota observed at shallower depths are present, the chemocline community is dominated by the filamentous green phototroph *Chloroflexus* sp. (with 0.6 μm diameter filaments) and the family *Beggiatoaceae* (two populations with filament diameters of 1 to 1.3 μm and 3 to 5 μm). At depths greater than few hundred micrometers beneath the chemocline, the populations of morphologically recognizable bacteria are greatly reduced, although scattered bundles of *M. chthonoplastes* and several varieties of unidentified nonphotosynthetic bacteria are apparent by light microscopy.

3. The Mat Microenvironment

In order to understand the function of a microbial mat community, it is first necessary to describe the physical and chemical microenvironment in which the microbes live. At Guerrero Negro, the environment within the mat differs substantially from that in the overlying water column. The community just beneath the mat surface typically experiences steep vertical gradients of light intensity and redox conditions; these change markedly due to daily variations in sunlight.

3.1. Mat–Water Column Interface

As the mat surface is approached from above, the dynamic convection that homogenizes the overlying water column diminishes. The internal friction of the water creates a viscous boundary water layer adjacent to the mat surface. The transport of solutes within this layer is dominated by diffusion; hence, it is called the diffusion boundary layer (DBL) (Jørgensen and Des Marais, 1990; Jørgensen, 1994). Although diffusion is typically considered to be a slow mechanism for transport, the DBL usually does not limit the rate of transport of dissolved compounds across the typical sediment–water interface. However, photosynthetic mats are very highly productive in intertidal and shallow subtidal environments, such as these hypersaline ponds. In these cases, transport of key species such as O_2 can become diffusion-limited. Furthermore, at night, the DBL creates a microaerophilic environment near the mat surface for key sulfide-oxidizing microorganisms such as purple sulfur bacteria and Beggiatoa. Thus, although Beggiatoa is killed by near-atmospheric O_2 concentrations, it can survive in the this DBL-sustained zone and obtain the O_2 it needs for respiration.

The effect of the DBL on transport was explored in the subtidal mats at Guerrero Negro (Jørgensen and Des Marais, 1990). The thickness of the DBL varies (from 0.2 to 0.8 mm) inversely as a function of water velocity above the mat (7.7 to 0.3 cm sec^{-1}, respectively). Thus higher flow rates favored more rapid transport of solutes across the mat–water interface. If the vertical relief of the mat surface (or surface roughness) is smaller than the thickness of the DBL, then this relief will have a negligible effect on both the fluid flows over the surface and the transport of solutes across the surface (Boudreau and Guinasso, 1982). However, for the subtidal mats studied at Guerrero Negro, the surface roughness is approximately 1 mm or greater and thus exceeds the DBL thickness, which is typically 0.5 mm. Thus the upper surface of the DBL closely follows the topography of the underlying mat, although the DBL is thinner over elevated areas and thicker over depressions. The DBL is thinner on the upstream sides of small mounds than on the downstream sides. Overall, the greater surface areas of rougher mats increase the rate of O_2 transport across the mat–water interface, even though the upper surface of the DBL is somewhat smoother than the solid surface of the mat.

Further studies of the mat–water interface should enhance our ability to interpret former mat surfaces that have been preserved as individual laminae in ancient stromatolites. The morphology of these laminae have long been interpreted to be characteristic of the microbiota that shaped them (e.g., Walter *et al.*, 1992); however, the potentially important biogeochemical factors that have governed lamina morphology are still largely unexplored.

3.2. Light Microenvironment

Below the mat surface, the internal structure absorbs and scatters incoming light. Gradients in light spectra and intensity were examined using a fiberoptic microsensor (Jørgensen and Des Marais, 1986, 1988; Jørgensen *et al.*, 1987).

The light flux penetrating the mat can be measured both as downward irradiance (the total downwelling light that passes through a horizontal plane) and as scalar irradiance (the sum of all light that converges on a given point within the mat). Due to the high density of photosynthetic organisms, bacterial mucilage, and mineral particles, light absorption is dominated by the light-harvesting pigments of the phototrophs and light is strongly scattered. Because absorption and scattering of light are quite substantial within the mat, scalar irradiance can differ substantially from downward irradiance (Jørgensen and Des Marais, 1988). Because scalar irradiance measures the total light actually available at a given location, it gives the most meaningful description of a microorganism's environment.

Measurements of scalar irradiance were obtained both from a microbial mat from pond 5 that was dominated by *M. chthonoplastes* and from a mat growing at

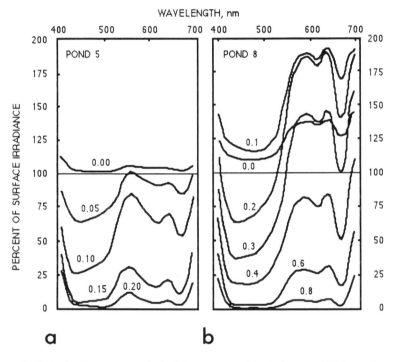

Figure 3. Scalar irradiance spectra obtained from a range of depths in two subtidal cyanobacterial mats. The scalar irradiances are expressed as the percent of the downward scalar irradiance measured at the mat surface. The depths (mm) below the mat surface are given adjacent to each spectral curve. The pond 5 mat was dominated by *M. chthonoplastes* filamentous cyanobacteria. The pond 8 mat harbored equivalent populations of unicellular and filamentous cyanobacteria (Des Marais *et al.*, 1992). These curves illustrate how the light available to the photosynthetic microorganisms can differ spectrally from the light source, even at the surface of the mat. (Modified from Jørgensen and Des Marais, 1988.)

higher salinity that was dominated by unicellular cyanobacteria (Fig. 3a) (Jørgensen and Des Marais, 1988). The pond 5 mat examined in the study did not have the typical surface colony of diatoms, and thus it was a dark green color. A strong decline in intensity and a marked change in spectral composition of the light is typically observed with depth in the mat (Fig. 3a). Minima in the spectra reflect the absorption maxima of the photosynthetic pigments of cyanobacteria. Chlorophyll a (Chl a) absorbs around 430 and 670 nm; phycocyanin absorbs around 620 nm; and various carotenoids absorb in the range 450 to 500 nm. In contrast, the mat that is dominated by unicellular cyanobacteria has a lower density of cells, a more gelatinous texture, and a light orange-tan color. Light penetrates more deeply, although blue light is strongly attenuated (Fig. 3b). Absorption is dominated by carotenoids, relative to Chl a. Longer-wavelength light, particularly longer than 900 nm, penetrates furthest into the mat.

These data illustrate the dramatic effects that the mat matrix can exert on the penetration of light. Light penetration within the photic zone is dominated by absorption effects. In the example shown in Fig. 3a for pond 5, light is virtually extinguished by 0.4 mm depth, although light can penetrate to more than 1 mm in other *M. chthonoplastes*-dominated mats (Canfield and Des Marais, 1993). In the unicellular cyanobacterial mat, light absorption is less substantial and scattered light was therefore more important. Thus scalar irradiance differs most substantially from incident irradiance with respect to its intensity and spectral composition (Fig. 3b). Maximum scalar irradiance actually significantly exceeds incident irradiance just below the mat surface, and, over a range of depths, blue light is substantially reduced and orange-red light enhanced, relative to the spectrum of incident irradiance. Such effects influence strongly the measurement of action spectra, which record the photosynthetic activity of microbes as a function of light intensity at discrete wavelengths. Action spectra are highly relevant to ecological studies of adaptation and evolution in mat communities. For example, benthic cyanobacteria that use light filtered by overlying diatom communities show greatest photosynthetic activity at wavelengths between 550 and 650 nm (Jørgensen *et al.*, 1987), a region that lies between the absorption maxima of Chl a. In contrast, planktonic cyanobacteria exposed to a broader spectrum of light show significant activity at wavelengths corresponding to the absorption maxima of Chl a (Jørgensen and Des Marais, 1988).

3.3. Chemical Gradients

The high rates of oxygenic photosynthesis that occur in the narrow photic zone of the mat create steep and variable gradients (Revsbech *et al.*, 1983) in the pH and in DIC and dissolved oxygen (DO) concentrations. The oxic zone is sustained by a highly dynamic balance between photosynthetic O_2 production and O_2 consumption by a host of sulfide-oxidizing and heterotrophic bacteria.

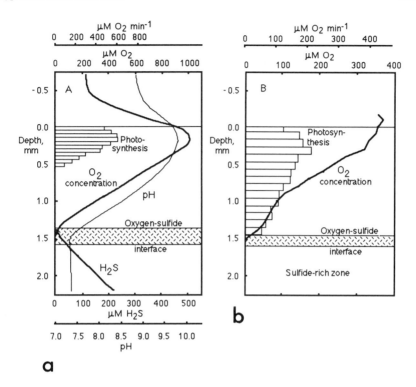

Figure 4. Depth profiles for key chemical constituents in the pond 5 mat: (a) photosynthetic O_2 production rates (bars), pH, and concentrations of O_2 and sulfide, measured by Jørgensen and Des Marais (1986); (b) photosynthetic O_2 production rates and O_2 concentrations measured during April 1990. (Data from Canfield and Des Marais, 1993.)

Using microelectrodes, the depth distribution of DO, sulfide concentrations, and pH was determined in a *M. chthonoplastes*-dominated mat from pond 5 (Fig. 4a) (Jørgensen and Des Marais, 1986). These depth profiles are typical of the profiles that have been obtained subsequently (Canfield and Des Marais, 1993). Extremely high rates of oxygenic photosynthesis create DO levels that are nearly five times the value of air-saturated brine, yet this O_2 has a residence time of only 2 min. Oxygen production can become negligible at a depth of only 0.55 mm, due to light limitation (Fig. 4a). However, O_2 diffuses further down to a point where, within a 100-μm-thick interval, it overlaps with sulfide diffusing up from below. This interval is inhabited typically by abundant *Chloroflexus* spp. and by Beggiatoa. As sunset approaches, the oxic zone collapses quickly and the oxic–anoxic boundary approaches the mat surface (Canfield and Des Marais, 1993). Thus the bacteria in the mat survive conditions that alternate between O_2 supersaturation and millimolar concentrations of sulfide.

4. The Microbial Mat Food Chain

The mat ecosystem depends on intimate interactions between key groups of bacteria. As already discussed, oxygenic photosynthesis by diatoms and cyanobacteria fulfills the role of primary production in this ecosystem. Organic decomposition and nutrient recycling sustain an additional vast consortium of bacteria.

4.1. Role of Dissolved Organic Matter

Organic compounds become available to the microbial food chain not only as dead cells, but also as dissolved organic matter (DOM) excreted by cyanobacteria and other microorganisms (see Paerl et al., 1993, for discussion). Indeed, DOM concentrations are observed to be highest during the daytime in the mat photic zone of M. chthonoplastes-dominated mats from pond 5 (Bauer et al., 1991). Autoradiography experiments with these mats demonstrate that populations of bacteria adhering to cyanobacterial sheaths do indeed incorporate DOM (Paerl et al., 1993). The radiolabeled organic substrates that are utilized include glucose, amino acids, and acetate.

4.2. Sulfate-Reducing Bacteria

Sulfate-reducing bacteria are quantitatively important consumers of DOM. Furthermore, the sulfide they produce sustains a wide variety of phototrophic and chemotrophic bacteria. The highest rates of sulfate reduction occur in the shallowest part of the mat, in close proximity to the photosynthetic source of fresh organic matter (Canfield and Des Marais, 1991, 1993). Although O_2 is typically an effective inhibitor of bacterial sulfate reduction, the highest reduction rates occur within the mat's aerobic zone during the daytime (Fig. 5) (Canfield and Des Marais, 1991). A thorough search was made for aerobic microenvironments within the aerobic zone that might serve as refugia for the sulfate-reducing bacteria, yet none were found. Any beneficial interactions between bacteria that might assist in neutralizing this O_2 inhibition of sulfate reduction are presently unknown.

4.3. Sulfide-Utilizing Bacteria

Sulfide is consumed both by chemotrophic and phototrophic bacteria. The absence of detectible sulfide concentrations in the aerobic zone of the mat, coupled with the high rates of sulfide production from aerobic sulfate-reducing bacteria (Canfield and Des Marais, 1991), indicate that a substantial population of sulfide-oxidizing bacteria must occupy the aerobic zone. Sulfide-oxidizing bacteria are most clearly visible under the microscope in the vicinity of the

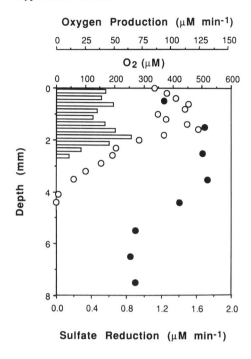

Figure 5. Depth distribution of O_2 production rates (bars), O_2 concentrations (open circles), and sulfate reduction rates (filled circles) for a subtidal mat from pond 5. (Adapted from Canfield and Des Marais, 1991.)

chemocline, the zone where detectible levels of O_2 and sulfide coexist (Fig. 4). In the chemocline of mats dominated by *M. chthonoplastes,* Beggiatoa, and *Chloroflexus* sp. are the most prominent of the sulfide-utilizing bacteria (Fig. 2) (D'Amelio *et al.*, 1989). Several varieties of *Chloroflexus*-like bacteria from a variety of mats in the Guerrero Negro salt ponds have recently been described (Pierson *et al.*, 1994). Mat layers dominated by these bacteria show sulfide-dependent photoautotrophic activity that depends on near-IR radiation. Their abundance at the chemocline indicates that these bacteria are important both as sulfide consumers and as secondary producers in the mat. Beggiatoa compete for this sulfide.

The relative population sizes of photoautotrophic and chemolithotrophic sulfide bacteria can be affected by the amount of light that reaches the chemocline (Jørgensen and Des Marais, 1986). Light levels as low as 1% of the incident near-IR radiation (800 to 900 nm) are sufficient for the phototroph *Chromatium* to dominate the chemocline. Thus the balance between the penetration of O_2 versus light into the sulfide-rich zone determines the balance in the populations of sulfide bacteria.

An unidentified filamentous purple bacterium has been described that actually lives inside the sheaths of viable *M. chthonoplastes* (D'Amelio *et al.*, 1987). This bacterium displays intracytoplasmic stacked photosynthetic lamellae that are similar to those of the purple sulfur bacteria of the genus *Ectothiorhodospira*. Its occurrence inside cyanobacterial sheaths is remarkable, because anoxygenic photosynthesis of the purple sulfur bacteria is inhibited by O_2. D'Amelio *et al.* (1987) proposed that, as light levels and DO vary during the day, this bacterium alternates between photoheterotrophic growth (at high light levels, using organic matter excreted by the cyanobacteria) and sulfide cometabolism with cyanobacteria (at relatively low light levels, where O_2 production is minimal). These preliminary observations illustrate the wealth of unexpected insights promised by further studies of the mat community.

4.4. Grazing and Bioturbation

Grazing of mats by metazoans is significant for several reasons (Farmer, 1992). First, metazoan activity can alter the primary productivity, structure, and biomass of mats. Grazing has even been invoked for causing the decline in stromatolite diversity and abundance in the late Proterozoic Eon, between 900 and 540 million years ago (e.g., Walter and Heys, 1985). Second, grazing strategies provide useful constraints for understanding the interactions and adaptations associated with the early evolution of Metazoa. Grazing has been examined in several marine hypersaline microbial mat ecosystems (e.g., Gerdes and Krumbein, 1987; Javor and Castenholz, 1984; Farmer, 1992).

In the subtidal *M. chthonoplastes*-dominated cyanobacterial mats at Guerrero Negro, small epifaunal and meiofaunal grazers are abundant, yet appear to have little impact on the distribution of these mats (Farmer, 1992). The meiofauna is dominated by several species of crustaceans and nematodes. In the mat's uppermost 0.5 to 1 mm, the harpacticoid copepod *Cletocampus dietersi* and an unidentified cladoceran species feed primarily on diatoms and organic detritus. At greater mat depths, up to 2 mm, the meiofauna is represented principally by the nematodes *Monhystera* sp., *Microlaimus* sp., and *Oncholaimus* sp., and by two species of chromadorids. *Microlaimus* sp. and the chromadorids appear to feed on diatoms, coccoid cyanobacteria, and organic detritus. *Monhystera* probably consumes smaller coccoid cyanobacteria. Nematodes are particularly abundant in the zone between the dense near-surface layer of filamentous cyanobacteria (*M. chthonoplastes*, etc.) and the deeper oxygen–sulfide interface. This nematode population has been estimated to maintain densities of 1600 individuals/cm^2 (Farmer, 1992).

The relatively small grazers observed in these hypersaline ponds do not inhibit the development of these subtidal mats, although they may affect mat fabric and surface texture (Farmer, 1992). Relatively few smaller species possess the necessary adaptations for consuming filamentous cyanobacteria. Intense grazing and bioturbation by macroinvertebrates with body sizes exceeding a few millimeters can prevent microbial mats from developing in other environments. However, mat development in Guerrero Negro ponds with salinities below 65 is limited instead by competition with the seagrass, *Ruppia maritima* (Des Marais *et al.*, 1992). Thus, factors such as the evolution of algal metaphytes might have been more important than grazing in promoting the decline of ancient microbial stromatolites (Farmer, 1992).

5. Microbial Motility

5.1. The Diel Cycle

In the *M. chthonoplastes*-dominated mats, Beggiatoa filaments can migrate to the surface at night as they follow the migration of the oxygen–sulfide interface (Jørgensen and Des Marais, 1986). These chemolithotrophs probably are important secondary producers. Furthermore, they might help to prevent regenerated nutrients from escaping into the overlying water column at night.

Motility was also examined (Garcia-Pichel *et al.*, 1994) in flat, laminated mats growing at higher salinities (approximately four times seawater salinity). Beggiatoa follows the oxygen–sulfide interface, traveling about 0.8 mm from a daytime position at depth to just below the mat surface at night. However, most of the Beggiatoa filaments remain at depth during the night. Light serves as a cue of paramount importance for stimulating and modulating the movements of *Oscillatoria* sp. and *Spirulina* cyanobacteria. During daylight, these filaments migrate to maintain a position where light intensities are sufficiently high to saturate photosynthesis but are insufficient to trigger photoinhibition. Thus motility allows the community to optimize both photosynthetic primary production and also the utilization of sulfide during secondary production.

5.2. Sedimentation Rate

The influx of detritus has both physical and chemical effects upon bacterial activity. Physically, the microflora must compete with detritus in order to populate the sediment surface at densities sufficient to maintain a coherent mat. The proportion of filamentous cyanobacteria increases at higher sedimentation rates, relative to unicellular cyanobacteria. This occurs presumably because gliding motility confers a selective advantage for the filamentous forms (Des Marais *et al.*, 1992). Filaments tend to be oriented more perpendicular to the growth surface (Farmer and Richardson, 1988), due to phototaxis and increased vertical migration.

Increases in the sedimentation rate of organic detritus increase the rate of organic decomposition, relative to that of oxygenic photosynthesis. The enhanced decomposition fuels higher rates of sulfate reduction, leading to higher rates of sulfide production. Lightly sheathed cyanobacteria of the genus *Oscillatoria* and sulfide-dependent bacteria such as Beggiatoa and anoxygenic phototrophs become more abundant, relative to the cyanobacteria that form thicker sheaths (Des Marais *et al.*, 1992). As inputs of organic detritus increase, these sulfur bacteria position themselves closer to the mat surface. The relatively lower contents of sheath material make the mat less stable against physical disruption.

6. Biogeochemical Cycles

The waters that host microbial mats are typically depleted in the basic nutrient elements (Javor, 1983a), yet microbial mats are among the most productive aquatic ecosystems. This apparent paradox can be explained if the recycling of key nutrients within the mat ecosystem is highly efficient. The biogeochemical cycling of carbon, oxygen, sulfur, and nutrients has been studied in *M. chthonoplastes*-dominated mats (e.g., Canfield and Des Marais, 1993; Bebout, 1992; Bebout *et al.*, 1994).

6.1. Carbon, Oxygen, and Sulfur

An overview of the key recycling processes is provided by examining the budgets of DIC and O_2 over a full 24-hour cycle (Fig. 6) (Canfield and Des Marais, 1993). During the daytime, O_2 is produced by oxygenic photosynthesis. This O_2 is removed by O_2 respiration and sulfide oxidation and by diffusion into the overlying brine. Thus, when the mat is functioning at steady state during the day, the processes of O_2 production and consumption are balanced, as follows:

$$O_2 \text{ photosynthesis} = O_2 \text{ respiration} + S \text{ oxidation} + \text{diffusion out}$$

In contrast, DIC is removed by oxygenic and anoxygenic photosynthesis, whereas it is produced by oxygen- and sulfate-based respiration and by diffusion both from the overlying brine and from deeper within the mat (Fig. 6). When the mat is at steady state, production balances consumption, as follows:

$$O_2 \text{ respiration} + \text{sulfate reduction} + \text{diffusion in} = \text{photosynthesis} + \text{chemosynthesis}$$

During the night, no O_2 is produced within the mat because of the absence of photosynthesis; thus the oxygen–sulfide interface moves up to just below the surface of the mat (Fig. 6). The mat becomes a net O_2 sink, as O_2 diffusing from the overlying brine is consumed by oxic respiration and sulfide oxidation. The following equation for O_2 applies at steady state:

$$O_2 \text{ diffusion in} = O_2 \text{ respiration} + S \text{ oxidation}$$

At night, the mat becomes a net source of DIC, which is produced by oxic respiration and sulfate reduction and then diffuses into the overlying brine (Fig.

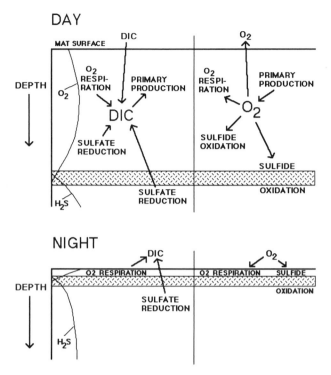

Figure 6. Schematic vertical cross-section of a microbial mat, depicting the various processes that control the budgets of dissolved inorganic carbon (DIC) and O_2, both during the daytime and at night. During the day, DIC, which either diffuses into the mat or is produced by oxic respiration or sulfate reduction, is subsequently consumed by photosynthesis. The O_2 produced by oxygenic photosynthesis can either diffuse out of the mat or be consumed by oxic respiration or sulfide oxidation. At night, DIC, produced either by sulfate reduction or oxic respiration, diffuses out of the mat, although some DIC can be reassimilated by chemoautotrophic bacteria. The O_2 diffusing into the mat oxidizes organic carbon or sulfide. (Adapted from Canfield and Des Marais, 1993.)

6). Some of this DIC may be consumed by chemoautotrophic carbon fixation sustained by sulfide oxidation near the mat–brine interface (e.g., by Beggiatoa and other sulfide oxidizers). The steady-state DIC balance is maintained as follows:

O_2 respiration + sulfate reduction = chemoautotrophy + diffusion out

In April 1990, the phototrophic community was dominated by *M. chthonoplastes,* and oxygenic photosynthesis extended to the oxygen–sulfide interface (Fig. 4b). Light therefore penetrated this interface and sustained a major population of the anoxygenic phototroph *Chloroflexus.* Beggiatoa was nearly absent.

David J. Des Marais

An example of the budgets for DIC, O_2, and S is given by data obtained in April 1990 (Table I and Fig. 6) (Canfield and Des Marais, 1993). In addition, about 57% of the daytime production of sulfide accumulated in the topmost 2 mm of the mat as elemental sulfur (S^o), but was quickly lost during the night. This daytime S^o accumulation reduced the demand for O_2 in sulfide oxidation.

Several key observations can be made regarding the O_2 and DIC budgets (Canfield and Des Marais, 1993). During the day, most of the O_2 produced is recycled within the mat by O_2 respiration and some sulfide oxidation. At night, O_2 is consumed principally by sulfide oxidation near the mat–water interface. Sulfate reduction is the principal source of DIC at night. Despite the abundant *Chloroflexus* filaments visible microscopically at the oxygen–sulfide interface, anoxygenic photosynthesis accounts for less than 10% of the total carbon fixation rate.

Table I. Diel O_2 and DIC Porewater Budget of a Subtidal Cyanobacterial Mat during April 1990[a]

	Flux[b] (mmoles m^{-2})			
	O_2		DIC	
Process	Day	Night	Day	Night
17 °C				
Oxygenic photosynthesis	+75	0	−76	0
Anoxygenic photsynthesis			−6	0
Oxic respiration	−53	c	+54	c
Sulfate reduction			+11	+14
Sulfide oxidation	−7	d		e
Porewater gain (+) or loss (−) due to exchange with overlying water	−15	+14	+35	−32
30 °C				
Oxygenic photosynthesis	+235	0	−236	0
Anoxygenic photsynthesis			−23	0
Oxic respiration	−155		+158	c
Sulfate reduction			+45	+60
Sulfide oxidation	−25	d		e
Porewater gain (+) or loss (−) due to exchange with overlying water	−55	+60	+98	−118

[a]Data from Canfield and Des Marais (1993).
[b]Positive values denote addition of O_2 or DIC to mat porewater.
[c]Not measured, but assumed to equal zero (Canfield and Des Marais, 1993).
[d]Not measured, but assumed to be responsible for virtually all of the O_2 consumed during the night (Canfield and Des Marais, 1993).
[e]Chemoautotrophy associated with sulfide oxidation by Beggiatoa removes a small amount of DIC from the porewater (Des Marais and Canfield, 1994) but this process was not quantified during this particular experiment.

A careful comparison of the relative O_2 and DIC fluxes across the mat–water interface reveals that, during the day, more DIC diffuses into the mat than O_2 diffuses out (Canfield and Des Marais, 1993). At night, more DIC diffuses out of the mat than O_2 diffuses in. However, the net O_2 and DIC fluxes balance over the full 24-hr cycle. This budget indicates that, during the day, carbon with an oxidation state greater than zero is incorporated into the mat, and carbon having a similarly high oxidation state is liberated by the mat at night. The chemical nature of this carbon is presently unknown.

Several general observations can be made regarding the cycling of carbon, oxygen, and sulfur. Although all of the key processes (Table I) are strongly influenced by temperature, their rates scale with temperature by roughly the same amount (Canfield and Des Marais, 1993). Over a 24-hr period, the net effect of these very high metabolic rates is that the net accumulation of carbon is low. Apparently this mat is a closely coupled system where high rates of photosynthetic carbon fixation fuel high rates of carbon oxidation. This efficient carbon oxidation is, in turn, required to maintain high rates of photosynthesis, because the high rates of oxidation regenerate the nutrients necessary for primary production.

6.2. Nitrogen

Direct measurements of nitrogen transformations and fluxes have been performed on these subtidal mats in parallel with measurements of the carbon, oxygen, and sulfur fluxes (e.g., Bebout et al., 1994; Canfield and Des Marais, 1994). Potentially key nitrogen sources for primary production include N_2 fixation, uptake of dissolved inorganic nitrogen (DIN) both from the overlying brine and from bacterial regeneration from deeper in the mat, dissolved organic nitrogen (DON) from the overlying brine or from bacterial sources, and recycling of recently produced nitrogen within the photic zone itself.

Observations made over the diel cycle illustrate several key aspects of the nitrogen budget. Rates of N_2 fixation are variable, with highest rates occurring at night (Bebout et al., 1994). Because the nitogenase enzyme is inhibited by O_2, the mat bacteria accomplish nitrogen fixation at night when no O_2 is being produced photosynthetically within the mat (Paerl et al., 1994). However, in these subtidal mats, the N_2 fixation rates are substantially less than those needed to sustain the observed rates of primary production. Bebout et al. (1994) observed uptake of DIN, NH_4^+ in particular, from the overlying brine. Large DIN (NH_4^+) concentrations exist below the photic zone; transport of this DIN to the photic zone could supply a major component of the fixed nitrogen required for primary production. Thus apparently the nitrogen demand of this actively photosynthesizing mat is met principally by the recycling of previously produced biomass. Relatively little net incorporation of nitrogen is observed in these subtidal mats.

These mats not only regenerate their own nitrogen, they also are quite efficient at retaining it (Canfield and Des Marais, 1994). Fluxes of NH_4^+, which is the principal form of dissolved nitrogen in mat porewaters, were monitored over a 24-hr period. Essentially no NH_4^+ was lost by the mat at any time during the 24-hr interval, even when regenerated DIC escapes into the overlying brine at night.

6.3. Carbon Isotopes

The distribution of stable carbon isotopes among various reservoirs of inorganic and organic carbon can offer useful insights regarding the structure and function of an ecosystem. The isotopic composition of organic matter in a microbial community is controlled by those enzymes that create, alter, and degrade organic compounds.

This isotopic control is modulated by fluxes of carbon, both within the community and between the community and its environment (Fig. 7) (Des Marais and Canfield, 1994). For example, if two processes competing for carbon at a branch point differ in their preference for ^{12}C over ^{13}C, then the $^{13}C/^{12}C$ value of their respective products will also be different (Monson and Hayes, 1980). In one case (see branch point A in Fig. 7), DIC diffusing in from the overlying brine is either fixed by the enzyme ribulose 1,5-bisphosphate carboxylase/oxygenase (RUBISCO) or it diffuses back out. Because RUBISCO discriminates strongly against ^{13}C (Roeske and O'Leary, 1984), the potential exists for the mat organic carbon to attain a substantially lower $^{12}C/^{13}C$ value, relative to the DIC in the overlying brine. However, essentially no isotopic discrimination was observed to accompany photosynthesis in these subtidal mats (Des Marais and Canfield, 1994). This is because virtually all of the DIC entering the mat was fixed; thus,

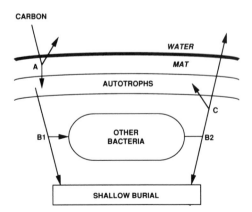

Figure 7. Pathways for inorganic and organic carbon in a microbial mat. Letters designate branch points in this flow and are discussed in the text. (Adapted from Des Marais *et al.*, 1992.)

no DIC flux existed to return unassimilated ^{13}C to the overlying brine. Thus point A was not operating as a true branch point; the rate of DIC uptake was limited by its rate of diffusion into the mat.

Subsequent transformations of organic carbon are achieved by various bacteria (see B1 and B2, Fig. 7). However, analyses of DOC and particulate organic carbon at various stages of diagenesis indicate that carbon isotopic discrimination is minimal in these subtidal mats (Des Marais and Canfield, 1994).

An additional opportunity for discrimination occurs at night, because part of the DIC stream that diffuses back to the overlying brine is assimilated by sulfide-dependent chemoautotrophs (point C, Fig. 7). Because sulfide-utilizing chemoautotrophs such as Beggiatoa employ RUBISCO, which is isotopically selective during DIC uptake, and because a significant fraction of the DIC indeed escapes into the overlying brine, a true branch point exists (point C, Fig. 7). Isotopic discrimination was indeed observed at night (Des Marais and Canfield, 1994). This discrimination is sufficient to explain the depletion observed in the $^{13}C/^{12}C$ value of the mat's organic matter, relative to the $^{13}C/^{12}C$ value of the DIC in the overlying brine. Thus the carbon isotopic composition of these subtidal mats derives from complex interactions between the carbon, oxygen, and sulfur cycles.

7. Diagenesis at Greater Burial Depths

7.1. Bacterial Activity

These subtidal mats typically achieve thicknesses of 5 cm or more, even though their photic zones are typically only 1 to 2 mm deep. Bacterial populations decline dramatically at depths below 2 mm, as evidenced by lower concentrations of total RNA content at depth in the mat (Risatti et al., 1994). Consistent with this trend, the populations of cyanobacteria and other bacteria observable in the light microscope decline markedly with depth in the permanently dark, anaerobic zone (D'Amelio et al., 1989). Occasional bundles of viable M. chthonoplastes filaments and numerous empty sheaths are the most abundant features observed below 2 mm. Organic decomposition reactions such as fermentation and bacterial sulfate reduction (Fig. 8) (Canfield and Des Marais, 1993) are perhaps the most important processes occurring at these depths. This sulfate reduction is sustained both by high sulfate concentrations in the hypersaline porewaters and by DOC, which either diffused down from the shallow, more active region of the mat (e.g., Jørgensen and Cohen, 1977) or was generated in situ by the decomposition of particulate organic matter.

A recent study of bacterial RNA indicated that specific populations of sulfate-reducing bacteria (SRB) are restricted to well-defined depth intervals within the mat community (Risatti et al., 1994). Desulfococcus SRB occur

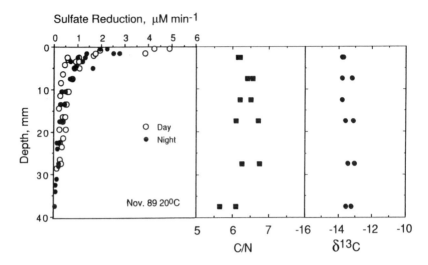

Figure 8. Trends with depth in the subtidal mat from pond 5, depicting some of the effects of organic diagenesis, as follows: (a) sulfate reduction rates (Canfield and Des Marais, 1993); (b) elemental carbon-to-nitrogen ratio of total organic carbon (Des Marais *et al.*, 1992); (c) isotopic composition of total organic carbon, where $\delta^{13}C = [(^{13}C/^{12}C_{sample}/(^{13}C/^{12}C_{standard} - 1] 1000$. The carbon isotope standard is the Peedee Belemnite carbonate (Des Marais *et al.*, 1992). Despite detectable rates of organic decomposition via sulfate reduction at depth in the mat, the parameters C/N and $\delta^{13}C$ still retain the values created by the photosynthetic community situated near the mat surface.

principally in the uppermost layers of the mat, including the photic zone. The *Desulfococcus* species apparently oxidize a variety of substrates, including lactate, ethanol, acetate (more slowly), and fatty acids (ranging from 3 to 16 carbon atoms). Thus *Desulfococcus* appears to be a nutritional generalist, a role consistent with its proximity to the dense bacterial populations and diverse organic substrates available within the mat's photic zone. Beneath the photic zone, sulfate reduction rates decline and the SRB population appears to be dominated by *Desulfovibrio* spp., which incompletely oxidize a limited array of organic substrates, typically lactate and ethanol, to produce acetate. The dominance of *Desulfovibrio* probably reflects its role as a consumer of substrates diffusing down from the shallowest mat layers with their photosynthesizing primary producers. At greater depths even further removed from the photic zone, the acetate-oxidizing *Desulfobacter* and *Desulfobacterium* become dominant. This depth-related pattern is consistent with the fact that acetate is situated further along the food chain from the substrates utilized by *Desulfovibrio*. Thus the phylogenetic structure of the mat's SRB populations appears to be consistent with depth-related trends in the availability of various organic substrates.

7.2. Survival of Organic Matter Produced in the Photic Zone

The composition of organic matter preserved at depth reflects that fraction of the organics from the surface photosynthetic community that survived diagenesis, plus those organics that were synthesized at depth, either by surviving photosynthesizers (e.g., Jørgensen and Cohen, 1987) or by anaerobic bacteria. Many photosynthetic carotenoid pigments survive well below the photic zone of those subtidal mats (Palmisano *et al.*, 1989) to depths corresponding to a sediment age greater than 8 years. However, the carotenoid/chlorophyll abundance ratio increases with depth, indicating that substantial decomposition of Chl *a* occurs over a timescale of months. The relative concentrations of the carotenoids generally reflect the relative abundances of their source organisms. For example, the carotenoid myxoxanthophyll is relatively abundant, and thus reflects the dominance of cyanobacteria.

Both the elemental C/N and $^{13}C/^{12}C$ values remain constant with depth (Fig. 8) (Des Marais *et al.*, 1992). The C/N value closely approximates the Redfield ratio, the ratio observed for organic matter in living marine plankton. These depth trends are consistent with light microscopic observations that reveal that major components of the mat's surface community (cyanobacteria and sheath material) survive at depth and retain the chemical and isotopic signatures of the surface community.

8. Future Research

Microbial mats offer more opportunities to increase our knowledge of microbial ecology than can possibly be summarized here. The work accomplished thus far with the mats in the Guerrero Negro salt ponds has created a useful overview of this enormously complex mat community. However, it is really just a prelude to achieving the central task of identifying fully the membership in that community as well as understanding how it functions with such extraordinary efficiency. A few examples of possible ecological investigations are given below.

Many of the studies described in this chapter have been performed only on subtidal hypersaline cyanobacterial mats dominated by *M. chthonoplastes*. Although intertidal and supratidal cyanobacterial mats have also received some attention, the level of effort summarized here must continue to be applied to other mat types. Examples of such mats include those dominated by unicellular cyanobacteria, eukaryotic algae, and nonphotosynthetic bacteria such as sulfide-oxidizing bacteria in deep-sea communities. Mats growing in low-sulfate environments such as lakes, streams, and thermal springs also merit more attention.

Studies of biogeochemical cycling in mats should be broadened to include additional populations of bacteria (e.g., heterotrophs, methanogens, and novel bacteria) that probably contribute substantially to the community. We must understand better how key nutrients such as nitrogen and phosphorus are regenerated and retained by the various types of mats. Mats that coexist with active mineral precipitation merit study in order to understand both the role of the microbes in the precipitation process as well as the impact of mineralization on mat biogeochemistry.

Because microenvironments within the mat sustain diverse populations of bacteria, the search for novel bacterial strains should continue. The observation of extraordinary O_2 tolerances among purple sulfur bacteria (D'Amelio et al., 1987) and SRBs (e.g., Canfield and Des Marais, 1991) demonstrates that fundamentally important principals of microbial physiology and ecology remain to be fully elucidated.

Perhaps most fundamental will be the use of genetic information to understand the ecology of microbial mat communities. Traditional phylogenetic methods based on morphology and physiology are not particularly effective when applied to microbial mats. However, methods for the identification and interpretation of 16s RNA and other macromolecules are improving rapidly and hold great promise. These phylogenic investigations should be done in combination with biogeochemical studies of naturally occurring microbial mats.

Finally, it is essential that consortia of specialists coordinate their efforts on a few "benchmark" microbial mats. Only through the synthesis of multiple perspectives on individual mats will we begin to comprehend the immense complexity and sophistication of these extraordinary, ancient ecosystems.

ACKNOWLEDGMENTS. The research summarized herein was supported by grants from NASA's Exobiology Program. I thank the continuing field support by Exportadora de Sal, S. A.

References

Bauer, J. E., Haddad, R. I., and Des Marais, D. J., 1991, Method for determining stable isotope ratios of dissolved organic carbon in interstitial and other natural marine waters, *Mar. Chem.* **33**:335–351.

Bebout, B. M., 1992, *Interactions of Nitrogen and Carbon Cycling in Microbial Mats and Stromatolites,* University of North Carolina at Chapel Hill, Ph.D. dissertation.

Bebout, B. M., Paerl, H. W., Bauer, J. E., Canfield, D. E., and Des Marais, D. J., 1994, Nitrogen cycling in microbial mat communities: The quantitative importance of N-fixation and other sources for N for primary productivity, in: *Microbial Mats: Structure, Development and Environmental Significance,* NATO ASI Series G, Vol. 35 (L. Stal and P. Caumette, eds.), Springer-Verlag, Heidelberg, pp. 265–271.

Boudreau, B. P., and Guinasso, N. L., 1982, The influence of a diffusive sublayer on accretion, dissolution, and diagenesis at the sea floor, in: *The Dynamic Environment of the Ocean Floor* (K. A. Fanning and F. T. Manheim, eds.), Lexington Books, Lexington, Massachusetts, pp. 115–145.

Cadée, G. C., and Hegeman, J., 1974, Primary production of the benthic microflora living on tidal flats in the Dutch Wadden Sea, *Neth. J. Sea Res.* **8**:260–291.

Canfield, D. E., and Des Marais, D. J., 1991, Aerobic sulfate reduction in microbial mats, *Science* **251:**1471–1473.

Canfield, D. E., and Des Marais, D. J., 1993, Biogeochemical cycles of carbon, sulfur, and free oxygen in a microbial mat, *Geochim. Cosmochim. Acta* **57:**3971–3984.

Canfield, D. E., and Des Marais, D. J., 1994, Cycling of carbon, sulfur, oxygen and nutrients in a microbial mat, in: *Microbial Mats: Structure, Development and Environmental Significance,* NATO ASI Series G, Vol. 35 (L. J. Stal and P. Caumette, eds.), Springer-Verlag, Heidelberg, pp. 255–263.

D'Amelio, E. D., Cohen, Y., and Des Marais, D. J., 1987, Association of a new type of gliding, filamentous, purple phototrophic bacterium inside bundles of *Microcoleus chthonoplastes* in hypersaline cyanobacterial mats, *Arch. Microbiol.* **147:**213–220.

D'Amelio, E. D., Cohen, Y., and Des Marais, D. J., 1989, Comparative functional ultrastructure of two hypersaline submerged cyanobacterial mats: Guerrero Negro, Baja California Sur, Mexico, and Solar Lake, Sinai, Egypt, in: *Microbial Mats, Physiological Ecology of Benthic Microbial Communities* (Y. Cohen and E. Rosenberg, eds.), American Society for Microbiology, Washington, D.C., pp. 97–113.

Des Marais, D. J., and Canfield, D. E., 1994, The carbon isotope biogeochemistry of microbial mats, in: *Microbial Mats: Structure, Development and Environmental Significance,* NATO ASI Series G, Vol. 35 (L. J. Stal and P. Caumette, eds.), Springer-Verlag, Heidelberg, pp. 289–298.

Des Marais, D. J., Cohen, Y., Nguyen, H., Cheatham, M., Cheatham, T., and Munoz, E., 1989, Carbon isotopic trends in the hypersaline ponds and microbial mats at Guerrero Negro, Baja California Sur, Mexico: Implications for Precambrian stromatolites, in: *Microbial Mats: Physiological Ecology of Benthic Microbial Communities* (Y. Cohen and E. Rosenberg, eds.), American Society for Microbiology, Washington, D.C., pp. 191–205.

Des Marais, D. J., D'Amelio, E., Farmer, J. D., Jorgensen, B. B., Palmisano, A. C., and Pierson, B. K., 1992, Case study of a modern microbial mat-building community: The submerged cyanobacterial mats of Guerrero Negro, Baja California Sur, Mexico, in: *The Proterozoic Biosphere: A Multidisciplinary Study* (W. J. Schopf and C. Klein, eds.), Cambridge University Press, New York, pp. 325–334.

Farmer, J. D. 1992, Grazing and bioturbation in modern microbial mats, in: *The Proterozoic Biosphere: A Multidisciplinary Study* (J. W. Schopf and C. Klein, eds.), Cambridge University Press, New York, pp. 295–297.

Farmer, J. D., and Richardson, L. L., 1988, Origin of microfabric in laminated microbial mats: Implications for interpreting stromatolites, *EOS; Transactions of the American Geophysical Union* **69:**1131.

Garcia-Pichel, F., Mechling, M., and Castenholz, R. W., 1994, Diel migrations of microorganisms within a benthic, hypersaline mat community, *Appl. Environ. Microbiol.* **60:**1500–1511.

Gerdes, G., and Krumbein, W. E., 1987, Faunal influence on the biolaminated deposits, in: *Lecture Notes in Earth Sciences,* Vol. 9 (S. Bhattacharji, G. M. Friedman, H. J. Neugebauer, and A. Seilacher, eds.), Springer-Verlag, Berlin, pp. 55–69.

Grøntved, J., 1960, On the productivity of micro-benthos and phytoplankton in some Danish Fjords, *Medd. Dan. Fisk. Havunders. New Ser.* **3:**55–91.

Javor, B., 1983a, Nutrients and ecology of the Western Salt and Exportadora de Sal Saltern brines, in: *Sixth International Symposium on Salt* (B. C. Schreiber and H. L. Harner, eds.), Salt Institute, Alexandria, Va., pp. 195–205.

Javor, B. J., 1983b, Planktonic standing crop and nutrients in a saltern ecosystem, *Limnol. Oceanogr.* **28:**153–159.

Javor, B. J., and Castenholz, R. W., 1984, Invertebrate grazers of microbial mats, Laguna Guerrero Negro, Mexico, in: *Microbial Mats: Stromatolites* (Y. Cohen, R. W. Castenholtz and H. O. Halvorson, eds.), Alan R. Liss, New York, pp. 85–94.

Jørgensen, B. B., 1994, Diffusion processes and boundary layers in microbial mats, in: *Microbial*

Mats: Structure, Development and Environmental Significance, NATO ASI Series G, Vol. 35 (L. Stal and P. Caumette, eds.), Springer-Verlag, Heidelberg, pp. 243–253.

Jørgensen, B. B., and Cohen, Y., 1977, Solar Lake (Sinai). 5. The sulfur cycle of the benthic cyanobacterial mats, *Limnol. Oceanogr.* **22**:657–666.

Jørgensen, B. B., and Cohen, Y., 1987, Photosynthetic potential and light-dependent oxygen consumption in a benthic cyanobacterial mat, *Appl. Environ. Microbiol.* **54**:176–182.

Jørgensen, B. B., and Des Marais, D. J., 1986, Competition for sulfide among colorless and purple sulfur bacteria in cyanobacterial mats, *FEMS Microbiol. Ecol.* **38**:179–186.

Jørgensen, B. B., and Des Marais, D. J., 1988, Optical properties of benthis photosynthetic communities: Fiber optic studies of cyanobacterial mats, *Limnol. Oceanogr.* **33**:99–113.

Jørgensen, B. B., and Des Marais, D. J., 1990, The diffusive boundary layer of sediments: Oxygen microgradients over a microbial mat, *Limnol. Oceanogr.* **35**:1343–1355.

Jørgensen, B. B., Cohen, Y., and Des Marais, D. J., 1987, Photosynthetic action spectra and adaptation to spectral light distribution of a benthic cyanobacterial mat, *Appl. Environ. Microbiol.* **53**:879–886.

Monson, K. D., and Hayes, J. M., 1980, Biosynthetic control of the natural abundance of carbon 13 at specific positions within fatty acids in *Escherichia coli,* evidence regarding the coupling of fatty acid and phospholipid synthesis, *J. Biol. Chem.* **255**:11435–11441.

Paerl, H. W., Bebout, B. M., Joye, S. B., and Des Marais, D. J., 1993, Microscale characterization of dissolved organic matter production and uptake in marine microbial mat communities, *Limnol. Oceanogr.* **38**:1150–1161.

Paerl, H. W., Bebout, B. M., Currin, C. A., Fitzpatrick, M. W., and Pinckney, J. L., 1994, Nitrogen fixation dynamics in microbial mats, in: *Microbial Mats: Structure, Development and Environmental Significance,* NATO ASI Series G, Vol. 35 (L. Stal and P. Caumette, eds.), Springer-Verlag, Heidelberg, pp. 325–337.

Palmisano, A. C., Cronin, S. E., D'Amelio, E. D., Munoz, E., and Des Marais, D. J., 1989, Distribution and survival of lipophilic pigments in a laminated microbial mat community near Guerrero Negro, Mexico, in: *Microbial Mats: Physiological Ecology of Benthic Microbial Communities,* (Y. Cohen and E. Rosenberg, eds.), American Society for Microbiology, Washington, D.C., pp. 138–152.

Pierson, B. K., Valdez, D., Larsen, M., Morgan, E., and Mack, E. E., 1994, *Chloroflexus*-like organisms from marine and hypersaline environments: Distribution and diversity, *Photosynth. Res.* **41**:35–52.

Revsbech, N. P., Jørgensen, B. B., and Blackburn, T. H., 1983, Microelectrode studies of the photosynthesis and O$_2$, H$_2$S and pH profiles of a microbial mat, *Limnol. Oceanogr.* **28**:1062–1074.

Risatti, J. B., Capman, W. C., and Stahl, D. A., 1994, Community structure of a microbial mat: The phylogenetic dimension, *Proc. Natl. Acad. Sci. USA* **91**:10173–10177.

Roeske, C. A., and O'Leary, M., 1984, Carbon isotope effects on the enzyme-catalyzed carboxylation of ribulose bisphosphate, *Biochemistry* **23**:6275–6284.

Walter, M. R. (ed.), 1976, *Stromatolites,* Elsevier, Amsterdam.

Walter, M. R., and Heys, G. R., 1985, Links between the rise of Metazoa and the decline of stromatolites, *Precambrian Res.* **29**:149–174.

Walter, M. R., Grotzinger, J. P., and Schopf, J. W., 1992, Proterozoic stromatolites, in: *The Proterozoic Biosphere: A Multidisciplinary Study* (J. W. Schopf and C. Klein, eds.), Cambridge University Press, New York, pp. 253–260.

7

Activity of Chemolithotrophic Nitrifying Bacteria under Stress in Natural Soils

H. J. LAANBROEK and J. W. WOLDENDORP

1. Introduction

Nitrification is an important process in the biogeochemical cycle of nitrogen, linking its reduced and oxidized parts. Since the conversion of ammonium to nitrate has a great impact on the environment, such as weathering of soils, production of greenhouse gases, and eutrophication of surface and ground waters (Van Breemen and Van Dijk, 1988), it is important to know the characteristics of the responsible organisms. Although many organotrophic microorganisms are able to produce oxidized nitrogenous compounds such as nitrite and nitrate (Focht and Verstraete, 1977; Killham, 1987; Kuenen and Robertson, 1987), chemolithotrophic nitrifying bacteria are considered to be the most important group producing these compounds from ammonia. A contribution to nitrate production by organotrophic microorganisms has only been observed in some acid coniferous forest soils (Schimel et al., 1984; Killham, 1986, 1990). According to *Bergey's Manual of Systematic Bacteriology* (Watson et al., 1989), the family of nitrifying bacteria is a diverse group of rods, vibrios, cocci, and spirilla, all having the ability to utilize ammonia or nitrite as a major source of energy and carbon dioxide as the chief source of carbon. With the exception of *Nitrobacter* species, all others are obligate chemolithotrophs, but some can grow mixotrophically on a mixture of CO_2 and small organic compounds. All strains are aerobic, but some may proliferate at low oxygen concentrations. Some *Nitrobacter* species might even grow at the expense of nitrate reduction in the absence of oxygen. Another important characteristic, especially of the chemolithotrophic

H. J. LAANBROEK • Centre for Limnology, Netherlands Institute of Ecology, 3631 AC Nieuwersluis, The Netherlands.   J. W. WOLDENDORP • Department of Soil Biology, Centre for Terrestrial Ecology, Netherlands Institute of Ecology, 6666 ZG Heteren, The Netherlands.
Advances in Microbial Ecology, Volume 14, edited by J. Gwynfryn Jones. Plenum Press, New York, 1995.

ammonia-oxidizing bacteria that have been isolated, is their sensitivity to low pH. The limited metabolic abilities of the chemolithotrophic nitrifying microorganisms will obviously restrict the number of suitable habitats. Nitrifying bacteria are commonly found in oxic environments with a neutral or slightly alkaline pH where ammonium is produced from organic matter by mineralization. However, chemolithotrophic ammonia-oxidizing bacteria are also observed in extreme environments such as acid tea soils (Walker and Wickramsinghe, 1979), sandstones of historic monuments (Meincke et al., 1989; Spieck et al., 1992), oceanic oxygen minimum zones (Ward, 1986), and the anaerobic hypolimnion of domestic wastewater reservoirs (Abeliovich, 1987).

Although sharing a specific and unique ability, i.e., the oxidation of ammonia without the necessity of organic compounds, the chemolithotrophic ammonia-oxidizing bacteria represent a phylogenetically diverse group of microorganisms belonging to different subdivisions of the Proteobacteria (Woese et al., 1984, 1985). Notwithstanding their highly different morphologies, Nitrosospira, Nitrosolobus, and Nitrosovibrio seem to be closely related phylogenetically on the basis of 16s rRNS analysis (Head et al., 1993). According to Macdonald (1986), Nitrosolobus, not the well-known Nitrosomonas, seems to be the most common ammonia-oxidizing genus in soils. In a stream receiving geothermal inputs of ammonium, Nitrosospira strains outnumbered the Nitrosomonas strains (Cooper, 1983). In spite of this, most experiments have been done with the latter genus. Hence, care must be taken with extrapolating ecophysiological knowledge of Nitrosomonas species to the in situ behavior of the whole ammonia-oxidizing community. With respect to the oxidation of nitrite by chemolithotrophic bacteria, the well-known and best-studied Nitrobacter is still regarded as the most dominant genus in soil.

The activity of chemolithotrophic nitrifying bacteria in natural soils characterized by conditions that might be regarded as marginal for these bacteria, i.e., acid heathland and forest soils as well as oxygen or ammonium-limited grassland soils, has been studied intensively by our department. The results of these studies will be discussed from an ecological perspective. Recently, the physiological characteristics of the chemolithotrophic nitrifying bacteria have been reviewed by Prosser (1989) and by Bock, Harms, Koops, and co-workers (Bock and Koops, 1992; Bock et al., 1992; Koops and Möller, 1992). In a former paper, we related the activity of nitrifying bacteria to the nitrogen nutrition of plants in natural ecosystems (Woldendorp and Laanbroek, 1989).

2. Nitrification at Low pH

2.1. Pure Culture Studies

In Nitrosomonas europaea, ammonia rather than ammonium appears to be the real substrate of ammonia mono-oxygenase, which is the first enzyme in volved in its chemolithotrophic energy generation (Suzuki *et al.*, 1974). In addition, bacterial membranes are permeable to ammonia but not to ammonium (Kleiner, 1985). However, this latter aspect might be of less importance when the ammonia mono-oxygenase is situated at the outside of the bacterial cell membrane as is hypothesized by Bock *et al.* (1992). The obligate use of ammonia and not of ammonium may explain why the activity of the chemolithotrophic ammonia-oxidizing bacteria is restricted to circumneutral conditions, where ammonia is more stable than under acid conditions ($pK_{NH3/NH4OH}$ = 9.25). The oxidation of hydroxylamine, the product of the enzyme ammonia mono-oxygenase, appeared to be less acid-sensitive in *N. europaea* strain ATCC 19178 compared to that of ammonia oxidation (Frijlink *et al.*, 1992a). Even at pH 5.0, a proton gradient across the cell membrane was produced during the oxidation of hydroxylamine, which led to ATP synthesis and energy-dependent transport of amino acids. The fact that hydroxylamine and not ammonia was oxidized at pH 5.0 suggests that ammonia hydroxylation is the predominant acid-sensitive process in *N. europaea*. Since *Nitrosomonas* cells failed to grow on hydroxylamine (Frijlink *et al.*, 1992a), as was also observed by others (compare Bock *et al.*, 1992), growth of these strictly chemolithotrophic organisms in acid environments is restricted to ammonia as a substrate. Notwithstanding the problem of growth on ammonia at low pH, ammonia-oxidizing bacteria have been isolated from acid soils (Allison and Prosser, 1991; De Boer *et al.*, 1989a; Hankinson and Schmidt, 1984; Martikainen and Nurmiaho-Lassila, 1985; Walker and Wickramasinghe, 1979). All of these strains turned out to be acid-sensitive. Therefore, ammonia oxidation by this type of acid-sensitive chemolithotrophic bacteria in acid soils is supposed to be restricted to microhabitats of circumneutral pH. In contrast, acid-tolerant or even acidophilic nitrite-oxidizing *Nitrobacter* strains have been isolated from acid soils (De Boer *et al.*, 1989a; Hankinson and Schmidt, 1988). Hence, the oxidation of ammonia in acid soils seems to be limited by the characteristics of the ammonia-oxidizing bacteria and accumulation of nitrite is hardly to be expected due to the acid-tolerance of the nitrite-oxidizing cells.

Besides the occurrence of nitrification in microhabitats of circumneutral pH, ammonia oxidation at low pH might also be stimulated by other factors such as the presence of surfaces or the hydrolysis of urea. A strain of *N. europaea*, which had a pH minimum of 7.0 for growth in liquid batch culture, was able to produce nitrite at a pH of 6.0 when attached to particles in continuous flow sand columns (Allison and Prosser, 1993). A mixed culture of *Nitrosospira* and *Nitrobacter*, both isolated from a fertilized, acid heathland soil, was only able to oxidize ammonia to nitrate at neutral pH (De Boer and Laanbroek, 1989). Nitrification by this mixed culture stopped at pH 5.5, unless urea was added. In the presence of urea, nitrate formation from ammonia was possible at a constantly low pH of 4.5 (Fig. 1). However, ammonia oxidation stopped as soon as urea was depleted. At a pH maintained at 5.0, the production of nitrate continued for another week after the urea was exhausted. Hence, the ratio between ammonium

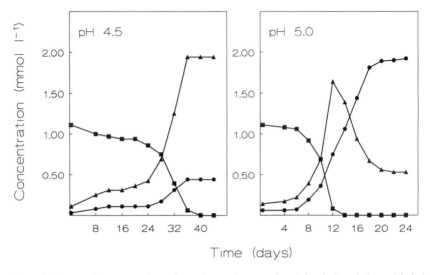

Figure 1. Changes in concentrations of urea (squares), ammonium (triangles), and nitrate (circles) during the incubation of mixed cultures of *Nitrosospira* strain AHB1 and *Nitrobacter* strain NHB1, at pH 4.5 and pH 5.0. Both strains had been isolated from a fertilized heathland soil. (After De Boer and Laanbroek, 1989.)

and nitrate produced at the end of the incubation period was dependent on the ambient pH.

The mechanism by which urea activates the cells is not yet understood. In contrast to ammonium, urea may provide the membrane-bound enzyme ammonia mono-oxygenase with substrate at a low pH. The enzyme urease in *N. europaea* is assumed to be located inside the bacterial cells (W. De Boer, personal communication). The ammonia produced by this enzyme might diffuse across the bacterial membrane where it is (partly) captured by the periplasmatic enzyme ammonia mono-oxygenase. The occurrence of chemodenitrification of nitrite at low pH as described by Van Cleemput and Baert (1984) could be excluded by the complete recovery of added urea in the form of ammonium and nitrate. The results with the mixed culture of the isolated *Nitrosospira* and *Nitrobacter* strains also underscored the acid-tolerant nature of the isolated nitrite-oxidizing strain. The ability to hydrolyse urea is quite common among the chemolithotrophic ammonia-oxidizing bacteria (Koops and Möller, 1992). Most of the *Nitrosomonas* and *Nitrosospira* strains isolated from acid soils in Scotland, appeared to be also ureolytic as well as acid-sensitive (Allison and Prosser, 1991).

2.2. Nitrification in Acid Heathland and Forest Soils

In Dutch acid heathland and forest soils characterized by atmospheric nitrogen inputs of 40 kg N ha^{-1} year^{-1} and more, nitrification is widespread and apparently chemolithotrophic as was indicated by inhibition experiments with acetylene (De Boer *et al.*, 1988, 1990, 1992, 1993; Martikainen and De Boer, 1993; Stams *et al.*, 1990; Tietema *et al.*, 1992; Troelstra *et al.*, 1990). Two types of chemolithotrophic nitrification could be distinguished in these soils: An acid-sensitive type, as defined by the absence of nitrate production at initial pH values below 6.0, and an acid-tolerant type, characterized by ammonium oxidation that started at pH values as low as pH 4.0 (De Boer *et al.*, 1989b, 1990). Enumerations of acid-tolerant ammonia-oxidizing bacteria by a most probable number (MPN) procedure, using mineral medium with ammonium as the sole energy substrate, only yielded numbers at pH 4.5 or 6.0, but not at pH 7.5. The acid-sensitive, but not the acid-tolerant type, could be stimulated by increasing the pH from 4.0 to 6.0 (De Boer *et al.*, 1989b). Nitrification in suspensions of degrading needles or the litter layer from a Douglas fir soil at pH 6.0 was several times higher than at pH 4.0, whereas pH had a variable effect on nitrification in suspensions of the fermentation layers (Fig. 2) (De Boer *et al.*, 1992). No effect on nitrification was observed in suspensions of the humus layers. In accordance with this result, MPN enumerations of (acid-sensitive) ammonia-oxidizing bacteria at pH 7.5 yielded larger numbers in the litter layer. This indicated a niche differentiation between acid-sensitive and acid-tolerant bacteria in the different forest soil layers due to environmental conditions. During active nitrate produc-

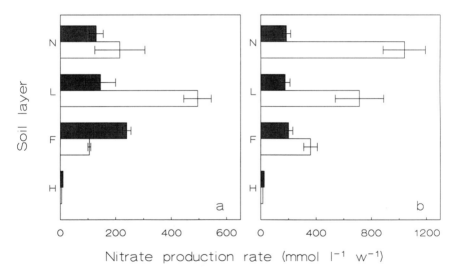

Figure 2. Nitrification rates in suspensions of overground degrading needles (N) and different soil layers of a Douglas fir forest soil: litter layer (L), fermentation layer (F), humus layer (H). Samples collected in (a) May 1989 and (b) January 1990 were incubated for 2 weeks at 20 °C and pH4 (open bars) or pH6 (filled bars). No suspensions of the mineral layer were incubated in January 1990. Data represent the mean of two replicates. (Data from De Boer *et al.*, 1992.)

tion in suspensions of heathland humus at constant pH 4.0, numbers of nitrite-oxidizing bacteria increased, whereas the MPN numbers of the ammonium-oxidizing cells determined at pH 6.0 remained at the same low level (De Boer et al., 1989b).

In accordance with the observations in Dutch ammonium-saturated forest soils, a different behavior of nitrifying bacteria with respect to pH could be observed on a pine forest site in Finland receiving high ammonium deposition rates (Martikainen et al., 1993): high nitrification potentials in the litter layer at pH 6.0 compared to pH 4.0, but no pH effect in the organic and upper mineral layers. In aerobic suspensions of the upper 5 cm of the soil profile of a drained, Finnish mire, the nitrification rate was higher at pH 6.0 than at pH 4.0 (Lang et al., 1993). However, deeper in the soil profile (5–25 cm), nitrification was greater at pH 4.0 than at pH 6.0. The effect of acetylene inhibition on nitrification was also dependent on the sampling layer as well as on pH. In the upper layer, nitrification was almost completely inhibited at pH 6.0, but practically not at pH 4.0. In the deeper layers, acetylene inhibited nitrification at pH 4.0 but not at pH 6.0. Hence, it could be concluded that in each layer the most dominant type of nitrification was autotrophic.

Several attempts to isolate the bacteria responsible for ammonia oxidation at pH 4.0 failed. Experiments with suspensions of the organic layer of a Douglas fir forest soil showed that filtration of the suspension stopped chemolithotrophic nitrification at pH 4.0 but not at pH 6.0 (De Boer et al., 1991). This suggests that the acid-tolerant ammonia oxidation was dependent on the presence of particles or aggregates. From electron microscopy, it appeared that small aggregates consisted of Nitrosospira-like cells embedded in an unknown biopolymer. In the larger aggregates, Nitrobacter-like bacteria, characterized by their peripheral cytomembrane systems, could be recognized at the outer side of the aggregates. The function of aggregate formation is not yet known. It is obvious that it does not protect the cells against low pH, since the pH in the aggregates will be even lower than in the surrounding medium due to the proton production by the ammonia-oxidizing bacteria. Protection of the cells against oxygen or nitric acid and concomitantly nitric oxide was not found (W. De Boer, personal communication). Cells of an acid-sensitive Nitrosospira strain isolated from acid heathland soil immobilized in alginate were also able to nitrify at a fixed pH value of 4.0 (De Boer et al., 1995). Daily fluctuations between pH 6.0 and 4.0 of an actively nitrifying mixed culture of the acid-sensitive Nitrosospira strain and an acid-tolerant Nitrobacter strain induced adaptation of single cells to be active at constant pH 4.0. Continuation of ammonia oxidation by these adapted cells after subculturing in fresh medium at pH 4.0 was dependent on the degree of dilution, since a high density of cells was apparently needed for a successful ammonia oxidation at low pH. These latter results suggest that acid-tolerant ammonium-oxidizing bacteria might be in fact adapted acid-sensitive cells.

3. Nitrification at Low Oxygen Concentrations

3.1. Pure Culture Studies

The first enzyme of the machinery involved in the oxidation of ammonia to nitrite by *N. europaea*, the copper-containing ammonia mono-oxygenase, is dependent on oxygen (Wood, 1987). Hence, chemolithotrophic nitrification cannot occur in the absence of oxygen. Oxygen is also the preferred electron acceptor for reoxidation of reduced compounds produced in excess during the oxidation of hydroxylamine, which is the product of the enzyme ammonia monooxygenase. However, during ammonia oxidation by *Nitrosomonas* strains at low oxygen tensions, nitrite itself might function as an alternative electron acceptor, with nitric oxide, nitrous oxide, or dinitrogen gas as end products (Bock *et al.*, 1992; Poth, 1986; Poth and Focht, 1985). Recently, it was shown by Anderson *et al.* (1993) that the size of nitrous and nitric oxide production by *N. europaea* was even more dependent on the presence of relatively high nitrite concentrations than on that of low partial oxygen concentrations. This suggests that nitrite and oxygen competed for electrons produced during the oxidation of hydroxylamine. Nitric oxide production by a *Nitrosomonas* strain isolated from sewage was stimulated by addition of organic materials (Poth, 1986). A slow consumption of pyruvate by *N. europaea* could be observed during anaerobic incubation in the presence of nitrite (Stüven *et al.*, 1992). Under these oxygen-limited conditions, nitrite was apparently reduced to nitric oxide and nitrous oxide. However, no increase in cell protein was observed. The statement of Abeliovich and Vonshak (1992) that *N. europaea* is able to grow anaerobically, using pyruvate as an electron donor and nitrite as an electron acceptor, is even more remarkable. Nitrite consumption by this organism was stimulated by the addition of ammonium. However, the role of ammonium with respect to anaerobic growth remained obscure. In addition to the production of nitric oxide by reduction of nitrite, this gas might also be formed directly during the oxidation of hydroxylamine (Hooper and Terry, 1979). The production of nitric oxide by *Nitrosomonas* strains was independent of the oxygen tension applied, whereas the production of nitrous oxide increased with decreasing pO_2 (Anderson and Levine, 1986). The ratio of nitric and nitrous oxide production during mixotrophic growth of *N. europaea* in a batch culture increased with increasing nitrite concentration (Stüven *et al.*, 1992). During growth of *N. europaea* on a mixture of ammonia and formate, hydroxylamine is released into the medium, where it subsequently reduces nitrite to nitric oxide and nitrous oxide by chemodenitrification. The production of nitric oxide and nitrous oxide at reduced oxygen tensions is not restricted to *Nitrosomonas* strains, but was also observed with *Nitrosolobus, Nitrosospira,* and *Nitrosococcus* isolates (Goreau *et al.*, 1980), as well as with a *Nitrosovibrio* strain (Stüven *et al.*, 1992).

Similar to the ammonia oxidation, oxygen is the preferred electron acceptor during the oxidation of nitrite by chemolithotrophic nitrite-oxidizing bacteria. In contrast to the ammonia-oxidizing bacteria, nitrite-oxidizing bacteria are not dependent on a mono-oxygenase for generation of energy. However, anaerobic growth of nitrite-oxidizing bacteria on nitrite with nitrate as an alternative electron acceptor is not feasible. Anaerobic growth on nitric oxide, which is an alternative inorganic growth substrate for *Nitrobacter winogradskyi* under oxic conditions (compare Bock *et al.*, 1992), has not yet been demonstrated. In anoxic environments, in the presence of simple organic compounds, *Nitrobacter* strains are able to grow anaerobically by reduction of nitrate to nitrite, ammonia, and nitrogen gases, in particular, nitrous oxide (Freitag *et al.*, 1987; Bock *et al.*, 1988). Nitrite-oxidizing cells of *Nitrobacter vulgaris* possess a nitrite reductase that generates nitric oxide (Ahlers *et al.*, 1990). This enzyme might have a function in NADH synthesis in *Nitrobacter* species at low nitrite concentrations (Freitag and Bock, 1990). Nitric oxide was also produced when nitrite-oxidizing cells of *Nitrobacter hamburgensis* grew mixotrophically at low oxygen tensions (Bock *et al.*, 1992). However, the *in situ* production of nitric oxide by nitrite-oxidizing bacteria in (semi)oxic environments is questionable. The production of this gas from corroding building stones was apparently caused by ammonia-oxidizing bacteria and not by *Nitrobacter* cells (Baumgärtner *et al.*, 1991). The same observation has been made with respect to nitrous oxide production during pure culture experiments (Goreau *et al.*, 1980).

In conclusion, the ammonia- as well as the nitrite-oxidizing bacteria can apparently cope with oxygen-limited conditions by using nitrite or nitrate as electron acceptors alternative to oxygen. However, due to the strict dependency on oxygen of the mono-oxygenase, chemolithotrophic oxidation of ammonia to nitrite cannot occur under strictly anoxic conditions. It seems unlikely that ammonia-oxidizing strains other than *Nitrosomonas* could oxidize ammonia without a monooxygenase.

3.2. Nitrification in Oxygen-Limited, Waterlogged Grassland Soils

In soils rich in organic material such as grassland soils, anoxic conditions are rapidly established after waterlogging due to a decreased oxygen diffusion in combination with oxygen consumption by soil organisms and plant roots (Ponnamperuma, 1984). According to the strict dependency on oxygen of the ammonia mono-oxygenase, as previously discussed, ammonia oxidation will stop in these waterlogged soils. However, radial oxygen loss from aerenchymatous plant roots creates a small oxidized layer around the roots in an otherwise anoxic environment (Laan *et al.*, 1989a). This radial oxygen loss might favor oxidation processes in the rhizosphere (Armstrong, 1964; Sand-Jensen *et al.*, 1982). One such oxidation process may be chemolithotrophic nitrification. The production of nitrate in the oxidized rhizosphere of waterlogged plants is indicated by the presence of nitrate in the root zone of the perennial macrophyte *Littorella uni-flora* (Christensen and Sørensen, 1986), the induction of the enzyme nitrate reductase in bog and rice plants (Blacquière, 1986; Uhel *et al.*, 1989), increased numbers of chemolithotrophic ammonia- and nitrite-oxidizing bacteria in the root zone of the bog plant *Glyceria maxima* compared to the reduced bulk soil (Fig. 3) (Both *et al.*, 1992a), and the preservation of potential nitrification activities in the waterlogged root zone of *Rumex palustris* (Engelaar *et al.*, in press). As only ammonium was supplied as inorganic nitrogen source, the high nitrate reductase activity measured in the leaves of the flooding-resistant, aerenchymatous plant species *R. palustris* indicated that there was also nitrate formation in its root zone. Nitrate consumption by *Rumex* species is well reflected by the level of nitrate reductase (Langelaan and Troelstra, 1992). Under waterlogged condi-tions, low nitrate reductase levels and low potential nitrification activities were observed in the leaves and root zones, respectively, of *Rumex thyrsiflorus*, which is also a *Rumex* species not resistant to flooding. In contrast to flooding-resistant *Rumex* species, *R. thyrsiflorus* produces hardly any new roots with aerenchyma

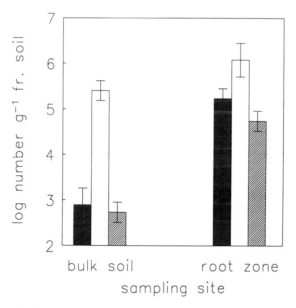

Figure 3. Most probable numbers of ammonium-oxidizing bacteria (filled bars) and nitrite-oxidizing bacteria enumerated in soil samples from out- and inside the root zone of *Glyceria maxima* in waterlogged soils. Numbers of nitrite-oxidizing bacteria were determined at 0.05 mM (open bars) and 5.0 mM (hatched bars), respectively. Bars indicate 95% confidence levels. (After Both *et al.*, 1992a.)

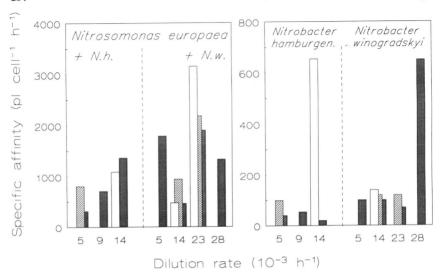

Figure 4. Specific affinities for oxygen consumption determined in mixed, steady-state cultures of *Nitrosomonas europaea* and either *Nitrobacter winogradskyi* (N.w.) or *Nitrobacter hamburgensis* (N.h.) grown in chemostats at 25 °C, pH 7.5, different dilution rates, and oxygen concentrations: 80% (filled bars), 10% (hatched bars), and 0% (open bars) oxygen. Notice the difference in scale used for the ammonia- and nitrite-oxidizing strains. (Modified after Laanbroek and Gerards, 1993; Laanbroek *et al.*, 1994.)

after flooding, and the radial oxygen release was consequently much lower (Laan *et al.*, 1989a,b). A prerequisite for the occurrence of ammonia oxidation in the root zone of flooded *R. palustris* is a continuous supply of sufficient ammonium (Engelaar *et al.*, 1991); otherwise, nitrification will be repressed by ammonium-assimilating organotrophic bacteria and plant roots, as will be discussed in Section 4.

The fact that increased numbers of chemolithotrophic nitrifying bacteria or at least nitrifying activity can be observed in the root zone of waterlogged plants indicates that these bacteria are able to consume a part of the oxygen released by the roots. The affinity of the chemolithotrophic ammonium-oxidizing bacteria for oxygen is mentioned to be much lower than the affinity of organotrophic bacteria for this electron acceptor (Prosser, 1989). However, half-saturation constants for oxygen found for *N. europaea* in mixed continuous cultures with either *N. hamburgensis* or *N. winogradskyi* (Fig. 4) were of the same magnitude as the K_m value for oxygen consumption measured with the organotrophic *Pseudomonas chlororaphis* (Laanbroek and Gerards, 1993; Laanbroek *et al.*, 1994; Bodelier and Laanbroek, unpublished results). In the small oxidized layer around plant roots, the outcome of a competition for oxygen will be determined by the differences in specific affinities of the competing organisms for oxygen. These differences are indicated by the ratio between the maximum oxygen consumption rate of the cells and the half-saturation constant for oxygen (Button, 1985). The average specific affinity for oxygen of the ammonia-oxidizing bacteria obtained with mixed continuous cultures of *N. europaea* and either *N. hamburgensis* or *N. winogradskyi* grown at different oxygen concentrations (Laanbroek *et al.*, 1994; Laanbroek and Gerards, 1993) are several orders of magnitude lower than those mentioned for organotrophic bacteria (Button, 1985). The fact that chemolithotrophic nitrification still occurs in the oxidized root zone of waterlogged plants suggests that the radial oxygen loss by the plants is proportionally larger than the release of usable carbon compounds. Hence, the organotrophic bacteria in the oxidized root zone appear to be carbon limited. Nitrogen limitation of organotrophic bacteria in the presence of actively ammonia-oxidizing cells is not very likely as is discussed in the next section.

3.3. Competition for Oxygen between Ammonia- and Nitrite-Oxidizing Bacteria

When at least a part of the available oxygen in the root zone is captured by the nitrifying bacteria, complete oxidation of ammonia to nitrate without accumulation of nitrite will be determined by the relative affinities of the ammonia- and nitrite-oxidizing bacteria for oxygen. When the oxygen supply was limited, nitrite accumulated in a mixed continuous culture of *N. europaea* and *N. winogradskyi* and cell numbers of the latter declined. This observation was supported by the specific affinities for oxygen measured for these species (Laanbroek and Gerards, 1993). The ammonia-oxidizing population had a greater specific affinity for oxygen than the nitrite-oxidizing bacteria (Fig. 4). Since the cell numbers of both species were of the same magnitude, most of the limiting amount of oxygen was used by the ammonium-oxidizing species. However, in a mixed continuous culture of *N. europaea* and mixotrophically growing *N. hamburgensis,* no nitrite accumulation was observed under oxygen-limited conditions, whereas the concentration of ammonium in the culture not used by the ammonia-oxidizing bacteria gradually increased and their numbers concomitantly decreased (Laanbroek *et al.*, 1994). Under conditions of limited oxygen supply, the specific affinities of the nitrite-oxidizing *N. hamburgensis* and the ammonia-oxidizing *N. europaea* were comparable and two to three times greater than the specific affinity of the nitrite-oxidizing *N. winogradskyi* under the same growing conditions (Fig. 4). Under conditions of sufficient oxygen supply, the specific affinities of *N. hamburgensis* for oxygen were much less than those of *N. europaea* and even less than those of *N. winogradskyi.* Cessation of the oxygen supply resulted in an immediate accumulation of nitrite in the mixed cultures of *N. europaea* and either *N. hamburgensis* or *N. winogradskyi* that had been growing at 80% oxygen saturation. The disadvantage of the low specific affinity of *N. hamburgensis* for oxygen under conditions of sufficient oxygen supply was not compensated for by its larger population size by means of mixotrophic growth on nitrite plus pyruvate. On the basis of its relatively great specific affinity under oxygen-limited conditions, it was hypothesized that in contrast to the more oxic *N. winogradskyi*, *N. hamburgensis* is adapted to growth in habitats that are constantly oxygen-limited (Both *et al.*, 1992c; Laanbroek *et al.*, 1994). If so, the similar numbers of *N. winogradskyi* and *N. hamburgensis* in peat bog soils, as has been observed by Laanbroek and Gerards (unpublished results) using specific antibody fluorescence enumerations, indicate that these water-logged soils contain oxic microsites possibly in the rhizosphere of aerenchymatous plants.

An alternative explanation of nitrite accumulation by mixed cultures of *N. europaea* and *Nitrobacter* species growing in the presence of organic compounds was presented by Stüven *et al.* (1992). These authors observed the accumulation of hydroxylamine during mixotrophic growth of *N. europaea* and *N. winogradskyi* or *N. vulgaris* on ammonium and pyruvate. A concomitant oxidation of ammonia and pyruvate by *N. europaea* may result in an overproduction of hydroxylamine, which may subsequently be released in the medium. Due to the extreme sensitivity of *Nitrobacter* species to this first intermediate of ammonia oxidation (Castignetti and Gunner, 1982), nitrite oxidation was apparently inhibited. The difference between *N. winogradskyi* and *N. hamburgensis* with respect to nitrite accumulation as observed in mixed oxygen-limited cultures (Laanbroek *et al.*, 1994; Laanbroek and Gerards, 1993) might then be explained by the different abilities for mixotrophic growth of these nitrite-oxidizing species. In contrast to the mixotrophic *N. hamburgensis*, the more chemolithotrophic *N. winogradskyi* would leave more organic compounds to be oxidized for the ammonia-oxidizing *N. europaea*. No pyruvate could be measured in mixed cultures of *N. europaea* and *N. hamburgensis*, whereas in mixed cultures of *N. europaea* and *N. winogradskyi* only a fraction of the pyruvate was consumed by the organisms (Laanbroek and Gerards, 1993; Laanbroek *et al.*, 1994).

4. Nitrification at Low Ammonium Concentrations

4.1. Pure Culture Studies

Half saturation concentrations for ammonia oxidation of *Nitrosomonas* cells are relatively high compared to *in situ* concentrations of ammonium (Prosser, 1989). There is a discrepancy between the measured low affinities of pure cultures of chemolithotrophic bacteria for ammonia and the observation that nitrification proceeds in oceanic waters at immeasurably low substrate concentrations. According to Ward (1986), this can be explained by at least two hypotheses: (1) pure cultures of bacteria, which constitute at least the major part of the natural population as indicated by immunofluorescent techniques, have lost the ability to nitrify at environmental low ammonia concentrations after isolation, and (2) ammonia-oxidizing bacteria possess complex enzyme systems capable of utilizing ammonia at concentration ranges of several orders of magnitude. Only the less sensitive system would then be measured under laboratory conditions. The K_m values for ammonium oxidation observed for *N. europaea* grown in mixed culture with either *N. hamburgensis* or *N. winogradskyi, i.e.,* 0.4–3.6 mM, were relatively independent of growth rate and oxygen concentration (Laanbroek *et al.*, 1994; Laanbroek and Gerards, 1993). The specific affinities of *N. europaea* for ammonia determined in these mixed culture experiments were several orders of magnitude lower than those mentioned for organotrophic bacteria, algae, or marine phytoplankton (Button, 1985).

Competition experiments in continuous cultures between *N. europaea* and the organotrophic *Arthrobacter globiformis,* performed in the presence of *N. winogradskyi* to prevent nitrite accumulation reaching toxic levels, showed that the chemolithotrophic bacteria were the weakest competitors for ammonium (Verhagen and Laanbroek, 1991). *Nitrosomonas europaea* could only oxidize the ammonium not used by the organotrophic bacteria (Fig. 5). However, when the organotrophic bacteria were both carbon- and nitrogen-limited, the nitrifying

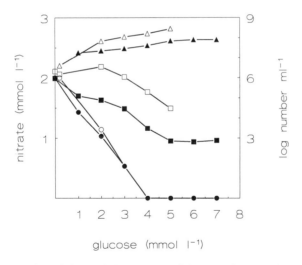

Figure 5. Concentrations of nitrate (circles), numbers of the ammonium-oxidizing bacterium *Nitrosomonas europaea* (squares), and numbers of the organotrophic bacterium *Arthrobacter globiformis* (triangles) in mixed, continuous cultures in the absence (open symbols) and presence (closed symbols) of the bacteriovorous flagellate *Adriamonas peritocrescens*. The culture was continuously supplied (D = 0.004 hr^{-1}) with mineral medium supplemented with 2 mM ammonium and variable amounts of glucose. (Modified after Verhagen and Laanbroek 1991, 1992.)

bacteria did not wash out and their numbers remained at a low but constant level in the chemostat. This phenomenon could only be explained by wall growth of the ammonia-oxidizing bacteria. The wall of the culture vessel might have offered a microhabitat where the competitive abilities of the chemolithotrophic bacteria with respect to ammonia-scavenging abilities may have been better than in the culture liquid itself. Introduction of the bacteriovorous flagellate *Adriamonas peritocrescens*, isolated from the same grassland soil as the *A. globiformis* strain (Verhagen *et al.*, 1994b), lowered the numbers of bacteria by grazing, but had no net effect on the production of nitrate from ammonia (Verhagen and Laanbroek, 1992). The flagellate apparently had a stimulating effect on the activity of the remaining ammonia-oxidizing cells, since fewer cells produced the same amount of nitrate per unit of time. Such a stimulation of nitrification in liquid cultures has also been shown by Griffiths (1989). The mechanism of this effect is not yet clear, but it is apparently not directly related to increased mineralization of organic nitrogen by the protozoa, since the total output of mineral nitrogen in the form of nitrate remained the same.

4.2. Nitrification in Ammonium-Limited Soils

In aquatic environments such as mixed continuous cultures, K_m values or specific affinities may be a useful tool to predict the outcome of competition for growth-limiting substrates between species. However, in more heterogenous environments such as soils, the utilization of these kinetic parameters might be less useful. Competition experiments between ammonia-oxidizing *N. europaea* and the organotrophic *A. globiformis*, in pot experiments and in soil columns continuously percolated with mineral medium supplemented with ammonium and glucose, confirmed the weak position of the chemolithotrophic bacteria with respect to ammonium consumption (Verhagen *et al.*, 1992, 1994a). Introduction of the bacteriovorous flagellate *A. peritocrescens* stimulated the activity of the ammonia-oxidizing cells again, whereas no effect on the total number of chemolithotrophic bacteria was observed (Verhagen *et al.*, 1993, 1994a). In addition to the stimulation of nitrification by a yet unknown factor as observed in homogenous continuous cultures, the flagellates may also have brought about a better mixing of the ammonia-oxidizing bacteria or their substrate in the soil. In the presence of growing plant roots, nitrification was completely repressed due to ammonium limitation (Verhagen *et al.*, 1994a). Here, a likely positive effect of protozoa on nitrification was overruled by the activity of the plant roots. Repression of ammonia-oxidizing bacteria as well as of potential nitrifying activities due to ammonium limitation were also observed in a pot experiment with the relatively fast-growing plant species *R. palustris* (Engelaar *et al.*, 1991). With the slower-growing *Rumex acetosa*, numbers of ammonia-oxidizing bacteria and potential nitrifying activities increased at the start of the incubation period. When

R. acetosa also reached a certain amount of biomass, numbers and activities of ammonia-oxidizing bacteria leveled off due to increasing ammonium limitation in the soils.

It had already been shown in agricultural soils by Jansson (1958), who applied labeled ammonium in the absence and presence of wheat straw, that nitrifying bacteria were regularly less effective than the organotrophic bacteria in the competition for mineralized ammonium. From pot experiments with mustard plants, Jansson (1958) concluded that the plants had only a very limited ability to compete effectively with the bacteria for nitrogen in the presence of farmyard manure. However, after nitrification had been well established in the presence of an excess of added ammonium, the nitrifying bacteria were able to use a considerable part of the ammonium chemically fixed in clay minerals, and consequently otherwise biologically unavailable ammonium became nitrified. However, it should be kept in mind that utilization of chemically fixed ammonium by nitrifying bacteria could only happen once in the same soil. In agreement with Jansson (1958), Zak *et al.* (1990) observed a decrease in nitrification in the presence of the plant species *Allium tricoccum* when compared to no-plant soils during short incubation experiments with ^{15}N-labeled ammonium. Most of the ^{15}N-labeled ammonium had been consumed by the organotrophic bacteria, since microbial bioimass contained 8.5 times as much nitrogen as *A. tricoccum* biomass 2 days after isotope addition. However, when plants and organotrophic bacteria are not nitrogen-limited, nitrification may occur (Riha *et al.*, 1986). On the basis of half-saturation values known for ammonium consumption by a soil community consisting of plant roots, organotrophic microorganisms, and ammonia-oxidizing bacteria, Rosswall (1982) predicted that ammonium-oxidizing bacteria would be the weakest competitors of them all and that nitrification would be repressed by the activity of plant roots and organotrophic bacteria under conditions of limiting ammonium supply.

4.3. Competition for Ammonium versus Allelopathy in Grassland Soils

Repression of nitrification in natural grassland soils themselves has often been observed, especially under climax vegetation. In New Zealand, for example, the nitrifying population of a low-fertility grassland soil was very small as indicated by MPN enumerations (Robinson, 1963). The population of nitrifying bacteria could be increased by field treatment with urea. In counts of chemolithotrophic nitrifying bacteria by a MPN method, Meiklejohn (1968) observed low numbers of ammonia- and nitrite-oxidizing bacteria in soils under a number of common indigenous African grasses. In contrast, numbers of nitrifying bacteria were about 100 times higher in heavily fertilized pastures. These observations all conform to the idea that nitrifying bacteria are weak competitors with respect to limiting amounts of ammonium. Studying one grassland and two different forest succession series, Rice and Pancholy (1972, 1973) also observed a decreasing number of nitrifying bacteria with increasing successional stages of the vegetation. However, the amount of ammonium also increased with increasing succession. In addition, the amounts of tannins in the soil of intermediate and late successional stages were high. So, it was hypothesized that nitrification in plant-nutrients-limited grassland soils is inhibited by allelochemical compounds produced by plants in order to conserve nitrogen and energy. This idea of inhibition of nitrifying bacteria by plant-derived allelochemicals, which had already been raised by Theron in 1951, was partly supported by a number of experiments (Moore and Waid, 1971; Munro, 1966; Rice and Pancholy, 1973). However, as can be seen from the experiments of Moore and Waid (1971), repeated addition of root washings reduced the repressive effect of these washings on the nitrification. Purchase (1974) observed that daily washings of living or decaying grass roots did not inhibit nitrification. The effects of tannins on pure cultures of ammonium-oxidizing bacteria were dependent on the genera studied (Bohlool *et al.*, 1977).

No inhibition of nitrification by allelochemicals was observed in the root zones of *Plantago* species. A positive correlation had been found between numbers of nitrifying bacteria in the rhizosphere of these plant species and the uptake of nitrate by the plant as indicated by the nitrate reductase activities in their leaves (Smit and Woldendorp, 1981). With an excess of ammonium, accumulation of *N. europaea* and of an organotrophic *Pseudomonas* species was observed around the roots of *Plantago lanceolata* in a small axenic growth chamber that was continuously percolated with mineral medium (Both, 1990). In this experiment, ammonium was always in excess. *Nitrosomonas europaea* might have been stimulated by ammonia production by the organotrophic bacteria growing on organic nitrogen compounds released by the plant roots. As described in Section 2.1, ammonia is likely to be a better substrate for the enzyme ammonia mono-oxygenase. However, the chemolithotrophic bacteria might also have been positively affected by these compounds themselves, since the *N. europaea* strain used in this experiment is able to incorporate exogenous amino acids as has been discussed above (Frijlink *et al.*, 1992b).

Stienstra *et al.* (1994), who studied nitrification in the root zone of *Holcus lanatus* originating from grassland soils, each with a different history of fertilizer application, observed decreasing potential ammonium-oxidizing activities with increasing periods of nonfertilization (Fig. 6). The grasslands studied constituted a series of different stages of plant succession in the vegetation resulting from decreasing nutrient availabilities (Olff, 1992). The potential nitrification activity in the root zones of *Holcus lanatus* and other plant species was similar in each field but differed significantly between the fields. In addition, MPN numbers of ammonium-oxidizing bacteria in the root zone of *H. lanatus* also decreased as the period of nonfertilization increased. So, these results are not in favor of the hypothesis of repressed nitrification by allelochemicals released by the plants

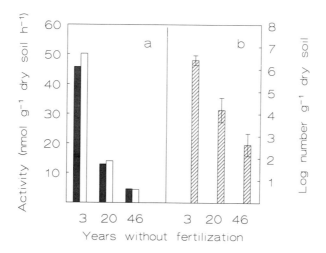

Figure 6. (a) Potential ammonia-oxidizing activities and (b) most probable numbers of ammonia-oxidizing bacteria measured in samples from extensively used grassland soils in which fertilization had been omitted since 3, 20, and 46 years, respectively. Samples for measurements of potential ammonia-oxidizing activities were collected in March 1991 from the root zones of *Holcus lanatus* (filled bars) and other dominant plant species (open bars) that are representative for a particular succession stage of the grassland vegetation: *Agrostis stolonifera* (3 years), *Anthoxanthum odoratum* (20 years), and *Agrostis capillaris* (46 years). Samples for determination of numbers were randomly taken in July 1991. Bars represent 95% confidence limits. (Modified after Stienstra *et al.*, 1994.)

under climax vegetation, but do support the idea of a diminished ammonium supply. However, it cannot be completely excluded that the production of nitrification-repressing organic compounds by *H. lanatus* may be stimulated by nutrient limitation.

4.4. Nitrification in Ammonium-Limited Grassland Soils in Relation to pH

Bramley and White (1989), who studied short-term nitrification activities in relation to pH in four pasture soils with a different history of liming, noticed that the optimum pH for an indigenous ammonium-oxidizing population was never far from the prevailing soil pH. In this study the upper 3 cm of soil, consisting of leaf litter and a dense root mat, was discarded to minimize any inhibitory effect that grass roots may have on the rate of nitrification. To study the pH optimum of the ammonium-oxidizing community in the series of grassland soils with a different history of fertilization, mentioned above (Stienstra *et al.*, 1994), the upper 10 cm, including the grass sod, was used. The pH_{KCl} values of the soils decreased from 4.9 to 4.4 as the period of nonfertilization increased. In the most recently fertilized grassland soils, the ammonium-oxidizing community showed a pH optimum of 7.0. As the periods of nonfertilization increased, the pH optimum decreased but became less distinct. In the grassland soil that had not been fertilized the longest time, no pH optimum was observed at all in the range of pH 5.0–8.0. It was hypothesized that the value of the optimum pH indicated the physiological condition of the ammonium-oxidizing bacteria, with the most active bacteria having the highest optimum pH. The grassland soil that had been fertilized relatively recently, i.e., only 3 years ago, also contained significantly higher concentrations of nitrate. In general, nitrate consumption by plants enhances the pH of the rhizosphere, whereas the opposite occurs during consumption of ammonium (e.g., Troelstra et al., 1992). Hence, nitrate consumption by the plant roots will create microsites of increased pH in an otherwise acid soil, hereby stimulating the acid-sensitive ammonium-oxidizing bacteria. So, a positive interaction might occur in these recently fertilized grassland soil between the nitrate-producing chemolithotrophic nitrifying bacteria and the nitrate-consuming plants. In addition, a shift in the ammonium-oxidizing community from acid-sensitive to acid-tolerant bacteria may also have occurred, with increased impoverishment of the soils due to the nonuse of fertilizers and annual removal of aboveground plant biomass. In contrast to acid-sensitive bacteria, acid-tolerant bacteria in heathland soils did not have a distinct pH optimum between pH 4.0 and pH 6.0 (De Boer et al., 1989b).

4.5. Effect of Substrate Concentrations on MPN Enumerations of Nitrite-Oxidizing Bacteria

A most remarkable observation was made by Both and co-workers when enumerating numbers of nitrite-oxidizing bacteria in natural grassland soils. In general, numbers of these bacteria counted by a MPN procedure were highly dependent on the nitrite concentration used in the enumeration medium. Usually, the highest numbers of nitrite-oxidizing bacteria were observed at a low nitrite concentration, i.e., 0.05 mM (Fig. 3) (Both et al., 1990, 1992a,b). It was only in a well-drained grassland soil on top of a former river bank that the highest numbers were observed for a relatively short period of time at the highest nitrite concentration applied, i.e., 5.0 mM (Fig. 7a) (Both et al., 1990, 1992b). In pot experiments with two grassland plant species, numbers of nitrite-oxidizing bacteria determined at 0.05 mM nitrite after 14–15 weeks of plant growth were significantly higher in the root zone compared to the nonrooted soil (Stienstra et al., 1993). In contrast to this, numbers determined at 5.0 mM nitrite after 11–12 weeks of incubations were significantly higher in the bulk soil. In arable soils characterized by different management practices, numbers of nitrite-oxidizing bacteria determined at 0.1 mM nitrite were always two orders of magnitude higher than those determined at 5.0 mM nitrite (Laanbroek and Gerards, 1991). Differences between numbers obtained at low and high nitrite concentrations

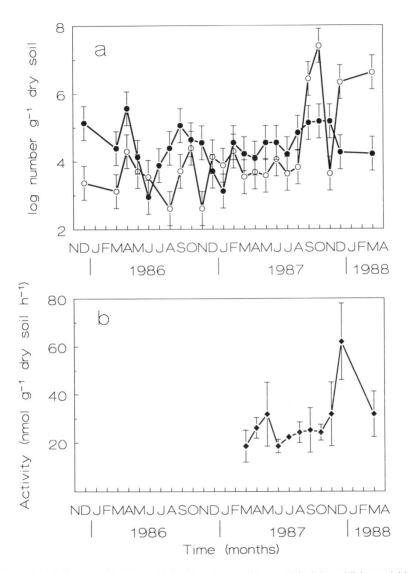

Figure 7. (a) Numbers of nitrite-oxidizing bacteria and (b) potential nitrite-oxidizing activities determined in samples collected from an extensively used grassland soil. Numbers of nitrite-oxidizing bacteria were determined in mineral medium supplemented with 0.05 mM (○) or 5.0 mM (●) nitrite. Bars represent 95% and 66% confidence limits for numbers and activities, respectively. (After Both *et al.*, 1990, 1992b.)

could be simulated by applying Monod and Haldane kinetics and different incubation times (Both and Laanbroek, 1991). According to the model they used, numbers of the nitrite-tolerant and -sensitive cells are underestimated at 0.05 mM and 5.0 mM nitrite, respectively, after applying an incubation period of 90 days. Therefore, low numbers of nitrite-oxidizing bacteria at 5.0 mM compared to 0.05 mM nitrite may indicate the predominance of nitrite-sensitive cells. It is not yet known what these cells represent. Aeration of anoxic waterlogged soils either by the presence of oxygen-releasing aerenchymatous plants or aerobic laboratory incubations in the presence of ammonium promoted the growth of nitrite-tolerant cells at 5.0 mM nitrite together with the ammonium-oxidizing cells (Both *et al.*, 1992a). No stimulation of nitrite-sensitive cells at 0.05 mM nitrite was observed in the presence of oxygen. Anoxic incubations decreased the numbers at 0.05 mM, but slightly stimulated the numbers at 5.0 mM nitrite. No effect of anoxic incubations was observed with respect to the numbers of ammonium-oxidizing bacteria. Since the behavior of the nitrite-tolerant bacteria and ammonium-oxidizing bacteria was similar toward oxic incubations in the presence of ammonium, this group of nitrite-oxidizing bacteria might represent the actively nitrifying part of the soil community, whereas the nitrite-sensitive cells enumerated at 0.05 mM nitrite may symbolize resting or organotrophically growing cells. To study the effect of starvation on the result of enumerations at low and high concentrations of nitrite, cells of different *Nitrobacter* species present in the waterlogged soils were enumerated after 18 months of starvation in the absence of nitrite (Laanbroek and Schotman, 1991). During these enumerations, no effect of the nitrite concentration on numbers was observed with nongrowing cells. Enumerations of *Nitrobacter* cells after subculturing for many generations on organic substrates were not affected by low or high nitrite concentrations (Stienstra *et al.*, 1993). So, the nature of the nitrite-sensitive cells obtained during enumerations at low nitrite concentrations is still obscure. But since these cells may have a relatively low half-saturation constant for nitrite oxidation as indicated by the model of Both and Laanbroek (1991), nitrite-sensitive cells might be responsible for scavenging traces of nitrite during low or nonnitrifying conditions in the soil.

4.6. Potential Nitrifying Activities in Grassland Soils

In addition to enumeration of chemolithotrophic nitrifying bacteria by MPN procedures described above, cell numbers have also been derived indirectly from determinations of potential nitrifying activities performed in soil slurries under optimal conditions in the laboratory according to the method of Schmidt and Belser (1982). These potential nitrifying activities do not represent *in situ* nitrate production rates, but indicate virtually the sizes of the enzyme systems able to oxidize ammonia or nitrite. In the soil of an extensively used pasture, determinations of potential nitrite-oxidizing activities were performed monthly, together with enumerations of nitrite-oxidizing bacteria by the MPN procedure (Both *et al.*, 1992b). Maximal values for the potential nitrite-oxidizing activities were obtained in early spring and early winter (Fig. 7b). However, these maximal values were only two to three times larger than the basic values measured in summer and midwinter. These relatively small differences between maximal and minimal values of potential nitrite-oxidizing activities were also observed by Berg and Rosswall (1986) in arable soils. Hence, the sizes of the nitrite-oxidizing enzyme systems present in the soil hardly fluctuated during seasons. Numbers of nitrite-oxidizing bacteria determined by the MPN method revealed larger seasonal fluctuations than the potential nitrite-oxidizing activities did (Fig. 7). This was also observed by Sarathchandra (1978) and Berg (1986). The latter suggested that the observed differences between MPN counts and potential nitrite-oxidizing activities may be due to the presence or absence of free-living bacteria. Under optimal conditions, a larger part of the nitrite-oxidizing community would be present as free-living cells in the soil, whereas more adverse conditions would force the larger part of the community to survive in microcolonies. Assuming a constant activity per cell, as was done by Berg, a concentration of cells in microcolonies would not effect the potential nitrite-oxidizing activity of the community, but would decrease the total number obtained in a MPN series. However, the activity per cell is dependent on the environmental conditions (Both *et al.*, 1992c; Laanbroek and Gerards, 1993; Laanbroek *et al.*, 1994). Low numbers of nitrite-oxidizing bacteria observed in a MPN series could also be due to the inability of cells from adverse soil conditions to proliferate in mineral medium, whereas these cells are still able to convert nitrite to nitrate during a measurement of potential nitrite-oxidizing activities.

Surprisingly, the potential nitrite-oxidizing activity determined monthly in the extensively used pasture soil correlated very well with the apparent saturation constant for nitrite oxidation (Both *et al.*, 1992b). Hence, the specific affinity for nitrite oxidation as indicated by the ratio between potential activity and apparent saturation constant (Button, 1985) was fairly constant throughout the year and averaged 295×10^6 liter (l) g^{-1} dry soil hr^{-1}. The average specific affinity for nitrite oxidation measured in 14 different soil samples from the same grassland location of 12×40 m, but taken in June a year later, amounted to 382×10^6 l g^{-1} dry soil hr^{-1}. This latter value was not significantly different from the annual average. Although astonishing, the meaning of this constancy in specific affinity throughout time is not obvious in a heterogenous system such as a soil where the ability of scavenging low amounts of substrates is likely to be more important than a high specific affinity.

5. Survival Mechanisms in the Absence of Mineral Nitrogen

As was previously discussed, chemolithotrophic ammonia-oxidizing bacteria are bad competitors with respect to oxygen and ammonium when compared with organotrophic bacteria. Which means that they will lose the competition from aerobic organotrophic bacteria under conditions of limiting supply of extracellular energy sources. So, nitrifying bacteria have to survive periods of deprivation of extracellular energy sources. Since *Nitrobacter* species can grow organotrophically (Bock *et al.*, 1992; Watson *et al.*, 1989), they may use organic compounds when there is no nitrite supply. Although observations of large numbers of *Nitrobacter* cells in soils in relation to ammonia-oxidizing cells are often explained by organotrophic growth of the former (e.g., Bock and Koops, 1992), nothing is known about their competitive ability with respect to organic compounds compared to other organotrophic bacteria. Under conditions of optimal energy supply, ammonia- as well as nitrite-oxidizing bacteria accumulate intracellular storage materials in the form of poly-β-hydroxybutyrate, polyphosphate, or glycogenlike granules (Bock and Heinrich, 1969; Bock and Koops, 1992; Gay *et al.*, 1983; Smith and Hoare, 1968; Terry and Hooper, 1970; Van Gool *et al.*, 1971). Chemolithotrophically grown cells of *N. winogradskyi* and *N. vulgaris* may possess an average poly-β-hydroxybutyrate content of 10–30% of the cell dry weight (compare Bock *et al.*, 1992). Even under growth-limiting conditions in the chemostat, the production of storage materials was very likely (Laanbroek and Gerards, 1993). Apparently, chemolithotrophic cells anticipate future energy-limited situations by production of intracellular reserve polymers under conditions of sufficient extracellular energy supply.

By endogenous respiration *N. europaea* cells were able to maintain a considerable proton gradient across the membrane in the absence of an extracellular substrate (Frijlink *et al.*, 1992a). Even at low pH ranges when ammonia oxidation was impossible, *N. europaea* cells could depend on endogenous substrates for the maintenance of a significant proton gradient. This gradient could be used to drive a secondary energy-dependent process such as active transport of alanine. The ability to actively accumulate a number of amino acids (Frijlink *et al.*, 1992b), which might include recapturing "lost" endogenous compounds, will have survival value under conditions of ammonia deprivation.

From experiments with specific inhibitors it was concluded that extra- and intracellular energy sources used two distinct systems for energy generation in *N. europaea* (Frijlink *et al.*, 1992a). Two different pathways for exo- and endogenous respiration are apparently also active in *N. winogradskyi* (Eigener, 1975; Eigener and Bock, 1975). According to these authors, nitrite oxidation in *N. winogradskyi* might even be inhibited by endogenous respiration, a characteristic that seems odd with respect to maximal utilization of extracellular energy sources and concomitant preservation of intracellular storage materials. However, a delay in switching from intra- to extracellular energy sources might be profitable in the longer term when the extracellular supply of mineral nitrogen is only available for a short period. Nitrification is inhibited in environments such as fish and oxidation ponds where relatively short cycles of oxic and anoxic conditions prevail (Diab *et al.*, 1993). Switching from anoxic to oxic conditions, nitrification started only after a 24 to 48 hr lag period. This observation may also be important with respect to the activity of nitrifying bacteria in the root zones of waterlogged, aerenchymatous plants, where oxygen concentrations will be subjected to diurnal changes due to plant activity (Buis, personal communication).

However, nothing is known about the exact survival mechanisms of chemolithotrophic nitrifying bacteria in soils, and hence this should be the subject of further studies. An adequate survival mechanism will be essential for these slowly growing bacteria with an extremely low cell yield due to growth on ammonia and nitrite as energy sources.

6. Future Perspectives

As indicated in the introduction of this chapter, the predominance of *Nitrosomonas* species in ammonia-oxidizing communities is often questioned. Nevertheless, most of the physiological studies concerning ammonia-oxidizing bacteria have been done with the species *N. europaea*. Therefore, hardly anything could be said about physiological differences between the various ammonium-oxidizing species or about possible differences in ecological niches in the soil. Differences between the ammonia-oxidizing species and genera might become apparent by studying their temporal and spatial distribution. This can be done by using specific antibodies (Bohlool and Schmidt, 1973, 1980; Schmidt, 1973; Gay *et al.*, 1983; Josserand *et al.*, 1981) or specific DNA probes (Head *et al.*, 1993; McCaig *et al.*, 1994).

The nitrite-oxidizing genus *Nitrobacter* contains several physiologically distinct species (Bock and Koops, 1992; Bock *et al.*, 1992). The distribution of physiologically different *Nitrobacter* species in an extensively used grassland soil was studied by using specific antibodies against these species (Both *et al.*, 1992b). No cross-reactions between antisera and antigenes were observed. The serotypes of *N. hamburgensis, N. winogradskyi* strain ATCC253g1, and *N. winogradskyi* strain *agilis* were simultaneously present throughout the year. Their numbers were always of the same magnitude. Coexistence of more than one *Nitrobacter* species in one soil sample is also reported by Josserand *et al.* (1981) and Laanbroek and Gerards (1991). On a yearly basis, numbers of *N. hamburgensis* and *N. winogradskyi* strain *agilis* in the extensively used grassland soil were significantly ($p < 0.05$) correlated with each other. During the study of spatial distribution of these nitrite-oxidizing bacteria, numbers of *N. hamburgensis* and both *N. winogradskyi* strains were correlated at a significance level of at least 5%. As revealed from principal component analyses, numbers of *N. hamburgensis* and *N. winogradskyi* strain *agilis* were always similarly affected by pairs of unknown factors, and in the case of spatial distribution numbers of the second *N. winogradskyi* strain were affected by these unknown factors in the same way as the other *Nitrobacter* species. This could indicate that these physiologically distinct species showed the same behavior in the soil, which seems to be unlikely. The corresponding performance of the different species might also be due to unreliability of the use of antibodies raised against laboratory strains grown under optimal conditions. In contrast to the DNA probes, specific antibodies are phenotypic markers that might be affected by the prevailing environmental conditions. Hence, antibodies raised in the laboratory may be less useful for enumerating *in situ* grown bacteria. The use of species-specific DNA markers seems to be more promising to determine possible niche differences between physiologically distinct chemolithotrophic nitrifying bacteria in soils, and more so when this enumeration method could be applied on a microscale.

References

Abeliovich, A., 1987, Nitrifying bacteria in wastewater reservoirs, *Appl. Environ. Microbiol.* **53:**754–760.

Abeliovich, A., and Vonshak, A., 1992, Anaerobic growth of *Nitrosomonas europaea*, *Arch. Microbiol.* **158:**267–270.

Ahlers, B., König, W., and Bock, E., 1990, Nitrite reductase activity in *Nitrobacter vulgaris*, *FEMS Microbiol. Lett.* **67:**121–126.

Allison, S. M., and Prosser, J.I., 1991, Urease activity in neutrophilic autotrophic ammonia-oxidizing bacteria isolated from acid soils, *Soil Biol. Biochem.* **23:**45–51.

Allison, S. M., and Prosser, J. I., 1993, Ammonia oxidation at low pH by attached populations of nitrifying bacteria, *Soil Biol. Biochem.* **25:**935–941.

Anderson, I. C., and Joel, J. S., 1986, Relative rates of nitric oxide and nitrous oxide production by nitrifiers, denitrifiers and nitrate respirers, *Appl. Environ. Microbiol.* **51:**938–945.

Anderson, I. C., Poth, M., Homstead, J., and Burdige, D., 1993, A comparison of NO and N$_2$O production by the autotrophic nitrifier *Nitrosomonas europaea* and the heterotrophic nitrifier *Alcaligenes faecalis*, *Appl. Environ. Microbiol.* **59:**3525–3533.

Armstrong, W., 1964, Oxygen diffusion from the roots of some British bog plants, *Nature* **204:**511–519.

Baumgärtner, M., Sameluck, F., Bock, E., and Conrad, R., 1991, Production of nitric oxide by amonium-oxidizing bacteria colonizing building stones, *FEMS Microbiol. Ecol.* **85:**95–100.

Berg, P., 1986, *Nitrifier Populations and Nitrification Rates in Agricultural Soil*, Swedish University of Agricultural Sciences, Uppsala, thesis.

Berg, P., and Rosswall, T., 1986, Seasonal variation in abundance and activity of nitrifiers in four arable cropping systems, *Microb. Ecol.* **13:**75–87.

Blacquière, T., 1986, Nitrate reduction in the leaves and numbers of nitrifiers in the rhizosphere of *Plantago lanceolata* growing in two contrasting sites, *Plant Soil* **91:**377–380.

Bock, E., and Heinrich, G., 1969, Morphologische und physiologische Untersuchungen an Zellen von *Nitrobacter winogradskyi* Buch, *Arch. Microbiol.* **69:**149–159.

Bock, E., and Koops, H.-P., 1992, The genus *Nitrobacter* and related genera, in: *The Prokaryotes. A Handbook on the Biology of Bacteria: Ecophysiology, Isolation, Identification, Applications*, 2nd ed., Vol. III (A. Balows, H. G. Trüper, M. Dworkin, W. Harder, and K.-H. Schleifer, eds.), Springer-Verlag, New York, pp. 2302–2309.

Bock, E., Wolderer, P. A., and Freitag, A., 1988, Growth of *Nitrobacter* in the absence of dissolved oxygen, *Water Res.* **22:**245–250.

Bock, E., Koops, H.-P., Ahlers, B., and Harms, H., 1992, Oxidation of inorganic nitrogen compounds as energy source, in: *The Prokaryotes. A Handbook on the Biology of Bacteria: Ecophysiology, Isolation, Identification, Applications,* 2nd ed., Vol. I (A. Balows, H. G. Trüper, M. Dworkin, W. Harder, and K.-H. Schleifer, eds.), Springer-Verlag, New York, pp. 414–430.

Bohlool, B. B., and Schmidt, E. L., 1973, A fluorescent antibody approach to determination of growth rates in soil, *Bull. Ecol. Res. Comm.* **17:**336–338.

Bohlool, B. B., and Schmidt, E. L., 1980, The immunofluorescence approach in microbial ecology, *Adv. Microb. Ecol.* **4:**203–241.

Bohlool, B. B., Schmidt, E. L., and Beasly, C., 1977, Nitrification in the intertidal zone: Influence of effluent type and effect of tannin on nitrifiers, *Appl. Environ. Microbiol.* **34:**523–528.

Both, G. J., 1990, *The Ecology of Nitrite-Oxidizing Bacteria in Grassland Soils,* State University of Groningen, the Netherlands, PhD thesis.

Both, G. J., and Laanbroek, H. J., 1991, The effect of the incubation period on the result of MPN enumerations of nitrite-oxidizing bacteria: Theoretical considerations, *FEMS Microbiol. Ecol.* **85:**355–344.

Both, G. J., Gerards, S., and Laanbroek, H. J., 1990, Enumeration of nitrite-oxidizing bacteria in grassland soils using a most probable number technique: Effect of nitrite concentration and sampling procedure, *FEMS Microbiol. Ecol.* **74:**277–286.

Both, G. J., Gerards, S., and Laanbroek, H. J., 1992a, The occurrence of chemolitho-autotrophic nitrifiers in water-saturated grassland soils, *Microb. Ecol.* **23:**15–26.

Both, G. J., Gerards, S., and Laanbroek, H. J., 1992b, Temporal and spatial variation in the nitrite-oxidizing bacterial community of a grassland soil, *FEMS Microbiol. Ecol.* **101:**99–112.

Both, G. J., Gerards, S., and Laanbroek, H.J., 1992c, Kinetics of nitrite oxidation in two *Nitrobacter* species grown in nitrite-limited chemostats, *Arch. Microbiol.* **157:**436–441.

Bramley, R. G. V., and White, R. E., 1989, The effect of pH, liming, moisture and temperature on the activity of nitrifiers in a soil under pasture, *Aust. J. Soil Res.* **27:**711–724.

Button, D. K., 1985, Kinetics of nutrient-limited transport and microbial growth, *Microbiol. Rev.* **49:**270–297.

Castignetti, D., and Gunner, H. B., 1982, Differential tolerance of hydroxylamine by an *Alcaligenes* sp., a heterotrophic nitrifier, and by *Nitrobacter agilis, Can. J. Microbiol.* **28:**148–150.

Christensen, P. B., and Sørensen, J., 1986, Temporal variation of denitrification activity in plant-covered, littoral sediment from Lake Hampen, Denmark, *Appl. Environ. Microbiol.* **51:**1171–1179.

Cooper, A. B., 1983, Population ecology of nitrifiers in a stream receiving geothermal inputs of ammonium, *Appl. Environ. Microbiol.* **45:**1170–1177.

De Boer, W., and Laanbroek, H. J., 1989, Ureolytic nitrification at low pH by *Nitrosospira* species, *Arch. Microbiol.* **152:**178–181.

De Boer, W., Duyts, H., and Laanbroek, H. J., 1988, Autotrophic nitrification in a fertilized acid heath soil, *Soil Biol. Biichem.* **20:**845–850.

De Boer, W., Duyts, H., and Laanbroek, H. J., 1989a, Urea stimulated autotrophic nitrification in suspensions of fertilized, acid heath soil, *Soil Biol. Biochem.* **21:**349–354.

De Boer, W., Klein Gunnewiek, P. J. A., Troelstra, S. R., and Laanbroek, H. J., 1989b, Two types of chemolithotrophic nitrification in acid heathland humus, *Plant Soil* **119:**229–235.

De Boer, W., Klein Gunnewiek, P. J. A., and Troelstra, S. R., 1990, Nitrification in Dutch heathland soils. II. Characteristics of nitrate production, *Plant Soil* **127:**193–200.

De Boer, W., Klein Gunnewiek, P. J. A., Veenhuis, M., Bock, E., and Laanbroek, H. J., 1991, Nitrification at low pH by aggregated chemolithotrophic bacteria, *Appl. Environ. Microbiol.* **57:**3600–3604.

De Boer, W., Tietema, A., Klein Gunnewiek, P. J. A., and Laanbroek, H. J., 1992, The chemolithotrophic ammonium-oxidizing community in a nitrogen-saturated acid forest soil in relation to pH-dependent nitrifying activity, *Soil Biol. Biochem.* **24**:229–234.

De Boer, W., Hunscheid, M. P. J., Schotman, J. M. T., Troelstra, S. R., and Laanbroek, H. J., 1993, *In situ* net N transformations in pine, fir and oak stands of different ages on acid sandy soil, 3 years after liming, *Biol. Fertil. Soils* **15**:120–126.

De Boer, W., Klein Gunnewiek, P. J. A., and Laanbroek, H. J., 1995, Nitrification at low pH by a bacterium belonging to the genus *Nitrosospira, Soil Biol. Biochem.* **27**:127–132.

Diab, S., Kochba, M., and Avnimelech, Y., 1993, Nitrification pattern in a fluctuating anaerobic-aerobic pond environment, *Water Res.* **27**:1469–1475.

Eigener, U., 1975, Adenine nucleotide pool variations in intact *Nitrobacter winogradskyi* cells, *Arch. Microbiol.* **102**:233–240.

Eigener, U., and Bock, E., 1975, Study on the regulation of oxidation and CO_2 assimilation in intact *Nitrobacter winogradskyi* cells, *Arch. Microbiol.* **102**:241–246.

Engelaar, W. M. H. G., Bodelier, P. L. E., Laanbroek, H. J., and Blom, C. W. P. M., 1991, Nitrification in the rhizosphere of a flooding-resistant and a flooding-non-resistant *Rumex* species under drained and waterlogged conditions, *FEMS Microbiol. Ecol.* **86**:33–42.

Engelaar, W. M. H. G., Symens, J. C., Laanbroek, H. J., and Blom, C. W. P. M., Preservation of nitrifying capacity and nitrate availability in waterlogged soils by radial oxygen loss from roots of wetland plants, *Biol. Fertil. Soils,* in press.

Focht, D. D., and Verstraete, W., 1977, Biochemical ecology of nitrification and denitrification, *Adv. Microb. Ecol.* **1**:135–214.

Freitag, A., and Bock, E., 1990, Energy conservation in *Nitrobacter, FEMS Microbiol. Lett.* **66**:157–162.

Freitag, A., Rudert, M., and Bock, E., 1987, Growth of *Nitrobacter* by dissimilatory nitrate reduction, *FEMS Microbiol. Lett.* **48**:105–109.

Frijlink, M. J., Abee, T., Laanbroek, H. J., de Boer, W., and Konings, W. N., 1992a, The bioenergetics of ammonia and hydroxylamine oxidation in *Nitrosomonas europaea* at acid and alkaline pH, *Arch. Microbiol.* **157**:194–199.

Frijlink, M. J., Abee, T., Laanbroek, H. J., de Boer, W., and Konings, W. N., 1992b, Secondary transport of amino acids in *Nitrosomonas europaea, Arch. Microbiol.* **157**:389–393.

Gay, G., Josserand, A., and Bardin, R., 1983, Growth of two serotypes of *Nitrobacter* in mixotrophic and chemoorganotrophic conditions, *Can. J. Microbiol.* **29**:394–397.

Goreau, T. J., Kaplan, W. A., Wofsy, S. C., McElroy, M. B., Valois, F. W., and Watson, S. W., 1980, Production of NO_2^- and N_2O by nitrifying bacteria at reduced concentrations of oxygen, *Appl. Environ. Microbiol.* **40**:526–532.

Griffiths, B. S., 1989, Enhanced nitrification in the presence of bacteriophagous protozoa, *Soil Biol. Biochem.* **21**:1045–1051.

Hankinson, T. R., and Schmidt, E. L., 1984, Examination of an acid forest soil for ammonia- and nitrite-oxidizing bacteria, *Can. J. Microbiol.* **30**:1125–1132.

Hankinson, T. R., and Schmidt, E. L., 1988, An acidophilic and a neutrophilic *Nitrobacter* strain isolated from the numerical predominant nitrite-oxidizing population of an acid forest soil, *Appl. Environ. Microbiol.* **54**:1536–1540.

Head, I. M., Hiorns, W. D., Embley, T. M., McCarthy, A. J., and Saunders, J. R., 1993, The phylogeny of autotrophic ammonia-oxidizing bacteria as determined by analysis of 16S ribosomal RNA gene sequences, *J. Gen. Microbiol.* **139**:1147–1153.

Hooper, A. B., and Terry, K. R., 1979, Hydroxylamine oxidoreductase of *Nitrosomonas europaea* production of nitric oxide from hydroxylamine, *Biochim. Biophys. Acta* **571**:12–20.

Jansson, S. L., 1958, Tracer studies on nitrogen transformations in soil with special attention to mineralization–immobilization relationships, *Kungl. Lantbr. Ann.* **24**:101–361.

Josserand, A., Gay, G., and Faurie, G., 1981, Ecological study of two *Nitrobacter* serotypes coexisting in the same soil, *Microb. Ecol.* **7**:275–280.

Killham, K., 1986, Heterotrophic nitrification, in: *Nitrification* (J. I. Prosser, ed.), IRL Press, Oxford, England, pp. 117–126.

Killham, K., 1987, A new perfusion system for measurement and characterization of potential rates of soil nitrification, *Plant Soil* **97**:267–272.

Killham, K., 1990, Nitrification in coniferous forest soils, *Plant Soil* **128**:31–44.

Kleiner, D., 1985, Bacterial ammonium transport, *FEMS Microbiol. Rev.* **32**:87–100.

Koops, H.-P., and Möller, U. C., 1992, The lithotrophic ammonia-oxidizing bacteria, in: *The Prokaryotes. A Handbook on the Biology of Bacteria: Ecophysiology, Isolation, Identification, Applications*, 2nd ed., Vol. III (A. Balows, H. G. Trüper, M. Dworkin, W. Harder, and K.-H. Schleifer, eds.), Springer-Verlag, New York, pp. 2625–2637.

Kuenen, J. G., and Robertson, L. A., 1987, Ecology of nitrification and denitrification, in: *The Nitrogen and Sulphur Cycles* (J. A. Cole and S. Ferguson, eds.), Cambridge University Press, Cambridge, England, pp. 162–218.

Laan, P., Smolders, A., Blom, C. W. P. M., and Armstrong, W., 1989a, The relative roles of internal aeration, radial oxygen losses, iron exclusion and nutrient balances in flood-tolerance of *Rumex* species, *Acta Bot. Neerl.* **38**:131–145.

Laan, P., Beerevoets, M. J., Lythe, S., Armstrong, W., and Blom, C. W. P. M., 1989b, Root morphology and aerenchyma formation as indicators of flood-tolerance of *Rumex* species, *J. Ecol.* **77**:693–703.

Laanbroek, H. J., and Gerards, S., 1991, Effects of organic manure on nitrification in arable soils, *Biol. Fert. Soils* **12**:147–153.

Laanbroek, H. J., and Gerards, S., 1993, Competition for limiting amounts of oxygen between *Nitrosomonas europaea* and *Nitrobacter winogradskyi* grown in mixed continuous cultures, *Arch. Microbiol.* **159**:453–459.

Laanbroek, H. J., and Schotman, J. M. T., 1991, Effect of nitrite concentration and pH on most probable number enumerations of non-growing *Nitrobacter* species, *FEMS Microbiol. Ecol.* **85**:269–278.

Laanbroek, H. J., Bodelier, P. L. E., and Gerards, S., 1994, Oxygen consumption kinetics of *Nitrosomonas europaea* and *Nitrobacter hamburgensis* grown in mixed continuous cultures at different oxygen concentrations, *Arch. Microbiol.* **161**:156–162.

Lang, K., Lehtonen, M., and Martikainen, P. J., 1993, Nitrification potentials at different pH values in peat samples from various layers of a drained mire, *Geomicrobiol. J.* **11**:141–147.

Langelaan, J. G., and Troelstra, S. R., 1992, Growth, chemical composition and nitrate reductase activity of *Rumex* species in relation to form and level of N supply, *Plant Soil* **145**:215–229.

Macdonald, R. M., 1986, Nitrification in soil: An introductory history, in: *Nitrification* (J. I. Prosser, ed.), IRL Press, Oxford, England, pp. 1–16.

Martikainen, P. J., and de Boer, W., 1993, Nitrous oxide production and nitrification in acidic soil from a Dutch coniferous forest, *Soil Biol. Biochem.* **25**:343–347.

Martikainen, P. J., and Nurmiaho-Lassila, E. L., 1985, *Nitrosospira*, an important ammonium-oxidizing bacterium in fertilized coniferous forest soil, *Can. J. Microbiol.* **31**:190–197.

Martikainen, P. J., Lehtonen, M., Lang, K., De Boer, W., Ferm, A., 1993, Nitrification and nitrous oxide production potentials in aerbic soil samples from the soil profile of a finnish coniferous site receiving high ammonium deposition, *FEMS Microbiol. Ecol.* **13**:113–121.

McCaig, A. E., Embley, T. M., and Prosser, J. I., 1994, Molecular analysis of enrichment cultures of marine ammonia oxidizers, *FEMS Microbiol. Ecol.* **120**:363–368.

Meiklejohn, J., 1968, Numbers of nitrifying bacteria in some Rhodesian soils under natural grass and improved pastures, *J. Appl. Ecol.* **5**:291–300.

Meincke, M., Krieg, E., and Bock, E., 1989, *Nitrosovibrio* spp., the dominant ammonia-oxidizing bacteria in building sandstone, *Appl. Environ. Microbiol.* **55**:2108–2110.

Moore, D. R. E., and Waid, J. S., 1971, The influence of washings of living plant roots on nitrification, *Soil Biol. Biochem.* **3**:69–83.

Munro, P. E., 1966, Inhibition of nitrifiers by grass root extracts, *J. Appl. Ecol.* **3**:231–238.

Olff, H., 1992, *On the Mechanisms of Vegetation Succession,* State University of Groningen, The Netherlands, Ph.D. thesis.

Ponnamperuma, F. N., 1984, Effects of flooding on soils, in: *Flooding and Plant Growth* (T. T. Kozlowaki, ed.), Academic Press, New York, pp. 10–45.

Poth, M., 1986, Dinitrogen production from nitrite by a *Nitrosomonas* isolate, *Appl. Environ. Microbiol.* **52**:957–959.

Poth, M., and Focht, D. D., 1985, ^{15}N kinetic analysis of N_2O production by *Nitrosomonas europaea:* An examination of nitrifier denitrification, *Appl. Environ. Microbiol.* **49**:1134–1141.

Prosser, J. I., 1989, Autotrophic nitrification in bacteria, *Adv. Microb. Physiol.* **30**:125–181.

Purchase, B. S., 1974, Evaluation of the claim that grass root exudates inhibit nitrification, *Plant Soil* **41**:527–539.

Rice, E. L., and Pancholy, S. K., 1972, Inhibition of nitrification by climax ecosystems, *Am. J. Bot.* **59**:1033–1040.

Rice, E. L., and Pancholy, S. K., 1973, Inhibition of nitrification by climax ecosystems. II. Additional evidence and possible role of tannins, *Am. J. Bot.* **60**:691–702.

Riha, S. J., Campbell, G. S., and Wolfe, J., 1986, A model of competition for ammonium among heterotrophs, nitrifiers and roots, *Soil Sci. Soc. Am. J.* **50**:1463–1466.

Robinson, J. B., 1963, Nitrification in a New Zealand grassland soil, *Plant Soil* **19**:173–182.

Rosswall, T., 1982, Microbiological regulation of the biogeochemical nitrogen cycle, *Plant Soil* **67**:15–34.

Sand-Jensen, K., Prahl, S., and Stockholm, H., 1982, Oxygen release from submerged aquatic macrophytes, *Oikos* **38**:349–354.

Sarathchandra, S. U., 1978, Nitrification activities of some New Zealand soils and the effect of some clay types on nitrification, *N. Z. J. Agr. Res.* **21**:615–621.

Schimel, J. P., Firestone, M. K., and Killham, K. S., 1984, Identification of heterotrophic nitrification in a Sierran forest soil, *Appl. Environ. Microbiol.* **48**:802–806.

Schmidt, E. L., 1973, Fluorescent antibody technique for the study of microbial ecology, *Bull. Ecol. Res. Comm.* **17**:67–76.

Schmidt, E. L., and Belser, L. W., 1982, Nitrifying bacteria, in: *Methods of Soil Analysis,* 2nd ed. (R. H. Miller and D. R. Keeney, eds.), American Society of Agronomy, Madison, Wis., pp. 1027–1042.

Smit, A. J., and Woldendorp, J. W., 1981, Nitrate production in the rhizosphere of *Plantago* species, *Plant Soil* **61**:43–52.

Smith, A. J., and Hoare, D. S., 1968, Acetate assimilation by *Nitrobacter agilis* in relation to its "obligate autotrophy," *J. Bacteriol.* **95**:844–855.

Spieck, E., Meincke, M., and Bock, E., 1992, Taxonomic diversity of *Nitrosovibrio* strains isolated from building sandstones, *FEMS Microbiol. Ecol.* **102**:21–26.

Stams, A. J. M., Flameling, E. M., and Marnette, E. C. L., 1990, The importance of autotrophic versus heterotrophic oxidation of atmospheric ammonium in forest ecosystems with acid soils, *FEMS Microbiol. Ecol.* **74**:337–344.

Stienstra, A. W., Both, G. J., Gerards, S., and Laanbroek, H. J., 1993, Numbers of nitrite-oxidizing bacteria in the root zone of grassland plants, *FEMS Microbiol. Ecol.* **12**:207–214.

Stienstra, A. W., Klein Gunnewiek, P. J. A., and Laanbroek, H. J., 1994, Repression of nitrification in soils under a climax grassland vegetation, *FEMS Microbiol. Ecol.* **12**:207–214.

Stüven, R., Vollmer, M., and Bock, E., 1992, The impact of organic matter on nitric oxide formation by *Nitrosomonas europaea,* *Arch. Microbiol.* **158**:439–443.

Suzuki, I., Dular, U., and Kwok, S. C., 1974, Ammonia or ammonium ion as substrate for oxidation by *Nitrosomonas europaea* cells and extracts, *J. Bacteriol.* **120**:556–558.

Terry, K. R., and Hooper, A. B., 1970, Polyphosphate and orthophosphate content of *Nitrosomonas europaea* as a function of growth, *J. Bacteriol.* **103**:199–206.

Theron, J. J., 1951, The influence of plants on the mineralization of nitrogen and the maintenance of organic matter in the soil, *J. Agr. Sci.* **41**:289–296.

Tietema, A., de Boer, W., Riemer, L., and Verstraten, J. M., 1992, Nitrate production in nitrogen-saturated acid forest soils: Vertical distribution and characteristics, *Soil Biol. Biochem.* **24**:235–240.

Troelstra, S. R., Wagenaar, R., and de Boer, W., 1990, Nitrification in Dutch heathland soils. I. General soil characteristics and nitrification in undisturbed soil cores, *Plant Soil* **127**:179–192.

Troelstra, S. R., Wagenaar, R., and Smant, W., 1992, Growth of actinorhizal plants as influenced by the form of N with special reference to *Myrica gale* and *Alnus incana, J. Exp. Botany* **43**:1349–1359.

Uhel, C., Roumet, C., and Salsac, L., 1989, Inducible nitrate reductase of rice plants as a possible indicator for nitrification in water-logged paddy soils, *Plant Soil* **116**:197–206.

Van Breemen, N., and Van Dijk, H. F. G., 1988, Ecosystems effects of atmospheric deposition of nitrogen in the Netherlands, *Environ. Poll.* **54**:249–274.

Van Cleemput, O., and Baert, L., 1984, Nitrite: A key compound in N loss processes under acid conditions? *Plant Soil* **76**:233–241.

Van Gool, A. P., Tobback, P. P., and Fischer, I., 1971, Autotrophic growth and synthesis of reserve polymers in *Nitrobacter winogradskyi, Arch. Microbiol.* **76**:252–264.

Verhagen, F. J. M., and Laanbroek, H. J., 1991, Competition for ammonium between nitrifying and heterotrophic bacteria in dual energy-limited chemostats, *Appl. Environ. Microbiol.* **57**:3255–3263.

Verhagen, F. J. M., and Laanbroek, H. J., 1992, Effects of grazing by flagellates on competition for ammonium between nitrifying and heterotrophic bacteria in chemostats. *Appl. Environ. Microbiol.* **58**:1962–1969.

Verhagen, F. J. M., Duyts, H., and Laanbroek, H. J., 1992, Competition for ammonium between nitrifying and heterotrophic bacteria in continuously percolated soil columns, *Appl. Environ. Microbiol.* **58**:3303–3311.

Verhagen, F. J. M., Duyts, H., and Laanbroek, H. J., 1993, Effects of grazing by flagellates on competition for ammonium between nitrifying and heterotrophic bacteria in soil columns, *Appl. Environ. Microbiol.* **59**:2099–2106.

Verhagen, F. J. M., Hageman, P. E. J., Woldendorp, J. W., and Laanbroek, H. J., 1994a, Competition for ammonium between nitrifying bacteria and plant roots in soil in pots; effects of grazing by flagellates and fertilization, *Soil Biol. Biochem.* **26**:89–96.

Verhagen, F. J. M., Zölffel, M., Bruggerolle, G., and Patterson, D. J., 1994b, *Adriamonas peritocrescens* gen. no., sp. nov., a new free-living soil flagellate (Protista incertae sedis), *Eur. J. Protozool.* **30**:295–308.

Walker, N., and Wickramasinghe, K. N., 1979, Nitrification and autotrophic nitrifying bacteria in acid tea soils, *Soil Biol. Biochem.* **11**:231–236.

Ward, B. B., 1986, Nitrification in marine environments, in: *Nitrification* (J. I. Prosser, ed.), IRL Press, Oxford, England, pp. 157–184.

Watson, S. W., Bock, E., Harms, H., Koops, H.-P., and Hooper, A. B., 1989, Nitrifying bacteria, in: *Bergey's Manual of Systematic Bacteriology*, Vol. 3 (J. T. Staley, M. P. Bryant, N. Pfennig, and J. G. Holt, eds.), Williams & Wilkins, Baltimore, pp. 1808–1834.

Woese, C. R., Weisburg, W. G., Hahn, C. M., Paster, B. J., Zahlen, L. B., Lewis, B. J., Macke, T. J., Ludwig, W., and Stackebrand, E., 1985, The phylogeny of purple bacteria: The gamma subdivision, *Syst. Appl. Microbiol.* **6**:25–33.

Woese, C. R., Weisburg, W. G., Paster, B. J., Hahn, C. M., Tanner, R. S., Krieg, N. R., Koops, H.-P., Harms, H., and Stackebrand, E., 1986, The phylogeny of purple bacteria: The beta subdivision, *Syst. Appl. Microbiol.* **5**:327–336.

Woldendorp, J. W., and Laanbroek, H. J., 1989, Activity of nitrifiers in relation to nitrogen nutrition of plants in natural ecosystems, *Plant Soil* **115**:217–228.

Wood, P. M., 1987, Monooxygenase and free radical mechanism for biological ammonia oxidation, in: *The Nitrogen and Sulphur Cycles* (J. A. Cole and S. Ferguson, eds.), Cambridge University Press, Cambridge, England, pp. 219–243.

Zak, D. R., Groffman, P. M., Pregitzer, K. S., Christensen, S., and Tiedje, J. M., 1990, The vernal dam: Plant–microbe competition for nitrogen in northern hardwood forests, *Ecology* **71:**651–656.

8

The Ecological and Physiological Significance of the Growth of Heterotrophic Microorganisms with Mixtures of Substrates

THOMAS EGLI

1. Introduction

It has been estimated that globally some 500×10^{12} kg of carbon dioxide are assimilated into biomass by autotrophic organisms annually. More than 99% of this assimilated carbon is remineralized, keeping the global biogeochemical carbon cycle roughly in balance (Hedges, 1992). In both terrestrial and aquatic ecosystems the majority of this primary biomass is not consumed directly by herbivorous animals, but decays to detritus and serves as a nutritional basis for the growth of consumers (for an extensive discussion, see Fenchel and Jørgensen, 1977). There is now substantial evidence suggesting that a large part of the energy and nutrients contained in this primary biomass is processed via the microbial detritus food chain, and this mineralizing ability makes heterotrophic microorganisms an important link in the global carbon cycle (Fenchel and Jørgensen, 1977; Paul and Voroney, 1980; Wetzel, 1984; Cole *et al.*, 1988; Mann, 1988). In addition, their ability to mineralize man-made xenobiotic organic chemicals has become increasingly important. This is illustrated by the fact that in industrialized countries the flux of synthetically produced organic material, much of which is ending up in the environment, has increased within the past two centuries to some 40 g C m^{-2} year^{-1}. This figure is equivalent to approximately 15% of the net primary biomass production in these regions (Egli, 1992).

THOMAS EGLI • Department of Microbiology, Swiss Federal Institute for Environmental Science and Technology (EAWAG), CH-8600 Dübendorf, Switzerland.

Advances in Microbial Ecology, Volume 14, edited by J. Gwynfryn Jones. Plenum Press, New York, 1995.

1.1. Detritus as the Basis for Heterotrophic Microbial Growth: There Is a Lot of Choice

Detritus present in ecosystems and its degradation products are traditionally divided into a particulate and a dissolved organic matter (carbon) fraction (POM, POC, DOM, DOC). In aquatic systems the two fractions are defined arbitrarily as the organic material either being retained by or passing a filter of 0.2- to 0.45-μm pore size (Williams, 1986; for soil, see Nedwell and Gray, 1987). However, one should keep in mind that the terms "particulate" and "dissolved" are relative, and that DOC also contains small particles and colloidal matter (Williams, 1986; Münster, 1993). In oceans the concentration of total organic carbon is typically in the range of one to a few milligrams per liter. Most of it is present as DOC, and the ratio between DOC:POC:living biomass is typically in the range of 100:10:1–2 (Fenchel and Jørgensen, 1977). Similar figures have been reported for oligo- and mesotrophic freshwaters (Münster and Chróst, 1990). In eutrophic aquatic systems the concentration of DOC is usually increased by a factor of 5–10 (Münster and Chróst, 1990).

Not only POC but also a large fraction of the DOC is present in a polymeric form. Some of the major components that have been identified include cellulose, lignin, chitin, hemicelluloses, lignocelluloses, proteins, inulin, nucleic acids, polyphenols, waxes, tannin, and melanin, but there are many others more, as well as many that have not yet been identified (Williams, 1986; Nedwell and Gray, 1987; Beauchamp *et al.*, 1989; Karl and Bailiff, 1989; Münster and Chróst, 1990). In lacustrine eutrophic waters a large fraction of the DOC, occasionally as much as 80% (Münster, 1993), can be attributed to aromatic compounds, carbohydrates, amino acids (most of them as polymers), and to long-chain fatty acids, urea, and organic acids. In contrast, in oligotrophic marine waters usually only some 10–15% of the organic carbon fraction can readily be identified (Williams, 1986). It should be added here that the chemical analysis of particulate, colloidal, and dissolved organic matter in soil or water is extremely complex; the problems in quantification and in determining its composition are many and beyond the scope of this chapter (for information, see, e.g., Williams, 1986; Koike *et al.*, 1990; Lee and Wakeham, 1992; Pakulski and Benner, 1992; Kaplan and Newbold, 1993).

Because microorganisms are only able to transport monomers and oligomers in the range of a few hundred daltons across the cell envelope (Cronan *et al.*, 1987; Lin, 1987; Furlong, 1987), only a small fraction of the total DOC (0.5–5%) can be readily utilized for microbial growth. In the literature this fraction of DOC is referred to as the "labile organic carbon pool" (Münster, 1993), "assimilable organic carbon" (van der Kooij, 1990), or "direct substrates" (Lancelot *et al.*, 1989). Hence, the major part of the potential carbon substrates in DOC initially has to be hydrolyzed outside of the cell into "digestible" portions before it can be taken up (Rogers, 1961; Azam and Cho, 1987; Hoppe *et al.*, 1988). [Directly utilizable compounds excreted by primary producers during photosynthetic activity or released during cell death or "sloppy" feeding seem to be a relevant source of carbon only in highly productive, eutrophic aquatic systems (Jackson, 1987; Williams, 1990; Münster and Chróst, 1990).] The processes that make particulate organic matter and polymeric dissolved organic matter available for the growth of heterotrophic microbes are essentially the same in aqueous systems, sediments, and soils (Azam and Cho, 1987; Nedwell and Gray, 1987; Ducklow and Carlson, 1992). They include enzymatic or (photo)chemical action and sometimes dissolution (Paul and Clark, 1989; Münster, 1991; Hedges, 1992; Lee and Wakeham, 1992). A substantial number of the hydrolytic enzymes that process polymeric carbon substrates are located at the surface of microbial cells, and the hydrolysis of, for example, proteins can be tightly coupled to uptake (Hollibaugh and Azam, 1983; Vives Rego *et al.*, 1985; Chróst, 1991; Hoppe, 1991; Martinez and Azam, 1993).

In aquatic environments the pool of readily utilizable monomeric compounds has been investigated extensively, with particular attention paid to carbohydrates, amino acids, and short- and long-chain fatty acids (Münster and Chróst, 1990; Münster, 1993). As indicated from their fast turnover times, dissolved free amino acids (DFAA) and carbohydrates (DFCHO) are considered to be especially important carbon, nitrogen, and energy sources for heterotrophic microorganisms in aquatic ecosystems (Münster, 1993). Concentrations of DFCHO and DFAA depend usually on the trophic state of the aquatic system. In the open ocean, average values for total DFCHO and DFAA have been reported to be in the range of 10 μg C liter^{-1} and 1 μg C liter^{-1}, respectively, whereas they are some five to ten times higher in more productive coastal marine waters, or mesotrophic and eutrophic freshwater samples. In the euphotic zone of a lake, glucose was frequently reported to contribute a significant fraction to the pool of DFCHO, with concentrations ranging between 1 to almost 200 μg liter^{-1} (Münster and Chróst, 1990). In this aquatic environment the concentrations of DFAA and DFCHO, and also of other substrates such as glycollate and glycine, were reported to exhibit a distinct day and night variation pattern, with peak concentrations observed in the late afternoon (Münster, 1991). Only recently it has been found that also the pool of nucleic acids is turned over quickly in the aquatic environment, suggesting that these compounds also contribute considerably to the diet of heterotrophic microbes (Karl and Bailiff, 1989). It should be pointed out, however, that there are many indications that not all of the analytically detectable monomeric substrates are readily available for microbial cells (Azam and Cho, 1987; Nedwell and Gray, 1987; Keil *et al.*, 1994).

1.2. Why Consider Growth with Mixtures of Substrates?

The scenario outlined for aquatic environments illustrates the conditions that heterotrophic microorganisms typically experience in nature with respect to their main growth-supporting substrates supplying carbon and energy (but partly also nitrogen and phosphorus). It is a pool consisting of hundreds of different carbonaceous compounds with concentrations of individual components that are extremely low, typically in the nanogram per liter or, at the most, microgram per liter range.

Early experiments suggested that the concentrations of individual utilizable carbon/energy substrates in aquatic ecosystems were too low to support growth of free-living microbes (Jannasch, 1967). This led to the assumption that most of the free-living bacteria in the sea and in freshwater were dormant and that perhaps only those attached to algal cells or particulate material, where they might be exposed to higher local substrate concentrations, were metabolically active and growing (see, e.g., Stevenson, 1978; Dawes, 1985). However, using microautoradiographic methods, it has now been demonstrated that, although extremely small and obviously starved (Morita, 1985), at least 30–80% of the free-living aquatic bacteria were assimilating organic substrates and actively dividing in oligotrophic environments (Hoppe, 1976; Meyer-Reil, 1978; Fuhrman and Azam, 1982; Tabor and Neihof, 1982; Douglas et al., 1987). For example, Tabor and Neihof (1982) demonstrated that 73% of bacterial cells in natural waters incorporated labeled thymidine and that 94% took up amino acids. Furthermore, it has been shown that mixed and pure cultures of free-living bacteria can be cultivated in batch and continuous culture at the expense of dissolved organic carbon present in filtered sea or lake water (Fuhrman and Azam, 1980; Ammerman et al., 1984; Tranvik and Höfle, 1987; Tranvik, 1990). Average generation times in the range of 12 to 48 hr have been measured in seawater, despite the extremely low concentrations of individual carbon substrates and grazing pressure (Azam and Cho, 1987). Hence, there is now no doubt that not only bacteria attached to organic particles but also the small free-living bacterial cells are growing and dividing (Kirchman, 1993; Kaprelyants et al., 1993). It has even been suggested that the free-living bacteria are mainly responsible for the mineralization of organic matter in the sea (Hoppe, 1976).

Considering the central role of carbon/energy sources for both maintenance and growth, it seems logical to presume that, at the low concentrations of carbon compounds found in highly oligotrophic environments, a successful growth strategy for a competitive heterotroph would be to take up as many as possible of the different available carbon/energy substrates at the same time. One can speculate that utilization of a mixture of different carbonaceous substrates simultaneously, instead of making use of only one of them, might give a competitive advantage to a cell. Surprisingly, microbial ecologists have paid little attention to this aspect of the "daily struggle for food," and defined studies considering growth with mixtures of substrates in the natural environment are scarce. How little this topic is considered to be of importance to microbial ecology may be seen from the fact that all major microbial ecology textbooks ignore this subject. Also, growth with substrate mixtures is hardly ever mentioned when the status of microbial ecology and the outstanding problems are discussed (see, e.g., Tate, 1986; Brock, 1987; Hobbie and Ford, 1993).

One may ask the question why growth with mixtures of substrates at low concentrations has received so little attention, despite the fact that its importance for microbial ecology has occasionally been pointed out in the literature. For example, it was emphasized by various researchers that laboratory media containing high concentrations of single carbon substrates do not reflect environmental growth conditions (e.g., ZoBell and Grant, 1942; Veldkamp and Jannasch, 1972). Already in the 1960s, the ability of pure cultures of bacteria and yeasts to utilize mixtures of two carbon sources simultaneously under carbon-limited growth conditions in continuous culture had been described (Harte and Webb, 1967; Mateles *et al.*, 1967; Silver and Mateles, 1969). Also, it had been realized that both the quantity and composition of readily available dissolved organic matter was one of the important parameters influencing both the extent and rates of the biodegradation of pollutants under environmental conditions (Simkins and Alexander, 1984; Schmidt and Alexander, 1985). Furthermore, it was recognized that the composition of microbial consortia in nature is a direct reflection of the composition and availability of the substrates for microbial growth (Brock, 1987); nevertheless, the composition of microbial consortia has attracted far more attention than the fact that they are feeding on substrate mixtures. Was it that the situation in nature with the presence of not only mixed substrates but also an unknown number of microbial strains was considered too complex to be approached experimentally? Or was the concept outlined by Monod (1942) that a single substrate controls the rate of growth, together with the elegant demonstration of the molecular regulatory mechanisms operating during catabolite repression (Jacob and Monod, 1961) and resulting in diauxic growth behavior (confirming that microbial cells also have economical sense and do not synthesize unnecessary proteins), too willingly applied to growth in the environment? This, despite the fact that it was pointed out that the concept of diauxic utilization of mixtures of substrates, commonly observed at high initial substrate concentrations in batch culture, can probably rarely be applied to conditions in nature (Veldkamp and Jannasch, 1972; Harder and Dijkhuizen, 1976). The former authors went as far as to state that diauxic growth was a "laboratory artifact."

All this does not mean that the utilization of mixtures of substrates has been totally neglected by microbiologists. After the first reports in the late 1960s, a considerable body of work has been published, especially on the simultaneous utilization of mixtures of carbon substrates by pure cultures of microorganisms under both batch and continuous cultivation conditions, indicating that this phenomenon occurs much more frequently than originally thought. A small number of these studies has also addressed ecological aspects. It is not the intention to review in depth the literature that has accumulated over the last two decades that confirms that pure cultures of microroganisms can utilize mixtures of substrates simultaneously under various cultivation conditions. There are a number of publications where the essential information has been compiled for interested readers (Harder and Dijkhuizen, 1976, 1982; Weide, 1983; Egli and Harder, 1984; Bull, 1985; Gottschal, 1986; Egli and Mason, 1991; Babel et al., 1993). Instead, this chapter will concentrate on aspects of mixed substrate utilization that have not been addressed in the above reviews, such as: (1) the physiological state of microbial cells under environmental conditions, in particular with respect to their nutritional versatility when growing carbon/energy-limited; (2) the kinetics of growth during the simultaneous utilization of mixtures of carbon sources; (3) the degradation of pollutants in the presence of other easily degradable "natural" carbon substrates; and (4) the interaction of substrates during growth with mixtures of compounds that do not fulfill the same nutritional (physiological) purpose.

Of course, with the techniques presently at hand, experimental studies in ecosystems on the use of mixtures of substrates at the single-cell level will be difficult if not impossible to perform. But this is not necessarily required for a better understanding of mixed substrate growth behavior of microorganisms in nature. The author is convinced that defined mixed substrate studies with pure microbial cultures also are a valuable way to understand the behavior of microorganisms in their natural environment. Therefore, by necessity rather than by free will, this chapter approaches the problem of the utilization of substrate mixtures from the pure culture side and tries to point out which information obtained in mixed substrate-pure culture studies might be usefully applied to the growth of microorganisms in the complex situation of ecosystems.

Growth of microbes in aquatic systems has been treated here with some preference. Obviously, such systems are less heterogeneous than soil. They are better defined with respect to substrate availability, turnover rates, and so on, and therefore are experimentally more easily accessible. Nevertheless, much of the information presented here on the physiological behavior of microorganisms during growth with and utilization of mixtures of substrates can also be applied to the situation in soil.

1.3. Mixed Substrates, Multiple Nutrients, and Limitation of Growth: Some Definitions

Several ways of expressing the nutritional role of compounds (or "resources") (Smith, 1993) that are required by microorganisms for growth and classifying this role in relation to other growth substrates are presently used in the literature (for a discussion, see Tilman, 1982; Baltzis and Fredrickson, 1988; Smith, 1993). If different compounds fulfill the same physiological function during growth, for example, as sources of carbon, they have been categorized as *homologous* (Harder and Dijkhuizen, 1976, 1982); other authors have chosen to call them *perfectly substitutable substrates* (León and Tumpson, 1975; Tilman, 1980). Compounds that serve different physiological requirements, for example, one acts as a source of carbon whereas the other supplies nitrogen, are called *heterologous* (Harder and Dijkhuizen, 1982), and synonyms proposed were *complementary* (Baltzis and Fredrickson, 1988; León and Tumpson, 1975) or *essential* (Tilman, 1980). These definitions can be applied not only for elemental components of growth substrates that are incorporated into biomass during growth but also to sources of energy or to terminal electron acceptors. In the present chapter, the nomenclature proposed by Harder and Dijkhuizen (1982), i.e., homologous and heterologous substrates, will be used.

The expressions "mixed substrates" and "mixed substrate growth," respectively, were introduced to indicate that a heterotrophic bacterium utilized two homologous (carbon/energy) substrates at the same time (Harte and Webb, 1967; Mateles *et al.*, 1967). Some authors also referred to this situation as *multiple substrate growth* (Kompala *et al.*, 1984). Mixed substrate growth has since been used in a wider context to embrace all types of homologous nutrients, including mixotrophic growth behavior (Harder and Dijkhuizen, 1982). It is in this sense that it will be used here as well.

In the case of mixtures of heterologous substrates serving different nutritional purposes, it is not so much the fact that they are consumed together that is of interest—this is a necessity anyway, as one cannot take over the physiological function of the other—but the effect of a simultaneous restricted availability of two or more such nutrients during growth and how this situation affects cell physiology. Such growth conditions were observed to occur in chemostat cultures fed with different ratios of two heterologous substrates, mainly mixtures of a carbon and a nitrogen source, and it was concluded that both nutrients appeared to simultaneously "limit" growth of the culture (for a summary, see Egli, 1991). To date, stable and reproducible conditions for such a nutritional state have only been described for growth in the chemostat, and it will be referred to as double- or multiple-nutrient-limited (or controlled) growth. But transiently, such conditions can also occur during other cultivation conditions, e.g., in batch or fed-batch cultures. The phenomenon will be described in Section 4.

Finally, a comment is necessary on the usage of the terms "nutrient limitation" and "nutrient-limited growth." These terms have been used to describe two different phenomena, i.e., first, the termination of growth by exhaustion of a particular nutrient in batch culture, and second, for nutrient-limited growth as it is encountered in continuous culture. As discussed recently (Mason and Egli, 1993), the two situations are fundamentally different for a cell and the occurring physiological processes. In this chapter the two terms will be used exclusively for growth conditions as they occur during cultivation in continuous culture.

2. Physiological State of Microbial Cells in the Environment

In contrast to higher organisms, microbial cells are directly exposed to the physical and chemical conditions of their environment and have only limited options to adjust it according to their needs. Therefore, in order to survive and compete successfully, most microorganisms are able to meet many of the environmental challenges by adjusting their cellular composition with respect to both structure and metabolic function. This adaptation can occur at both the phenotypic and the genotypic level, and the limits within which it can take place vary, of course, for different genospecies. Consequently, for a microbial cell with a given genotype, the structural composition and metabolic capacity, i.e., its physiological state, is intimately linked to the environmental conditions in which it is growing (e.g., Kluyver, 1956; Herbert, 1961b; Tempest, 1970b; Koch, 1976; Tempest and Neijssel, 1978).

In our attempt to understand what strategies heterotrophic microorganisms growing in nature are following to cope with the presence of mixtures of substrates, it is important to know what the physiological states of such cells are with respect to their ability to utilize different nutritional compounds. Furthermore, one should know how this ability is affected by environmental growth conditions. Answers to these questions require information at the level of single cells and less at the level of populations. Ideally, information obtained by directly studying individual microbial cells *in situ* and, for example, measuring their uptake potential for different substrates or even investigating their pattern of expressed versus repressed catabolic enzymes is necessary. However, the methods presently available for such studies are still seriously limited. Although one can obtain information on specific properties or activities of special groups of organisms or even of individual cells (see, e.g., White, 1986; Karl, 1986; Mills and Bell, 1986; Paul, 1993), it is not yet possible to obtain a more general picture of the physiological state and metabolic potential of a microbial cell growing in the environment (Hobbie and Ford, 1993; Paul, 1993).

Due to this dilemma, we have to return to the wealth of information that has been obtained in the laboratory with pure cultures of heterotrophic microorganisms mostly grown on media with a single growth-limiting substrate and to try and carefully examine which of the data could be applied and extrapolated to conditions as they occur in ecosystems. In the following section, some of the important information will be reviewed that might give us an idea of the nutritional versatility and potential of heterotrophic microbial cells growing in the environment. It should be kept in mind that most of this information originated from cells growing in suspended culture, and, whereas the link to the growth of free-living microbes in aquatic ecosystems is obvious, extrapolation to soil systems has to be made with care.

2.1. Experimental Approaches to Nutrient-Limited Growth and the Importance of Limitation of Heterotrophic Growth by Carbon/Energy Substrates

In addition to physicochemical factors, such as temperature, light, or pH, the availability of different nutrients is known to be a key parameter determining the physiological state of microbial cells. Two fundamentally different approaches are presently used to experimentally study the physiological events that take place in microorganisms when the availability of a particular substrate (nutrient) becomes restricted. A first technique is to deprive (starve) cells of specific nutrients and follow as a function of time the changes that take place in the cells after exposure to this challenge. The usual procedures are that cells are either cultivated in batch culture where a specific nutrient runs out during growth or by transferring exponentially growing cells from batch cultures to a starvation medium lacking a particular component. The state of the art of this approach has been documented in a recent book (Kjelleberg, 1993; see, e.g., Östling et al., 1993; Hengge-Aronis, 1993; Siegele et al., 1993; Nyström, 1993). The second approach is that of continuous cultivation where the rate of growth of a microbial culture can be maintained at any value below the maximum specific growth rate (μ_{max}) by the constant addition of fresh medium containing a particular substrate at limiting concentrations (Herbert, 1961a; Tempest, 1970a). When in equilibrium (steady state), culture parameters such as growth environment and cellular composition are constant (Herbert, 1961a). Much information has been collected from continuous cultures on the influence of different nutrient limitations and growth rate on the physiological state of microorganisms (reviewed in, e.g., Tempest, 1970b; Tempest and Neijssel, 1978; Harder and Dijkhuizen, 1983; Tempest et al., 1983; Harder et al., 1984).

It is appropriate to recall that the word "starvation" is presently used to describe two different processes: On one hand, it means exposure of a cell to a complete absence of one or more specific nutrients (as done in batch starvation experiments); on the other hand, it is used to describe the state of slow (or even no) growth as it occurs in nature, where cells remain active and are adapted to the low nutrient fluxes that are available for them (which corresponds roughly to growth in continuous culture at low dilution rates). The two processes are physiologically completely different: whereas the former is a transient state phenomenon [intracellular resources are mobilized within the cell and reallocated as necessary (see Wanner and Egli, 1990)], the latter results in a physiological state that is virtually stable for extended periods of time.

Of course, neither of the two systems is able to exactly reproduce the growth and starvation conditions that microorganisms experience in ecosystems. However, as recently pointed out by Morita (1993), from an ecological viewpoint one must ask the critical question of whether the right experiments were in fact carried out when using the first experimental approach and exposing microbes to starvation conditions. (This does not detract from the relevance of the fast and exciting progress that is presently made with respect to the molecular understanding of adaptation processes when using this technique nor does it imply that these are ecologically meaningless.) Morita argued that, although it is now known with considerable certainty that most cells in aquatic (and most probably also in terrestrial) ecosystems are viable and metabolically active, much of the ecological research concentrates on growing bacteria in laboratory media and subjecting them to starvation conditions, whereas from an ecological viewpoint one should study the response of bacterial cells growing under oligotrophic conditions (their normal growth environment) and investigate their response to nutrient fluctuations (including eutrophic conditions). Morita pointed out that slow growth in continuous culture represents a method that probably most closely resembles the growth conditions bacteria encounter in the natural environment and that this system allows to control growth conditions to a better degree than any other cultivation system available.

There is still much debate about which nutrient controls the growth of heterotrophic microbes in different ecosystems and to what extent they are in a state of starvation (discussed in, e.g., van Es and Meyer-Reil, 1982; Morita, 1988; Moriarty and Bell, 1993; Kirchman, 1993). However, a common view seems to be that, in addition to temperature, the slow rates of hydrolysis of organic matter and the ability of heterotrophic bacteria to utilize organic substrates down to extremely low concentrations leads to a situation where the availability of carbon/energy either alone or in combination with other nutrients, such as nitrogen and phosphorus, restricts growth. The observations that (1) turnover times of monomeric carbohydrates and amino acids are usually short in aquatic systems and that (2) added utilizable energy substrates are consumed immediately (Novitsky and Morita, 1978; van Es and Meyer-Reil, 1982) both support the idea that the carbon flux through the pool of dissolved organic utilizable substrates from unavailable particulate or dissolved organic matter into bacterial biomass is fast and not saturated. This suggests that the decomposition of particulate and dissolved polymeric organic material is the rate-limiting step over extended periods of time. This applies not only to aquatic ecosystems (Azam and Cho, 1987; Cole *et al.*, 1988; Hoppe *et al.*, 1988; Morita and Moyer, 1989; Kirchman, 1990; Münster, 1991; Ducklow and Carlson, 1992; Lee and Wakeham, 1992), but also to soils and aquatic sediments, despite the presence of large amounts of organic matter (Williams, 1985; Nedwell and Gray, 1987).

The conclusion that growth of heterotrophic microbes in the marine environment is limited by carbon/energy more than by any other nutrient has recently obtained support from laboratory investigations with pure cultures of a marine gram-negative bacterium (Kjelleberg *et al.*, 1993; Östling *et al.*, 1993). These authors have investigated the processes occurring when cells of *Vibrio* sp. S14 were starved for different nutrients in an artificial seawater microcosm. When this bacterium was subjected to carbon starvation, miniaturization of the cells occurred and at the same time they remained viable and developed resistance to various environmental stresses. Similar results were obtained when carbon in combination with nitrogen and phosphorus was omitted from the starvation medium. However, when cells were transferred into nitrogen- or phosphorus-deficient media, the *Vibrio* sp. S14 used did not develop into minicells, had a low survival rate, and exhibited reduced resistance to stresses. This finding correlates well with the reports that free-living bacteria found in the sea and in freshwater are usually very small in size (termed dwarf cells, ultramicrocells, minicells). Such cells have recently been isolated and grown in pure culture (Schut *et al.*, 1993, 1995), and others were reported to increase in size and multiply when fed properly and again undergo "miniaturization" when exposed to nutrient-restricted conditions (Torella and Morita, 1981; Morita, 1985; Kjelleberg *et al.*, 1987).

Hence, there is now good evidence that growth of heterotrophic microbes in aquatic and terrestrial ecosystems is to a large part restricted by the availability of carbon/energy substrates. Nevertheless, there are indications that the rate of heterotrophic microbial growth might transiently be controlled in many ecosystems—ranging from the ultraoligotrophic open ocean to heavily eutrophic aquatic environments—also by other nutrients, especially nitrogen or phosphorus, either alone or in combination with carbon (e.g., Paerl, 1982, 1993; Lancelot and Billen, 1985). A new concept of why and under what conditions such a limitation of microbial growth by several nutrients simultaneously might occur is discussed in more detail in Section 4.

2.2. Nutritional Versatility of Microbial Cells Growing Carbon/Energy-Limited

2.2.1. Strategies Followed in the Expression of Enzymes Involved in Carbon Catabolism

Only a few of the proteins encoded on a microbial genome are constantly expressed; a considerable part of them are synthesized only when required for growth or survival (Koch, 1976; Magasanik, 1976; Neidhardt *et al.*, 1990). Indeed, it has been shown that the synthesis of many catabolic enzymes is regulated as a function of both nutrient availability and environmental growth conditions. Because this subject has received much attention in the past years (Tempest and Neijssel, 1978; Harder and Dijkhuizen, 1983; Tempest *et al.*, 1983; Harder *et al.*, 1984; Poindexter, 1987), only the most essential information will be outlined.

2.2.1a. High Substrate Concentrations. At high concentrations, many carbon/energy substrates supporting fast growth have the ability to inhibit the utilization of carbon sources that allow only lower growth rates. The classic example is that of *Escherichia coli* growing in batch culture with a mixture of glucose and lactose, where glucose represses the expression of enzymes involved in the metabolism of lactose (Monod, 1942). As a result of this regulation, a diauxic growth pattern is observed where lactose is utilized only after all the glucose has been consumed. Such regulation patterns have also been observed for many other microorganisms and substrate combinations (Monod, 1942; Harder and Dijkhuizen, 1982; Wanner and Egli, 1990). The molecular mechanisms involved in this regulation at the level of gene expression have been studied extensively for different systems and carbon catabolite repression, induction, and inducer exclusion have been identified as the main regulatory strategies (see, e.g., Postma, 1986; Beckwith, 1987; Saier, 1989; Lengeler, 1993). However, repression-induction is not the only mechanism controlling the synthesis of catabolic enzymes. Synthesis of catabolic enzymes that are specific for the breakdown of carbon sources that were not supplied in the growth medium has frequently been observed. This phenomenon where a protein is synthesized in the absence of an "inducing" compound has been called *derepression* (Gorini, 1960). This definition is at the phenomenological level and does not imply a certain mechanistic way of regulation at the molecular level. It also differs from the definition of derepression usually found in the genetic literature where it is used as a synonym for *induced*, i.e., synthesis of an enzyme in response to the presence of an inducing compound (Lewin, 1994). Furthermore, it contrasts with the definition of *constitutivity*, which implies that an enzyme is present always at the same level, irrespective of the growth conditions.

2.2.1b. *Low Substrate Concentrations.* Various strategies are known for adaptation of growth at low substrate concentrations. For instance, a cell's scavenging ability for a particular substrate can be enhanced by increasing the amount of transport enzyme or substrate-specific porins in the outer membrane (i.e., increasing V_{max}), by changing from a low-affinity uptake system to one of higher affinity (i.e., decreasing K_m), or by supporting a transport enzyme by the synthesis of a periplasmic substrate capturing (binding) protein. A further option is to keep the intracellular concentration of a substrate low by enhancing either V_{max} or decreasing K_m of the initial metabolic steps. Examples for all these strategies are well documented and have been discussed in detail in several excellent reviews (Tempest and Neijssel, 1978; Tempest *et al.*, 1983; Harder *et al.*, 1984; Harder and Dijkhuizen, 1983; Kjelleberg *et al.*, 1987; Lengeler, 1993). A less well-documented but highly interesting strategy might be the use of multiphasic transport systems, i.e., transport enzymes that change their kinetic behavior depending on substrate concentration (Nissen *et al.*, 1984). In addition to optimizing the efficiency of transport and subsequent metabolism of individual substrates, a cell also has the option to enhance the rate of carbon supply into the central pathways by either increasing the number of different catabolic enzyme systems or by synthesizing transport proteins with a broader specificity but still high affinity for the different substrates.

2.2.2. The Cells Are Ready: Evidence for Derepression of Catabolic Enzymes

The restricted availability of nutrients, especially that of carbon/energy sources, has led to the selection of heterotrophic microorganisms with physiological properties that allow them to efficiently cope and survive in "unhospitable" low-nutrient environments. One strategy employed is to shut down metabolism, form a resting state, and wait for better times (see, e.g., Dow et al., 1983; Sonenshein, 1989). If this is not possible, a strategy has to be adopted where the most is made out of the low concentrations of available substrates and to take up these carbon compounds as efficiently as possible. Many gram-negative heterotrophic bacteria, especially those living in the open ocean, seem to be successful with the latter option. Considering the complex mixture of carbonaceous compounds present at low concentrations in oligotrophic ecosystems, it would make perfect sense to express a wide range of catabolic enzymes at a low level, rather than to specialize on metabolizing only one or a few carbon substrates. Such expression of a set of enzymes for carbon catabolism would enable a cell, on the one hand, to maximize the simultaneous uptake of a multiplicity of substrates and, on the other hand, to immediately utilize a compound when it becomes available and reduce the time required to react toward a transient availability of a particular substrate (Matin, 1979; Harder and Dijkhuizen, 1982; Poindexter, 1987). Hence, we can ask the question: How is the expression of potentially available catabolic enzymes regulated at the genetic level when cells are growing under carbon/energy-limited conditions in the environment, and what is the resulting capacity of such a cell to utilize different carbon substrates?

Indeed, there is now considerable experimental evidence for the derepression of catabolic enzymes. This has been shown in cells growing under carbon/energy-limited conditions at low dilution rates in continuous culture (Pfennig and Jannasch, 1962; Dean, 1972; Matin, 1979; de Vries et al., 1988), as well as in cultures growing under carbon-sufficient conditions of a batch culture. Derepression was characteristically found when cells were supplied with substrates that supported medium-to-low maximum specific growth rates. A well-investigated example is that of methylotrophic yeasts. Derepression of enzymes involved in the dissimilation of methanol was observed at low dilution rates in glucose-, sorbitol-, glycerol-, or xylose-limited cultures of the methylotrophic yeasts Hansenula polymorpha and Kloeckera sp. 2201 (Egli et al., 1980; Eggeling and Sahm, 1981). Similarly, in the exponential phase of batch cultures, growth with a range of different sugars (except glucose) resulted in derepression of methanol-related enzymes to various degrees (Eggeling and Sahm, 1978; de Koning et al., 1987). Further evidence for an extensive derepression of catabolic enzyme systems was reported for two unidentified heterotrophic bacterial strains isolated from freshwater (Sepers, 1984). One of the strains was grown in carbon-limited chemostat culture at different dilution rates (D) with aspartate as the only carbon/energy source. Cells collected from the chemostat were tested in an oxygen probe for their capacity to respire 39 different carbon sources. When grown at $D = 0.001$ hr^{-1}, the strain was able to oxidize 29 out of the 39 substrates. In contrast, only nine of these substrates stimulated the oxygen uptake rate of cells grown at $D = 0.34$ hr^{-1}.

These observations suggest that microbial cells growing in carbon/energy-limited chemostat culture with a particular carbon substrate should in principle be able to immediately utilize and grow with a range of alternative carbon substrates. This was tested recently by examining cells of *E. coli* grown in glucose-limited chemostat culture at different dilution rates for their ability to immediately utilize and grow with six sugars other than glucose (Lendenmann and Egli, 1995). Cells collected from the chemostat were exposed under batch culture conditions to excess concentrations of individual sugars, and both uptake and growth were recorded (Table I). When cultivated at $D = 0.2$ hr^{-1}, the cells were able to immediately utilize mannose, maltose, ribose, and galactose at rates that were two- to threefold higher than the actual glucose consumption rate of the culture in the chemostat at this dilution rate. In addition, the cells also were able

Table I. Potential for Immediate Substrate Consumption $(q_{s(excess)})$ and Growth of *Escherichia coli*[a,b]

Sugar supplied	$q_{s(excess)}$	lag($q_{s(excess)}$)	μ (hr^{-1})	lag(μ)
Glucose	−1.7	<10 sec	0.64	<3 min
Fructose[c]	−0.01 (−0.4)	<10 sec (3 min)	0.58	<3 min
Mannose	−0.9	<10 sec	0.45	<3 min
Maltose	−1.6	<10 sec	0.58	<3 min
Ribose	−1.4	<10 sec	0.49	<3 min
Galactose	−0.9	<10 sec	0.49[d]	~10 min
Arabinose	0	~6 min	0.53[e]	~20 min

[a] Adapted from Lendenmann (1994).
[b] Grown in carbon-limited chemostat culture at a dilution rate of 0.2 hr^{-1} with glucose as the only source of carbon and energy, when exposed to excess concentrations of other sugars in batch culture. Specific consumption rates q_s are given in hr^{-1} [g sugar (g cell dry wt.)$^{-1}$ hr^{-1}]. $q_{s(glu)}$ in the chemostat was −0.42 hr^{-1}. lag(q), lag time before consumption of the sugar was observed; lag(μ), lag time before growth of cells was observed; μ, specific growth rate (hr^{-1}) after transfer of cells from glucose-limited chemostat culture to batch cultures with the different sugars indicated.
[c] The rate of fructose uptake during the first minute was low, but it increased and became constant after 3 or more min. Biomass concentration used in uptake assays was 45 mg liter^{-1}, and during the assay it increased typically to 47 mg liter^{-1}.
[d] Achieved after 30 min.
[e] Achieved after 60 min.

to grow immediately at a specific growth rate that was at least twice as high as the dilution rate at which they were cultivated in the glucose-limited chemostat. Although some exceptions to this pattern were observed (e.g., a small lag before fructose was consumed, but no visible effect on growth, and a considerable lag before growth with galactose or arabinose started), the data clearly demonstrate that cells of *E. coli* growing on glucose were able to replace glucose and grow virtually instantaneously with several other sugars with at least the same specific growth rate. Similar results were obtained for cells cultivated at $D = 0.3$ hr^{-1} and 0.6 hr^{-1}, although lag times observed for galactose and arabinose became longer before growth with these substrates commenced. From these experiments, one can conclude that even at a specific growth rate equivalent to two thirds of μ_{max} (which was 0.92 hr^{-1}), *E. coli* did not "suffer" significantly from carbon catabolite repression by glucose when cultivated under carbon-limited conditions.

A similar observation has been reported for the facultatively autotrophic *Thiobacillus* A2 (Gottschal *et al.*, 1981a). During autotrophic growth in chemostat culture, this organism always retained a low capacity to oxidize various heterotrophic substrates (acetate, formate, fructose, lactate, butyrate, glycollate, and malate), indicating at least a partial derepression of catabolic pathways. As a result, when a culture of this bacterium was shifted in continuous culture from purely autotrophic conditions to a heterotrophic medium containing acetate as the only source of carbon, the cells continued growing with no apparent lag.

An increase in the capacity to scavenge nutrients was also reported for cells of *Vibrio* sp. S14 when subjected to starvation conditions (Mårdén *et al.*, 1987; Albertson *et al.*, 1990a,b). In this gram-negative bacterium, exposure to multiple nutrient (C, N, P) starvation resulted in an increase in the affinity of glucose transport, the synthesis of a high-affinity leucine uptake system with broad substrate specificity, as well as increased exoprotease activity. In addition, the synthesis of three periplasmic proteins was reported as a response to carbon starvation; the fact that they were of a similar size as that normally found for binding proteins suggests that they might be responsible for the cell's increased capacity to scavenge nutrients from the environment.

Further evidence for derepression comes from several reports that demonstrate that marine *Vibrio* spp. retain the capacity to metabolize carbon/energy substrates immediately, even after long-term starvation. For example, starved *Vibrio* sp. S14 was able to immediately respond to glucose addition by a 25-fold increase in the rate of protein synthesis (Albertson *et al.*, 1990c). Similarly, in the marine *Vibrio* strain ANT300 after 7 weeks of starvation in the ultramicrocell state, addition of glucose resulted in growth occurring almost instantaneously (Amy *et al.*, 1983; Morita, 1993). The same research group reported that 16 isolates of marine bacteria were able to immediately utilize glutamate after 8 months of starvation (Amy and Morita, 1983). This ability to respond instantaneously to the addition of nutrients indicates that starving marine *Vibrio* spp. maintain a set of catabolic enzymes, although the cells were not exposed to these substrates during starvation. Also, cells of *Pseudomonas aeruginosa* were reported to maintain the capacity to immediately transport virtually all of the common amino acids when they were starved for either carbon or nitrogen (Kay and Gronlund, 1969).

All of these examples clearly indicate that under carbon/energy-limited conditions, the derepression of catabolic enzymes is not at all wasteful but allows an organism to respond quickly to changes in substrate availability in the environment. Unfortunately, to date only a few studies are available where the metabolic potential of cells was investigated at different (especially low) growth rates in the chemostat. However, one can envisage that derepression of catabolic pathways under carbon/energy-limited growth conditions is a general phenomenon that confers an important advantage to a microbial cell growing in an oligotrophic environment.

Recently, Decker *et al.* (1993) have provided an explanation for the phenomenon of derepression of maltose-utilizing enzymes in *E. coli*. The synthesis of these enzymes requires maltose or maltotriose as the inducing compound, and the utilization of maltose is affected by both inducer exclusion and catabolite repression. Nevertheless, it was observed that enzymes for maltose utilization can be induced with exogenous glucose in the absence of maltose. It turned out that both maltose and maltotriose were formed intracellularly from glucose and glucose 1-phosphate, and that this resulted in a partial induction (some 10–20% of the maximum induction) of the maltose system, on condition that glucose did not exert catabolite repression and reduced intracellular cAMP levels. The author is not aware of other cases where such a mechanism leading to derepression has been investigated in such detail, but it might well turn out that the intracellular formation of inducers is the basis for derepression of other catabolic enzymes when the inducing substrate is not available in the environment. This raises the important question of the possible occurrence of the derepression of enzymes involved in the degradation of xenobiotic compounds where expression requires the presence of an inducing compound. One must assume that in most cases the normal cellular metabolism does not allow the formation of the inducing compound intracellularly. Hence, in the construction of strains for the degradation of xenobiotic compounds, it might be a sensible strategy to put the degradative enzymes under control of a derepressible system, such as the one described above for maltose. This would ensure a constant low degradative capacity and at the same time the ability for a fast response in case the compound becomes available in the environment, without conferring a strong competitive disadvantage to the cell due to the constitutive expression of an unnecessary enzyme.

2.2.3. Further Evidence for an Increased or Altered Microbial Catabolic Versatility during Growth under Carbon/Energy-Limited Conditions

There are several additional reports that provide either direct or indirect evidence that cells growing slowly under nutrient-limited conditions might be able to grow with a wider (or a different) range of substrates than cells cultivated with an excess of nutrients at high growth rates. Although much of this information is certainly equivocal and not necessarily connected to the phenomenon of derepression, it seems appropriate to refer to it here. For example, it was already pointed out by ZoBell and Grant (1943), who investigated 70 marine bacterial isolates with respect to their nutritional versatility, that many of the isolates grew with glucose when it was supplied at low concentrations, but failed to produce acid from glucose in the standard glucose utilization test.

Evidence for derepression of catabolic enzymes has also been presented by Pedersen et al. (1978). Using two-dimensional gel electrophoresis, these authors have quantified the amount of 140 individual proteins in E. coli cultivated at different growth rates in carbon-sufficient batch cultures. Together with the ribosomal proteins, they accounted for 5% of the genome capacity and two thirds of the total cellular protein. Thirty-eight of these proteins exhibited an inverse relationship with growth rate. At a specific growth rate of 0.38 hr^{-1}, this group of proteins accounted for almost 25% of the total cellular protein. One of these proteins seemed to have a role in acetate utilization, whereas the others remained unidentified. The authors mentioned that such an expression pattern is typical for catabolic enzymes that are subject to cAMP control.

Further evidence for the hypothesis that an important part of the adaptation of microbial cells to a carbon/energy-limited growth environment occurs at the level of nutrient transport was reported recently for E. coli (Kurlandzka et al., 1991; Death et al., 1993). In both studies it was found that cells exposed to prolonged cultivation in a glucose-limited chemostat exhibited increased levels of outer-membrane porins, which are known to participate in the diffusion of low-molecular-weight compounds such as mono- and disaccharides, amino acids, and nucleosides. In particular, the concentration of the porin lamB was enhanced by approximately a factor of ten, and Death and co-workers (1993) reported that this increase was paralleled by a decrease in the cells' affinity for glucose. Considering that the cells were grown with glucose as the only source of carbon and energy and that apparently primarily lamB is involved in the translocation of glucose through the outer membrane (Death et al., 1993), the synthesis of increased amounts of other porins suggests a derepression of various other nutrient transport systems.

In order to answer the question of whether microorganisms growing in low-nutrient environments were more or less sensitive to pollutants, Horowitz and colleagues (1983) isolated and characterized some 600 bacterial strains from the subarctic sea. Initially, colonies were isolated from seawater and sediment using both high- and low-nutrient agar plates. Bacterial colonies were subsequently characterized for over 300 different properties, including their nutritional versatility (i.e., their ability to use carbohydrates, alcohols, carboxylic acids, amino acids, or hydrocarbons as their only source of carbon and energy). Significant differences were found with respect to the nutritional versatility of the different isolates. Isolates from low-nutrient media were nutritionally far more versatile than those isolated on high-nutrient media. Typically, they were able to utilize two to three times the number of carbon substrates than isolates from high-nutrient media. The authors concluded that an important characteristic of bacteria of the subarctic marine microbial population was euryheterotrophy, i.e., the ability to utilize a wide range of different carbon/energy compounds for growth, and that organisms with this property were favored under the oligotrophic conditions encountered in seawater. In a similar study, where the nutritional flexibility of oligotrophic and copiotrophic bacterial isolates from Antarctic freshwater lake sediments was investigated, it was found that those strains isolated and kept on low-nutrient media were nutritionally much more versatile than the copiotrophic strains isolated and maintained on rich media (Upton and Nedwell, 1989). Whereas the copiotrophs were able to grow with 22% (mean value) of the substrates tested, oligotrophic isolates utilized on the average 89% of the carbon sources for growth.

Finally, when discussing the nutritional versatility of microbes, it is worthwhile pointing out that the spectrum of growth-supporting carbon substrates may change for a particular microorganism as a function of both substrate concentration and growth conditions. Martin and MacLeod (1984) reported that their marine isolates were able to grow on some carbon sources only when they were supplied at low but not at high concentrations; hence, one would classify these isolates as "oligotrophs" (compare also ZoBell and Grant, 1943). However, with other carbon substrates the strains exhibited "eutrophic" behavior and were able to grow both at low and high concentrations. Similarly, some carbon substrates supported growth of eutrophic strains at low concentrations, whereas others did not. Also, van der Kooij and co-workers (van der Kooij et al., 1982; van der Kooij and Hijnen, 1983) observed that for both P. aeruginosa and a Flavobacterium sp., the spectrum of carbon substrates able to support growth was dependent on the concentration. There are also various reports in the literature that the terminal electron acceptor that has to be used for growth influences the spectrum of utilizable carbon sources (Bryan, 1981; Beauchamp et al., 1989), or that some substrates can be utilized only in combination with others (Lüttgens and Gottschalk, 1980; Cogan, 1987; Bitzi et al., 1991).

All of this indicates that the ability to use particular carbon/energy substrates and the nutritional versatility and flexibility of a microbial cell growing in a carbon-restricted environment are most probably higher than that determined under carbon-sufficient conditions in the laboratory. Also, as suggested by ZoBell and Grant (1943), the spectrum of carbon substrates utilizable under environmental conditions is not necessarily identical to that determined with standard laboratory test procedures used in taxonomical investigations. Even microorganisms considered to be highly nutritionally specialized, such as obligate methano- and methylotrophic bacteria and obligate autotrophs, have the ability to incorporate a considerable fraction of their cellular carbon from non-C_1 compounds (Rittenberg, 1969; Egli and Harder, 1984).

2.3. Influence of Other Nutrient Limitations on Nutritional Versatility

In analogy to the derepression of carbon catabolic enzyme systems observed in heterotrophic microbes during carbon/energy-limited growth, one should expect that a restriction of other nutrients would also cause the cell to express enzyme systems that might be useful for the acquisition of this nutrient. In the case of enzyme systems involved in the transport and assimilation of nitrogen, this has indeed been demonstrated for a number of microorganisms. For example, in *E. coli* limitation by leucine leads to the derepression of a high-affinity-binding protein transport system (LIV-I). In comparison to the nonrepressible membrane-bound transport system that exhibits low affinity to all branched amino acids, LIV-I not only has an increased affinity to leucine, but also has a wider substrate spectrum and can transport a range of branched amino acids efficiently (Rahmanian and Oxender, 1972; Wood, 1975; Anderson and Oxender, 1978; Furlong, 1987). The LIV-I system consists of a common transport protein that is fed from the periplasm by either a leucine-specific binding protein or a binding protein that accepts various branched amino acids. Also, in cells of *Hyphomicrobium* ZV620 growing in ammonium-limited chemostat cultures derepression of NADPH-dependent glutamate dehydrogenase (GDH), even in the presence of the high-affinity ammonium assimilation system [glutamine synthetase/glutamine oxoglutarate aminotransferase (GS/GOGAT)], was observed at low dilution rates (Gräzer-Lampart *et al.*, 1986). Whereas the synthesis of GDH was repressed at $D > 0.05$ hr^{-1}, its specific activity reached at $D = 0.01$ hr^{-1} approximately 20% of that found in carbon-limited cells grown at the same growth rate.

Several studies have reported the derepression of nitrogenase in phototrophic microorganisms in response to nitrogen-limited growth conditions. In anaerobic cultures of *Rhodospirillum rubrum* cultivated in a chemostat with malate as the carbon source, limitation of growth by either ammonium or glutamate resulted in the derepression of nitrogenase (Munson and Burris, 1969). Because no N_2 was needed for the expression of N_2-fixing activity, the authors concluded that the cells expressed (derepressed) nitrogenase in response to the restricted availability of nitrogen rather than to the presence of the substrate N_2.

Unfortunately, it was not tested whether the culture was able to simultaneously reduce dinitrogen and utilize externally supplied ammonia under N-limited conditions. Derepression of nitrogenase in cells grown photoheterotrophically with C_4 acids and amino acids as a source of nitrogen was also reported for *Rhodopseudomonas capsulata* (Hillmer and Gest, 1977) and for *R. rubrum* (Hoover and Ludden, 1984). In the latter bacterium, growth under carbon-limited conditions with glutamate as the only source of carbon and nitrogen resulted in excretion of excess ammonia and repression of nitrogenase. When the culture was shifted to (most likely) nitrogen-limited growth conditions by addition of carbon (malate) to both the culture and the medium reservoir, the simultaneous presence of three different nitrogen-assimilating enzyme systems, GDH, GS, and nitrogenase was observed. One can conclude that the cells expressed all of the nitrogen-assimilating enzyme systems in order to get access to all possible forms of nitrogen. Not only phototrophic bacteria but also *Clostridium pasteuranium* were reported to derepress the synthesis of nitrogenase during nitrogen-limited growth with ammonia (Daesch and Mortenson, 1968).

High concentrations of orthophosphate tightly control the expression of many proteins involved in the hydrolysis and/or transport of organic phosphorus-containing compounds. In batch cultures this results in a diauxic growth pattern as, for example, that reported for *Comamonas* (formerly *Pseudomona testosteron*) in the presence of a mixture of orthophosphate (utilized first) and methylphosphonate (Daughton *et al.*, 1979). However, limited availability of phosphorus leads to the synthesis of a number of enzyme systems able to process, transport, and metabolize potential phosphorus sources such as inorganic forms of phosphate or organically bound phosphorus. Among them are a phosphate-binding protein, a high-affinity phosphate carrier, a porin involved in the diffusion of the phosphate-anion through the outer membrane, and alkaline and acid phosphatase that allow hydrolysis of organically bound phosphorus (Wanner 1987a,b; Torriani-Gorini, 1987).

From the information compiled in this section for the bulk nutrients carbon and nitrogen, it appears to be a common microbial strategy to react to the restricted availability or even absence of a particular nutrient with the synthesis of enzyme systems that will allow access to other possible forms of this nutrient. Although the regulatory mechanisms that lead to the synthesis of these systems are in most cases not yet known in detail, the response usually seems not to require the extracellular presence of an inducing substrate. Rather, it seems to be triggered by the general limited availability of this particular nutrient. To what extent the different regulatory proteins involved in the response to carbon starvation in different organisms (Matin, 1991; Kjelleberg *et al.*, 1993; Nyström, 1993; Östling *et al.*, 1993; Hengge-Aronis, 1993) are involved in the derepression of catabolic pathways remains to be investigated.

3. Mixtures of Substrates Serving the Same Physiological Purpose

A wealth of information is now available that demonstrates that a wide range of different microorganisms are able to simultaneously utilize mixtures of substrates that serve the same physiological function. All reliable information concerning mixed substrate growth originates from pure culture work. This is not surprising, considering the experimental problems one would encounter to prove that a particular member in a mixed culture is utilizing a number of homologous substrates simultaneously. In the majority of studies, the growth of heterotrophic microorganisms with mixtures of carbon/energy sources was investigated. Nevertheless, there are also a few of examples where pure cultures were shown to simultaneously assimilate two sources of nitrogen or to use two terminal electron acceptors (see Section 3.4).

Over the past years, several reviews on mixed substrate growth of microorganisms have been published that either stressed the regulation of enzyme synthesis (Harder and Dijkhuizen, 1976, 1982), or biotechnological aspects (Weide, 1983; Bull, 1985; Babel et al., 1993), or were focused on special groups of microorganisms (Egli and Harder, 1984; Egli and Mason, 1991). Instead of retracing old ground, a selection of well-documented examples for the growth with mixtures of homologous substrates is given in Table II. In addition, the principal phenomena at the level of substrate utilization, biomass production, and enzyme regulation that are usually observed during the growth of microbial cultures with mixtures of homologous substrates will be briefly outlined. Subsequently, important aspects that were either not treated in earlier reviews or where considerable information has accumulated over the last few years, such as the degradation of pollutants in the presence of easily degradable carbon substrates and mixed substrate kinetics, will be discussed in more detail in separate sections.

3.1. Simultaneous Utilization of Mixtures of Carbon Substrates

3.1.1. Substrate Excess Conditions in Batch Culture

Under substrate excess conditions of batch cultures, sequential utilization of mixtures of carbon substrates (frequently resulting in a diauxic growth pattern) is considered to be the rule rather than the exception. However, the information collected in Table II clearly demonstrates that the simultaneous utilization of different carbon sources is also commonly observed under substrate excess conditions. This phenomenon has been reported for both bacteria and yeasts under aerobic, anaerobic, or thermophilic growth conditions. Especially combinations of substrates that support medium- or low-maximum specific growth rates are utilized simultaneously. For example, on a mixture of glucose and methanol the

Table II. Simultaneous Utilization of Mixtures of Carbon Substrates in Batch and Chemostat Cultures[a]

Organism	Conditions	Substrates	Remarks	Reference
Acinetobacter calcoaceticus	Chemostat	Ac+Glc	Glc energy source only; oxidized to Gluc	Müller and Babel (1986)
Arthrobacter P1	Chemostat	Methylamine+Ac	Depending on ratio, Ac as energy source only	Levering and Dijkhuizen (1985)
Azotobacter vinelandii	Batch	Glc + Ac	Ac first; high Glc conc. overrides repression by Ac	Tauchert et al. (1990)
Azotobacter vinelandii	Batch	Gal + Glc	Glc co-utilized, dependent on pregrowth of inoculum	Wong et al. (1995)
Beneckea natriegens	Chemostat	Glu+For	For as energy source only yield increased	Linton et al. (1981)
Bifidobacterium breve	Batch	Glc+Ara	0.8–1.0g/liter, each; fermenting conditions	Degnan and MacFarlane (1993)
Butyrivibrio fibrisolvens	Batch	Glc+Xyl	Anaerobic, 0.18g/liter, each	Russel and Baldwin (1978)
Butyrivibrio fibrisolvens	Batch	Glc+Xyl	Anaerobic, 2g/liter, each	Marounek and Kopecný (1994)
Candida boidinii	Batch	Xyl+MeOH	2–3g/liter, each	Volfová et al. (1988)
Candida boidinii	Chemostat	Man+MeOH		Kyslíková and Volfová (1990)
Candida boidinii	Chemostat	Oleate+MeOH		Waterham et al. (1992)
Candida boidinii (Kloeckera sp. 2201)	Chemostat	Glc+MeOH	Additive growth yields	Egli et al. (1982a,b)
Candida boidinii (Kloeckera sp. 2201)	Chemostat	Glc+MeOH	Steady-state MeOH conc. reduced on mixtures	Egli et al. (1983)
Candida utilis	Batch	Ac+Xyl	Both at pH 6.0, but at pH 4,5 Ac first	Pöhland et al. (1966)
Candida utilis	Chemostat	Ac+Xyl	At pH 4.5 both utilized up to D = 0.2/hr	Pöhland et al. (1966)
Candida utilis	Batch	Glc+EtOH	10g/liter, each	Weusthuis et al. (1994)
Candida utilis	Batch	Glc+Fru+Suc	Fermenting conditions	Aker and Robinson (1987)
Candida shehatae	Batch	Xyl+Glc		Kastner and Roberts (1990)
Caulobacter crescentus	Batch	Glc+Glu	Glu alone slowly utilized, but fast with Glc	Poindexter (1987)

Organism	Mode	Substrate	Notes	Reference
Chelatobacter heintzii	Batch	Glc+NTA	Increased μ_{max}, excess ammonia present	Egli *et al.* (1988)
Chelatobacter heintzii	Chemostat	Glc+NTA	Additive growth yields, excess ammonia present	Bally *et al.* (1994)
Chelatococcus asaccharolyticus	Batch	Ac+NTA	Increased μ_{max}, excess ammonia present	Egli *et al.* (1988)
Clostridium thermohydrosulfuricum	Chemostat	Glc+Cel	In batch Glc first; in chemostat simultaneously at $D < 0.2$/hr; D_{crit} for celobiose 0.3/hr	Slaff and Humphrey (1986)
Clostridium thermohydrosulfuricum	Batch	Glc+Xyl	Anaerobic conditions, thermophilic (70°C)	Cook *et al.* (1993)
Comamonas testosteroni	Batch	pTS+Ac	Increased μ_{max}, excess sulfate present	M. Witschel (unpublished)
Comamonas testosteroni	Chemostat	pTS+Ac	Excess sulfate present	M. Witschel (unpublished)
Corynebacterium sp.	Chemostat	Glc+Arg	Steady-state Glc conc. measured	Law and Button (1977)
Corynebacterium sp.	Chemostat	Glc+AA mixture	Steady-state Glc conc. measured	Law and Button (1977)
Escherichia coli ML30	Batch	Glc+Gal	Below 5 mg/liter Glc	Lendenmann (1994)
Escherichia coli ML30	Batch	Glc+Fru	Below 5 mg/liter Glc; rate of Fru utilization dependent on previous conditions	Lendenmann (1994)
Escherichia coli ML30	Batch	Glc+pyruvate	0.3–1.0g/liter	Hegewald and Knorre (1978)
Escherichia coli	Chemostat	Glc+Lac	At low D, at high D only Glc utilized	Mateles *et al.* (1969)
Escherichia coli	Chemostat	Glc+Asp	At low D, at high D only Glc utilized; μ_{max} on Asp 0.2/hr, but still utilized at $D = 0.9$/hr	Silver and Mateles (1969)
Escherichia coli	Chemostat	Glc+Xyl	At low and high D, close to μ_{max}	Standing *et al.* (1972)
Escherichia coli	Chemostat	Glc+Gal	At low and high D, close to μ_{max}	Standing *et al.* (1972)
Escherichia coli ML30	Chemostat	Glc+Gal+Rib+Fru+Mal+Ara	Constant D; various mixtures with up to six sugars, additive growth yields, steady-state conc. measured	Lendenmann (1994)

(*continued*)

Table II. (Continued)

Organism	Conditions	Substrates	Remarks	Reference
Eubacterium limosum	Batch	Glc+MeOH	At low concentrations	Loubière *et al.* (1992b)
Hansenula polymorpha MH30	Batch	Xyl+MeOH	Increased μ_{max}	Brinkmann and Babel (1992)
Hansenula polymorpha MH30	Chemostat	GlyOH+Xyl	With mixtures reduced steady-state GlyOH conc.	Babel *et al.* (1993)
Hansenula polymorpha MH30	Chemostat	Xyl+Glc	Xyl conc. given as $f(D)$	Babel *et al.* (1993)
Hansenula polymorpha MH 30	Chemostat	Xyl+MeOH	$D_{crit}(Xyl) = 0.175/hr; D_{crit}(MeOH) = 0.21/hr; D_{crit}$ on mixture $= 0.36/hr$	Brinkmann and Babel (1992)
Hansenula polymorpha MH 30	Chemostat	Xyl+MeOH	With mixtures reduced MeOH steady-state conc.	Babel *et al.* (1993)
Hansenula polymorpha CBS 4732	Chemostat	Glc+MeOH	MeOH utilized at $D>\mu_{max}$(MeOH); D where repression of MeOH utilization started dependent on ratio	Egli *et al.* (1982a, 1983) Egli *et al.* (1986)
Hyphomicrobium X	Batch	EtOH+methylamine	At low concentrations only	Brooke and Attwood (1983)
Klebsiella aerogenes NCIB 8021	Chemostat	Glc+Mal	μ_{max} on Mal 0.6/hr, together with Glc still utilized at $D = 1.0$/hr	Harte and Webb (1967)
Klebsiella aerogenes	Chemostat	Glc+Lac	Except at very high D	Baidya *et al.* (1967)
Lactobacillus sanfrancisco	Batch	Glc+Mal	Fermenting conditions	Stolz *et al.* (1993)
Lactobacillus xylosus	Batch	Xyl+Glc	Fermenting conditions, at low concentrations	Tyree *et al.* (1990)
Leuconostoc sp.	Batch	Glc+Cit	Anoxic, 2–4g/liter	Cogan (1987)
Leuconostoc mesenteroides	Batch	Lac+Cit	Anoxic, products formed and again utilized	Schmitt and Divies (1991)
Leuconostoc oenos	Batch	Glc+malate	Anoxic, improved growth energetics	Loubière *et al.* (1992a)
Leuconostoc oenos	Batch	Glc+Cit	Anoxic, increased μ_{max} and yield, S_0 ca. 10g/liter, each	Salou *et al.* (1991) Salou *et al.* (1994)
Leuconostoc oenos	Batch	Glc+Fru	Anoxic, increased μ_{max} and yield, S_0 ca. 10g/liter, each	Salou *et al.* (1994)

Organism	Culture	Substrates	Notes	Reference
Methylobacterium	Chemostat	Ac+methylene-chloride		A. Tien (unpublished)
Nocardia sp. 239	Batch	MeOH+Glc	1.8g/liter, each	de Boer *et al.* (1990)
Neocallimastix sp.	Batch	Xyl+other reducing sugar(s)	Anaerobic rumen fungus, up to 1 g/liter	Lowe *et al.* (1987)
Paracoccus denitrificans	Chemostat	Man+MeOH	Depending on ratio, MeOH only energy source	van Verseveld *et al.* (1979)
Paracoccus denitrificans	Chemostat	Man+For	For only energy source	van Verseveld and Stouthamer (1980)
Pseudomonas aeruginosa	Chemostat, N–limitation	Cit+Glc	Threshold concentration of Cit needed for utilization	Ng and Dawes (1973)
Pseudomonas fragi	Batch	Glc+Lact+Cit+Asp+Glu+creatine+creatinine	Conc. range 5–80 mM	Molin (1985)
Pseudomonas fragi	Chemostat	Glc+Lact+Cit+Asp+Glu+creatine+creatinine		Molin (1985)
Pseudomonas oxalaticus	Chemostat	Ac+Ox	At $D>0.2$ Ox utilization repressed, mixotrophy	Harder and Dijkhuizen (1982)
Pseudomonas oxalaticus	Chemostat	Ac+For	Depending on ratio. For only energy source	Dijkhuizen and Harder (1979)
Pseudomonas putida	Batch	Phenol+p-cresol	50–100 µg/liter	Hutchinson and Robinson (1988)
Pseudomonas putida P1	Batch	Glc+phenol	μ_{max} with phenol alone 0.1/hr; but utilized up to 0.41/hr with Glc	Reber and Kaiser (1981)
Pseudomonas sp. KI	Batch	Aniline+Lact	1.4g/liter Lact + 80mg/liter aniline	Konopka *et al.* (1989)
Rhizobium trifolii	Batch	Man+Ara+Glc	0.7g/liter, each	de Hollander and Stouthamer (1979)
Ruminococcus albus	Batch	Cel+Xyl	Fermenting conditions	Thurston *et al.* (1994)
Ruminococcus albus	Batch	Glc+Xyl	Fermenting conditions	Thurston *et al.* (1994)
Saccharomyces cerevisiae	Batch	Rib+Xyl	Xyl not used alone, oxidized mainly to xylitol	van Zyl *et al.* (1989)
Saccharomyces cerevisiae	Batch	Glc+Xyl	(as above)	van Zyl *et al.* (1989)

(continued)

Table II. (Continued)

Organism	Conditions	Substrates	Remarks	Reference
Saccharomyces cerevisiae	Batch	EtOH+Xyl	(as above)	van Zyl et al. (1989)
Saccharomyces cerevisiae	Batch	Man+Glc+Gal		Pineault et al. (1977)
Saccharomyces cerevisiae	Chemostat	Glc+EtOH	Additive growth yields	Geurts et al. (1980)
				de Jong-Gubbels et al. (1995)
Saccharomyces cerevisiae	Chemostat	Glc+Prop	Prop energy source only, increased yield	Pronk et al. (1994)
Saccharomyces cerevisiae	Chemostat	Glc+Mal	Constant D, steady-state conc.	Häggström and Cooney (1984)
Saccharomyces cerevisiae	Chemostat	Mal+Glc	measured	Weusthuis et al. (1993)
Selenomonas ruminantium HD4	Batch	Glc+Xyl+Suc	Anaerobic culture, 180 mg/liter, each	Russel and Baldwin (1978)
Streptococcus sanguis	Chemostat	Glc+19 AA	All, except Asp and threonine at high D	Rogers and Reynolds (1990)
Thiobacillus versutus	Batch	Glc+Gal	Increase of μ_{max}	Wood and Kelly (1977)
Unidentified gram-negative strain	Batch	i-Prop+ acetone+MeOH	i-Prop as initial substrate, MeOH and acetone produced from i-Prop	Bitzi et al. (1991)
Unidentified gram-negative strain	Batch	EtOH+i-Prop +n-But	All utilized, acetone produced from i-Prop, acetone was utilized in the late exponential phase	Al-Awadhi (1989)
Unidentified ultramicrobacterium	Batch	Ala+Leu+Pro+ Phe+Ser+His+ Arg+Glu+ Asp+Gln	50–100 mg/liter, each	Schut et al. (1995)
Unidentified ultramicrobacterium	Chemostat	Glc+Ala		Schut et al. (1995)

[a]Note that in most cases it has not been proven experimentally that a carbon compound served both as a carbon and an energy source. If not indicated otherwise, cultivation in chemostat culture was limited by carbon. Abbreviations of substrates: acetate (Ac), amino acid (AA), arabinose (Ara), arginine (Arg), asparagine (Asp), cellobiose (Cel), citrate (Cit), ethanol (EtOH), formate (For), fructose (Fru), galactose (Gal), gluconate (Gluc), glucose (Glc), glutamate (Glu), glycerol (GlyOH), lactate (Lact), lactose (Lac), maltose (Mal), mannose (Man), methanol (MeOH), nitrilotriacetate (NTA), oxalate (Ox), para-toluenesulfonate (pTS), ribose (Rib), sucrose (Suc), xylose (Xyl).

methylotrophic yeasts exhibit a clear diauxic growth pattern (Sahm and Wagner, 1973; Sakai *et al.*, 1987), whereas when supplied with mixtures of glycerol/methanol (Wanner and Egli, 1990) or xylose/methanol (Volfová *et al.*, 1988; Brinkmann and Babel, 1992) the two carbon sources were utilized simultaneously. It should be added that in neither of these cases was it tested whether the cultures separated into two specialized subpopulations (see discussion in Section 3.5.1). However, the fact that yeast cells grown in batch culture with either xylose, sorbitol, or glycerol as the only substrate exhibited high activities (derepression) of all enzymes required for methanol metabolism makes this possibility highly unlikely (Eggeling and Sahm, 1978; de Koning *et al.*, 1987). Of course, more than just two carbon sources may be utilized simultaneously, and this was demonstrated for batch growth of both bacteria and yeasts (de Hollander and Stouthamer, 1979; Aker and Robinson, 1987; Schut *et al.*, 1995). For example, *P. fragi* consumed glucose, lactate, citrate, aspartate, and glutamate at the same time, when these were present at concentrations of 0.5 to 1 g liter^{-1} each (Molin, 1985). Similarly, an unidentified gram-negative thermophilic isolate NA 17, when supplied with a mixture of the three solvents ethanol, isopropanol, and *n*-butanol, utilized them all simultaneously. At the same time, it produced acetone from isopropanol in the early phase of growth, which was reutilized in the later phase of growth (Al-Awadhi, 1989). These two examples indicate that microorganisms are capable of handling quite complex combinations of substrates simultaneously. It has also been observed that environmental conditions may influence the utilization pattern for mixtures of carbon substrates. Such a case was reported for *Candida utilis* (Pöhland *et al.*, 1966), which utilized acetate and xylose simultaneously at pH 6, but preferred acetate at low pH (4.5). In comparison to growth with the individual carbon sources, a stimulation in the maximum specific growth rate was observed in many cases when mixtures of carbon substrates were used simultaneously (see Section 3.2). However, from the limited data available, it is not yet possible to judge whether the extent of stimulation in growth rate might be predictable.

Incompatible combinations of substrates, leading to a diauxic growth behavior when supplied at high concentrations, are frequently consumed simultaneously in batch culture when the initial concentrations are lowered (Brooke and Attwood, 1983; Senn, 1989; Tyree *et al.*, 1990; Loubière *et al.*, 1992b; Lendenmann, 1994). This is shown in Fig. 1. for a culture of *E. coli* growing with glucose and galactose. This combination of sugars is utilized in a diauxic manner when they are supplied in the gram per liter range (Monod, 1942; Standing *et al.*, 1972); however, they are utilized simultaneously when the initial substrate concentrations are reduced to 5 mg liter^{-1} or lower (Senn, 1989; Lendenmann, 1994). The same authors found similar results for mixtures of glucose plus fructose. Likewise, *Eubacterium limosum* grew mixotrophically with mixtures of glucose plus methanol when initial glucose concentrations were

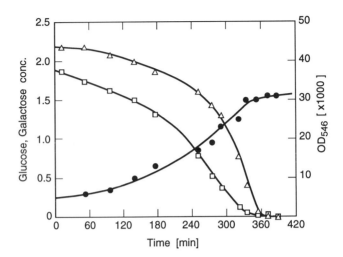

Figure 1. Batch growth of *Escherichia coli* in a synthetic medium supplied with a mixture of glucose (□) and galactose (△) as the sole source of carbon and energy. Inoculum was taken from a glucose-limited chemostat culture growing at $D = 0.20$ hr^{-1}. Growth is given as OD$_{546}$ (●) × 1000; concentrations of sugars are given in mg liter^{-1}. (Adapted from Senn, 1989.)

below 6 mM, whereas at higher glucose concentrations diauxic growth was observed (Loubière *et al.*, 1992b). Unfortunately, there are few data available in the literature where the transition from diauxic growth pattern to simultaneous utilization has been studied systematically.

In contrast to the many studies that can be found in the literature where the utilization of individual compounds by microbial cultures was investigated at low, "environmentally" realistic concentrations in batch cultures (such conditions are probably mostly carbon/energy-limited and not sufficient), there are only few where the utilization of mixtures of carbon sources was considered. For example, in the microgram per liter concentration range, acetate and phenol (13 and 2 μg liter^{-1}, respectively) or glucose and aniline (3 μg liter^{-1}, each) were utilized in combination by pure cultures of pseudomonads (Schmidt and Alexander, 1985). *Pseudomonas aeruginosa* was reported to grow with a mixture of 45 carbonaceous compounds, each added to tap water at a concentration of 1 μg liter^{-1} of C, whereas none of these compounds supported growth on their own at this concentration (van der Kooij *et al.*, 1982). (Note that in this case it was not tested how many of the 45 compounds were used and to what extent.) It should be kept in mind that such studies are usually difficult to interpret because of the interference of uncharacterized DOC, which is usually present in the cultivation media. As pointed out recently by Alexander (1994), under environmental conditions the effects of one substrate on the utilization of another are largely unknown.

3.1.2. Carbon-Limited Conditions in Continuous Culture

There are numerous examples that have demonstrated the simultaneous utilization of mixtures of diauxic carbon compounds by pure microbial cultures under carbon-limited growth conditions in the chemostat (see Table II). Some of the best-studied microorganisms with respect to substrate utilization, biomass formation, and enzyme regulation are the facultatively autotrophic bacteria *P. oxalaticus* and *Thiobacillus versutus* (formerly A2) under mixotrophic growth conditions (reviewed in Harder and Dijkhuizen, 1982; Gottschal, 1986). In both cases interest was focused on the simultaneous utilization of a heterotrophic substrate and CO_2 as carbon sources. With respect to growth with mixtures of heterotrophic substrates, methylotrophic yeasts and *E. coli* have been studied in most detail.

It seems logical, and the author is not aware of any contradictory report, that mixtures of carbon substrates utilized together in batch culture by an organism will also be consumed together during carbon-limited growth in continuous culture (for examples, see Table II). Therefore, the more interesting case is that of carbon compounds that are incompatible in batch culture and provoke distinct diauxic growth. The utilization of mixtures of the diauxic substrate combination glucose and methanol by methylotrophic yeasts will be used here as an example to explain some of the observations that are commonly made during cultivation of microorganisms with mixed carbon compounds. In methylotrophic yeasts the synthesis of methanol-utilizing enzymes is repressed during growth with glucose in batch culture. Repression is also observed at high dilution rates in a glucose-limited chemostat; however, reducing the dilution rate results in increasing derepression of both dissimilatory and assimilatory methanol enzymes (Egli *et al.*, 1980, 1983). Derepression levels of most enzymes were usually between a few and 20% of those observed in methanol-grown, fully induced cells. For a culture of *H. polymorpha* growing in a glucose-limited chemostat at a fixed dilution rate of 0.14 hr^{-1}, addition of methanol in increasing amounts to the glucose-medium resulted, first, in the simultaneous and complete utilization of methanol together with glucose independent of the mixture composition in the feed and, second, in increasing induction of methanol metabolic enzymes, until a plateau was reached (see Fig. 4). Depending on the enzyme, proportions of 10 to 50% methanol in the mixture were sufficient to achieve maximum expression of enzyme-specific activity, corresponding to that found in methanol-grown cells at this dilution rate. Virtually identical patterns for methanol-related enzymes were reported for this yeast during growth at a constant dilution rate with mixtures of xylose/methanol (de Koning *et al.*, 1990). Applying the same experimental strategy, similar regulatory patterns consisting of derepression (although sometimes low in the absence of the inducing substrate) combined with induction to a variable degree (depending on the composition of the substrate mixture supplied) were observed for *Chelatobacter heintzii* growing with mixtures of glucose plus nitrilotriacetic acid (Bally *et al.*, 1994) and *C. testosteroni* cultivated with acetate/paratoluenesulfonate mixtures (M. Witschel, unpublished data). The relevance of these regulatory patterns for both growth kinetics and pollutant degradation will be discussed in more detail.

As a rule, mixtures of diauxic carbon compounds are utilized simultaneously at low dilution rates in a C-limited chemostat. This is demonstrated in Fig. 2 for growth of the yeast *H. polymorpha* (Egli *et al.*, 1982b, 1986). Supplied as the only source of carbon, methanol and glucose supported maximum

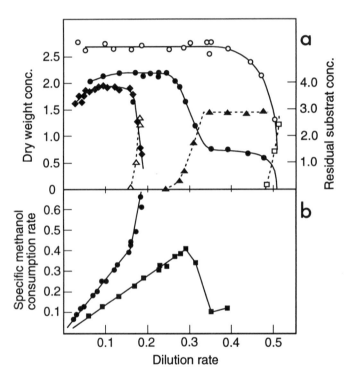

Figure 2. Growth of *Hansenula polymorpha* in carbon-limited chemostat culture supplied with either methanol, a mixture of glucose/methanol (38.2%/61.8%, wt/wt), or glucose as the source(s) of carbon and energy, as a function of dilution rate. (a) Concentrations of cell dry weight (in g liter^{-1}) formed during growth with methanol alone (♦), with the mixture (●), or with glucose only (○); steady-state residual substrate concentrations (in g liter^{-1}) of methanol during growth with methanol (△), or with the mixture (▲), and of glucose during growth with both the mixture and glucose alone (□). (b) Specific methanol consumption rates for methanol [in grams methanol (g cell dry wt)$^{-1}$ hr^{-1}] during growth with methanol alone (●) or with the glucose/methanol mixture (■). The total concentration of methanol and/or glucose in the inflowing medium was always 5 g liter^{-1}. (Adapted from Egli *et al.*, 1986.)

specific growth rates in the chemostat of roughly 0.19 hr^{-1} and 0.51 hr^{-1}, respectively, where washout was observed (D_{crit}). When the yeast was fed with a mixture of glucose:methanol (approx. 3:2, wt:wt), both compounds were utilized simultaneously up to a dilution rate of approximately 0.26 hr^{-1}, where methanol started to accumulate. At a dilution rate higher than 0.33 hr^{-1}, only glucose was utilized by the culture. Hence, during growth with a mixture of the "good" carbon source glucose, *H. polymorpha* was able to utilize the "slow" carbon source methanol at dilution rates exceeding its μ_{max}. This phenomenon has been observed frequently for both bacteria and yeasts growing with mixed carbon sources (Harte and Webb, 1967; Silver and Mateles, 1969; Eggeling and Sahm, 1981; Brinkmann and Babel, 1992; Lendenmann, 1994), and only in one instance was the opposite effect found (Slaff and Humphrey, 1986) (Table II). For *H. polymorpha,* it was found that the dilution rate at which the culture stopped utilizing methanol was strongly dependent on the composition of the substrate mixture in the feed. The lower the proportion of methanol in the mixture supplied, the higher was the dilution rate at which mixed substrate utilization ceased; e.g., when methanol contributed only 20% of the total carbon, it was utilized up to dilution rates of almost 0.4 hr^{-1} (Egli *et al.,* 1986).

This surprising effect can be explained in the following way. Consumption of a carbon/energy substrate(s) at a certain specific rate (q_{tot}) supports a particular cellular specific growth rate (μ). This can be achieved by either consuming one substrate at a high rate (q_{S1}) as indicated in Eq. (1) or by consuming several substrates at a reduced rate [Eq. (2), assuming unchanged yield coefficients for the different substrates ($Y_{X/S1...n}$) during single and mixed substrate growth]:

$$\mu = Y_{X/S} \cdot q_{tot} = Y_{X/S1} \cdot q_{S1} \tag{1}$$

$$\mu = Y_{X/S1} \cdot q_{S1} + Y_{X/S2} \cdot q_{S2} + \ldots + Y_{X/Sn} \cdot q_{Sn} \tag{2}$$

Thus, during growth in continuous culture at a certain dilution rate, the consumption of several substrates simultaneously leads to a reduced flux of carbon through the catabolic pathways of the individual substrates. The effect is illustrated in Fig. 2B for the case of methanol in *H. polymorpha* growing with either methanol alone or a mixture of glucose plus methanol. During growth with methanol as the sole carbon source, the maximum methanol consumption capacity ($q_{MeOH(max)}$) of this yeast is in the range of 0.40 to 0.45 g methanol (gram cell dry weight)$^{-1}$ hr^{-1} at a dilution rate of approximately 0.17 hr^{-1}. Increased q_{MeOH} at higher D were due to accumulation of unutilized methanol in the culture, resulting in excessive oxidation of methanol until the system collapsed. However, when the culture was cultivated with the mixture, q_{MeOH} at $D = 0.17$ was only approximately half $q_{MeOH(max)}$ and, theoretically, the pathway for methanol catabolism had still spare capacity to handle methanol. When the dilution rate was increased in this experiment, the cells continued to utilize methanol until they consumed methanol at the maximum possible rate (in this case, at $D = 0.28$ hr^{-1}). Only when they were pushed over this limit, by further increasing the dilution rate, did the cells cease methanol consumption and switch to growth on glucose only. This indicates that the capacity of the catabolic pathway to handle methanol was determining the point (dilution rate) up to which mixed substrate growth was possible, which, in turn, was determined in combination by dilution rate and the proportion of methanol in the mixture supplied. In this way also an increase in μ_{max} during growth with compatible substrate mixtures above the μ_{max} supported by either of the substrates on their own (as observed by Brinkmann and Babel, 1992) can be explained.

Concerning the efficiency of carbon assimilation from different substrates during mixed substrate growth, it was found in several cases that the formation of biomass was approximately additive over a wide range of mixture compositions and growth rates (Egli et al., 1982a; Bally et al., 1994; Lendenmann, 1994; de Jong-Gubbels et al., 1995). However, there are many examples in the literature where the growth yields were not additive during mixed substrate growth, and for details the reader is referred to some comprehensive reviews (Gommers et al., 1988; Babel et al., 1993). In many cases the nonadditive behavior is due to a change in metabolic function of one of the substrates. For example, for *Paracoccus denitrificans* growing on mixtures of mannitol and methanol, methanol was used as an energy source only (depending on mixture ratio), whereas the bacterium was perfectly able to use this substrate as a carbon/energy source on its own (van Verseveld et al., 1979). Similar observations were made for the growth of *P. oxalaticus* with acetate/formate mixtures (Dijkhuizen and Harder, 1979), or for *Arthrobacter* P1 with acetate/methylamine mixtures, where surprisingly acetate took the role as an energy source (Levering and Dijkhuizen, 1985). It should be added here that many organisms are able to utilize reduced carbon substrates, which on their own do not support growth, as additional energy sources (especially C_1 compounds) and exploit the derived energy to improve the assimilation efficiency of their growth substrate (Egli and Harder, 1983; Bitzi et al., 1991; Babel et al., 1993).

The present discussion included only examples of pure cultures fed with binary carbon substrate mixtures. But more complex substrate mixtures also were shown to be degraded simultaneously by pure microbial cultures under carbon-limited conditions in the chemostat. For example, Molin (1985) reported simultaneous utilization of glucose, lactate, citrate, aspartate, glutamate, creatine, and creatinine by *P. fragi;* Aker and Robinson (1987) reported that of *C. utilis* with glucose, fructose, and sucrose; Rogers and Reynolds (1990) reported that of 19 amino acids plus glucose by *Streptococcus sanguis;* and Lendenmann (1994) reported that of six sugars by *E. coli.*

3.2. Kinetics of Growth with Mixed Carbon Substrates

Traditional kinetic descriptions of bacterial growth are based on the assumption that the growth rate is dependent on the concentration of a single, growth-limiting substrate. In contrast to this, it was demonstrated in the previous section that it is highly likely that under environmental growth conditions microorganisms will utilize a variety of carbon/energy substrates simultaneously. This, of course, raises the question of whether and how kinetic models based on growth rate control by one substrate only can be applied to a situation where a cell is utilizing several substrates simultaneously. Is it possible to describe mixed substrate growth by combining kinetic relationships determined for individual substrates during single-substrate-limited growth, or is it necessary (and possible) to develop alternative mathematical models for this situation?

Before discussing kinetics of mixed substrate growth, it seems appropriate at this point to give a short overview on the "state of the art" with respect to microbial growth kinetics with single substrates. Kinetics of single-substrate-limited growth have been investigated extensively, but a closer look at the published data shows that even for a particular microorganism–substrate combination the results are not consistent. Indeed, it is well-known that kinetic investigations have always been hampered by two fundamental experimental problems (Powell, 1967; Koch, 1982; Button, 1985; Owens and Legan, 1987; Rutgers *et al,* 1991; Jannasch and Egli, 1993). The first was the difficulty to measure reliably the low concentrations of substrates in the growth-limiting range; because of this difficulty, the dependence of the specific growth rate (μ) on the extracellular concentration of a substrate (s) has usually been determined by measuring the specific growth rate of a microorganism in batch culture at substrate concentrations estimated by calculation, rather than at either analytically determined concentrations or by measuring steady-state concentrations in a chemostat at different dilution rates. Second, it was observed that kinetic properties of microbial strains were not constant and that they were affected during adaptation from high to low substrate concentrations and vice versa (Jannasch, 1968; Shehata and Marr, 1971; Koch and Wang, 1982; Höfle, 1983; Rutgers *et al.,* 1987; Senn *et al.,* 1994).

Probably the best-studied example is that of growth of *E. coli* with glucose for which large variations for the Monod affinity constant K_s are found in the literature, ranging from approximately 100 mg liter^{-1} down to 50 μg liter^{-1} (Senn *et al.,* 1994). Generally, K_s for glucose measured for *E. coli* adapted to low growth rates (usually cells grown for an extended time in a chemostat) were in the range of 50 to 100 μg liter^{-1}, whereas the values measured for unadapted cells (i.e., cells first cultivated in batch culture and subsequently exposed to low substrate concentrations) were more than two orders of magnitude higher (Koch and Wang, 1982; Senn *et al.,* 1994). In a detailed analysis of experimental data from chemostat-grown *E. coli* with either glucose (Senn *et al.,* 1994) or galactose (Lendenmann, 1994) it was found that due to the differences in kinetic properties exhibited by low- and high-substrate-concentration-adapted cells, neither of several kinetic models described the data satisfactorily. Nevertheless, the Monod equation described well the experimental data obtained in the ecologically important growth range, i.e., at low- and medium-dilution rates.

It should be added that the case of *E. coli* discussed above does not stand alone, and that a critical evaluation of single-substrate kinetic data published indicates that more often than not they were obtained from cultures adapted to high laboratory substrate concentrations. Because it has been demonstrated for several bacteria that this adaptation is a slow process and needs in the order of 100–200 generations (Höfle, 1983; Rutgers *et al.*, 1987; Senn *et al.*, 1994), the data must therefore be interpreted with caution. Needless to say, that the kinetic properties of cells adapted to low-dilution rates are more relevant for growth in the environment.

3.2.1. Experimental Data Available

In contrast to single-substrate-limited growth, few defined studies have been published containing experimental data on concentrations of individual limiting substrates during mixed substrate growth in either batch or chemostat cultures and their influence on growth rate. Mostly the macromolecular or biochemical cellular composition or competitive aspects have been studied under such conditions (Harder and Dijkhuizen, 1982; Egli and Mason, 1991; Gottschal, 1993).

Kinetic studies employing mixtures of carbon substrates in batch culture are presently restricted to positive effects on the specific maximum growth rate. Such an increase in μ_{max} when exposed to mixtures of carbon sources in comparison to growth with either of these substrates as single carbon sources has been reported for several bacteria (Table II). This effect seems not restricted to the utilization of mixed carbon substrates, but has also been observed during the simultaneous utilization of molecular oxygen and nitrate as terminal electron acceptors by *Thiosphera pantotropha* (Robertson and Kuenen, 1990). Similar effects have been observed in chemostat culture, resulting in increased critical dilution rates (D_{crit}) during mixed substrate growth (Slaff and Humphrey, 1986; Brinkmann and Babel, 1992; Loubière *et al.*, 1992b). For example, the methylotrophic yeast *H. polymorpha* MH30 exhibited a D_{crit} during growth with either xylose or methanol alone of $0.18\ hr^{-1}$ and $0.21\ hr^{-1}$, respectively, whereas a mixture of the two substrates supported growth in the chemostat up to a D_{crit} of $0.36\ hr^{-1}$ (Brinkmann and Babel, 1992). However, in neither of these examples was the range investigated where growth rate was influenced by the concentrations of the substrates.

The very first thorough kinetic investigation was that of Law and Button (1977), who studied the growth of a marine *Corynebacterium* in carbon-limited chemostat culture with various mixtures of glucose and amino acids at a constant D of $0.03\ hr^{-1}$. They found that the steady-state concentration of glucose

($\bar{s}_{glucose}$) was reduced from 210 μg liter^{-1}, with glucose as the only carbon substrate to less than 10 μg liter^{-1} when cells were cultivated with a mixture of amino acids (concentration in the feed, $S_0 = 1$ mg liter^{-1}) and glucose ($S_0 = 1.7$ mg liter^{-1}). A similar observation was made for a methylotrophic yeast growing at a low dilution rate in continuous culture with mixtures of glucose plus methanol (Egli et al., 1983); also in this case, $\bar{s}_{methanol}$ was reduced compared to growth at the same dilution rate with methanol alone, and the reduction was dependent on the composition of the glucose/methanol mixture supplied in the feed as shown in Fig. 4. Experimental evidence for this phenomenon has been published recently also for growth of the acetogenic anaerobe *Eubacterium limosum* on mixtures of glucose and methanol (Loubière et al., 1992b). During growth in carbon-limited chemostat culture at $D = 0.2$ hr^{-1}, glucose was detectable, whereas when this bacterium was grown with mixtures of glucose plus methanol the concentration of glucose was below the detection limit. Unfortunately, in all of these examples only the concentration of one component of the substrate mixture provided in the medium could be measured. Steady-state concentrations of carbon substrates during mixed substrate growth were also recently measured during cultivation of the methylotrophic yeast *H. polymorpha* MH 30 in carbon-limited chemostat culture with a mixture of either glycerol/xylose or xylose/methanol (Babel et al., 1993). For glycerol there was clear evidence that the steady-state concentration during growth with mixtures was reduced in comparison to single-substrate growth. It should be added that this effect has not only been observed for pure suspended cultures but also for the more complex systems of a biofilm (Namkung and Rittman, 1987b).

It seems, of course, appealing to speculate that the lowering of steady-state concentrations during simultaneous utilization of mixed carbon substrates by heterotrophic microorganisms might be a general phenomenon, applicable to all of the substrates, and not just one, and valid for growth with even more complex mixtures. If this were the case, it could considerably change our perception of microbial growth under environmental conditions. However, it is obvious that in order to deduce general kinetic principles for microbial growth with mixed substrates, the concentrations of all individual substrates used by the cells should be known as a function of the composition of the substrate mixture supplied; and this, preferably, not only for steady-state but also during transient growth conditions. Only recently, two chemostat studies have been published providing for the first time steady-state concentrations for all substrate components (Lendenmann et al., 1992; Egli et al., 1993; Weusthuis et al., 1993; Lendenmann, 1994). Although in one study a yeast was used, whereas in the second *E. coli* was employed, the information obtained in both studies demonstrates that the reduction of steady-state concentrations during mixed substrate growth is observed for all substrates involved.

The most extensively studied example is that for the growth of *E. coli* in carbon-limited chemostat culture with defined mixtures of up to six different sugars (Lendenmann et al., 1992; Egli et al., 1993; Lendenmann, 1994). In a first series of experiments, steady-state sugar concentrations were measured in cultures growing with different mixtures of glucose and galactose at a fixed dilution rate of 0.3 hr^{-1}, and the results are given in Fig. 3. As anticipated for the low growth rate employed (μ_{max} with either glucose or galactose was 0.92 hr^{-1}), both sugars were utilized simultaneously, independent of the composition of the glucose/galactose mixture supplied in the feed. The data clearly demonstrate that the steady-state concentrations of glucose and galactose were always lower during growth with sugar mixtures than during growth with a single sugar, and furthermore that the concentrations were roughly proportional to their ratio in the medium feed. The steady-state concentrations of the two sugars were always the same, irrespective of whether the experiment was carried out with a total concentration of sugars in the feed of 1, 10, or 100 mg liter^{-1}, or whether the culture was first grown with glucose before stepwise increasing the proportion of galactose in the feed or vice versa. This demonstrates that the pattern observed was independent of the biomass concentration in the culture medium and the growth history of the cells. Furthermore, experiments performed between $D = 0.3$ and 0.7 hr^{-1} with a glucose:galactose mixture of constant composition (1:1) indicated that the phenomenon was not dependent on growth rate.

The results obtained in the experiment described for a mixture of two sugars

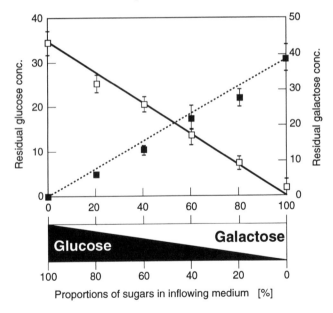

Figure 3. Steady-state concentrations of glucose ($\square$) and galactose ($\blacksquare$) during growth of *Escherichia coli* at a constant dilution rate of 0.30 hr^{-1} in carbon-limited chemostat culture with different mixtures of the two sugars. The proportion of the sugars in the mixture fed to the culture is given as weight percentages. The total sugar concentration in the feed was always 100 mg liter^{-1}. Steady-state sugar concentrations are given in μg liter^{-1}. (Adapted from Egli *et al.*, 1993.)

(A and B) suggest that during growth at a constant rate the observed steady-state concentration of a particular sugar ($\bar{s}_A$ and $\bar{s}_B$) was proportional to its contribution to the total sugar concentration in the medium reservoir ($S_{0(A)} + S_{0(B)}$). Hence, a first simple model that describes the observed steady state concentration of sugar ($\bar{s}_{A(mix)}$) during growth of *E. coli* with sugar mixtures can be put forward:

$$
\bar{s}_{A(mix)} = \bar{s}_{A(100\%)} \frac{S_{0(A)}}{S_{0(A)} + S_{0(B)}} = \bar{s}_{A(100\%)} \frac{S_{0(A)}}{S_{0(tot)}}, \tag{3}
$$

where $\bar{s}$ (100%) is the steady- state concentration of *A* during growth with *A* only. It seems reasonable to assume that *E. coli* should be able to grow not only with two but with three or even more carbon substrates simultaneously. It was of course tempting to assume that this principle might be equally applicable to growth with more complex sugar mixtures. Equation (3) implies that the steady-state concentration of a sugar contributing a minor fraction to the total substrate utilized would be reduced correspondingly, and furthermore that one that contributes always the same proportion of carbon to the mixture would stay constant, irrespective of the relative proportions of other sugars in the medium.

To confirm the validity of this simple model, *E. coli* was grown in carbon-limited chemostat culture at a fixed dilution rate of 0.6 hr^{-1} with mixtures of six different sugars that all supported growth when supplied as single substrates. In this experiment the composition of mixtures supplied for growth was such that one sugar contributed 50% of the total carbon, whereas each of the other five sugars contributed 10%. As expected, all of the six sugars were utilized simultaneously. The biomass concentration measured was constant (47.3 mg liter^{-1} dry wt) during growth with all six mixtures tested (the total sugar concentration in the feed was always 100 mg. liter), indicating that the growth yields of the individual sugars were additive and identical (Lendenmann, 1994). The steady-state concentrations measured in this experiment for the different sugars are shown in Table III. The results clearly support the concept that in the case of the six sugars tested the observed steady-state concentration for a particular sugar is a reflection of its contribution to the overall substrate utilized by the cell. For example, the average steady-state concentration of glucose was 12 ± 4 µg liter^{-1} whenever this sugar contributed 10% to the total substrate concentration in the feed, and this steady-state concentration was not influenced by the relative proportion of other sugars. The same pattern was also observed for the other sugars. It should be noted that *E. coli* was able to utilize sugars such as glucose and galactose down to concentrations of some 10 µg liter^{-1} at a dilution rate that corresponded to $^2/_3$ of μ_{max}. Concentrations of glucose in this range are frequently found in freshwater and marine ecosystems (Münster and Chróst, 1990). Another interesting observation was made with respect to growth with ribose: Although the maximum specific growth rate for this sugar when supplied as the

Table III. Average Steady-State Concentrations of Individual Sugars during Growth of *Escherichia coli*[a,b]

Contribution in mixture	Steady-state concentration (μg liter^{-1})					
	Maltose	Glucose	Galactose	Ribose	Arabinose	Fructose
10%	23	12	11	32	52	69
50%	95	50	70	182	256	254
100%	214	138	209	c	d	e

[a] Adapted from Lendenmann (1994).
[b] Grown in carbon limited chemostat culture at $D = 0.6$ hr^{-1} with different mixtures of six sugars as the carbon/ energy source, as a function of their contribution to the total substrate in the feed (given in %, wt/wt). The composition of the mixtures was such that one sugar contributed 50% of the total sugar in the feed, whereas the other five sugars contributed 10% each. For comparison, the steady-state concentrations during growth with a single sugar (indicated as 100%) at the same dilution rate are also given, The total sugar concentration in the feed was in all experiments 100 mg liter^{-1}.
[c] μ_{max} of *E. coli* with ribose was 0.57 hr^{-1}; therefore, no steady-state concentration can be given for growth with ribose only.
[d] Cells flocculated during growth with arabinose as the only carbon source, making the determination of steady-state substrate concentration impossible.
[e] Steady-state concentration not reported because of adaptation process observed during growth with fructose as the major substrate (Lendenmann, 1994).

only source of carbon and energy was lower (0.57 hr^{-1}) than the dilution rate applied (0.6 hr^{-1}), ribose was utilized simultaneously during growth with all mixtures and the steady-state concentrations measured were approximately proportional to the relative contribution of this sugar in the feed. It was subsequently shown that at this dilution rate overproportionally enhanced steady-state concentrations of ribose were observed only when the proportion of ribose to the total sugar concentration exceeded 50% (Lendenmann, 1994).

3.2.2. Effect of Inducible Enzyme Systems on Growth Kinetics

Concentrations of transport proteins and catabolic enzymes present in cells are known to influence their kinetic properties, and this implies that steady-state concentrations of substrates in chemostat culture are strongly connected with the extent of expression of such proteins (for reviews, see, e.g., Button, 1985, 1993; Rutgers *et al.*, 1991; van Dam and Jansen, 1991). A good example to illustrate the interaction between enzyme concentration of a catabolic pathway and the steady-state concentration of its substrate in the culture medium during mixed substrate growth is that of the methylotrophic yeast *Kloeckera* sp. 2201 (Egli *et al.*, 1982a, 1983). It should be added that for interpretation of the data it is important to know that methanol enters the cell by diffusion and that the yeast has no possibility to replace the first enzyme in the pathway—alcohol oxidase—with an enzyme of different affinity. When this yeast was cultivated in chemostat

culture at a constant dilution rate with different mixtures of glucose and methanol, the pattern observed for the steady-state concentration of methanol exhibited
two distinct regions: As long as methanol contributed less than approximately
50% (wt/wt) to the total substrate in the medium feed, its steady-state concentration in the culture stayed constant, whereas it increased dramatically when the

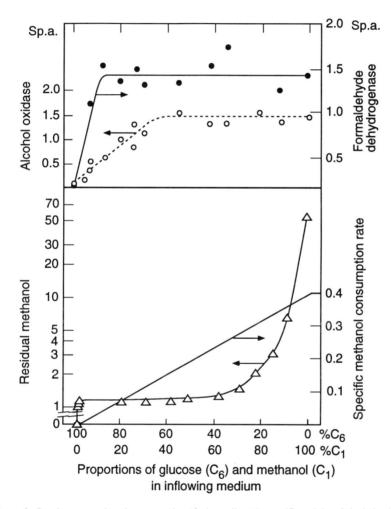

Figure 4. Steady-state methanol concentration ($\triangle$, in mg liter^{-1}), specific activity of alcohol oxidase
($\bigcirc$) and formaldehyde dehydrogenase ($\bullet$), and specific methanol consumption rate (—) in a culture
of *Kloeckera* sp. 2201 during simultaneous utilization of mixtures of glucose and methanol in carbon-
limited chemostat culture at a constant dilution rate of 0.14 hr^{-1}. The specific activity patterns shown
for alcohol oxidase and formaldehyde dehydrogenase [in μmole (mg protein/min)$^{-1}$] are representative for all other dissimilatory and assimilatory enzymes in the metabolic pathway for methanol.
(Adapted from Egli *et al.*, 1982a, 1983.)

proportion of methanol in the feed was higher than 50% (Fig. 4). This pattern, in combination with that observed for specific activities of enzymes involved in the dissimilation and assimilation of methanol, indicates that this yeast followed two different control strategies in regulating the metabolic flux of methanol ($q_{methanol}$) in the cell [Eq. (4); where it is assumed that the Monod model can be used to describe the relationship between $q_{methanol}$ and the steady-state methanol concentration $\bar{s}$]. As long as the proportion of methanol flowing through the pathway was lower than 50%, they adjusted the amount of enzyme to the flux, i.e., cells changed $q_{max(methanol)}$. This strategy resulted in a constant extracellular steady-state concentration of methanol ($\bar{s}$). However, when the flux of methanol exceeded 50%, a further increase of the amount of enzyme was obviously not possible; hence, in order to increase $q_{methanol}$ through the pathway, the concentration of the substrate had to increase from approximately 1.2 mg liter^{-1} to 70 mg liter^{-1} to sustain growth at the set dilution rate:

$$q_{methanol} = q_{max(methanol)} \frac{\bar{s}}{K_s + \bar{s}} \tag{4}$$

This example indicates that the resulting patterns of extracellular substrate concentrations are intimately linked to the concentrations of enzymes in a metabolic pathway. Therefore, especially for compounds whose utilization requires induction of specific enzymes, a detailed knowledge of the regulatory enzyme pattern is probably needed to interpret and predict substrate concentrations and link them with growth kinetics. Furthermore, even if the concentration of enzymes in a pathway stayed constant, one can expect considerably different patterns of steady-state substrate concentrations in experiments as discussed above for *E. coli* and *Kloeckera* sp. 2201, dependent on whether the pathway operates far from or close to saturation (Egli *et al.*, 1993). In this respect, the almost linear behavior of the steady-state concentration patterns obtained for the growth of *E. coli* at a constant dilution rate with mixtures of glucose and galactose, but also those with six sugars, suggests that, independent of the sugar mixture supplied, all enzyme systems involved were more or less fully induced and operating far from saturation.

3.2.3. Kinetic Models for Mixed Substrate Growth

Despite much evidence for mixed substrate growth, the kinetic models used today to describe microbial growth assume that the concentration of a single, limiting substrate controls the rate of growth. Although for this situation a range of different mathematical models have been proposed for the relationship between specific growth rate (μ) and the concentration of the growth-limiting substrate (s) (reviewed in Jannasch and Egli, 1993; Lendenmann, 1994), Eq. (5), originally proposed by Monod (1942), is by far the one that is most frequently used to describe kinetics of microbial growth, competition, or biodegradation in the environment (Bazin, 1982; Simkins and Alexander, 1984; Gottschal, 1993; Senn *et al.*, 1994):

$$\mu = \mu_{max} \frac{s}{K_s + s} \tag{5}$$

Nevertheless, several authors have extended Monod's model—usually by combining two or more (modified) Monod terms—to describe the utilization of mixtures of both homologous (Yoon *et al.*, 1977; Yoon and Blanch, 1977; Gondo *et al.*, 1978; Bell, 1980; Kompala *et al.*, 1984; Slaff and Humphrey, 1986; Namkung and Rittman, 1987a; Bley and Babel, 1992) and heterologous combinations of substrates (Megee *et al.*, 1972; Bader, 1978, 1982; Lee *et al.*, 1984; Baltzis and Fredrickson, 1988; Mankad and Bungay, 1988). Although most of these models were originally designed to describe diauxic growth patterns in batch culture (therefore most of them include inhibition terms), some of them also were applied to growth in continuous culture. Whereas these models could be easily tested for batch culture growth, it was not possible to prove their validity for mixed substrate growth in the chemostat because of the lack of experimental data on steady-state concentrations of growth-limiting substrates.

The experimental data presented for *E. coli* growing with mixtures of sugars in the chemostat certainly give an answer to the question of how microorganisms are able to reduce concentrations of carbon substrates in ecosystems to the low levels observed and still grow reasonably fast. However, they are bound to make the description of microbial growth kinetics under environmental conditions not much easier. Using these data, different mixed substrate models were evaluated for their ability to predict the growth rate from the steady-state concentrations measured in chemostat cultures of *E. coli* with different sugar mixtures (Lendenmann, 1994). Several assumptions were made in this evaluation: (1) kinetics of mixed substrate growth are determined by the concentrations of the individual growth-limiting components and not by a lumped parameter (e.g., DOC); (2) as for growth with a single limiting substrate, the kinetics of a particular substrate during mixed substrate growth can still be described by a Monod relationship [which, based on data obtained for a mixture of glucose:galactose (1:1) seems to be the case (Lendenmann, 1994)]; (3) the Monod kinetic parameters determined for single-substrate growth can also be applied to mixed substrate growth; and last but not least, (4) it was assumed that all substrates were utilized simultaneously. Of the models listed above none was able to describe all the experimental data (Lendenmann, 1994). The growth rate predicted from the steady-state concentrations measured in the chemostat during cultivation with glucose:galactose mixtures from Fig. 3 fit well ($0.28 \text{ hr}^{-1} < \mu < 0.31 \text{ hr}^{-1}$) with the experimental dilution rate of 0.3 hr^{-1} when the model of Bell (1980) was used, based on a summation of Monod terms for the different substrates [Eq. (6)]. However, this model considerably overestimated the growth rates ($1.07\,hr^{-1} < \mu < 1.32\,hr^{-1}$) when applied to the data from the six sugar mixtures at $D = 0.6\,hr^{-1}$ (Table III).

$$\mu = \sum \mu_{max,i} \times \frac{s_i}{K_i + s_i} \qquad (6)$$

For the experiment performed with six different sugars at $D = 0.6\,hr^{-1}$, best estimates for the growth rate ($0.57\,hr^{-1} < \mu < 0.67\,hr^{-1}$) from measured steady-state concentrations were obtained with the mathematical relationship shown in Eq. (7) (Lendenmann, 1994):

$$\mu = \frac{\mu_{max} \times \sum a_i \times s_i}{\mu_{max} + \sum a_i \times s_i} \qquad (7)$$

where a_i is the specific affinity μ_{max}/K_i^{-1} (note that a_i was originally defined as v_{max}/K_i; Button, 1991) and μ_{max} is the maximum specific growth rate for glucose ($0.92\,hr^{-1}$). When applied to the data from the glucose : galactose experiment, this model slightly underestimated the actual growth rate of the culture ($0.25\,hr^{-1} < \mu < 0.29\,hr^{-1}$).

Both models were able to predict the dilution rate of the culture within a range of approximately $\pm 15\%$, whereas with other models proposed for the description of growth kinetics with mixtures of homologous substrates the deviations in the predicted values for the specific growth rate from the experimentally set dilution rate were considerably higher (U. Lendenmann, unpublished data). The first model has the unfortunate property that the upper limit of the maximum specific growth rate that theoretically can be achieved is not defined and increases with the number of substrates used. This is reflected in the overestimation of growth rates when applied to the experimental data (Table III) obtained for *E. coli* growing with mixtures of six sugars. On the other hand, at very low substrate concentrations, well below K_s for the different substrates, the specific growth rate depends approximately linearly on the sugar concentration. Therefore this model provides a reasonable prediction of the growth rate in our experiments. A serious disadvantage of both models is that they assume constant kinetic properties of the cells and do not consider that the utilization of certain substrates depends on the induction of the relevant enzymes, as discussed in the example given above for the growth of methylotrophic yeasts with mixtures of glucose and methanol.

It is obvious that a good estimate for the growth rate of a microbial cell would depend on the knowledge of all (or at least the major) growth-limiting substrates utilized under natural conditions together with their concentrations. Furthermore, for either of the two models, a good estimate of microbial growth rates in nature is very much dependent on the use of environmentally relevant values for substrate affinities, K_s. As discussed above for the case of *E. coli* with glucose, this parameter can change over a range of two to three orders of magnitude and is dependent on culturing conditions and the degree of adaptation of the cells. In contrast, differences in maximum specific growth rates exhibited by adapted and nonadapted cells vary at the most by a factor of approximately two to three (Jannasch, 1968; Harvey, 1970; Lendenmann, 1994).

It remains to be seen whether or not it will be possible at a later stage for certain microorganisms to use lumped parameters such as DOC (or a utilizable fraction thereof) in combination with average substrate affinities and maximum specific growth rates to predict growth rates in nature. However, as it is known that the quality of DOC changes during degradation, i.e., "older" DOC is less well degradable than "young" DOC (e.g., Azam and Cho, 1987; Ducklow and Carlson, 1992; Hobbie and Ford, 1993), this will probably be a difficult task. It should also be pointed out that the summation of growth rate contributions from different substrates might not always follow a linear pattern, since it is known that some substrates can be utilized only at low rates when presented on their own, whereas when present in combination with other substrates they may contribute to a considerable part of the growth rate (see Table III) (e.g., Silver and Mateles, 1969; Reber and Kaiser, 1981; Poindexter, 1987).

3.3. Degradation of Pollutants in the Presence of Readily Utilizable Carbon Compounds

3.3.1. Pollutant Degradation at Low Concentrations

A central question concerning the fate of pollutants in the environment is that of their degradation at low concentrations. Concentrations of (xenobiotic) pollutants in aquatic environments are usually in the low microgram or nanogram per liter range; even organic compounds used in large amounts, such as laundry detergents, only reach peak concentrations of a few milligrams per liter in municipal wastewater treatment plants (e.g., Meijers and van der Leer, 1976; Howard, 1989; Alexander, 1994). This is a concentration range at which carbon sources of natural origin, such as sugars or amino acids, are present (Münster and Chróst, 1990) and usually the concentration of DOC available for microbes in such systems is 10- to 100-times higher than that of pollutants (Schmidt and Alexander, 1985). It is almost certain that under such conditions pollutants are utilized simultaneously together with other available carbon substrates.

It has been demonstrated for many natural and xenobiotic carbon sources that they can be utilized and assimilated by microorganisms when they are present at concentrations of a few nanogram per liter or even femtogram per liter (Boethling and Alexander, 1979a,b; Subba-Rao *et al.*, 1982; Fuhrman and Ferguson, 1986; van der Kooij, 1990; Morita, 1993), i.e., far below the "threshold" levels for growth (see also Section 3.5.2). It has been suggested that this ability to metabolize substrates at extremely low concentrations is a consequence of the fact that they are simultaneously utilized together with uncharacterized DOC that can support either growth or supply energy required for maintenance. This explanation is supported by the results reported by van der Kooij and co-workers (van der Kooij *et al.*, 1980, 1982; van der Kooij and Hijnen, 1988; van der Kooij, 1990), who found that the use of mixtures of substrates enabled several bacteria, including *Pseudomonas, Aeromonas,* and *Flavobacterium* strains, to more efficiently utilize certain substrates at low concentrations or even to co-utilize substrates that did not support growth on their own. This implies that in ecosystems both the rate and the extent of biodegradation of xenobiotic organic compounds at low concentrations are probably controlled by the presence of other organic compounds and that the patterns of stimulation or inhibition of the degradation of pollutants observed at a particular time are to a large part a result of the particular mixture, concentration, and availability of additional (carbon) compounds of natural origin.

Clearly, the data for the growth kinetics of *E. coli* with mixtures of sugars suggest that the simultaneous utilization of a pollutant in combination with other carbon sources should result in reduced residual concentrations of the pollutant compared to conditions where the pollutant is used as the only carbon substrate. Unfortunately, there are only a few studies in which this question was addressed at all (LaPat-Polasko *et al.*, 1984; Schmidt and Alexander, 1985), and the author is not aware of any systematic experimental approaches. The many reports on die-away studies of pollutants using water or soil samples are essentially blackbox systems and therefore difficult to interpret, although the observed degradation for some compounds seems to follow a rather simple kinetic pattern (see, e.g., Rubin *et al.*, 1982). Thus, many of the fundamental questions concerning kinetics and extent of (pollutant) biodegradation under realistic conditions are still unanswered.

3.3.2. The Case of Nitrilotriacetic Acid: A Pollutant Degraded by an Inducible Enzyme System

Derepression (or even constitutivity) of a metabolic pathway responsible for the degradation of a particular pollutant under environmental (carbon-limited) growth conditions would ensure the presence of a certain degradation capacity even in the absence of that pollutant. Hence, the cells should be able to immediately utilize the pollutant when it becomes available (let us not consider the rates that can be achieved and the interference of other available carbon substrates for the time being). In contrast, if the synthesis of enzymes of a degradation pathway requires the pollutant as an inducing compound, then the question arises as to how much pollutant is required to stimulate induction, what the time course of this induction process looks like, and how rates and extent of degradation are influenced by the presence of natural carbon substrates?

In an attempt to understand the degradation of the complexing agent nitrilotriacetic acid (NTA) in wastewater treatment plants and in the environment, this question has been investigated in our laboratory for the NTA-degrading bacterium *Chelatobacter heintzii* ATCC 29600 using carbon-limited mixed substrate continuous cultures. The study has revealed some interesting regulatory features that are probably not only valid for this particular system but may be a first step toward a more refined understanding of the regulation of pollutant-degrading enzymes under complex environmental conditions.

In the gram-negative bacterium *Ch. heintzii*, NTA is degraded via two key enzymes: NTA mono-oxygenase (NTA-Mo) and iminodiacetate dehydrogenase (IDA-DH) (Uetz *et al.*, 1992; Uetz and Egli, 1993; Egli, 1994). The synthesis of both enzymes is repressed during growth with glucose as the only source of carbon and energy (Uetz and Egli, 1993). Neither in glucose-limited chemostat culture at low dilution rates nor in the exponential or stationary phase of glucose batch cultures was significant derepression of NTA-Mo and IDA-DH observed in the absence of NTA. Nevertheless, low levels of the two enzymes always appeared to be present under such conditions (Bally *et al.*, 1994; Bally and Egli, submitted). Induction of the two enzymes requires the presence of NTA or IDA, and so far no other substrates are known to be able to induce their synthesis. However, if there is a repression effect by glucose, it is easily overruled, because when presented with mixtures of glucose and NTA, both substrates were utilized simultaneously in both the exponential growth phase of batch cultures (Egli *et al.*, 1988) and in carbon-limited (ammonia in excess) chemostat culture (Bally *et al.*, 1994).

3.3.2a. Substrate Mixture and Enzyme Induction. Utilization of NTA and the regulation of NTA-specific enzymes have been investigated in carbon-limited chemostat culture at a constant dilution rate during growth with mixtures of glucose and NTA of different composition (Bally *et al.*, 1994). During growth with all mixtures tested, both NTA and glucose were consumed to completion. It was found that the synthesis of NTA-Mo and IDA-DH became induced above the low constitutive level only when the carbon from NTA contributed more than approximately 1–3% to the total carbon in the substrate mixture supplied in the feed (Fig. 5). Although NTA-Mo and IDA-DH were not induced during growth with glucose–NTA mixtures containing less than 3.6% NTA carbon, NTA was utilized simultaneously with glucose; for example, when a mixture consisting of 0.26 mg NTA-C liter^{-1} plus 727 glucose-C liter^{-1} (99.964% glucose-C/0.036% NTA-C) was supplied in the inflowing medium, the resulting NTA concentration in the culture was 12 μg liter^{-1}. Whether or not this degradation of NTA was due to a low expression of the established NTA-degrading enzymes NTA-Mo and IDA-DH or by a yet unknown enzyme system is presently not known.

These results suggest that under environmental conditions NTA-degrading bacteria will most likely utilize NTA not as an exclusive carbon source, but that they will metabolize it simultaneously with other available carbon substrates. Of

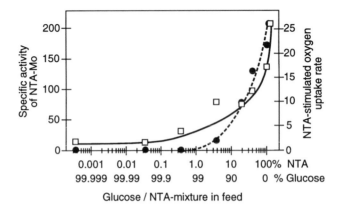

Figure 5. Regulation of NTA metabolism in *Chelatobacter heintzii* ATCC 29600 growing at a constant dilution rate of 0.06 hr^{-1} in carbon-limited chemostat culture supplied with either different mixtures of glucose plus NTA, or with NTA alone as the source(s) of carbon and energy. Irrespective of the mixture supplied in the feed, both substrates were consumed to completion (see text). Specific activity of NTA mono-oxygenase [in μmole (g protein × min)$^{-1}$] in cell-free extracts (□) and oxygen uptake rate of washed cells [μmole O_2 (g dry wt × min)$^{-1}$] in the presence of excess NTA (●). The total concentration of carbon from both NTA and glucose in the medium feed was always 60.6 mM. (Adapted from Bally *et al.*, 1994.)

course, this raises the question as to what fraction NTA contributes to the total carbon substrate consumed during growth of NTA-degrading bacteria in treatment plants or surface waters. Furthermore, one would like to know whether the cells are induced or whether they will operate at the low constitutive enzyme level. A rough estimate for the ratio of NTA to total carbon available for NTA-utilizers can be made from measured environmental concentrations. (One should be aware, however, of the fact that the concentrations of substrates in ecosystems do not necessarily reflect the flux and consumption rates of substrates by individual cells. Such an estimate also assumes that under environmental conditions kinetic constants are in the same range for the different substrates.) Data from monitoring programs indicate that NTA concentrations in river water are usually in the range of 0.4 to 4 μg C liter^{-1}, a value that can increase to some 10–20 μg C liter^{-1} in rivers receiving a high load of treated sewage (Bally *et al.*, 1994). Considering average concentrations of DOC in surface waters of a few milligrams per liter and taking into account that usually only some 1–5% of this carbon is available for microbial growth suggests that under these conditions NTA may contribute to approximately 0.1–1% of the total carbon for growth of NTA-utilizing bacteria. If NTA degraders only utilized glucose from the available carbon pool (average concentrations in the environment are in the range of 5–20 μg C liter^{-1}), NTA would contribute some 10% of the carbon utilized for growth. Based on the chemostat experiment presented above and assuming that induction of enzymes involved in NTA degradation is regulated also in ecosystems via the flux of carbon from NTA, one can speculate that under environmental conditions induction of NTA enzymes should be triggered if glucose was the main carbon source for these bacteria, whereas no induction can be expected under conditions where the bacteria utilize a wider range of carbon sources.

3.3.2b. Level of Induction in the Environment. To investigate in more detail the question of whether NTA-degrading bacteria are induced and to what extent and whether or not degradation in wastewater treatment plants occurs via expression of NTA-enzymes or enrichment of NTA-degraders, both the number of NTA-degrading bacteria and the expression of NTA mono-oxygenase components were determined in two different wastewater treatment plants and a laboratory model treatment system consisting of porous pots fed with artificial sewage containing different concentrations of NTA (Bally, 1994). In all systems the number of cells reacting immunopositively toward the antiserum raised against the NTA-degrading bacterium *Ch. heintzii* ATCC 29600 was in the range of approximately 1% of the total population (Table IV), which corresponds well with the results obtained in an earlier investigation (Wilberg *et al.*, 1993). In one municipal wastewater treatment plant (Glatt), the concentrations of NTA and DOC measured in the aeration tank indicated that NTA contributed well below 1% of the total dissolved carbon available for microbial growth. As expected from the pure culture experiment shown in Fig. 5, both components of NTA-Mo were below the detection limit, although the plant degraded NTA with a high efficiency. A similar response was obtained for a porous pot model treatment system that was fed with synthetic sewage containing 1.1% NTA-carbon (Bally, 1994). However, when the proportion of NTA-C in the synthetic sewage fed to the porous pots was increased to some 5%, an increased level of component A of NTA-Mo was observed, whereas component B remained below the detection

Table IV. Presence of NTA–Mono-oxygenase Components and of Immunopositive Bacterial Cells in Wastewater Treatment Plants (WWTP) and a Porous Pot System Fed with Different Ratios of NTA–Carbon to DOC[a,b]

| | WWTP Glatt | Porous pots[c] | | WWTP Säntis |
		3 mg NTA	15 mg NTA	
NTA-C : DOC	0.06%	1.1%	5.4%	6.9%
Immunopositive cells (% of total cells)	4.2%	0.5%	1.0%	0.9%
NTA–Mo				
cA [mg (g dry wt)$^{-1}$]	<0.0005	<0.0005	0.023	0.045
cB [mg (g dry wt)$^{-1}$]	<0.0005	<0.0005	<0.0005	0.007
NTA-elimination efficiency	98.9%	98.9%	98.8%	99.9%

[a]Adapted from Bally (1994).
[b]Cells immunopositive for α-*Chelatobacter heintzii* ATCC 29600. Not detectable, <0.0005 mg protein (g dry wt)$^{-1}$.
[c]Porous pots were fed with NTA plus synthetic sewage.

limit. At the same time, the proportion of immunopositive bacterial cells in the mixed culture increased only slightly from 0.5 to 1.0%. In addition, the presence of immunopositive cells and NTA-Mo was studied in a small wastewater treatment plant on Mount Säntis. This plant is fed with wastewater from three tourist restaurants all of which use a dishwasher detergent containing NTA as the main chelating agent. In this plant the contribution of NTA to the total carbon was, on busy days, between some 5–10%. Also in this plant no obvious enrichment of immunopositive NTA-degrading bacteria was observed, although both components A and B of the NTA-Mo were detected in considerable amounts. Of course, one has to be aware of the fact that interpretation of these data is affected by a number of uncertainties. For example, it is not known whether or not the number of immunopositive cells gives a fair estimate for the total number of NTA-degrading bacteria in the activated sludge microbial community [certainly not all NTA-degraders reacted immunopositively and earlier results indicate that the proportion of *Chelatococcus* spp. is similar to that of *Chelatobacter* spp. in activated sludge (Wilberg *et al.*, 1993; Bally, 1994)]. Similarly, it is not known whether all NTA-degrading cells were employing the NTA-Mo system described from *Ch. heintzii* ATCC 29600 against which the antibodies used in this study had been raised (Uetz *et al.*, 1992). Nevertheless, the information obtained suggests that regulation of NTA degradation at the level of enzyme expression is an important factor in activated sludge plants, and that the response of the NTA-degrading bacteria under complex environmental conditions to the availability of NTA as a growth substrate might be similar to that described above for the pure culture growing with mixtures of glucose plus NTA in the chemostat.

One would expect that the level of induction of pollutant-degrading cells will influence the behavior of a treatment facility especially during transient exposure to increased concentrations of NTA. Activated sludge containing an induced population of NTA-degrading bacteria should be able to cope better with a pulse of NTA than uninduced cultures. Indications for such a behavior have indeed been obtained for porous pot plants that were run with different concentrations of NTA in the synthetic sewage feed (Bally, 1994). Plants run with synthetic sewage containing 3 mg liter^{-1} NTA exhibited a lower transient accumultation of NTA when the concentration of the complexing agent was increased in the medium feed to 15 mg liter^{-1} NTA than plants previously adapted to the degradation of 0.3 mg liter^{-1} NTA.

3.3.3. Importance of Additional Carbon/Energy Sources for Enzyme Synthesis under Transient Conditions

3.3.3a. Effect on Induction. If induction is an important process in the regulation of the degradation capacity of activated sewage sludge or natural microbial populations in surface waters, it is of course of interest to know how fast the enzymes required for the degradation of a particular pollutant can be synthesized in case it becomes available in considerable amounts in such a mixed substrate environment. Vice versa, one is interested to know how quickly they will be lost in the absence of the inducing substrate. Several aspects concerning the dynamics of induction of NTA-degrading enzymes during mixed substrate growth have been studied recently for *Ch. heintzii* (Bally, 1994; Bally and Egli, submitted). When a culture growing at a constant dilution rate in carbon-limited chemostat culture with glucose as the only source of carbon and energy was subjected to a medium shift where carbon from glucose in the medium was completely replaced by NTA carbon, the cells required almost 24 hr before they were able to metabolize NTA, as indicated by the oxygen uptake rate exhibited by cells taken from the chemostat and exposed to excess NTA (Fig. 6). Almost complete washout of the culture occurred, and accumulation of NTA was observed during the time before induction started. The influence of supplying mixtures of glucose–NTA instead of NTA only was surprising. A shift from glucose to a mixture consisting of 1% glucose-C/99% NTA-C resulted in a reduction of the time required until induction of NTA-Mo started from 24 to approximately 9 hr. A further increase of the proportion of glucose-C in the mixture again reduced the time required until, using a mixture of 90% glucose-C/10% NTA-C, induction started after 30 min and no accumulation of NTA was detected during this shift experiment (Bally, 1994). The fact that for mixtures containing up to 80% NTA-C the time required for induction was approximately

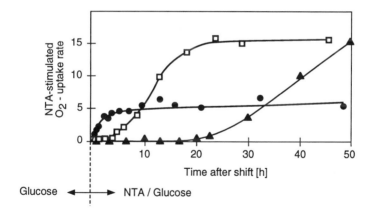

Figure 6. Induction of NTA metabolism in *Chelatobacter heintzii* ATCC 29600 growing in continuous culture after switching the feed from a medium containing glucose as the only carbon source to media containing either only NTA (▲), or to mixtures of glucose plus NTA containing a proportion of either 50% (□) or 10% (●) of NTA. Induction of enzymes involved in the metabolism of NTA was assessed by measuring the specific oxygen consumption rate of washed cells exposed to excess concentrations of NTA in a Clark oxygen electrode [given in μmole O_2 (g dry wt $\times$ min)$^{-1}$]. Throughout the experiment the dilution rate was kept constant at $D = 0.06$ hr^{-1}. The total carbon concentration in the feed was always 60.0 mM. (Adapted from Bally and Egli, submitted.)

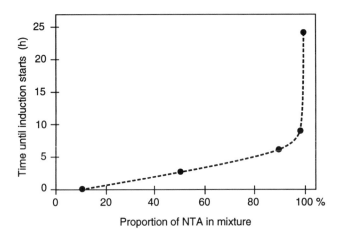

Figure 7. Time needed to initiate induction of NTA-degrading enzymes in *Chelatobacter heintzii* ATCC 29600 grown in continuous culture with glucose as the only source of carbon after switching the feed to media containing either NTA only, or mixtures of NTA plus glucose of different composition (indicated as % of carbon from NTA). Experimental conditions as specified in Fig. 6. (Adapted from Bally and Egli, submitted.)

linearly related to the amount of glucose-C in the mixture (Fig. 7) was striking. The data demonstrate that glucose was not inhibiting the induction but actually supported expression of NTA-metabolizing enzymes, probably by supplying the energy and building blocks needed for the necessary rearrangement of the cellular metabolism and particularly for the synthesis of NTA-degrading enzymes. The overproportionally high accelerating effect of as little as 1% of glucose-carbon, which reduced the time needed for induction from 23 to 9 hr, suggests that the level of carbon flux needed to rearrange metabolism without sacrificing (degrading) much of its own cell constituents is in the range of 0.004 g glucose (g dry cell wt)$^{-1}$ hr^{-1}. It is interesting that this substrate specific consumption rate is approximately one order of magnitude below the range of maintenance coefficients given by Pirt (1975) for different microbial cells.

Similar evidence for the importance of energy availability for cells on the ability to adapt to new growth conditions was reported for the facultative chemolithotrophic *Thiobacillus* A2 (now *T. versutus*) by Gottschal and co-workers (Gottschal *et al.*, 1981a). This bacterium can grow either heterotrophically or autotrophically using thiosulfate as an energy source. Furthermore, on condition that these substrates are supplied at growth-limiting concentrations, cells are able to grow mixotrophically in the chemostat, i.e., simultaneously using a heterotrophic carbon source and thiosulfate–CO_2 (Smith and Kelly, 1979; Gottschal and Kuenen, 1980a). During growth in acetate-limited chemostat culture at a dilution rate of 0.05 hr^{-1}, *Thiobacillus* A2 exhibited no autotrophic capacity, measured as the ability to both oxidize thiosulfate and to fix carbon dioxide via the Calvin cycle. When the medium fed to such a culture was shifted from acetate as the growth-limiting substrate to one containing thiosulfate–CO_2 (at the same time D was reduced to 0.02 hr^{-1}), only a very slow increase of the capacity to oxidize thiosulfate and ribulose-1,5-bisphosphate carboxylase (RubPC) was observed during the first 7–10 hr. No growth occurred during this time, and washout of the cells was found at a rate equal to the dilution rate. The authors suspected that this slow induction of autotrophic capacity might be due to the lack of energy for the synthesis of new enzymes. When the same shift was repeated at a dilution rate of 0.05 hr^{-1} using a mixture of thiosulfate/acetate (16 mM/6 mM), an increase of thiosulfate-oxidizing capacity was observed after half an hour (RubPC activity was not followed in this experiment). It should be added that the rate of addition of acetate to the culture had to be so low that acetate remained "limiting" throughout the shift. This was demonstrated by pulsing excess acetate to the culture, which led to an immediate repression of the synthesis of autotrophic enzymes and initiated their degradation. Unfortunately, the authors did not investigate the response of the culture to different rates of acetate addition. However, starting out from cells grown under a range of other conditions, Gottschal and colleagues investigated the time course of adaptation of this *Thiobacillus* strain during shifts to chemoautotrophic growth conditions (Gottschal *et al.*, 1981a,b). They generally observed that cells provided with a utilizable energy source, even if they had to degrade intracellularly stored reserve materials such as polyhydroxybutyrate (PHB), were able to rapidly induce the enzymes necessary to oxidize thiosulfate and/or RubPC. In contrast, shift conditions under which the cells had no possibility to gain energy in one way or another resulted in considerable lag phases before they were able to grow autotrophically.

3.3.3b. Influence on Degradation of Inducible Enzymes. In analogy to the experiments discussed above, one can expect that the presence of an additional carbon/energy substrate might influence the rate of turnover and degradation of a particular inducible catabolic enzyme after depletion of its substrate. For both *Ch. heintzii* and *Thiobacillus* A2, this process has been investigated in some detail and the transient response of the two organisms turned out to be quite different. When cells of *Ch. heintzii* were grown in chemostat culture with NTA as the only source of carbon/energy and NTA was replaced in the inflowing medium by glucose, the initial loss of NTA-Mo (both specific activity and NTA-Mo components A and B) was in the range of 5% per hour. However, if the feed of NTA was interrupted and the cells were starved, this led to a rapid loss of NTA-Mo components and activity. Within the first few hours following the shift, the calculated rate of degradation of this enzyme was in the range of 30% per hour for component A and as high as 90% for component B (Bally and Egli, submitted; Bally, 1994). This observation suggests that the availability of an alternative carbon/energy source helped the cell to avoid excess degradation of intracellular resources during adaptation to the new growth conditions. In contrast, the activity of both the thiosulfate-oxidizing capacity and RubPC did not decrease significantly during starvation of autotrophically grown cells of *Thiobacillus* A2 (Gottschal *et al.*, 1981a). However, when cells of *Thiobacillus* A2 were shifted from thiosulfate–CO_2 to acetate or to fructose, a rapid inactivation of both RubPC and thiosulfate-oxidizing capacity was observed. In the case of RubPC, the authors confirmed that the loss of enzyme activity was due to proteolysis. Whereas the rate of inactivation of RubPC seemed independent of the dilution rate after the shift, the rate of loss of thiosulfate-oxidizing activity was clearly proportional to the dilution rate.

3.4. Mixtures of Homologous Nutrients Other than Carbon

There are very few reports where the potential for simultaneous utilization of homologous mixtures of nutrients other than carbon have been investigated. The simultaneous utilization of mixtures of amino acids as nitrogen (and possibly also carbon) sources that probably always occurs during growth in complex medium is an obvious example. This has been demonstrated for a variety of different microorganisms and growth conditions, such as growth under nitrogen excess, carbon excess, or carbon-limited conditions (see, e.g., Molin, 1985; Rogers and Reynolds, 1990; Schut *et al.*, 1995). It should be pointed out that in no case the exact nutritional roles of individual amino acids were traced, e.g., whether and to what extent they served as a either a nitrogen or a carbon source, or what fraction was incorporated directly into protein.

As mentioned earlier, several nitrogen-fixing bacteria were shown to derepress nitrogenase under nitrogen-limited growth conditions in the chemostat (Daesch and Mortenson, 1968; Munson and Burris, 1969; Hillmer and Gest, 1977; Hoover and Ludden, 1984). However, none of these authors tested whether these cultures were able to use the growth-limiting nitrogen source and at the same time fix dinitrogen. Similarly, this author is aware of only one report demonstrating convincingly the simultaneous utilization of, for example, inorganic phosphate plus an organic phosphorus-containing compound (Schowanck and Verstraete, 1990). In this work, a strain of *Pseudomonas paucimobilis* was found to simultaneously utilize inorganic phosphate and methylphosphonate under both P-sufficient conditions in batch culture and P-limited conditions in continuous culture.

An exception is the report by Robertson and Kuenen (1984), who investigated the growth of *Thiosphera pantotropha* with mixtures of oxygen and nitrate as terminal electron aceptors in batch culture. When this bacterium was grown with acetate and ammonia as the carbon/energy and nitrogen sources and it was supplied with either molecular oxygen or nitrate (in the absence of molecular oxygen) as the only terminal electron acceptor, the cultures exhibited a doubling time of approximately 2.2 and 2.8 hr, respectively. However, when nitrate was added to an aerobic culture, the doubling time was reduced to approximately 1.5 hr. It was reported that when growing in a mixed culture with other denitrifying bacteria unable to utilize either electron acceptors at the same time, such as *Pc. denitrificans*, this ability led to a competitive advantage of *T. pantotropha* (Robertson and Kuenen, 1990).

3.5. Some Conclusions, Implications, and Speculations

3.5.1. Simultaneous Utilization or Segregation of Culture into Subpopulations?

Earlier in this chapter the question was addressed of whether or not a culture growing with a mixture of substrates separates into two subpopulations, each specialized for growth on a particular substrate in the mixture. Although it has been demonstrated that cultures of microorganisms may segregate into metabolically different subpopulations, especially during very slow growth (see Kaprelyants *et al.*, 1993), there is much indirect evidence that in the experiments presented in Sections 3.2 and 3.3, several carbon substrates were utilized simultaneously by each individual cell in the microbial population.

For example, in the case of methylotrophic yeasts, high levels of derepression of enzymes involved in the metabolism of methanol was found in cells growing in either batch or chemostat culture in the absence of methanol. The specific activities of some of the methanol enzymes occasionally exceeded 50% of that found in methanol-grown cells (Eggeling and Sahm, 1978, 1981; Egli *et al.*, 1980; de Koning *et al.*, 1987, 1990). In addition, in all the cells of a culture of *H. polymorpha*, growing either in a glucose-limited chemostat or with glycerol in batch culture, two to four peroxisomes were found containing both alcohol oxidase and catalase (Egli *et al.*, 1980). This all suggests that the yeast population was rather homogeneous and not that only a fraction of the population expressed enzymes required for the utilization of methanol.

Further indirect evidence against segregation of the population comes from the pattern of specific activities for methanol-dissimilating and assimilating enzymes in the yeasts *H. polymorpha* and *Kloeckera* sp. 2201 during growth in continuous culture at a constant dilution rate with different mixtures of methanol plus glucose (Egli *et al.*, 1982a, 1983). Let us assume that two subpopulations exist under such growth conditions, one growing on glucose only and one growing exclusively on methanol. This would imply that the subpopulation growing with methanol would be fully induced because it is growing with methanol as the only substrate at the constant dilution rate imposed. This implies again that the level of induction of individual cells would be always the same, independent of the mixture composition supplied in the feed, but that the fraction of methanol-consuming cells would be proportional to the fraction of methanol fed to the culture. Consequently, the total activity of methanol-related enzymes in the culture should be proportional to the fraction of methanol in the mixture, and, because the biomass concentration in the culture was kept approximately constant in this experiment (Egli et al., 1982a), the specific activity would increase linearly. As this is obviously not the case, one must draw the conclusion that all the cells were utilizing methanol in conjunction with glucose.

Additional evidence supporting simultaneous utilization is the observation made for pure cultures of both bacteria and yeasts that carbon substrates that support low growth rates can be utilized in the chemostat at dilution rates that exceed by far the μ_{max} they can support on their own in either batch or chemostat culture (see Fig. 2, and Egli et al., 1986). Last but not least, the patterns of steady-state substrate concentrations measured for carbon-limited growth of E. coli with mixtures of sugars presented in Section 3.3.1 strongly point to a simultaneous utilization. In theory, chemostat cultivation with a single growth-limiting substrate at a particular dilution rate results in a steady-state concentration of this substrate that is independent of its concentration in the feed and therefore the culture density (Monod, 1950; Tempest, 1970a), and in the case of E. coli this has been demonstrated recently for growth in a glucose-limited chemostat (Senn et al., 1994). This implies that in the experiment shown in Fig. 3, segregation of the culture into a glucose- and a galactose-utilizing subpopulation would result in a constant steady-state concentration of both glucose and galactose, and this is obviously not the case.

3.5.2. Threshold Concentrations

The threshold concentration for a compound that serves as a nutrient for a microorganism is usually defined as the lowest concentration at which this compound still allows growth and multiplication of the cell (for a summary, see, e.g., Schmidt et al., 1985; Alexander, 1994). This does not imply that the compound is not utilized at lower concentrations. In the case of xenobiotic compounds the term is also frequently used to indicate that a compound is not consumed below a certain concentration (i.e., threshold for utilization). In the case of substrates of natural origin the evidence supporting the existence of threshold concentrations is weak because compounds, such as amino acids or sugars, are turned over quickly even at the very low environmental concentrations (see Section 2.1). The low, fairly stable concentrations of such compounds observed in ecosystems represent a steady-state between their utilization and continuous formation from particulate material rather than a threshold concentration for utilization. Nevertheless, for pure cultures of various bacteria, a threshold concentration for growth has been demonstrated in the laboratory when they were supplied with single-carbon substrates. Depending on organism and carbon source, the threshold concentrations for growth are typically in the range of 1 to 100 μg liter^{-1} (Jannasch, 1967; Shehata and Marr, 1971; Law and

Button, 1977; van der Kooij *et al.*, 1980; Alexander, 1994). However, it should be pointed out that experimental demonstration of such threshold concentrations—even in aqueous systems—is fraught with difficulties (not to speak of the problems encountered in soil). On one hand, the influence of contaminating DOC in media or from the air is difficult to exclude in such experiments. On the other hand, adaptation of cells to low nutrient levels seems to influence threshold concentrations. For example, adaptation of *Aeromonas hydrophila* and other bacteria to grow in drinking water resulted in reduced threshold values for growth below 1 μg/liter (van der Kooij, 1990; van der Kooij *et al.*, 1980; van der Kooij and Hijnen, 1988).

The experiments presented in Section 3.2.1 for the growth of *E. coli* in chemostat culture with mixtures of sugars suggest that an individual sugar should theoretically be consumed even if it would contribute only a minute fraction of the total carbon substrate utilized by the culture (i.e., assuming linear interpolation in Fig. 3 down to 0%), on condition that at least one other growth-supporting carbon substrate is present. This indicates that during simultaneous utilization of mixtures of carbon substrates the energetic reasoning, i.e., that a threshold amount of energy is needed to maintain cell function (see, e.g., Morita, 1993), should probably not be extended to the utilization of individual substrates, but should be kept at the level of the sum of carbon/energy substrates consumed by a heterotrophic cell. Consequently, under such conditions, cells should be able to utilize individual substrates down to indefinitely low concentrations, on condition that the enzymes required for transport and utilization remain expressed and that other substrates will not interfere.

There is now considerable experimental evidence to support the idea that threshold concentrations for the utilization of (and growth with) particular carbon compounds can be lowered by simultaneous utilization of other carbon substrates. For example, *Salmonella typhimurium* failed to grow in batch culture in a synthetic medium when supplied with 1 μg liter^{-1} glucose (Schmidt and Alexander, 1985). However, in the presence of growth-supporting concentrations of arabinose (5 mg liter^{-1}), glucose was still utilized at concentrations of 0.5 μg liter^{-1} and lower. A similar reduction of the threshold concentration for glucose was observed in chemostat culture for a marine coryneform bacterium when cultivated with mixtures of glucose plus amino acids in comparison to glucose as the only substrate (Law and Button, 1977). For *E. coli*, Shehata and Marr (1971) have reported a threshold concentration of 18 μg liter^{-1} for growth with glucose as the only carbon source, whereas in the experiments reported by Lendenmann (1994), this bacterium was able to utilize and grow with significantly lower glucose concentrations when mixtures of sugars were supplied (see Fig. 3 and Table III). Similarly, van der Kooij *et al.* (1982) found that a mixture of 45 compounds, each present at a concentration of 1 μg liter^{-1}, supported growth of *P. aeruginosa*, whereas the bacterium was not able to grow with the individual compounds present at this concentration.

All this information gives a reasonable explanation for the observation that many compounds can be utilized by microorganisms even if they are present at very low concentrations. Indeed, the utilization in the nanogram per liter or even the femtogram per liter concentration range is documented for many naturally occurring and xenobiotic compounds, far below the threshold concentrations typically observed for the utilization of individual compounds (see Alexander, 1994). It seems justified to conclude that the presence and simultaneous utilization of mixed substrates influences both the kinetics and the extent of the degradation of individual compounds. However, a more systematic experimental approach is required in order to obtain a more general picture as to why threshold concentrations for utilization of certain compounds but not for others are observed in ecosystems. The author is convinced that the method of continuous culture of both pure and mixed microbial populations in combination with mixed substrates will be an important tool to approach this problem in a rational way.

3.5.3. Mixed Substrates and Competition

The phenomenon that simultaneous utilization of carbon sources results in lowered steady-state concentrations of individual substrates compared to single substrate growth has now been demonstrated for a number of bacteria and yeasts, although complete data sets are available only in the case of *E. coli* and *Saccharomyces cerevisiae*. Nevertheless, it seems reasonable to propose that this principle is generally applicable for the growth of heterotrophs under carbon-limited conditions.

With respect to competition in the environment, this implies that in principle the ability of a cell to grow simultaneously with a wide range of different carbon substrates present should result in enhanced competitiveness over one that is only able to utilize a smaller number of substrates for growth (assuming that the two strains differ just in the number of substrates they are able to utilize but not in their kinetic properties with respect to the common substrates). There are indeed several examples reported in the literature where nutritionally versatile strains were reported to outcompete specialist strains under mixed substrate limited conditions (Gottschal *et al.*, 1979; Smith and Kelly, 1979; Dykhuizen and Davis, 1980; see also Gottschal, 1993). However, although competition at the level of substrate concentrations seems the most plausible explanation for the outcome of these experiments, one should recall that in no case have the actual residual substrate concentrations in the growth medium been determined. Therefore, other effects, such as interaction between the strains at the physiological level resulting from altered substrate fluxes through the different species, cannot be excluded (discussed by Gottschal, 1993).

This indicates that one is still far from understanding microbial competition in the environment (except perhaps in some clear-cut cases where different species are competing for a single nutrient under excess conditions). Certainly, the present understanding of competition will change considerably in the future as more information becomes available on growth under environmental conditions, including not only the utilization of mixed substrates, but also multiple-nutrient-limited and transient situations. For an excellent discussion of many of these points, interested readers are referred to Gottschal (1993).

3.5.4. Transient Growth Behavior

The ability to rapidly respond to changing environmental conditions is certainly an important physiological property for a microorganism in all natural (and technical) environments. The examples discussed above all point to the crucial role of energy availability for the degree of flexibility. Consequently, one must assume that a heterotrophic nutritional generalist thriving on a number of different substrates simultaneously has a better chance to make some energy available to support a rapid response to a substrate that becomes suddenly available in its environment in significant amounts than does an organism that is restricted to the utilization of only one or a few energy sources.

The two examples discussed above indicate that the role of alternative carbon/energy substrates in ecosystems should be seen in a different light, especially that of glucose. It is normally assumed that the presence of glucose results in the repression of other catabolic enzyme systems (see Section 2.2.1). However, the results suggest that when consumed at growth-limiting rates, this catabolite repression provoking carbon source also is able to support the induction of other required proteins and helps to cut down the time required for the process of adaptation, e.g., when new substrates become available in the environment or when the cell is challenged by an environmental stress.

The discussion here indicates that in aquatic ecosystems versatile heterotrophic microorganisms will rarely experience "black-and-white" transient conditions where they are forced to suddenly replace its primary source of carbon/energy by another compound (although, within limits, *E. coli* seems to be able to cope with such a situation as well) (see Section 2.2.2 and Lendenmann and Egli, 1995). Probably much more realistic are smooth transitions with respect to the proportions of the different substrates contributing to the overall carbon/energy consumption. In terrestrial ecosystems the situation might be more dramatic in the way that within a short time substrates might become available that before were inaccessible, as indicated by bursts of denitrification in soil after rainfall (Davidson *et al.*, 1993).

4. Multiple-Nutrient-Limited Growth

In laboratory systems the composition of media is usually designed such that from all the different essential nutrients only one of them is "limiting" microbial growth, whereas all other nutrients are present in excess (Liebig's law). In the decelerating and early stationary phase of a batch culture, exhaustion of this nutrient terminates growth and determines the maximum amount of biomass that can be produced (Monod, 1942; Stephenson, 1949; Mason and Egli, 1993). In a chemostat the continuous addition of this nutrient controls the specific rate of growth and the steady-state concentration of biomass in the culture (Monod, 1950; Tempest, 1970a). Applying this concept to ecosystems, identification of the growth-limiting nutrient is usually done by addition of specific nutrients, e.g., glucose, ammonia, or inorganic phosphate, to samples taken from the environment and measuring the stimulation in growth rate (Gerhart and Likens, 1975; Paerl, 1982; Ducklow and Carlson, 1992). However, this method rarely gives a straightforward answer and it is a common observation that the best stimulation of growth is obtained when a combination of nutrients is supplied and not a single nutrient (e.g., Morris and Lewis, 1992; Coveney and Wetzel, 1992; Wang *et al.*, 1992; Zweifel *et al.*, 1993; Schweitzer and Simon, 1995). In fact, it was pointed out long ago that the assumption that growth in nature is limited by a single nutrient or even a single substrate is probably rarely true (Veldkamp and Jannasch, 1972; Paerl, 1982).

As shown above, there is now a considerable body of literature on the simultaneous utilization of homologous substrates, especially on mixtures of carbon sources. This contrasts sharply with the few studies in which the simultaneous limitation of growth by two or more nonhomologous nutrients has been addressed. Consequently, most microbiologists still support the view that only one compound can "limit," i.e., control, the rate of growth at any one time.

4.1. Experimental Observations

The first experimental indications for the fact that more than one nutrient may limit growth were obtained for combined C/vitamin, C/O_2-, C/N-, C/K-, C/P-, C/Zn-, and N/P-limited conditions (Harrison, 1972; Megee *et al.*, 1972; Sinclair and Ryder, 1975; Cooney *et al.*, 1976; Hueting and Tempest, 1979; Panikov, 1979; Lawford *et al.*, 1980; Egli, 1982; Lee *et al.*, 1984). In addition, a number of studies were published where the phenomenon was treated on a purely theoretical basis (Yoon and Blanch, 1977; Sykes, 1973; Bader, 1978, 1982; Chen and Christensen, 1985; Thingstad and Pengerud, 1985; Thingstad, 1987; Baltzis and Fredrickson, 1988). Clear experimental proof for the existence of a dual-nutrient-limited (C/N or C/P) growth regimen in continuous culture was given for both bacteria and yeasts on the basis of culture parameters and the composition and physiological properties of the cells (Egli and Quayle, 1984, 1986; Gräzer-Lampart *et al.*, 1986; Pengerud *et al.*, 1987; Minkevich *et al.*, 1988; Rutgers *et al.*, 1990; Duchars and Attwood, 1989). Also, experimental evidence for the existence of a double (N/P)-limited growth regimen for the growth of algae was reported by Ahlgren (1980) and Rhee (1978) during cultivation at a constant dilution rate in continuous culture (although the latter author concluded that the transition from nitrogen- to phosphorus-limited growth occurred in a stepwise manner).

The typical pattern observed in all of these experiments is shown in Fig. 8 for the growth of *Hyphomicrobium* ZV620 in a chemostat at a constant dilution rate, as a function of the carbon:nitrogen ratio in the medium feed (Gräzer-Lampart *et al.*, 1986). Steady-state concentrations of residual methanol and ammonia, as well as the cellular composition and activities of ammonia-assimilating enzymes, indicate the existence of three distinct growth regimens: (1) carbon limitation with nitrogen in excess (C:N < 7.1); (2) nitrogen limitation with carbon in excess (C:N > 12.6); and (3) at intermediate C:N ratios (7.1 < C:N < 12.6), a growth regimen where steady-state conditions could be established with neither carbon nor nitrogen present in excess. When growth was limited by either carbon or nitrogen only, the cellular composition was essentially constant. Within the transition growth regimen, however, the composition of the cells depended on the C:N ratio in the medium feed. Both the overall composition (N, protein, and PHB content) and the enzymic pattern of the cells (enzymes involved in ammonia assimilation) were adjusted according to the availability of the two nutrients restricting the synthesis of cellular material. It is remarkable to see the different regulation strategies for GDH, GS, and GOGAT. Virtually identical patterns with respect to either culture parameters or cellu-

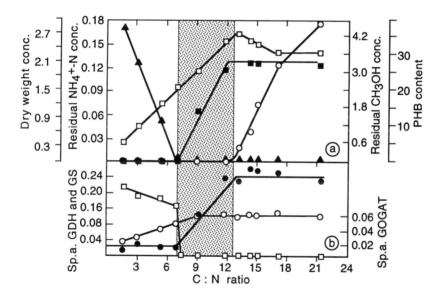

Figure 8. Culture parameters and cellular and enzymic composition of *Hyphomicrobium* ZV620 growing in chemostat culture with methanol and ammonium at a constant dilution rate of $D = 0.054$ hr^{-1}, as a function of the carbon:nitrogen ratio in the inflowing medium. [$S_0(NH_4^+ - N)$ was held constant, S_0 (methanol) was increased]. (a) Cell dry weight (□), residual methanol (○), and $NH_4^+ - N$ (▲) concentration in culture (all in g liter^{-1}), cellular PHB content (■, in % of dry wt). (b) Specific activities [all in μmole (mg protein × min)$^{-1}$] of glutamate dehydrogenase (GDH) (□), glutamine synthetase (GS) (●), glutamine oxoglutarate aminotransferase (GOGAT) (○). (Adapted from Gräzer-Lampart *et al.*, 1986.)

lar composition have been observed under carbon/nitrogen-limited growth for a number of other microorganisms, including the yeasts *H. polymorpha* and *C. valida* (Egli and Quayle, 1984, 1986; Minkevich *et al.*, 1988), *Hyphomicrobium* X (Duchars and Attwood, 1989), a thermotolerant *Bacillus* sp. (Al-Awadhi *et al.*, 1990), *Klebsiella pneumoniae* (Rutgers *et al.*, 1990), or for carbon/phosphorus-limited growth of *Acinetobacter johnsonii* (Bonting *et al.*, 1992). All of this experimental evidence indicates that in experiments such as the one presented for *Hyphomicrobium* ZV620, the existence of a dual-nutrient-limited growth regimen between two distinctly single-nutrient-limited zones of growth is a general phenomenon.

An interesting experiment with respect to the positioning of the dual-limited zone was performed by Pengerud *et al.* (1987). These authors grew a pure culture of *Ps. putida* in a chemostat at a dilution rate of $0.03 \, hr^{-1}$ with different C:P ratios in the medium (the phosphate concentration was kept constant in the feed while the glucose concentration was increased). With increasing glucose concentrations in the feed, the biomass increased linearly and, as in the case of *Hyphomicrobium* ZV620 (Fig. 8), it only became constant when clear phosphorus limitation was reached at C:P ~ 300. A C/P-limited zone was observed between $140 < C:P < 300$. When the same experiment was done with a bacteriovorous nanoflagellate added to the culture, the number of flagellates increased with increasing C:P feed ratios up to approximately 300, whereas the concentration of bacteria in the culture remained roughly constant at a very low density. Although the bacterial density was kept low by the flagellates, glucose was consumed to completion up to C:P feed ratios of 500. In this experiment the zone of double-limited growth shifted to considerably higher C:P ratios (approximately between $350 < C:P < 500$), and distinct phosphorus limitation was observed only at C:P ratios exceeding 500. This must be interpreted in the way that the number of bacteria was controlled at a low level by predation by the flagellates, but at the same time the specific glucose consumption rate of the bacterial biomass increased, recycling of phosphorous occurred, and the specific growth rate of the bacteria must have been faster than the dilution rate set (Thingstad and Pengerud, 1985).

4.2. Dependence of Dual-Nutrient-Limited Zone on Growth Rate

Based on their experimental data, Egli and Quayle (1984, 1986) proposed that the extension of this zone where growth is limited by two nutrients at the same time can be predicted from the growth yields for the individual substrates, and they speculated that its extension should depend on growth rate (a similar suggestion was put forward by Thingstad, 1987). An empirical formula was proposed for calculating the bounderies of the dual-nutrient-limited zone at a fixed dilution rate. This can be illustrated for the example of *Hyphomicrobium* ZV620 shown in Fig. 8. In this case, the steady-state concentration of biomass ($\bar{x}$) at the border between C-limited and the C/N-limited zone can be calculated from the carbon utilized for growth ($C_0 - \bar{c}$) and the corresponding growth yield measured under C-limited conditions ($Y_{X/C}$) from Eq. (8), or alternatively, from the nitrogen utilized ($N_0 - \bar{n}$) and the nitrogen-based growth yield measured under C-limited growth ($Y_{X/N}$). Because both $\bar{c}$ and $\bar{n}$ are very small compared to C_0 and N_0, respectively, Eq. (8) can be rearranged to yield the C:N ratio in the medium feed [Eq. (9)]:

$$\bar{x} = (C_0 - \bar{c})\, Y_{X/C} = (N_0 - \bar{n})\, Y_{X/N} \tag{8}$$

$$\frac{C_0}{N_0} \approx \frac{Y_{X/N}}{Y_{X/C}} \tag{9}$$

Similarly, the boundary between C/N-limited and N-limited growth can be calculated from the growth yields for carbon and nitrogen measured under N-limited conditions. In this particular case, the $Y_{X/C}$ and $Y_{X/N}$ measured under C-limited conditions were 0.99 and 7.19, and those measured under N-limited growth were 0.80 and 10.3, respectively (Gräzer-Lampart et al., 1986). From these yield factors the predicted zone for dual-nutrient-limited growth should be observed for growth with media of a C:N ratio between 7.3 and 12.9, which corresponds nicely with the experimentally determined values of 7.1 and 12.6, respectively. Eq. 9 implies that any change in growth yield should lead to a zone of dual-nutrient-limited growth. Furthermore, the dual-nutrient-limited zone should become more extended the wider the range is within the growth yield coefficients, which can vary. Because the differences in cellular composition, and hence the growth yields, are known to be more pronouced at low growth rates (Herbert, 1976), one should expect that slow growth would result in quite extended transition growth regimens.

This concept was applied to predict the boundaries of the carbon–nitrogen-limited zone as a function of growth rate for *Klebsiella pneumoniae* growing with glycerol and ammonia from yield constants that are available in the literature (Egli, 1991). The result of this calculation predicts that the extension of the carbon–nitrogen-limited growth regimen in a chemostat should be strongly dependent on the growth rate of the culture (Fig. 9). At high dilution rates close to μ_{max}, the boundary should be very narrow and difficult to detect experimentally, whereas at low dilution rates, a rather extended zone of dual limitation must be expected. The zone of dual limitation not only became wider as the dilution rate decreased, but it also moved to higher C:N ratios. This suggests, for example, that cultivation of the bacterium with a medium of a C:N ratio of 8, growth would be carbon-limited at low dilution rates; however, with the same medium, growth should become nitrogen-limited at high dilution rates. Clearly, the position and shape of the C/N-limited zone is determined by the growth yield coefficients and the plasticity of the cell composition with respect to carbonaceous and nitrogenous components. This includes that for highly oxidized substrates the zone should be expected at higher C:N ratios, or that at low dilution rates the rate of endogenous (maintenance) activity will determine the shifting of the zone toward higher C:N ratios. Experimental data reported by Minkevich and co-workers (1988) for the growth of the yeast *C. valida*

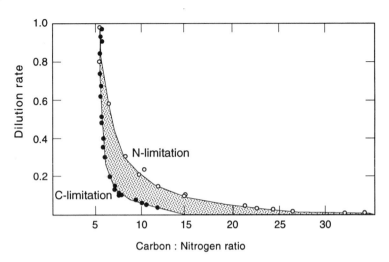

Figure 9. Predicted extension of the double (carbon/nitrogen) limited growth regimen (shaded area), as a function of the carbon:nitrogen ratio supplied in the growth medium and the dilution rate (in hr^{-1}), for a culture of *Klebsiella pneumoniae* cultivated in the chemostat with glycerol and NH_4^+ as the only sources of carbon/energy and nitrogen. Calculated C:N ratios for the transition between carbon-limited to double-limited growth (●); calculated C:N ratios for the transition between nitrogen-limited and double-limited growth (○). (Adapted from Egli, 1991.)

with ethanol and ammonia support such a shape and position of the C/N-limited zone in the C:N ratio–dilution rate plane.

The evidence presented indicates that zones of dual-nutrient-limited growth as described here for carbon and nitrogen should exist also for other combinations of nutrients, and that simultaneous limitation by more than two nutrients should be possible (Egli, 1991). Unfortunately, experimental data reported in the literature are still scarce, and information is usually restricted to a single medium nutrient ratio and a particular dilution rate. Nevertheless, the variability observed in the data for the cellular composition of microorganisms suggests that extended dual-nutrient-limited zones should be observed for nutrient combinations such as C/P, C/Mg or C/K (Herbert, 1961b, 1976; Wanner and Egli, 1990). Since growth yields also change with the degree of oxidation of a substrate, for example, for carbon (Linton and Stephenson, 1978) but also of nitrogen (e.g., during growth with either ammonia or nitrate), one would expect a shift in the position of the multiple-nutrient-limited zone as a function of the type of substrate supplied for growth. The exact position of a multiple-nutrient-limited area will certainly be influenced by the simultaneous utilization of mixtures of substrates as it will occur under environmental conditions. However, such effects remain to be investigated because the simultaneous utilization of the different homologous substrates might be affected by the extent of control the different nutrients have over the growth rate. This has been observed, for example, during growth of a methylotrophic yeast fed with mixtures of glucose and methanol where repression of the simultaneous utilization of methanol together with glucose occurred within the C/N-limited growth regimen (Egli and Quayle, 1986).

4.3. Control of Growth Rate under Multiple-Nutrient-Limited Growth Conditions

The fact that growth regimens exist where two or more nutrients are consumed to completion and therefore seem to "limit" growth simultaneously raises the question of whether only one of the different nutrients or perhaps several of them simultaneously exert control over the cellular growth rate. This question has been examined by Rutgers *et al.* (1990) for the simultaneous limitation of *K. pneumoniae* during growth in chemostat culture with glucose and ammonia (these authors prefer to speak of "multiple-nutrient-controlled" growth rather than of "multiple-nutrient-limited" growth). For growth at three different dilution rates they measured steady-state concentrations of the two nutrients throughout the three growth regimens, i.e., glucose limitation, dual (glucose/ammonia) limitation, and ammonia limitation. Their data established that under distinctly single-nutrient-limited growth conditions, steady-state concentrations of the growth-controlling nutrient remained constant and independent from the C:N ratio in the feed, whereas within the dual-limited zone, steady-state concentrations of both glucose and ammonia were changing with the C:N ratio in the inflowing medium. Applying the metabolic flux control theory (Kell and Westerhoff, 1986; Rutgers *et al.*, 1991), the results were analyzed to quantify the relative control of the concentrations of the two "limiting" nutrients (glucose and ammonia) on the growth rate of the culture. The analysis gives a first strong indication that, similar to the cellular composition that constantly adapts within this zone (compare Fig. 8), the control of growth rate gradually shifted from glucose to ammonia within the dual-limited zone and vice versa, and did not distinctly switch from one to the other via a step function at a distinct C/N–feed ratio. This finding that microbial growth rate can be controlled by two (or even more) nutrients simultaneously may explain the observation that better stimulation of bacterial growth is obtained in environmental samples upon addition of a combination rather than single nutrients. It also gives further support to the idea that growth in ecosystems is frequently controlled (limited) by more than one nutrient at the same time.

Although there is now convincing evidence that stable multiple-nutrient-limited growth regimens exist and that the rate of growth within this region seems to be controlled by two (or more) nutrients at the same time, one should be aware of the fact that the kinetic relationship between the concentrations of growth-controlling nutrients and growth rate have not yet been worked out. Whether one of the many models that have been proposed in the literature (MeGee *et al.*, 1972; Bader, 1982;

Lee *et al.*, 1984; Baltzis and Fredrickson, 1988; Mankad and Bungay, 1988; Haas, 1994) is able to describe the specific growth rate under such conditions requires experimental confirmation. It should be added that the existence of an extended zone of double-nutrient-limited growth between growth regimens that are clearly limited by a single nutrient are in contrast to the abrupt changes proposed according to Liebig's principle (see, e.g., Smith, 1993). And the shape of the double-nutrient-limited zone as shown in Fig. 9 might also explain why under some conditions an abrupt and under other conditions a gradual transition have been observed between transitions from the limitation by one nutrient to another nutrient.

4.4. Multiple-Nutrient-Limited Growth in Ecosystems

One must ask the question whether or not the concept outlined above for the growth of pure cultures under steady-state conditions in continuous culture may also be applied to growth of heterotrophic microorganisms in natural environments where the availability of different nutrients is known to fluctuate, resulting in transiently changing growth rates and perhaps also temporary changes in the nature of the nutritional growth regimen. Unfortunately, only a few controlled dynamic studies where microbial cultures were subjected to defined transients between different nutritional growth regimens have been carried out. However, the data obtained for a culture of the yeast *H. polymorpha* shifted from a carbon- to a nitrogen-limited growth medium (Egli, 1982) demonstrate that "double-nutrient-limited" growth can also be observed under transient growth conditions. A similar observation was made only recently in our laboratory for a culture of *Ps. oleovorans* that was shifted from carbon- to nitrogen-limited growth conditions in a chemostat (M. Zinn and R. Durner, unpublished data). The fact that several authors reported large variations in the cellular carbon and nitrogen ratio of both natural algal and microbial populations (see, e.g., Goldman, 1980; Lancelot and Billen, 1985; Goldman *et al.*, 1987; Currie, 1990) indicates also that growth of cells in ecosystems might oscillate between single- and dual- or multiple-nutrient-limited conditions. However, the fact that the positioning of the zones of double- or multiple-nutrient-limited growth is dependent not only on the growth yield [i.e., quality of the carbon utilized, especially with respect to its degree of reduction (Egli, 1991)] but also on the presence of predators (Pengerud *et al.*, 1987) adds a considerable degree of complexity.

The information presently available indicates that the zones of multiple-nutrient-limited growth should generally become broader with decreasing growth rates. (Thus, as indicated in Fig. 9 for the two nutrients carbon and nitrogen, extremely high C:N ratios are needed at low growth rates to make a culture growing distinctly nitrogen-limited.) This, in combination with the observation that growth rates of microorganisms in ecosystems are usually very low (Moriarty, 1986) and that excess concentrations of utilizable carbon compounds are virtually never (or only transiently) observed in the environment suggest that heterotrophic growth should oscillate somewhere between distinct limitation by carbon and limitation by carbon in combination with one or several other essential nutrients, such as nitrogen or phosphorus. Because multiple-nutrient-limited growth has rarely been considered, virtually nothing is known on physiological and kinetic properties of both pure cultures and natural populations growing under such conditions. The author hopes that the concept outlined above will help engender a more structured experimental approach to the problem of nutrient limitation of growth in nature.

5. Concluding Remarks

This chapter indicates that we are just beginning to understand the different strategies employed by heterotrophic microorganisms and their consequences for growth in the dilute and complex mixture of nutrients available in the natural environment. It seems timely that the traditional single substrate approach—which has (admittedly successfully) determined during the past decades the strategies used for isolation of microorganisms from nature and their subsequent cultivation and characterization in the laboratory, and as a consequence, also our understanding of microbial growth in nature—is extended toward a more realistic scenario. For example, the mixed substrate enrichment and cultivation strategy has demonstrated the competitive advantages and importance of facultatively chemolithotrophic thiobacilli in nature, a group of bacteria that was only occasionally enriched by the single substrate approach, and was therefore considered to be of little importance in nature (Gottschal and Kuenen, 1980b).

Despite the fact that mixed substrate growth has not yet been demonstrated *in situ,* the information presently available demonstrates that, compared to growth with a single carbon source, the ability to simultaneously utilize mixtures of carbon substrates confers some important advantages to nutritionally versatile heterotrophic microorganisms (and to a certain degree probably also to organisms that are known today as nutritional specialists). I also hope that the examples discussed here have demonstrated convincingly that many of the questions of heterotrophic growth in nature with dissolved organic carbon can be approached from the pure (or defined mixed) culture side by systematically increasing the complexity of the growth medium supplied. It should be stressed that the technique of continuous culture is crucial in this approach.

There is a range of questions that can be identified as important for a better understanding of microbially mediated transformations (including cell growth) in nature and where we are seriously lacking information. These include among many others: (1) growth and degradation kinetics of mixtures of compounds present at low concentrations, especially of xenobiotic compounds in the presence of substrates of natural origin; (2) regulation of catabolic enzymes involved in the degradation of xenobiotic compounds under environmental conditions; (3) intimately linked to the former question is the knowledge of fluxes of important natural substrates in ecosystems and not only their concentrations; and (4) hardly anything is known on the interaction of different nutrients (multiple nutrient limitation) and the physiological consequences for microbial growth, especially under dynamic conditions. A deeper knowledge of the ability and the strategies employed by heterotrophic microorganisms to deal with mixtures of substrates will not just be of crucial interest for microbial ecologists. It can also be applied to both environmental and industrial biotechnological processes (see, e.g., Egli and Mason, 1991) to manipulate and direct metabolic fluxes at the physiological level into product-forming routes by using specific mixtures of substrates.

ACKNOWLEDGEMENTS. I would like to thank Urs Lendenmann, Matthias Bally, and Karin Kovářová for many stimulating discussions. I am indebted to Roland Durner, Albert Tien, Margarete Witschel, and Manfred Zinn, who have generously permitted me to use unpublished results; to Meinhard Simon and Hans van Dijken for supplying manuscripts in advance of publication; and to Gwynfryn Jones for giving me the time to write. Special thanks go to Tony Mason for many suggestions and for careful reading of the various drafts of this manuscript. The financial support for the work on NTA came from both the Research Commission of ETH Zürich (Project No. 0.330.089.86) and Lever (Switzerland)/Unilever Merseyside (England). The research on mixed substrate kinetics was financed by the Swiss National Science Foundation (Project Nos. NF-3.474.-083, 31-9210.87, 31-30004.90, 31-30004.90/2).

References

Ahlgren, G., 1980, Effects on algal growth rates by multiple nutrient limitation, *Arch. Hydrobiol.* **89**:43–53.

Aker, K. C., and Robinson, C. W., 1987, Growth of *Candida utilis* on single- and multicomponent-sugar substrates and on waste banana pulp liquors for single-cell protein production, *MIRCEN J.* **3**:255–274.

Al-Awadhi, N., 1989, *Characterization and Physiology of Some Thermotolerant and Thermophilic Solvent-Utilizing Bacteria,* Swiss Federal Institute of Technology, Zürich, Switzerland, Ph.D. thesis No. 8118.

Al-Awadhi, N., Egli, T., Hamer, G., and Mason, C. A., 1990, The process utility of thermotolerant methylotrophic bacteria: I. An evaluation in chemostat culture, *Biotechnol. Bioeng.* **36**:816–820.

Albertson, N. H., Nyström, T., and Kjelleberg, S., 1990a, Exoprotease activity in two marine bacteria during starvation, *Appl. Environ. Microbiol.* **56**:218–223.

Albertson, N. H., Nyström, T., and Kjelleberg, S., 1990b, Starvation-induced modulations in binding protein-dependent glucose transport in the marine *Vibrio* sp. S14, *FEMS Microbiol. Lett.* **70**:205–210.

Albertson, N. H., Nyström, T., and Kjelleberg, S., 1990c, Macromolecular synthesis during recovery of the marine *Vibrio* sp. S14 from starvation, *J. Gen. Microbiol.* **136**:2201–2207.

Alexander, M., 1994, *Biodegradation and Bioremediation,* Academic Press, San Diego, Calif.

Ammerman, J. W., Fuhrman, J. A., Hagström, Å., and Azam, F., 1984, Bacterioplankton growth in seawater: I. Growth kinetics and cellular characteristics in seawater cultures, *Mar. Ecol. Prog. Ser.* **18**:31–39.

Amy, P. S., and Morita, R. Y., 1983, Starvation-survival patterns of sixteen freshly isolated open-ocean bacteria, *Appl. Environ. Microbiol.* **45**:1109–1115.

Amy, P. S., Pauling, C., and Morita, R. Y., 1983, Recovery from nutrient starvation by a marine *Vibrio* sp., *Appl. Environ. Microbiol.* **45**:1685–1690.

Anderson, J. J., and Oxender, D. L., 1978, Genetic separation of high- and low-affinity transport systems for branched-chain amino acids in *Escherichia coli, J. Bacteriol.* **136**:168–174.

Azam, F., and Cho, B. C., 1987, Bacterial utilization of organic matter in the sea, in: *Ecology of Microbial Communities* (M. Fletcher, T. R. G. Gray, and J. G. Jones, eds.), Cambridge University Press, Cambridge, England, pp. 261–281.

Babel, W., Brinkmann, U., and Müller, R. H., 1993, The auxiliary substrate concept—an approach for overcoming limits of microbial performance, *Acta Biotechnol.* **13**:211–242.

Bader, F. G., 1978, Analysis of double-substrate limited growth, *Biotechnol Bioeng.* **20**:183–202.

Bader, F. G., 1982, Kinetics of double-substrate limited growth, in: *Microbial Population Dynamics* (M. J. Bazin, ed.), CRC Press, Boca Raton, Fla., pp. 1–32.

Baidya, T. K. N., Webb, F. C., and Lilly, M. D., 1967, The utilization of mixed sugars in continuous fermentation. I, *Biotechnol. Bioeng.* **9**:195–204.

Bally, M., 1994, *Physiology and Ecology of Nitrilotriacetate Degrading Bacteria in Pure Culture, Activated Sludge and Surface Waters,* Swiss Federal Institute of Technology Zürich, Switzerland, Ph.D. thesis No. 10821.

Bally, M., and Egli, T., 1995, Dynamics of substrate consumption and enzyme synthesis in *Chelatobacter heintzii* sp. ATCC 29600 during growth in carbon-limited chemostat culture with different mixtures of glucose and nitrilotriacetate (NTA), submitted.

Bally, M., Wilberg, E., Kühni, M., and Egli, T., 1994, Growth and enzyme synthesis in the nitrilotriacetic acid (NTA) degrading *Chelatobacter heintzii* ATCC 29600, *Microbiology* **140**:1927–1936.

Baltzis, B. C., and Fredrickson, A. G., 1988, Limitation of growth rate by two complementary nutrients: Some elementary but neglected considerations, *Biotechnol. Bioeng.* **31**:75–86.

Bazin, M. J. (ed.), 1982, *Microbial Population Dynamics.* CRC Series in Mathematical Models in Microbiology, CRC Press, Boca Raton, Fla.

Beauchamp, E. G., Trevors, J. T., and Paul, P. W., 1989, Carbon sources for bacterial denitrification, *Adv. Soil Sci.* **10**:113–142.

Beckwith, J., 1987, The lactose operon, in: *Escherichia coli and Salmonella typhimurium,* Vol. 2 (F. C. Neidhardt, J. L. Ingraham, K. Brooks, B. Magasanik, M. Schaechter, and H. E. Umbarger, eds.), American Society for Microbiology, Washington, D.C., pp. 1444–1452.

Bell, W. H., 1980, Bacterial utilization of algal extracellular products. 1. The kinetic approach, *Limnol. Oceanogr.* **25**:1007–1020.

Bitzi, U., Egli, T., and Hamer, G., 1991, The biodegradation of mixtures of organic solvents by mixed and monocultures of bacteria, *Biotechnol. Bioeng.* **37**:1037–1042.

Bley, T., and Babel, W., 1992, Calculating affinity constants of substrate mixtures in a chemostat, *Acta Biotechnol.* **12**:13–15.

Boethling, R. S., and Alexander, M., 1979a, Effect of concentration of organic chemicals on their biodegradation by natural microbial communities, *Appl. Environ. Microbiol.* **37**:1211–1216.

Boethling, R. S., and Alexander, M., 1979b, Microbial degradation of organic compounds at trace levels, *Environ. Sci. Technol.* **13**:989–991.

Bonting, C. F. C., van Veen, H. W., Taverne, A., Kortstee, G. J. J., and Zehnder, A. J. B., 1992, Regulation of polyphosphate metabolism in *Acinetobacter* strain 210A grown in carbon- and phosphate-limited continuous culture, *Arch. Microbiol.* **158**:139–144.

Brinkmann, U., and Babel, W., 1992, Simultaneous utilization of heterotrophic substrates by *Hansenula polymorpha* results in enhanced growth, *Appl. Microbiol. Biotechnol.* **37**:98–103.

Brock, T. D., 1987, The study of microorganisms *in situ:* Progress and problems, in: *Ecology of Microbial Communities* (M. Fletcher, T. R. G. Gray, and J. G. Jones, eds.), Cambridge University Press, Cambridge, England, pp. 3–17.

Brooke, A. G., and Attwood, M. M., 1983, Regulation of enzyme synthesis during the growth of *Hyphomicrobium* X on mixtures of methylamine and ethanol, *J. Gen. Microbiol.* **129**:2399–2404.

Bryan, B. A., 1981, Physiology and biochemistry of denitrification, in: *Denitrification, Nitrification and Atmospheric Nitrous Oxide* (C. C. Delwiche, ed.), Wiley, New York, pp. 67–84.

Bull, A. T., 1985, Mixed culture and mixed substrate systems, in: *Comprehensive Biotechnology* (M. Moo-Young, ed.), Vol. 1, *The Principles of Biotechnology: Scientific Fundamentals* (A. T. Bull and H. Dalton, eds.), Pergamon Press, Oxford, England, pp. 281–299.

Button, D. K., 1985, Kinetics of nutrient-limited transport and microbiol growth, *Microbial. Rev.* **49**:270–297.

Button, D. K., 1991, Biochemical basis for whole-cell uptake kinetics: Specific affinity, oligotrophic capacity, and the meaning of the Michaelis constant, *Appl. Environ. Microbiol.* **57**:2033–2038.

Button, D. K., 1993, Nutrient-limited microbial growth kinetics: Overview and recent advances, *Antonie van Leeuwenhoek* **63**:225–235.

Chen, C. Y., and Christensen, E. R., 1985, A unified theory for microbial growth under multiple nutrient limitation, *Water Res.* **19**:791–798.

Chróst, R. J., 1991, Environmental control of the synthesis and activity of aquatic microbial ectoenzymes, in: *Microbial Enzymes in the Aquatic Environments* (R. J. Chróst, ed.), Springer-Verlag, New York, pp. 29–59.

Cogan, T. M., 1987, Co-metabolism of citrate and glucose by *Leuconostoc* spp.: Effects on growth, substrate and products, *J. Appl. Bacteriol.* **63**:551–558.

Cole, J. J., Findlay, S., and Pace, M. L., 1988, Bacterial production in fresh and saltwater ecosystems: A cross-system overview, *Mar. Ecol. Prog. Ser.* **43**:1–10.

Cook, G. M., Janssen, P. H., and Morgan, H. W., 1993, Simultaneous uptake and utilisation of glucose and xylose by *Clostridium thermohydrosulfuricum, FEMS Microbiol. Lett.* **109**:55–62.

Cooney, C., Wang, D. C. I., and Mateles, R. I., 1976, Growth of *Enterobacter aerogenes* in a chemostat with double nutrient limitations, *Appl. Environ. Microbiol.* **31**:91–98.

Coveney, M. F., and Wetzel, R. G., 1992, Effects of nutrients on specific growth rate of bacterioplankton in oligotrophic lake water cultures, *Appl. Environ. Microbiol.* **58**:150–156.

Cronan, Jr., J. E., Gennis, R. B., and Maloy, S. R., 1987, Cytoplasmic membrane, in: *Escherichia coli* and *Salmonella typhimurium*, Vol. 1 (F. C. Neidhardt, J. L. Ingraham, K. Brooks, B. Magasanik, M. Schaechter, and H. E. Umbarger, eds.), American Society for Microbiology, Washington, D.C., pp. 31–55.

Currie, D. J., 1990, Large scale variability and interactions among phytoplankton, bacterioplankton and phosphorus, *Limnol. Oceanogr.* **35**:1437–1455.

Daesch, G., and Mortenson, 1968, Sucrose catabolism in *Clostridium pasteuranium* and its relation to N_2 fixation, *J. Bacteriol.* **96**:346–351.

Daughton, C. G., Cook, A. M., and Alexander, M., 1979, Phosphate and soil binding: Factors limiting bacterial degradation of ionic phosphorus-containing pesticide metabolites, *Appl. Environ. Microbiol.* **37**:175–184.

Davidson, E. A., Matson, R. A., Vitousek, P. M., Riley, R., Dunkin, K., Garcïa-Mendez, G., and Maass, J. M., 1993, Processes regulating soil emission of NO and N_2O in a seasonally dry tropical forest, *Ecology* **74**:130–139.

Dawes, E. A., 1985, Starvation, survival and energy reserves, in: *Bacteria in their Natural Environments* (M. Fletcher and G. D. Floodgate, eds.), Academic Press, London, pp. 43–79.

Dean, A. C. R., 1972, Influence of environment on the control of enzyme synthesis, *J. Appl. Chem. Biotechnol.* **22:**245–259.

Death, A., Notley, L., and Ferenci, T., 1993, Derepression of LamB protein facilitates outer membrane permeation of carbohydrates into *Escherichia coli* under conditions of nutrient stress, *J. Bacteriol.* **175:**1475–1483.

de Boer, L., Euverink, G. J., van der Vlag, J., and Dijkhuizen, L., 1990, Regulation of methanol metabolism in the facultative methylotroph *Nocardia* sp. 239 during growth on mixed substrate in batch and continuous culture, *Arch. Microbiol.* **153:**33–343.

Decker, K., Peist, R., Reidl, J., Kossmann, M., Brand, B., and Boos, W., 1993, Maltose and maltotriose can be formed endogenously in *Escherichia coli* from glucose and glucose 1-phosphate independently of enzymes of the maltose system, *J. Bacteriol.* **175:**5655–5665.

Degnan, B. A., and Macfarlane, G. T., 1993, Transport and metabolism of glucose and arabinose in *Bifidobacterium breve*, *Arch. Microbiol.* **160:**144–151.

de Hollander, J. A., and Stouthamer, A. H., 1979, Multicarbon-substrate growth of *Rhizobium trifolii*, *FEMS Microbiol. Lett.* **6:**57–59.

de Jong-Gubbels, P., Vanrollehem, P., Heijnen, S., van Dijken, J. P., and Pronk, J. T., 1995, Regulation of carbon metabolism in chemostat cultures of *Saccharomyces cerevisiae*, *Yeast* **11:**407–418.

de Koning, W., Gleeson, M. A. G., Harder, W., and Dijkhuizen, L., 1987, Regulation of methanol metabolism in the yeast *Hansenula polymorpha*. Isolation and characterization of mutants blocked in methanol assimilatory enzymes, *Arch. Microbiol.* **147:**375–382.

de Koning, W., Weusthuis, R. A., Harder, W., and Dijkhuizen, L., 1990, Metabolic regulation in the yeast *Hansenula polymorpha*. Growth of dihydroxyacetone kinase/glycerol kinase-negative mutants on mixtures of methanol and xylose in continuous cultures, *Yeast* **6:**107–115.

de Vries, G. E., Harms, N., Maurer, K., Papendrecht, A., and Stouthamer, A. H., 1988, Physiological regulation of *Paracoccus denitrificans* methanol dehydrogenase synthesis and activity, *J. Bacteriol.* **170:**3731–3737.

Dijkhuizen, L., and Harder, W., 1979, Regulation of autotrophic and heterotrophic metabolism in *Pseudomonas oxalaticus* OX1: Growth on mixtures of acetate and formate in continuous culture, *Arch. Microbiol.* **123:**47–53.

Douglas, D. J., Novitsky, J. A., and Fournier, R. O., 1987, Microautoradiography-based enumeration of bacteria with estimates of thymidine-specific growth and production rates, *Mar. Ecol. Prog. Ser.* **36:**91–99.

Dow, C. S., Whittenbury, R., and Carr, N. G., 1983, The "shut down" or "growth precursor" cell—an adaptation for survival in a potentially hostile environment, in: *Microbes in Their Natural Environments* (J. H. Slater, R. Whittenbury, and J. W. T. Wimpenny, eds.), Cambridge University Press, Cambridge, England, pp. 187–247.

Duchars, M. G., and Attwood, M. M., 1989, The influence of the carbon:nitrogen ratio of the growth medium on the cellular composition and regulation of enzyme activity in *Hyphomicrobium* X, *J. Gen. Microbiol.* **135:**787–793.

Ducklow, H. W., and Carlson, C. A., 1992, Oceanic bacterial production, *Adv. Microb. Ecol.* **12:**113–181.

Dykhuizen, D., and Davies, M., 1980, An experimental model: Bacterial specialists and generalists competing in chemostats, *Ecology* **61:**1213–1227.

Eggeling, L., and Sahm, H., 1978, Derepression and partial insensitivity to carbon catabolite repression of the methanol-dissimilating enzymes in *Hansenula polymorpha*, *Eur. J. Appl. Microbiol. Biotechnol.* **5:**197–202.

Eggeling, L., and Sahm, H., 1981, Enhanced utilization-rate of methanol during growth on mixed substrate: A continuous culture study with *Hansenula polymorpha*, *Arch. Microbiol.* **130:**362–365.

Egli, T., 1982, Regulation of protein synthesis in methylotrophic yeasts: Repression of methanol dissimilating enzymes by nitrogen limitation, *Arch. Microbiol.* **131**:95–101.

Egli, T., 1991, On multiple-nutrient-limited growth of microorganisms, with special reference to carbon and nitrogen substrates, *Antonie van Leeuwenhoek* **60**:225–234.

Egli, T., 1992, General strategies in the biodegradation of pollutants, in: *Metal Ions in Biological Systems. Degradation of Environmental Pollutants by Microorganisms and Their Metalloenzymes*, Vol. 28 (H. Sigel and A. Sigel, eds.), Marcel Dekker, New York, pp. 1–39.

Egli, T., 1994, Biochemistry and physiology of the degradation of nitrilotriacetic acid and other metal complexing agents, in: *Biochemistry of Microbial Degradation* (C. Ratledge, ed.), Kluwer Academic Publishers, Dordrecht, pp. 179–195.

Egli, T., and Harder, W., 1984, Growth on methylotrophs on mixed substrates, in: *Microbial Growth on C1 Compounds* (R. L. Crawford and R. S. Hanson, eds.), American Society for Microbiology, Washington, D.C., pp. 330–337.

Egli, T., and Mason, C. A., 1991, Mixed substrates and mixed cultures, in: *Biology of Methylotrophs* (I. Goldberg and J. S. Rokem, Eds.), Butterworth-Heinemann, Boston, pp. 173–201.

Egli, T., and Quayle, J. R., 1984, Influence of the carbon:nitrogen ratio on the utilization of mixed carbon substrates by the methylotrophic yeast *Hansenula polymorpha*, in: Proc. 100th Ann. Meeting Soc. Gen. Microbiol. (abstract booklet), Warwick, p. M13.

Egli, T., and Quayle, J. R., 1986, Influence of the carbon:nitrogen ratio of the growth medium on the cellular composition and the ability of the methylotrophic yeast *Hansenula polymorpha* to utilize mixed carbon sources, *J. Gen. Microbiol.* **132**:1779–1788.

Egli, T., van Dijken, J. P., Veenhuis, M., Harder, W., and Fiechter, A., 1980, Methanol metabolism in yeasts: Regulation of the synthesis of catabolic enzymes, *Arch. Microbiol.* **124**:115–121.

Egli, T., Käppeli, O., and Fiechter, A., 1982a, Regulatory flexibility of methylotrophic yeasts in chemostat cultures: Simultaneous assimilation of glucose and methanol at a fixed dilution rate, *Arch. Microbiol.* **131**:1–7.

Egli, T., Käppeli, O., and Fiechter, A., 1982b, Mixed substrate growth of methylotrophic yeasts in chemostat culture: Influence of dilution rate on the utilisation of a mixture of glucose and methanol, *Arch. Microbiol.* **131**:8–13.

Egli, T., Lindley, N. D., and Quayle, J. R., 1983, Regulation of enzyme synthesis and variation of residual methanol concentration during carbon-limited growth of *Kloeckera* sp. 2201 on mixtures of methanol and glucose, *J. Gen. Microbiol.* **129**:1269–1281.

Egli, T., Bosshard, C., and Hamer, G., 1986, Simultaneous utilization of methanol-glucose mixtures by *Hansenula polymorpha* in chemostat: Influence of dilution rate and mixture composition on utilization pattern, *Biotechnol. Bioeng.* **28**:1735–1741.

Egli, T., Weilenmann, H.-U., El-Banna, T., and Auling, G., 1988, Gram-negative, aerobic, nitrilotriacetate-utilizing bacteria from wastewater and soil, *Syst. Appl. Microbiol.* **10**:297–305.

Egli, T., Lendenmann, U., and Snozzi, M., 1993, Kinetics of microbial growth with mixtures of carbon sources, *Antonie van Leeuwenhoek* **63**:289–298.

Fenchel, T. M., and Jørgensen, B. B., 1977, Detritus food chains of aquatic ecosystems: The role of bacteria, *Adv. Microb. Ecol.* **1**:1–58.

Fuhrman, J. A., and Azam, F., 1980, Bacterioplankton secondary production estimates for coastal waters of British Columbia, Antarctica, and California, *Appl. Environ. Microbiol.* **39**:1085–1095.

Fuhrman, J. A., and Azam, F., 1982, Thymidine incorporation as a measure of heterotrophic bacterioplankton production in marine surface waters: Evaluation and field results, *Mar. Biol.* **66**:109–120.

Fuhrman, J. A., and Ferguson, R. L., 1986, Nanomolar concentrations and rapid turnover of dissolved free amino acids in seawater: Agreement between chemical and microbiological measurements, *Mar. Ecol. Prog. Ser.* **33**:237–242.

Furlong, C. E., 1987, Osmotic-shock-sensitive tranport systems, in: *Escherichia coli and Salmonella typhimurium*, Vol. 1 (F. C. Neidhardt, J. L., Ingraham, K. Brooks, B. Magasanik, M. Schaechter, and H. E. Umbarger, eds.), American Society for Microbiology, Washington, D.C., pp. 768–796.

Gerhart, D. W., and Likens, G. E., 1975, Enrichment experiments for determining nutrient limitation: Four methods compared, *Limnol. Oceanogr.* **20**:649–653.

Geurts, T. G. E., de Kok, H. E., and Roels, J. A., 1980, A quantitative description of the growth of *Saccharomyes cerevisiae* CBS426 on a mixed substrate of glucose and ethanol, *Biotechnol. Bioeng.* **22**:2031–2043.

Goldman, J. C., 1980, Physiological processes, nutrient availability and the concept of relative growth rate in marine phytoplankton ecology, in: *Primary Productivity in the Sea* (P. G. Falkowski, ed.), Plenum Press, New York, pp. 179–194.

Goldman, J. C., Caron, D. A., and Dennett, M. R., 1987, Regulation of gross growth efficiency and ammonium regeneration in bacteria by substrate C:N ratio, *Limnol. Oceanogr.* **32**:1239–1252.

Gommers, P. J. F., van Schie, B. J., van Dijken, J. P., and Kuenen, J. G., 1988, Biochemical limits to microbial growth yields: an analysis of mixed substrate utilization, *Biotechnol. Bioeng.* **32**:86–94.

Gondo, S., Kaushik, K., and Venkatasubramanian, K., 1978, Two carbon-substrate continuous culture: Multiple steady states and their stability, *Biotechnol. Bioeng.* **20**:1479–1485.

Gorini, L., 1960, Antagonism between substrate and repression in controlling the formation of a biosynthetic enzyme, *Proc. Nat. Acad. Sci. USA* **46**:682–690.

Gottschal, J. C., 1986, Mixed substrate utilization by mixed cultures, in: *Bacteria in Nature* (J. S. Poindexter and E. R. Leadbetter, eds.), Plenum Press, New York, pp. 261–292.

Gottschal, J. C., 1993, Growth kinetics and competition—some contemporary comments, *Antonie van Leeuwenhoek* **63**:299–313.

Gottschal, J. C., and Kuenen, J. G., 1980a, Mixotrophic growth of *Thiobacillus* A2 on acetate and thiosulfate as growth limiting substrates in the chemostat, *Arch. Microbiol.* **126**:33–42.

Gottschal, J. C., and Kuenen, J. G., 1980b. Selective enrichment of facultatively chemolithotrophic thiobacilli and related organisms in continuous culture, *FEMS Microbiol. Lett.* **7**:241–247.

Gottschal, J. C., de Vries, S. and Kuenen, J. G., 1979, Competition between the facultatively chemolithotrophic *Thiobacillus* A2, and obligately chemolithotrophic *Thiobacillus* and a heterotrophic *Spirillum* for inorganic and organic substrates, *Arch. Microbiol.* **121**:241–249.

Gottschal, J. C., Pol, A., and Kuenen, J. G., 1981a, Metabolic flexibility of *Thiobacillus* A2 during substrate transitions in the chemostat, *Arch. Microbiol.* **129**:23–28.

Gottschal, J. C., Nanninga, H. J., and Kuenen, J. G., 1981b, Growth of *Thiobacillus* A2 under alternating growth conditions in the chemostat, *J. Gen. Microbiol.* **126**:85–96.

Gräzer-Lampart, S. D., Egli, T., and Hamer, G., 1986, Growth of *Hyphomicrobium* ZV620 in the chemostat: Regulation of NH_4^+-assimilating enzymes and cellular composition, *J. Gen. Microbiol.* **132**:3337–3347.

Haas, C. N., 1994, Unified kinetic treatment for growth on dual nutrients, *Biotechnol. Bioeng.* **44**:154–164.

Häggström, M. H., and Cooney, C. L., 1984, α-Glucosidase synthesis in batch and continuous culture of *Saccharomyces cerevisiae, Appl. Biochem. Biotechnol.* **9**:475–481.

Harder, W., and Dijkhuizen, L., 1976, Mixed substrate utilization, in: *Continuous Culture 6. Applications and New Fields* (A. C. R. Dean, D. C. Ellwood, C. G. T. Evans, and I. Melling, eds.), Ellis Horwood, Chichester, England, pp. 297–314.

Harder, W., and Dijkhuizen, L., 1982, Strategies of mixed substrate utilization in microorganisms, *Phil. Trans. R. Soc. Lond. B.* **297**:459–480.

Harder, W., and Dijkhuizen, L., 1983, Physiological responses to nutrient limitation, *Annu. Rev. Microbiol.* **37**:1–23.

Harder, W., Dijkhuizen, L., and Veldkamp, H., 1984, Environmental regulation of microbial metabolism, in: *The Microbe 1984* (D. P. Kelly and N. G. Carr, eds.), Cambridge University Press,
Cambridge, England, pp. 51–95.

Harrison, D. E. F., 1972, Physiological effects of dissolved oxygen tension and redox potential on
growing populations of microorganisms, *J. Appl. Chem. Biotechnol.* **22:**417–440.

Harte, M. J., and Webb, F. C., 1967, Utilization of mixed sugars in continuous fermentation. II,
Biotechnol. Bioeng. **9:**205–221.

Harvey, R. J., 1970, Metabolic regulation in glucose-limited chemostat cultures of *Escherichia coli*,
J. Bacteriol. **104:**698–706.

Hedges, J. I., 1992, Global biogeochemical carbon cycles: Progress and problems, *Mar. Chem.*
39:67–93.

Hegewald, E., and Knorre, W. A., 1978, Kinetics of growth and substrate consumption of
Escherichia coli ML30 on two carbon sources, *Z. Allg. Mikrobiol.* **18:**415–426.

Hengge-Aronis, R., 1993, The role of *rpoS* in early stationary-phase gene regulation in *Escherichia
coli* K12, in: *Starvation in Bacteria* (S. Kjelleberg, ed.), Plenum Press, New York, pp. 171–
200.

Herbert, D., 1961a, A theoretical analysis of continuous culture systems, *Soc. Chem. Ind. Monogr.
(London)* **12:**21–53.

Herbert, D., 1961b, The chemical composition of micro-organisms as a function of their growth
environment, in: *Microbial Reaction to the Environment* (C. G. Meynell and H. Gooder, eds.),
Cambridge University Press, Cambridge, England, pp. 391–416.

Herbert, D., 1976, Stoichiometric aspects of microbial growth, in: *Continuous Culture 6: Applications and New Fields* (A. C. R. Dean, D. C. Ellwood, C. G. T. Evans, and J. Melling, eds.),
Ellis Horwood, Chichester, England, pp. 1–30.

Hillmer, P., and Gest, H., 1977, H_2 metabolism in the photosynthetic bacterium *Rhodopseudomonas
capsulata:* Production and utilization of H_2 by resting cells, *J. Bacteriol.* **129:**732–739.

Hobbie, J. E., and Ford, T. E., 1993, A perspective on the ecology of aquatic microbes, in: *Aquatic
Microbiology. An Ecological Approach* (T. E. Ford, ed.), Blackwell Scientific Publications,
Boston, Oxford, pp. 1–14.

Höfle, M. G., 1983, Long-term changes in chemostat cultures of *Cytophaga johnsonae, Appl.
Environ. Microbiol.* **46:**1045–1053.

Hollibaugh, J. T., and Azam, F. 1983, Microbial degradation of dissolved proteins in seawater,
Limnol. Oceanogr. **28:**1104–1116.

Hoover, T. R., and Ludden, P. W., 1984, Derepression of nitrogenase by addition of malate to
cultures of *Rhodospirillum rubrum* grown with glutamate as the carbon and nitrogen source, *J.
Bacteriol.* **159:**400–403.

Hoppe, H.-G., 1976, Determination and properties of actively metabolizing heterotrophic bacteria in
the sea, investigated by means of micro-autoradiography, *Mar. Biol.* **36:**291–302.

Hoppe, H.-G., 1991, Microbial extracellular enzyme activity: A new key parameter in aquatic
ecology, in: *Microbial Enzymes in the Aquatic Environments* (R. J. Chróst, ed.), Springer-
Verlag, New York, pp. 60–83.

Hoppe, H.-G., Kim. S.-J., and Gocke, K., 1988, Microbial decomposition in aquatic environments:
Combined process of extracellular enzyme activity and substrate uptake, *Appl. Environ. Microbiol.* **54:**784–790.

Horowitz, A., Krichevsky, M. I., and Atlas, R. M., 1983, Characteristics and diversity of subarctic
marine oligotrophic stenoheterotrophic and euryheterotrophic bacterial populations, *Can. J.
Microbiol.* **29:**527–535.

Howard, P. H., 1989, *Handbook of Environmental Fate and Exposure Data for Organic Chemicals,*
Vol. I, *Large Production and Priority Pollutants,* Lewis Publishers, Chelsea, Mich.

Hueting, S., and Tempest, D. W., 1979, Influence of the glucose input concentration on the kinetics of metabolite production by *Klebsiella aerogenes* NCTC418: Growing in chemostat culture in potassium- or ammonia-limited environments, *Arch. Microbiol.* **123**:189–199.

Hutchinson, D. H., and Robinson, C. W., 1988, Kinetics of the simultaneous batch degradation of p-cresol and phenol by *Pseudomonas putida*, *Appl. Microbiol. Biotechnol.* **29**:599–604.

Jacob, F., and Monod, J., 1961, Genetic regulatory mechanisms in the synthesis of proteins, *J. Molec. Biol.* **3**:318–356.

Jackson, G. A., 1987, Physical and chemical properties of aquatic environments, in: *Ecology of Microbial Communities* (M. Fletcher, T. R. G. Gray, and J. G. Jones, eds.), Cambridge University Press, Cambridge, England, pp. 213–233.

Jannasch, H. W., 1967, Growth of marine bacteria at limiting concentrations of organic carbon in seawater, *Limnol. Oceanogr.* **12**:264–271.

Jannasch, H. W., 1968, Growth characteristics of heterotrophic bacteria in seawater, *J. Bacteriol.* **95**:722–723.

Jannasch, H. W., and Egli, T., 1993, Microbial growth kinetics: A historical perspective, *Antonie van Leeuwenhoek* **63**:213–224.

Kaplan, L. A., and Newbold, J. D., 1993, Biogeochemistry of dissolved organic carbon entering streams, in: *Aquatic Microbiology. An Ecological Approach* (T. E. Ford, ed.), Blackwell Scientific Publications, Boston, Oxford, pp. 139–165.

Kaprelyants, A. S., Gottschal, J. C., and Kell, D. B., 1993, Dormancy in non-sporulating bacteria, *FEMS Microbiol. Lett.* **104**:271–286.

Karl, D. M., 1986, Determination of *in situ* microbial biomass, viability, metabolism and growth, in: *Bacteria in Nature*, Vol. 2 (J. S. Poindexter and E. R. Leadbetter, eds.), Plenum Press, New York, pp. 85–176.

Karl, D. M., and Bailiff, M. D., 1989, The measurement and distribution of dissolved nucleic acids in aquatic environments, *Limnol. Oceanogr.* **34**:543–558.

Kastner, J. R., and Roberts, R. S., 1990, Simultaneous fermentation of D-xylose and glucose by *Candida shehatae*, *Biotechnol. Lett.* **12**:57–60.

Kay, A. A., and Gronlund, A. F., 1969, Influence of carbon or nitrogen starvation on amino acid transport in *Pseudomonas aeruginosa*, *J. Bacteriol.* **100**:276–282.

Keil, R. G., Montluçon, D. B., Prahl, F. G., and Hedges, J. I., 1994, Sorptive preservation of labile organic matter in marine sediments, *Nature* **370**:549–552.

Kell, D. B., and Westerhoff, H. V., 1986, Metabolic control theory: Its role in microbiology and biotechnology, *FEMS Microbiol. Rev.* **39**:305–320.

Kirchman, D. L., 1990, Limitation of bacterial growth by dissolved organic matter in the subarctic Pacific, *Mar. Ecol. Prog. Ser.* **62**:47–54.

Kirchman, D. L., 1993, Particulate detritus and bacteria in marine environments, in: *Aquatic Microbiology, An Ecological Approach* (T. E. Ford, ed.), Blackwell Scientific Publications, Boston, Oxford, pp. 1–14.

Kjelleberg, S. (ed.), 1993, *Starvation in Bacteria*, Plenum Press, New York.

Kjelleberg, S., Hermansson, M., Mårdén, P., and Jones, G. W., 1987, The transient phase between growth and nongrowth of heterotrophic bacteria, with emphasis on the marine environment, *Annu. Rev. Microbiol.* **41**:25–49.

Kjelleberg, S., Albertson, N., Flärdh, K., Holmquist, L., Jouper-Jaan, Å., Marouga, R., Östling, J., Svenblad, B., and Weichart, D., 1993, How do non-differentiating bacteria adapt to starvation? *Antonie van Leeuwenhoek* **63**:333–341.

Kluyver, A. J., 1956, Life's flexibility: Microbial adaptation, in: *The Microbe's Contribution to Biology* (A. J. Kluyver and C. B van Niel, eds.), Harvard University Press, Cambridge, Mass., p. 93.

Koch, A. L., 1976, How bacteria face depression, recession and derepression, *Persp. Biol. Med.* **20**:44–63.

Koch, A. L., 1982, Multistep kinetics: Choice of models for the growth of bacteria, *J. Theor. Biol.* **98**:401–417.

Koch, A. L., and Wang, C. H., 1982, How close to the theoretical diffusion limit do bacterial uptake systems function, *Arch. Microbiol.* **131**:36–42.

Koike, I., Hara, S., Terauchi, K., and Kogure, K., 1990, Role of sub-micrometer particles in the ocean, *Nature* **345**:242–244.

Kompala, D. S., Ramakrishna, D., and Tsao, G. T., 1984, Cybernetic modeling of microbial growth on multiple substrates, *Biotechnol. Bioeng.* **26**:1272–1281.

Konopka, A., Knight, D., and Turco, R. F., 1989, Characterization of a *Pseudomonas* sp. capable of aniline degradation in the presence of secondary carbon sources, *Appl. Environ. Microbiol.* **55**:385–389.

Kurlandzka, A., Rosenzweig, R. F., and Adams, J., 1991, Identification of adaptive changes in an evolving population of *Escherichia coli:* The role of changes with regulatory and highly pleiotrophic effects, *Mol. Biol. Evol.* **8**:261–281.

Kyslíková, E., and Volfová, O., 1990, Simultaneous utilization of methanol and mannose in a chemostat culture of *Candida boidinii* 2, *Folia Microbiol. (Praha)* **35**:484.

Lancelot, C., and Billen, G., 1985, Carbon–nitrogen relationships in nutrient metabolism of coastal marine ecosystems, *Adv. Aquat. Microbiol.* **3**:263–321.

Lancelot, C., Billen, G., and Mathot, S., 1989, Ecophysiology of phyto- and bacterioplankton growth in the Southern Ocean, in: *Plankton Ecology,* Vol. 1 (S. Caschetto, ed.), Science Policy Office, Brussels, pp. 4–92.

LaPat-Polasko, L. T., McCarty, T. L., and Zehnder, A. J. B., 1984, Secondary substrate utilization of methylene chloride by an isolated strain of *Pseudomonas* sp., *Appl. Environ. Microbiol.* **47**:825–830.

Law, A. T., and Button, D. K., 1977, Multiple-carbon-source-limited growth kinetics of a marine coryneform bacterium, *J. Bacteriol.* **129**:115–123.

Lawford, H. G., Pik, J. R., Lawford, G. R., Williams T., and Kligerman, A., 1980, Physiology of *Candida albicans* in zinc-limited chemostats, *Can. J. Microbiol.* **26**:64–70.

Lee, A. L., Ataai, M. M., and Shuler, M. L., 1984, Double-substrate-limited growth of *Escherichia coli, Biotechnol. Bioeng.* **26**:1398–1401.

Lee, C., and Wakeham, S. G., 1992, Organic matter in the water column: Future research challenges, *Mar. Chem.* **39**:95–118.

Lendenmann, U., 1994, *Growth Kinetics of* Escherichia coli *with Mixtures of Sugars,* No. 10658, Swiss Federal Institute of Technology Zürich, Switzerland, Ph.D. thesis.

Lendenmann, U., and Egli, T., 1995, Is *Escherichia coli* growing in glucose-limited chemostat culture able to utilise other sugars without lag? *Microbiology* **141**:71–78.

Lendenmann, U., Snozzi, M., and Egli, T., 1992, Simultaneous utilization of diauxic sugar mixtures by *Escherichia coli,* in: 6th Intern. Symp. Microb. Ecol. Abstracts (abstract booklet), p. 254, Barcelona, Spain.

Lengeler, J. W., 1993, Carbohydrate transport in bacteria under environmental conditions, a black box? *Antonie van Leeuwenhoek* **63**:275–288.

León, J. A., and Tumpson, D. B., 1975, Competition between two species for two complementary or substitutable resources, *J. Theor. Biol.* **50**:185–201.

Levering, P. R., and Dijkhuizen, L., 1985, Regulation of methylotrophic and heterotrophic metabolism in *Arthrobacter* P1. Growth with mixtures of methylamine and acetate in batch and continous cultures, *Arch. Microbiol.* **142**:113–120.

Lewin, B., 1994, *Genes V,* Oxford University Press, Oxford, England.

Lin, E. C. C., 1987, Dissimilatory pathways for sugars, polyols, and carboxylates, in: *Escherichia coli and Salmonella typhimurium,* Vol. 1 (F. C. Neidhardt, J. L. Ingraham, K. Brooks, B. Magasanik, M. Schaechter, and H. E. Umbarger, eds.), American Society for Microbiology, Washington, D.C., pp. 244–284.

Linton, J. D., and Stephenson, R. J., 1978, A preliminary study on growth-yields in relation to the carbon and energy-content of various organic growth substrates, *FEMS Microbiol. Lett.* **3**:95–98.

Linton, J. D., Griffiths, K., and Gregory, M., 1981, The effect of mixtures of glutamate and formate on the yield and respiration of a chemostat culture of *Beneckea natriegens, Arch. Microbiol.* **129**:119–122.

Loubière, P., Salou, P., Leroy, M. J., Lindley, N. D., and Parreilleux, A., 1992a, Electrogenic malate uptake and improved growth energetics of the malolactic bacterium *Leuconostoc oenos* grown on glucose-malate mixtures, *J. Bacteriol.* **174**:5302–5308.

Loubière, P., Gros, E., Paquet, V., and Lindley, N. D., 1992b, Kinetics and physiological implications of the growth behavior of *Eubacterium limosum* on glucose/methanol mixtures, *J. Gen. Microbiol.* **138**:979–985.

Lowe, W. E., Theodorou, M. K., and Trinci, A. P. J., 1987, Growth and fermentation of an anaerobic rumen fungus on various carbon sources and effect of temperature on development, *Appl. Environ. Microbiol.* **53**:1210–1215.

Lütgens, M., and Gottschalk, G., 1980, Why a co-substrate is required for growth of *Escherichia coli* on citrate, *J. Gen. Microbiol.* **119**:63–70.

Magasanik, B., 1976, Classical and postclassical modes of regulation of the synthesis of degradative bacterial enzymes, *Prog. Nucl. Acids Res. Molec. Biol.* **17**:99–115.

Mankad, T., and Bungay, H. R., 1988, Models for microbial growth with more than one limiting nutrient, *J. Biotechnol.* **7**:161–166.

Mann, K. H., 1988, Production and use of detritus in various freshwater, estuarine, and costal marine ecosystems, *Limnol. Oceanogr.* **33**:910–930.

Mårdén, P., Nyström, T., and Kjelleberg, S., 1987, Uptake of leucine by a marine gram-negative heterotrophic bacterium exposed to starvation conditions, *FEMS Microbiol. Ecol.* **45**:233–241.

Marounek, M., and Kopecný, J., 1994, Utilization of glucose and xylose in ruminal strains of *Butyrivibrio fibrisolvens, Appl. Environ. Microbiol.* **60**:738–739.

Martin, P., and MacLeod, R. A., 1984, Observations on the distinction between oligotrophic and eutrophic marine bacteria, *Appl. Environ. Microbiol.* **47**:1017–1022.

Martinez, J., and Azam, F., 1993, Perplasmic aminopeptidase and alkaline phosphatase activities in a marine bacterium: Implications for substrate processing in the sea, *Mar Ecol. Prog. Ser.* **92**:89–97.

Mason, C. A., and Egli, T., 1993, Dynamics of microbial growth in the decelerating and stationary phase of batch culture, in: *Starvation in Bacteria* (S. Kjelleberg, ed.), Plenum Press, New York, pp. 81–102.

Mateles, R. I., Chian, S. K., and Silver, R., 1967, Continuous culture on mixed substrates, in: *Microbial Physiology and Continuous Culture* (E. O. Powell, C. G. T. Evans, R. E. Strange, and D. W. Tempest, eds.), H. M. S. O., London, pp. 232–239.

Matin, A., 1979, Microbial regulatory mechanisms at low nutrient concentrations as studied in chemostat, in: *Strategies of Microbial Life in Extreme Environments* (Berlin: Dahlem Konferenzen, M. Shilo, ed.), Verlag Chemie, Weinheim, pp. 323–339.

Matin, A., 1991, The molecular basis of carbon-starvation-induced general resistance in *Escherichia coli, Molec. Microbiol.* **5**:3–10.

Megee, R. D., Drake, J. F., Fredrickson, A. G., and Tsuchiya, H. M., 1972, Studies in inter-microbial symbiosis, *Saccharomyces cerevisiae* and *Lactobacillus casei, Can. J. Microbiol.* **18**:1733–1742.

Meijers, A. P., and van der Leer, R. Chr., 1976, The occurrence of organic micropollutants in the River Rhine and the River Maas in 1974, *Water Res.* **10**:597–604.

Meyer-Reil, L. A., 1978, Autoradiography and epifluorescence microscopy combined for the determination of number and spectrum of actively metabolizing bacteria in natural waters, *Appl. Environ. Microbiol.* **36**:506–512.

Mills, A. L., and Bell, P. E., 1986, Determination of individual organisms and their activities in situ, in: *Microbial Autecology* (R. L. Tate III, ed.), Wiley, New York, pp. 27–60.

Minkevich, I. G., Krynitskaya, A. Y., and Eroshin, V. K., 1988, A double substrate limitation zone of continuous microbial growth, in: *Continuous Culture* (P. Kyslik, E. A. Dawes, V. Krumphanzl, and M. Novak, eds.), Academic Press, London, pp. 171–189.

Molin, G., 1985, Mixed carbon source utilization of meat-spoiling *Pseudomonas fragi* 72 in relation to oxygen limitation and carbon dioxide inhibition, *Appl. Environ. Mirobiol.* **49**:1442–1447.

Monod, J., 1942, *Recherches sur la Croissance des Cultures Bactériennes*, Hermann and Cie, Paris.

Monod, J., 1950, La technique de culture continue; Théorie et application, *Ann. Inst. Pasteur* **79**:390–410.

Moriarty, D. J. W., 1986, Measurement of bacterial growth rates in aquatic systems from rates of nucleic acid synthesis, *Adv. Microb. Ecol.* **9**:245–292.

Moriarty, D. J. W., and Bell, R. T., 1993, Bacterial growth and starvation in aquatic environments, in: *Starvation in Bacteria* (S. Kjelleberg, ed.), Plenum Press, New York, pp. 25–33.

Morita, R. Y., 1985, Starvation and miniaturisation of heterotrophs, with special emphasis on maintenance of the starved viable state, in: *Bacteria in Their Natural Environments* (M. Fletcher and G. D. Floodgate, eds.), Academic Press, London, pp. 111–130.

Morita, R. Y., 1988, Bioavailability of energy and its relationship to growth and starvation survival in nature, *J. Can. Microbiol.* **43**:436–441.

Morita, R. Y., 1993, Bioavailability of energy and the starvation state, in: *Starvation in Bacteria* (S. Kjelleberg, ed.), Plenum Press, New York, pp. 1–23.

Morita, R. Y., and Moyer, C. L., 1989, Bioavailability of energy and the starvation state, in: *Recent Advances in Microbial Ecology* (T. Hattori, Y. Ishida, Y. Maruyama, R. Y. Morita, and A. Uchida, eds.), Japan Scientific Societies Press, Tokyo, pp. 75–79.

Morris, D. R., and Lewis, Jr., W. M., 1992, Nutrient limitation of bacterioplankton growth in Lake Dillon, Colorado, *Limnol. Oceanogr.* **37**:1179–1192.

Müller, R. H., and Babel, W., 1986, Glucose as an energy donor in acetate growing *Acinetobacter calcoaceticus*, *Arch. Microbiol.* **144**:62–66.

Munson, T. O., and Burris, R. H., 1969, Nitrogen fixation by *Rhodospirillum rubrum* grown in nitrogen-limited chemostat culture, *J. Bacteriol.* **97**:1093–1098.

Münster, U., 1991, Extracellular enzyme activity in eutrophic and polyhumic lakes, in: *Microbial Enzymes in the Aquatic Environments* (R. J. Chróst, ed.), Springer-Verlag, New York, pp. 95–121.

Münster, U., 1993, Concentrations and fluxes of organic carbon substrates in the aquatic environment, *Antonie van Leeuwenhoek* **63**:243–264.

Münster, U., and Chróst, R. J., 1990, Origin, composition, and microbial utilization of dissolved organic matter, in: *Aquatic Microbial Ecology, Biochemical and Molecular Approaches* (J. Overbeck and R. J. Chróst, eds.), Springer, New York, pp. 8–46.

Namkung, E., and Rittman, B. E., 1987a, Modeling bisubstrate removal by biofilms, *Biotechnol. Bioeng.* **29**:269–278.

Namkung, E., and Rittman, B. E., 1987b, Evaluation of bisubstrate secondary utilization kinetics by biofilms, *Biotechnol. Bioeng.* **29**:335–342.

Nedwell, D. B., and Gray, T. R. G., 1987, Soils and sediments as matrices for microbial growth, in: *Ecology of Microbial Communities* (M. Fletcher, T. R. G., Gray, and J. G. Jones, eds.), Cambridge University Press, Cambridge, England, pp. 21–54.

Neidhardt, F. C., Ingraham, J. L., and Schaechter, M., 1990, *Physiology of the Bacterial Cell, A Molecular Approach*, Sinauer, Sunderland, Mass.

Ng, F. M.-W., and Dawes, E. A., 1973, Chemostat studies on the regulation of glucose metabolism in *Pseudomonas aeruginosa* by citrate, *Biochem. J.* **132**:129–140.

Nissen, H., Nissen, P., and Azam, F., 1984, Multiphasic uptake of D-glucose by an oligotrophic marine bacterium, *Mar. Ecol. Prog. Ser.* **16**:155–160.

Novitsky, J. A., and Morita, R. Y., 1978, Possible strategy for the survival of marine bacteria under starvation conditions, *Marine Biol.* **48**:289–295.

Nyström, T., 1993, Global systems approach to the physiology of the starved cell, in: *Starvation in Bacteria* (S. Kjelleberg, ed.), Plenum Press, New York, pp. 129–150.

Östling, J., Holmquist, L., Flärdh, K., Svenblad, B., Jouper-Jaan, Å., and Kjelleberg, S., 1993, Starvation and recovery of *Vibrio,* in: *Starvation in Bacteria* (S. Kjelleberg, ed.), Plenum Press, New York, pp. 103–127.

Owens, J. D., and Legan, J. D., 1987, Determination of the Monod substrate saturation constant for microbial growth, *FEMS Microbiol. Rev.* **46**:419–432.

Paerl, H. W., 1982, Factors limiting productivity of freshwater ecosystems, *Adv. Microb. Ecol.* **6**:75–110.

Paerl, H. W., 1993, Interaction of nitrogen and carbon cycles in the marine environment, in: *Aquatic Microbiology. An Ecological Approach* (T. E. Ford, ed.), Blackwell Scientific Publications, Boston, Oxford, pp. 343–381.

Pakulski, J. D., and Benner, R., 1992, An improved method for hydrolysis and MBTH analysis of dissolved and particulate carbohydrates in seawater, *Mar. Chem.* **40**:143–160.

Panikov, N., 1979, Steady-state growth kinetics of *Chlorella vulgaris* under double substrate (urea and phosphate) limitation, *J. Chem. Tech. Biotechnol.* **29**:442–450.

Paul, E. A., and Clark, F. E., 1989, *Soil Microbiology and Biochemistry,* Academic Press, New York.

Paul, E. A., and Voroney, R. P., 1980, Nutrient and energy flows through soil microbial biomass, in: *Contemporary Microbial Ecology* (D. C. Ellwood, J. N. Hedger, M. J. Latham, J. M. Lynch, and J. H. Slater, eds.), Academic Press, London, pp. 215–237.

Paul, J. H., 1993, The advances and limitations of methodology, in: *Aquatic Microbiology. An Ecological Approach* (T. E. Ford, ed.), Blackwell Scientific Publications, Boston, Oxford, pp. 15–46.

Pedersen, S., Bloch, P. L., Reeh, S., and Neidhardt, F. C., 1978, Patterns of protein synthesis in *Escherichia coli:* A catalog of the amount of 140 individual proteins at different growth rates, *Cell* **14**:179–190.

Pengerud, B., Skjoldal, E., and Thingstad, F., 1987, The reciprocal interaction between degradation of glucose and ecosystem structure. Studies in mixed chemostat cultures of marine bacteria, algae, and bacteriovorous nanoflagellates, *Mar. Ecol. Prog. Ser.* **35**:111–117.

Pfennig, N., and Jannasch, H., 1962, Biologische Grundfragen bei der homokontinuierlichen Kultur von Mikroorganismen, *Ergebnisse der Biologie* **25**:93–135.

Pineault, G., Pruden, B. B., and Loutfi, H., 1977, The effects of mixing, temperature, and nutrient concentration on the fermentation of a mixed sugar solution simulating the hexose content of waste sulfite liquor, *Can. J. Chem. Eng.* **55**:333–340.

Pirt, 1975, *Principles of Microbe and Cell Cultivation,* Blackwell, London.

Pöhland, D., Ringpfeil, M., and Behrens, U., 1966, Die Assimilation von Glucose, Xylose und Essigsäure durch *Candida utilis, Z. Allg. Mikrobiol.* **6**:387–395.

Poindexter, J. S., 1987, Bacterial responses to nutrient limitation, in: *Ecology of Microbial Communities* (M. Fletcher, T. R. G., Gray, and J. G. Jones, eds.), Cambridge University Press, Cambridge, England, pp. 283–317.

Postma, P. W., 1986, Catabolite repression and related processes, in: *Regulation of Gene Expression—25 Years On* (I. R. Booth and C. F. Higgins, eds.), Cambridge University Press, Cambridge, England, pp. 21–49.

Powell, E. O., 1967, The growth rate of micro-organisms as a function of substrate concentration, in: *Microbial Physiology and Continuous Culture* (P. O. Powell, C. G. T. Evans, R. E. Strange, and D. W. Tempest, eds.), H. M. S. O., London, pp. 34–56.

Pronk, J. T., van der Linden-Beuman, A., Verduyn, C., Scheffers, A. W., and van Dijken, J. P.,

1994, Propionate metabolism in *Saccharomyces cerevisiae:* Implications for the metabolon hypothesis, *Microbiology* **140**:717–722.

Rahmanian, M., and Oxender, D. L., 1972, Depressed leucine transport activity in *Escherichia coli, J. Supramolec. Struct.* **1**:55–59.

Reber, H. H., and Kaiser, R., 1981, Regulation of the utilization of glucose and aromatic substrates in four strains of *Pseudomonas putida, Arch. Microbiol.* **130**:243–247.

Rhee, G.-Y., 1978, Effects of N:P atomic ratios and nitrate limitation on algal growth, cell composition, and nitrate uptake, *Limnol. Oceanogr.* **23**:10–25.

Rittenberg, S. C., 1969, The roles of exogenous organic matter in the physiology of chemolithotrophic bacteria, *Adv. Microb. Physiol.* **3**:151–196.

Robertson, L. A., and Kuenen, J. G., 1984, Aerobic denitrification, a controversy revived, *Arch. Microbiol.* **139**:351–354.

Robertson, L. A., and Kuenen, J. G., 1990, Mixed terminal electron acceptors (oxygen and nitrate), in: *Mixed and Multiple Substrates and Feedstocks* (G. Hamer, T. Egli, and M. Snozzi, eds.), Hartung-Gorre, Konstanz, pp. 97–106.

Rogers, A. H., and Reynolds, E. C., 1990, The utilization of casein and amino acids by *Streptococcus sanguis* P_4A_7 in continuous culture, *J. Gen. Microbiol.* **136**:2535–2550.

Rogers, H. J., 1961, The dissimilation of high molecular weight organic substrates, in: *The Bacteria,* Vol. 2 (I. C. Gunsalus and R. Y. Stanier, eds.), Academic Press, New York, pp. 261–318.

Rubin, H. E., Subba-Rao, R. V., and Alexander, M., 1982, Rates of mineralization of trace concentrations of aromatic compounds in lake water and sewage samples, *Appl. Environ. Microbiol.* **43**:1133–1138.

Russel, J. B., and Baldwin, R. L., 1978, Substrate preferences in rumen, bacterial evidence of catabolite regulatory mechanisms, *Appl. Environ. Microbiol.* **36**:319–329.

Rutgers, M., Teixeira de Mattos, M. J., Postma, P. W., and van Dam, K., 1987, Establishment of the steady state in glucose-limited chemostat cultures of *Klebsiella pneumoniae, J. Gen. Microbiol.* **133**:445–453.

Rutgers, M., Balk, T. A., and van Dam, K., 1990, Quantification of multiple-substrate controlled growth: Simultaneous ammonium and glucose limitation in chemostat cultures of *Klebsiella pneumoniae, Arch. Microbiol.* **153**:478–484.

Rutgers, M., van Dam, K., and Westerhoff, H. V., 1991, Control and thermodynamics of microbial growth: Rational tools for bioengineering, *CRC Crit. Rev. Biotechnol.* **11**:367–395.

Sahm, H., and Wagner, F., 1973, Mikrobielle Verwertung von Methanol. Eigenschaften der Formaldehyddehydrogenase und Formiatdehydrogenase aus *Candida boidinii, Arch. Mikrobiol.* **90**:263–268.

Saier, Jr., M. H., 1989, Protein phosphorylation and allosteric control of inducer exclusion and catabolite repression by bacterial phosphoenolpyruvate: sugar phosphotransferase system, *Microbiol. Rev.* **53**:109–120.

Sakai, Y., Sawai, T., and Tani, Y., 1987, Isolation and characterization of a catabolite repression-insensitive mutant of a methanol yeast *Candida boidinii* A5 producing alcohol oxidase in glucose-containing medium, *Appl. Environ. Microbiol.* **53**:1812–1818.

Salou, P., Leroy, M. J., Goma, G., and Pareilleux, A., 1991, Influence of pH and malate–glucose ratio on the growth kinetics of *Leuconostoc oenos, Appl. Microbiol. Biotechnol.* **36**:87–91.

Salou, P., Loubière, P. and Pareilleux, A., 1994, Growth and energetics of *Leuconostoc oenos* during cometabolism of glucose with citrate or fructose, *Appl. Environ. Microbiol.* **60**:1459–1466.

Schmidt, S. K., and Alexander, M., 1985, Effects of dissolved organic carbon and second substrates on the biodegradation of organic compounds at low concentrations, *Appl. Environ. Microbiol.* **49**:822–827.

Schmidt, S. K., Alexander, M., and Schuler, M. L., 1985, Predicting threshold concentrations of organic substrates for bacterial growth, *J. Theor. Biol.* **114**:1–8.

Schmitt, P., and Divies, C., 1991, Co-metabolism of citrate and lactose by *Leuconostoc mesenteroides* subsp. *cremoris, J. Ferment. Bioeng.* **71**:72–74.

Schowanek, D., and Verstraete, W., 1990, Phosphonate utilization by bacteria in the presence of alternative phosphorus sources, *Biodegradation* **1**:43–53.

Schut, F., de Vries, E. J., Gottschal, J. C. Robertson, B. R., Harder, W., Prins, R. A., and Button, D., 1993, Isolation of typical marine bacteria by dilution culture: Growth, maintenance, and characteristics of isolates under laboratory conditions, *Appl. Environ. Microbiol.* **59**:2150–2160.

Schut, F., Jansen, M., Pedro Gomes, T. M., Gottschal, J. C., Harder, W., and Prins, R. A., 1995, Substrate uptake and utilization by a marine ultramicrobacterium, *Microbiology* **141**:351–361.

Schweitzer, B., and Simon, M., 1995, Growth limitation of planktonic bacteria in a large mesotrophic lake, *Microb. Ecol.* **30**:89–104.

Senn, H., 1989, *Kinetik und Regulation des Zuckerabbaus von* Escherichia coli *ML 30 bei tiefen Zuckerkonzentrationen,* Swiss Federal Institute of Technology, Zürich, Switzerland, Ph.D. thesis No. 8831.

Senn, H., Lendenmann, U., Snozzi, M., Hamer, G., and Egli, T., 1994, The growth of *Escherichia coli* in glucose-limited chemostat culture: A re-examination of the kinetics, *Biochim. Biophys. Acta* **1201**:424–436.

Sepers, A. B. J., 1984, The uptake capacity for organic compounds of two heterotrophic bacterial strains at carbon-limited growth, *Z. Allg. Mikrobiol.* **24**:261–267.

Shehata, T. E., and Marr, A. G., 1971, Effect of nutrient concentration on the growth of *Escherichia coli, J. Bacteriol.* **107**:210–216.

Siegele, D., Almirón, M., and Kolter, R., 1993, Approaches to the study of survival and death in stationary-phase *Escherichia coli,* in: *Starvation in Bacteria* (S. Kjelleberg, ed.), Plenum Press, New York, pp. 151–169.

Silver, R. S., and Mateles, R. I., 1969, Control of mixed-substrate utilization in continuous culture of *Escherichia coli, J. Bacteriol.* **97**:535–543.

Simkins, S., and Alexander, M., 1984, Models for mineralization kinetics with the variables of substrate concentration and population density, *Appl. Environ. Microbiol.* **47**:1299–1306.

Sinclair, C. G., and Ryder, D. N., 1975, Models for the continuous culture of microorganisms under both oxygen and carbon limiting conditions, *Biotechnol. Bioeng.* **17**:375–398.

Slaff, G. F., and Humphrey, A. E., 1986, The growth of *Clostridium thermohydrosulfuricum* on multiple substrates, *Chem. Eng. Commun.* **45**:33–51.

Smith, A. L., and Kelly, D. P., 1979, Competition in the chemostat between an obligately and a facultatively chemolithotrophic *Thiobacillus, J. Gen. Microbiol.* **115**:377–384.

Smith, V. H., 1993, Implication of resource-ratio theory for microbial ecology, *Adv. Microb. Ecol.* **13**:1–37.

Sonenshein, A. L., 1989, Metabolic regulation of sporulation and other stationary-phase phenomena, in: *Regulation of Prokaryotic Development* (I. Smith, A. Slepecky, and P. Setlow, eds.), American Society for Microbiology, Washington, D.C., pp. 109–130.

Standing, C. N., Fredrickson, A. G., and Tsuchiya, H. M., 1972, Batch and continuous culture transients for two substrate systems, *Appl. Microbiol.* **23**:354–359.

Stephenson, M., 1949, Growth and nutrition, in: *Bacterial Metabolism,* 3rd ed., Longmans, Green and Co., London, pp. 159–178.

Stevenson, L. H., 1978, A case for bacterial dormancy in aquatic systems, *Microb. Ecol.* **4**:127–133.

Stolz, P., Böcker, G., Vogel, R. F., and Hammes, W. P., 1993, Utilisation of maltose and glucose by lactobacilli isolated from sourdough, *FEMS Microbiol. Lett.* **109**:237–242.

Subba-Rao, R. V., Rubin, H. E., and Alexander, M., 1982, Kinetics and extent of mineralization of organic chemicals at trace levels in freshwater and sewage, *Appl. Environ. Microbiol.* **43**:1139–1150.

Sykes, R. M., 1973, Identification of the limiting nutrient and specific growth rate, *J. Water Poll. Control Fed.* **45**:888–895.

Tabor, P. S., and Neihof, R. A., 1982, Improved microautoradiographic method to determine individual microorganisms active in substrate uptake in natural waters, *Appl. Environ. Microbiol.* **44**:945–953.

Tate III, R. L., 1986, Environmental microbial autecology: Practicality and future, in: *Microbial Autecology* (R. L. Tate III, ed.), Wiley, New York, pp. 249–259.

Tauchert, K., Jahn, A., and Oelse, J., 1990, Control of diauxic growth of *Azotobacter vinelandii*, *J. Bacteriol.* **172**:6447–6451.

Tempest, D. W., 1970a, The continuous cultivation of microorganisms. 1. Theory of the chemostat, *Meth. Microbiol.* **2**:259–276.

Tempest, D. W., 1970b, The place of continuous culture in microbiological research, *Adv. Microb. Physiol.* **4**:223–250.

Tempest, D. W., and Neijssel, O. M., 1978, Eco-pysiological aspects of microbial growth in aerobic nutrient-limited environments, *Adv. Microb. Ecol.* **2**:105–153.

Tempest, D. W., Neijssel, O. M., and Zevenboom, W., 1983, Properties and performance of microorganisms in laboratory culture: Their relevance to growth in natural ecosystems, in: *Microbes in Their Natural Environment* (J. H. Slater, R. Whittenbury, and J. W. T. Wimpenny, eds.), Cambridge University Press, Cambridge, England, pp. 119–152.

Thingstad, T. F., 1987, Utilization of N, P, and organic C by heterotrophic bacteria. I. Outline of a chemostat theory with a consistent concept of "maintenance" metabolism, *Mar. Ecol. Prog. Ser.* **35**:99–109.

Thingstad, T. F., and Pengerud, B., 1985, Fate and effect of allochthonous organic material in aquatic microbial ecosystems. An analysis based on chemostat theory, *Mar. Ecol. Prog. Ser.* **21**:47–62.

Thurston, B., Dawson, K. A., and Strobel, H. J., 1984, Pentose utilization by the ruminal bacterium *Ruminococcus albus*, *Appl. Environ. Microbiol.* **60**:1087–1092.

Tilman, D., 1980, Resources: A graphical–mechanistic approach to competition and predation, *Am. Nat.* **116**:362–393.

Tilman, D., 1982, *Resource Competition and Community Structure*, Princeton University Press, Princeton, N.J.

Torella, F., and Morita, R. Y., 1981, Microcultural study of bacterial size changes and microcolony and ultramicrocolony formation by heterotrophic bacteria in seawater, *Appl. Environ. Microbiol.* **41**:518–527.

Torriani-Gorini, A., 1987, The birth and growth of the Pho regulon, in: *Phosphate Metabolism and Cellular Regulation in Microorganisms* (A. Torriani-Gorini, F. G. Rothman, S. Silver, A. Wright, and E. Yagil, eds.), American Society for Microbiology, Washington, D.C., pp. 12–19.

Tranvik, L. J., 1990, Bacterioplankton growth on fractions of dissolved organic carbon of different molecular weights from humic and clear waters, *Appl. Environ. Microbiol.* **56**:1672–1677.

Tranvik, L. J., and Höfle, M., 1987, Bacterial growth in mixed cultures on dissolved organic carbon from humic and clear waters, *Appl. Environ. Microbiol.* **53**:482–488.

Tyree, R. W., Clausen, E. C., and Gaddy, J. L., 1990, The fermentative characteristics of *Lactobacillus xylosus* on glucose and xylose, *Biotechnol Lett.* **12**:51–56.

Uetz, T., and Egli, T., 1993, Characterization of an inducible, membrane-bound iminodiacetate dehydrogenase from *Chelatobacter heintzii* ATCC 29600, *Biodegradation* **3**:423–434.

Uetz, T., Schneider, R., Snozzi, M., and Egli, T., 1992, Purification and characterization of a two component monooxygenase that hydroxylates nitrilotriacetate from "*Chelatobacter*" strain ATCC 29600, *J. Bacteriol.* **174**:1179–1188.

Upton, A. C., and Nedwell, D. B., 1989, Nutritional flexibility of oligotrophic and copiotrophic antarctic bacteria with respect to organic substrates, *FEMS Microbiol. Ecol.* **62**:1–6.

van Dam, K., and Jansen, N., 1991, Quantification of control of microbial metabolism by substrates and enzymes, *Antonie van Leeuwenhoek* **60:**209–223.

van der Kooij, D., 1990, Assimilable organic carbon (AOC) in drinking water, in: *Drinking Water Microbiology: Progress and Recent Developments* (G. A. McFeters, ed.), Springer-Verlag, New York, pp. 57–87.

van der Kooij, D., and Hijnen, W. A. M., 1983, Nutritional versatility of a starch-utilizing *Flavobacterium* at low substrate concentrations, *Appl. Environ. Microbiol.* **45:**804–810.

van der Kooij, D., and Hijnen, W. A. M., 1988, Nutritional versatility and growth kinetics of an *Aeromonas hydrophila* strain isolated from drinking water, *Appl. Environ. Microbiol.* **54:**2842–2851.

van der Kooij, D., Visser, A., and Hijnen, W. A. M., 1980, Growth of *Aeromonas hydrophila* at low concentrations of substrates added to tap water, *Appl. Environ. Microbiol.* **39:**1198–1204.

van der Kooij, D., Oranje, J. P., and Hijnen, W. A. M., 1982, Growth of *Pseudomonas aeruginosa* in tap water in relation to utilization of substrates at concentrations of a few micrograms per liter, *Appl. Environ. Microbiol.* **44:**1086–1095.

van Es, F. B., and Meyer-Reil, L.-A., 1982, Biomass and metabolic activity of heterotrophic marine bacteria, *Adv. Microb. Ecol.* **6:**111–170.

van Verseveld, H. W., and Stouthamer, A. H., 1980, Two-(carbon) substrate-limited growth of *Paracoccus denitrificans* on mannitol and formate, *FEMS Microbiol. Lett.* **7:**207–211.

van Verseveld, H. W., Boon, J. P., and Stouthamer, A. H., 1979, Growth yields and the efficiency of oxidative phosphorylation of *Paracoccus denitrificans* during two-(carbon) substrate-limited growth, *Arch. Microbiol.* **121:**213–223.

van Zyl, C., Prior, B. A., Kilian, S. G., and Kock, J. L. F., 1989, D-Xylose utilization by *Saccharomyces cerevisiae, J. Gen. Microbiol.* **135:**2791–2798.

Veldkamp, H., and Jannasch, H. W., 1972, Mixed culture studies with the chemostat, *J. Appl. Chem. Biotechnol.* **22:**105–123.

Vives–Rego, J., Billen, G., Fontigny, A., and Somville, M., 1985, Free and attached proteolytic activity in water environments, *Mar. Ecol. Prog. Ser.* **21:**245–249.

Volfová, O., Korínek, V., and Kyslíková, E., 1988, Alcohol oxidase in *Candida boidinii* on methanol–xylose mixtures, *Biotechnol. Lett.* **10:**643–648.

Wang, L., Miller, T. D., and Priscu, J. C., 1992, Bacterioplankton nutrient deficiency in a eutrophic lake, *Arch. Hydrobiol.* **125:**423–439.

Wanner, B., 1987a, Phosphate regulation of gene expression, in: *Escherichia coli and Salmonella typhimurium,* Vol. 2 (F. C. Neidhardt, J. L. Ingraham, K. Brooks, B. Magasanik, M. Schaechter, and H. E. Umbarger, eds.), American Society for Microbiology, Washington, D.C., pp. 1326–1333.

Wanner, B., 1987b, Bacterial alkaline phosphatase gene regulation and the phosphate response in *Escherichia coli,* in: *Phosphate Metabolism and Cellular Regulation in Microorganisms* (A. Torriani-Gorini, F. G. Rothman, S. Silver, A. Wright, and E. Yagil, eds.), American Society for Microbiology, Washington, D.C., pp. 12–19.

Wanner, U., and Egli, T., 1990, Dynamics of microbial growth and cell composition in batch culture, *FEMS Microbiol. Rev.* **75:**19–44.

Waterham, H. R., Keizer-Gunnink, I., Goodman, J. M., Harder, W., and Veenhuis, M., 1992, Development of multipurpose peroxysomes in *Candida biodinii* growing in oleic acid–methanol limited continuous cultures, *J. Bacteriol.* **174:**4057–4063.

Weide, H., 1983, Mikrobielle Verwertung von Mischsubstraten, *Z. Allg. Mikrobiol.* **23:**37–40.

Wetzel, R. G., 1984, Detrital dissolved and particulate organic carbon functions in aquatic ecosystems, *Bull. Mar. Sci.* **35:**503–509.

Weusthuis, R. A., Adams, H., Scheffers, W. A., and van Dijken, J. P., 1993, Energetics and kinetics of maltose transport in *Saccharomyces cerevisiae:* A continuous culture study, *Appl. Environ. Microbiol.* **59:**3102–3109.

Weusthuis, R. A., Luttik, M. A. H., Scheffers, W. A., van Dijken, J. P., and Pronk, J. T., 1994, Is the Kluyver effect caused by product inhibition? *Microbiology* **140**:1723–1729.

White, D. C., 1986, Quantitative physiochemical characterization of bacterial habitats, in: *Bacteria in Nature*, Vol. 2 (J. S. Poindexter and Leadbetter, E. R., eds.), Plenum Press, New York, pp. 177–203.

Wilberg, E., El-Banna, T., Auling, G., and Egli, T., 1993, Serological studies on nitrilotriacetic acid (NTA)-utilising bacteria: Distribution of *Chelatobacter heintzii* and *Chelatococcus asaccharovorans* in sewage treatments plants and aquatic ecosystems, *System. Appl. Microbiol.* **16**:147–152.

Williams, P. J. le B., 1990, The importance of losses during microbial growth: Commentary on the physiology, measurement and ecology of the release of dissolved organic material, *Mar. Microb. Food Webs* **4**:175–206.

Williams, P. M., 1986, Chemistry of the dissolved and particulate phases in the water column, in: *Plankton Dynamics of the Southern California Bight. Lecture Notes on Coastal and Estuarine Studies*, Vol. 15 (R. W. Eppley, ed.), Springer-Verlag, Berlin, pp. 53–83.

Williams, S. T., 1985, Oligotrophy in soil: Fact or fiction? in: *Bacteria in the Natural Environment* (M. Fletcher and G. D. Floodgate, eds.), Academic Press, London, pp. 81–110.

Wong, T. Y., Pei, H., Bancroft, K., and Childers, G. W., 1995, Diauxic growth of *Azotobacter vinelandii* on galactose and glucose: Regulation of glucose transport by another hexose, *Appl. Environ. Microbiol.* **61**:430–433.

Wood, A. P., and Kelly, D. P., 1977, Heterotrophic growth of *Thiobacillus* A2 on sugars and organic acids, *Arch. Microbiol.* **113**:257–264.

Wood, J. M., 1975, Leucine transport in *Escherichia coli*. The resolution of mulitple transport systems and their coupling to metabolic energy, *J. Biol. Chem.* **250**:4477–4485.

Yoon, H., and Blanch, H. W., 1977, Competition for double growth-limiting nutrients in continuous culture, *J. Appl. Chem. Biotechnol.* **27**:260–268.

Yoon, H,. Klinzing, G., and Blanch, H., 1977, Competition for mixed substrates by microbial populations, *Biotechnol. Bioeng.* **19**:1193–1210.

ZoBell, C. E., and Grant, C. W., 1942, Bacterial activity in dilute nutrient solutions, *Science* **96**:189.

ZoBell, C. E., and Grant, C. W., 1943, Bacterial utilization of low concentrations of organic matter, *J. Bacteriol.* **45**:555–564.

Zweifel, U. L., Norman, B., and Hagström, A., 1993, Consumption of dissolved organic carbon by marine bacteria and demand for inorganic nutrients, *Mar. Ecol. Prog. Ser.* **101**:23–32.

Index

In the index that follows, *f* indicates that the information can be found in a figure on the page indicated; *t* signifies that the information is found in a table.

DATE DUE

DEMCO, INC. 38-2971